Microplastics: Ecological and Food Security Implications

Adnan Abubakr • Adnan Amin • Monisa M Malik
Editors

Microplastics: Ecological and Food Security Implications

Springer

Editors
Adnan Abubakr
Division of Aquatic Environmental Management
Faculty of Fisheries Rangil
Ganderbal SKUAST-Kashmir
Jammu and Kashmir, India

Adnan Amin
Division of Aquatic Environmental Management
Faculty of Fisheries Rangil
Ganderbal SKUAST-Kashmir
Jammu and Kashmir, India

Monisa M Malik
Division of Aquatic Environmental Management
Faculty of Fisheries Rangil
Ganderbal SKUAST-Kashmir
Jammu and Kashmir, India

ISBN 978-3-031-94479-6 ISBN 978-3-031-94480-2 (eBook)
https://doi.org/10.1007/978-3-031-94480-2

This Springer imprint is published by the registered company Springer Nature Switzerland AG
The registered company address is: Gewerbestrasse 11, 6330 Cham, Switzerland

Abbreviations

ACR	Acrylate Copolymer
AFM	Atomic Force Microscopy
AOPs	Advanced Oxidation Processes
APC	Antigen-Presenting Cell
BBB	Blood-Brain Barrier
BPA	Bisphenol A
CAT	Catalase
CDL	Container Deposit Legislation
CIEL	Center for International Environmental Law
COPD	Chronic Obstructive Pulmonary Disease
COX	Cyclooxygenases
CP	Cellulose Propionate
DRS	Deposit Return Scheme
DSP	Diarrhetic Shellfish Poisoning
EDCs	Endocrine-Disrupting Chemicals
EDS	Energy Dispersive Spectroscopy
EPFRs	Environmentally Persistent Free Radicals
EPR	Extended Producer Responsibility
EPS	Expanded Polystyrene
ERA	Environmental Risk Assessment
FPA	Focal Plane Array
GC–MS	Gas Chromatography-Mass Spectrometry
GESAMP	Group of Experts on the Scientific Aspects of Marine Environmental Protection
GHG	Greenhouse Gas
GPML	Global Partnership on Marine Litter
GPx	Glutathione Peroxidase
GR	Glutathione Reductase

HDPE	High-Density Polyethylene
INC	Intergovernmental Negotiating Committee
Keap1	Kelch-Like ECH-Associated Protein 1
KOH	Potassium Hydroxide
LCA	Life Cycle Assessment
LCC	Leaf-branch Compost Cutinase
LDIR	Laser Direct Infrared
LDPE	Low-Density Polyethylene
LMS	Lysosomal Membrane Stability
LOX	Lipoxygenases
MDA	Malondialdehyde
MEAs	Multilateral Environmental Agreements
MMTPA	Million Metric Tons Per Annum
MNPs	Microplastics and Nanoparticles
MPD	Microplastic Debris
MPs	Microplastics
MSW	Municipal Solid Waste
NAFLD	Nonalcoholic Fatty Liver Disease
NPs	Nanoplastics
Nrf2	Nuclear Factor Erythroid 2-Related Factor 2
PAHs	Polycyclic Aromatic Hydrocarbons
PAHs	Polycyclic Aromatic Hydrocarbons
PAN	Polyacrylonitrile
PBDEs	Polybrominated Diphenyl Ethers
PCBs	Polychlorinated Biphenyls
PCCPs	Personal-Care and Cosmetic Products
PE	Polyethylene
PET	Polyethylene Terephthalate
PhAC	Pharmaceutically Active Compounds
PHAs	Polyhydroxyalkanoates
PLA	Polylactic Acid
PLOM	Polarized-Light Optical Microscopy
POPs	Persistent Organic Pollutants
PP	Polypropylene
PPE	Personal Protective Equipment
PS	Polystyrene
PSP	Paralytic Shellfish Poisoning
PUR	Polyurethane
PVC	Polyvinyl Chloride
PWM	Plastic Waste Management
Py-GC–MS	Pyrolysis Gas Chromatography-Mass Spectrometry
ROS	Reactive Oxygen Species
SBA	Swachh Bharat Abhiyan
SDS	Sodium Dodecyl Sulfate
SEM	Scanning Electron Microscopy

SOD	Superoxide Dismutase
SUPs	Single-Use Plastics
TBMP	Tire and Bitumen Microplastic Particles
TBT	Tributyl Tin
TED-GC–MS	Thermal Extraction Desorption Gas Chromatography-Mass Spectrometry
TEM	Transmission Electron Microscopy
TGA	Thermogravimetric Analysis
TGA-DSC	Thermogravimetry Coupled with Differential Scanning Calorimetry
TLR-6	Toll Like Receptor 6
TNF-α	Tumour Necrosis Factor Alpha
ULBs	Urban Local Bodies
μ-FTIR	Micro-Fourier Transform Infrared Spectroscopy
WWMPs	Wastewater Microplastics
WWTPs	Wastewater Treatment Plants

Contents

Editors and Contributors

About the Editors

Adnan Abubakr is a distinguished academic and researcher currently serving as Professor of Limnology and Head of the Division of Aquatic Environmental Management, Faculty of Fisheries, Sher-e-Kashmir University of Agricultural Sciences and Technology of Kashmir (SKUAST-K). He holds both MSc and PhD degrees in Limnology from the Department of Limnology, Barkatullah University, Bhopal, with expertise in water and wastewater analysis, including analytical and bacteriological techniques applicable to freshwater and brackish water systems.

He specializes in the design and development of artificial wetland systems for wastewater treatment. With over 15 years of extensive experience in teaching, research, and extension in the field of aquatic ecology, he has significantly contributed to the advancement of aquatic environmental sciences. He played a pivotal role in establishing the MFSc and PhD degree programs in Aquatic Environmental Management within the Division of AEM at SKUAST-K.

A prolific researcher, he has published more than 60 research papers in peer-reviewed journals. He has successfully completed two externally funded and three university-funded research projects and has supervised more than 20 postgraduate and doctoral students in their research pursuits.

Adnan Amin is an Assistant Professor cum Junior Scientist in the Division of Aquatic Environmental Management at the Faculty of Fisheries, Rangil, Ganderbal, Sher-e-Kashmir University of Agricultural Sciences and Technology—Kashmir (SKUAST-K). He earned his undergraduate degree from the College of Fishery Science Muthukur, Nellore, under Sri Venkateswara Veterinary University (SVVU), Tirupati, Andhra Pradesh. He completed his master's at the College of Fisheries, Mangalore, specializing in Aquatic Environment Management, and obtained his doctorate from the College of Fisheries, Ratnagiri, in the same field.

His research interests encompass aquatic toxicology, limnology, and the impact of emerging contaminants on aquatic ecosystems. His significant contributions include studies on heavy metal and pesticide toxicity in aquatic organisms and the effects of climate change on aquatic biodiversity. He has published over 35 research papers in national and international peer-reviewed journals, along with 5 book chapters, 10 review papers, and 18 magazine articles.

In recognition of his academic excellence, he received a merit fellowship during his master's program. He holds life membership in the Society of Fisheries and Life Science, College of Fisheries, Mangalore. His editorial contributions include co-editing books such as *Xenobiotics in Aquatic Animals: Reproductive and Developmental Impacts*, *Food Security, Nutrition and Sustainability Through Aquaculture Technologies*, and *Coldwater Fisheries and Aquaculture Management: Technology for Sustainable Food Production*. His work is widely cited, with his publications receiving numerous citations, reflecting his impact in the field of aquatic environmental management

Monisa M. Malik is currently serving as an Assistant Professor cum Junior Scientist in the Division of Aquatic Environmental Management, Faculty of Fisheries, Rangil, Ganderbal, Sher-e-Kashmir University of Agricultural Sciences and Technology – Kashmir (SKUAST-K). She holds a bachelor's degree in Fisheries Science from Dr. Balasaheb Sawant Konkan Krishi Vidyapeeth (DBSKKV), Dapoli, where she was recognized as the Best Academic Performer and honored with the Best Student Award.

She completed both her master's and PhD in Fisheries Science from SKUAST-Kashmir, earning a Gold Medal for her outstanding performance during her master's program. She is also a recipient of the prestigious National INSPIRE Fellowship for her doctoral research.

Her research expertise includes Limnology, Wastewater Treatment, Aquatic Ecology, and Aquatic Environmental Management. She is an active member of several editorial boards, including the *International Journal of Energy and Environmental Science*, *Journal of Zoo Biology*, and *Nature Science e-Magazine*. In addition, she holds life memberships in reputed organizations such as the *Journal of Fisheries and Life Sciences* and the *Journal of Environmental Biology*.

She has contributed over 22 research articles to national and international peer-reviewed journals and is the author of a book published by an international publisher. She has also written several impactful book chapters. Her scholarly contributions have been recognized with a Best Poster Presentation Award from a national organization.

Contributors

Zaheer Abass Division of Fisheries Resources Management, Faculty of Fisheries, Sher-e-Kashmir University of Agricultural Science and Technology of Kashmir, Shalimar, Srinagar, J&K UT, India

Waseem Abbas Department of Zoology, Division of Science and Technology, University of Education, Township, Lahore, Pakistan

A. P. Abeygunawardana Department of Animal Science, Faculty of Animal Science and Export Agriculture, Uva Wellassa University, Badulla, Sri Lanka

Adnan Abubakr Division of Aquatic Environmental Management, Faculty of Fisheries Rangil, Ganderbal, SKUAST-Kashmir, Jammu and Kashmir, India

S. Abuthagir Iburahim ICAR – Central Institute of Fisheries Education, Mumbai, India

Ishtiyaq Ahmad Division of Fish Genetics and Biotechnology, Faculty of Fisheries, SKUAST-K, Rangil, Ganderbal, Jammu and Kashmir, India

Aiman Institute of Zoology, Bahauddin Zakariya University, Multan, Pakistan

Sahina Akter Centre for Marine Living Resources and Ecology (CMLRE), Kochi, India

Shigufta Ali Fish Nutrition and Research Laboratory, Department of Zoology, Aligarh Muslim University, Aligarh, India

Adnan Amin Division of Aquatic Environmental Management, Faculty of Fisheries Rangil, Ganderbal, SKUAST-Kashmir, Jammu and Kashmir, India

Sonam Angmo ICAR-Central Institute of Fisheries Education (CIFE), Mumbai, Maharashtra, India

Hudisa Banoo Division of Fisheries Resources Management, Faculty of Fisheries, Sher-e-Kashmir University of Agricultural Science and Technology of Kashmir, Shalimar, Srinagar, J&K UT, India

Naila M. Bhat Faculty of Fisheries, Sher-e- Kashmir University of Science and Technology, Baramulla, Kashmir, India

Balasaheb Ramdas Chavan College of Fisheries, Dr. Balasaheb Sawant Konkan Agricultural University, Dapoli, Ratnagiri, India

Dechen Chuskit Division of Fisheries Resources Management, Faculty of Fisheries, Sher-e-Kashmir University of Agricultural Science and Technology of Kashmir, Shalimar, Srinagar, J&K UT, India

Joseph Joy Cindrella KSCSTE – Centre for Water Resources Development and Management (CWRDM), Kozhikode, Kerala, India

Viraj Yeshwant Daunde Center of Excellence in Wildlife, Exotic, and Aquatic Animal Pathology, Faculty of Veterinary Science, Chulalongkorn University, Bangkok, Thailand

International Graduate Course of Veterinary Science and Technology (VST), Faculty of Veterinary Science, Chulalongkorn University, Bangkok, Thailand

Ndakalimwe Naftal Gabriel Department of Fisheries and Ocean Sciences, Faculty of Agriculture, Engineering and Natural Science, Sam Nujoma Campus, University of Namibia, Henties Bay, Namibia

Rohit Kumar Gautam Department of Zoology, Babasaheb Bhimrao Ambedkar University, Lucknow, UP, India

Faculty of Fisheries Science, Department of Fisheries Resources Management, Kerala, India

University of Fisheries and Ocean Studies, Cochin, Kerala, India

Keisham Geenita ICAR-Central Institute of Fisheries Education (CIFE), Mumbai, Maharashtra, India

Ifra Tabassum Giri Division of Aquatic Environmental Management, Faculty of Fisheries Rangil, Ganderbal, SKUAST-Kashmir, Jammu and Kashmir, India

K. P. G. K. P. Guruge Department of Animal Science, Faculty of Animal Science and Export Agriculture, Uva Wellassa University, Badulla, Sri Lanka

Ishfaq Hamid Division of Aquatic Environmental Management, Faculty of Fisheries Rangil, Ganderbal, SKUAST-Kashmir, Jammu and Kashmir, India

Muzammal Hoque ICAR-Central Institute of Fisheries Education (CIFE), Mumbai, Maharashtra, India

Gowhar Iqbal ICAR-Central Institute of Fisheries Education, Mumbai, India

Saman Iram Institute of Zoology, Bahauddin Zakariya University, Multan, Pakistan

Manoj Tukaram Kamble Center of Excellence in Wildlife, Exotic, and Aquatic Animal Pathology, Faculty of Veterinary Science, Chulalongkorn University, Bangkok, Thailand

Aasia Karim Department of Zoology, Sardar Bahadur Khan Women's University, Quetta, Pakistan

Gurjeet Kaur ICAR-Central Institute of Post-Harvest Engineering and Technology, Ludhiana, Punjab, India

Simran Kaur College of Fisheries, Guru Angad Dev Veterinary and Animal Sciences University, Ludhiana, Punjab, India

Muhammad Sabtain Khan Department of Zoology, Division of Science and Technology, University of Education, Township, Lahore, Pakistan

Saurav Kumar Aquatic Environmental Management, AEHM Division, ICAR-Central Institute of Fisheries Education, Versova, Mumbai, Maharashtra, India

Vikas Kumar ICAR-Central Institute of Post-Harvest Engineering and Technology, Ludhiana, Punjab, India

Fariha Latif Institute of Zoology, Bahauddin Zakariya University, Multan, Pakistan

Kenyum Lollen ICAR-Central Institute of Fisheries Education (CIFE), Mumbai, Maharashtra, India

Samiya Majeed Division of Aquaculture, Faculty of Fisheries Rangil Ganderbal, SKUAST-Kashmir, Srinagar, India

Mohd Ashraf Malik ICAR-Central Institute of Fisheries Education, Mumbai, India

Monisa M. Malik Division of Aquatic Environmental Management, Faculty of Fisheries Rangil, Ganderbal, SKUAST-Kashmir, Jammu and Kashmir, India

Hafsa Maqbool Division of Fish Post Harvest Technology, Faculty of Fisheries Rangil, Ganderbal, SKUAST-Kashmir, Jammu & Kashmir, India

Pranali Prabhakar Marbade Late Shri Punaram Nishad College of Fisheries Kawardha, Sawaikachhar, Chhattisgarh, India

Seema Vijay Medhe ASEAN Institute for Health Development, Mahidol University, Salaya, Putthamonthon, Nakhon Pathom, Thailand

Shakir Ahmad Mir ICAR-Central Institute of Fisheries Education, Mumbai, India

Abha Mishra Department of Zoology, Babasaheb Bhimrao Ambedkar University, Lucknow, UP, India

Saba Nabi College of Fisheries, BASU, Kishanganj, Bihar, India

Kurapati Nagendrasai Aquatic Environmental Management, AEHM Division, ICAR-Central Institute of Fisheries Education, Versova, Mumbai, Maharashtra, India

Bejawada Chanikya Naidu ICAR – Central Institute of Fisheries Technology, Kochi, India

Ubair Nisar Business and Policy Research (IBPR), Sher-e- Kashmir University of Science and Technology, Baramulla, Kashmir, India

Gargi Rajesh Palekar Center of Excellence in Wildlife, Exotic, and Aquatic Animal Pathology, Faculty of Veterinary Science, Chulalongkorn University, Bangkok, Thailand

International Graduate Course of Veterinary Science and Technology (VST), Faculty of Veterinary Science, Chulalongkorn University, Bangkok, Thailand

Nopadon Pirarat Center of Excellence in Wildlife, Exotic, and Aquatic Animal Pathology, Faculty of Veterinary Science, Chulalongkorn University, Bangkok, Thailand

Gora Shiva Prasad Department of Aquatic Environmental Management, Faculty of Fishery Sciences, West Bengal University of Animal and Fishery Sciences, Kolkata, India

Neha Wajahat Qureshi ICAR-Central Institute of Fisheries Education, Mumbai, India

Karankumar Kishorkumar Ramteke ICAR-Central Institute of Fisheries Education (CIFE), Mumbai, Maharashtra, India

Ifrah Rashid Faculty of Fisheries, Sher-e- Kashmir University of Science and Technology, Baramulla, Kashmir, India

Sajad Ahmad Rather Division of Aquatic Environmental Management, Faculty of Fisheries Rangil, Ganderbal, SKUAST-Kashmir, Jammu and Kashmir, India

Nikheel Bhojraj Rathod Department of Post-Harvest Management of Meat, Poultry and Fish, PG Institute of Post-Harvest Management (Dr. Balasaheb Sawant Konkan Krishi Vidyapeeth, Dapoli) Killa-Roha, Raigad, Maharashtra, India

Vijay Kumar Reddy College of Fisheries, Guru Angad Dev Veterinary and Animal Sciences University, Ludhiana, Punjab, India

K. R. Renjith KSCSTE – Centre for Water Resources Development and Management (CWRDM), Kozhikode, Kerala, India

M. D. Sahana ICAR – Central Institute of Fisheries Education, Mumbai, India

Ashish Sahu Faculty of Fisheries Science, Department of Fisheries Resources Management, Kerala University of Fisheries and Ocean Studies, Panangad, Cochin, Kerala, India

ICAR-National Bureau of Fish Genetic Resources, Lucknow, Uttar Pradesh, India

J. D. M. Senevirathna Department of Animal Science, Faculty of Animal Science and Export Agriculture, Uva Wellassa University, Badulla, Sri Lanka

Tasaduq Hussain Shah Division of Fisheries Resources Management, Faculty of Fisheries, Sher-e-Kashmir University of Agricultural Science and Technology of Kashmir, Shalimar, Srinagar, J&K UT, India

Muhammad Mudassar Shahzad Department of Zoology, Division of Science and Technology, University of Education, Township, Lahore, Pakistan

Siddhnath College of Fisheries, Guru Angad Dev Veterinary and Animal Sciences University, Ludhiana, Punjab, India

Mahender Singh ICAR-National Bureau of Fish Genetic Resources, Lucknow, Uttar Pradesh, India

Mohammad Talib ICAR-Central Institute of Fisheries Education, Mumbai, India

Sangharsh Himmat Tayade Center of Excellence in Wildlife, Exotic, and Aquatic Animal Pathology, Faculty of Veterinary Science, Chulalongkorn University, Bangkok, Thailand

International Graduate Course of Veterinary Science and Technology (VST), Faculty of Veterinary Science, Chulalongkorn University, Bangkok, Thailand

G. G. N. Thushari Department of Animal Science, Faculty of Animal Science and Export Agriculture, Uva Wellassa University, Badulla, Sri Lanka

Muhammad Umer Department of Zoology, Division of Science and Technology, University of Education, Township, Lahore, Pakistan

Muhammad Usman Department of Zoology, Division of Science and Technology, University of Education, Township, Lahore, Pakistan

Rinkesh Nemichand Wanjari Department of Fisheries Resource Management, School of Fisheries, Centurion University of Technology and Management, Paralakhemundi, Odisha, India

Fatima Yasin Department of Zoology, Division of Science and Technology, University of Education, Township, Lahore, Pakistan

Microplastics: A Rising Global Concern Impacting Food Safety and Human Well-Being

Bejawada Chanikya Naidu, M. D. Sahana, and S. Abuthagir Iburahim

1 Introduction

Plastics, initially valued for their versatility, durability, and cost-effectiveness, have become a major environmental concern due to their persistent accumulation in natural ecosystems (Rukmangada et al., 2023). Among these pollutants, microplastics (MPs), defined as synthetic polymer particles ranging from 1 μm to 5 mm, pose a substantial threat due to their widespread distribution and complex interactions with biological systems (Palanisamy et al., 2024). MPs are classified as primary MPs, which are intentionally manufactured for commercial and industrial applications, and secondary MPs, which result from the fragmentation of larger plastic debris through physicochemical and biological degradation (Venkatesh et al., 2024). Their ubiquitous dispersal is evident from their detection in remote regions such as the Mariana Trench and Mount Everest (Naidu et al., 2025). The escalation of global plastic production, from two million tonnes in the 1950s to 368 million tonnes in 2019, with projections reaching 712 million tonnes by 2040, further exacerbates this crisis. Each year, an estimated 5–13 million metric tonnes of plastic waste enter the marine environment, significantly contaminating both aquatic and terrestrial ecosystems (Megha et al., 2025). The small size of MPs enhances their bioavailability, making them easily ingested by a wide range of aquatic organisms, including zooplankton, bivalves, crustaceans, and fish (Mallik et al., 2021). Their resemblance to natural food particles facilitates bioaccumulation and trophic transfer within food webs. Additionally, MPs have a high affinity for persistent organic pollutants (POPs), heavy metals, and antibiotic residues, acting as vectors for toxic contaminants (Naidu et al., 2022a). This has profound implications for human

B. C. Naidu (✉)
ICAR-Central Institute of Fisheries Technology, Kochi, India

M. D. Sahana · S. Abuthagir Iburahim
ICAR-Central Institute of Fisheries Education, Mumbai, India

A. Abubakr et al. (eds.), *Microplastics: Ecological and Food Security Implications*,
https://doi.org/10.1007/978-3-031-94480-2_1

health, particularly through seafood consumption, drinking water, and packaged food products. The COVID-19 pandemic has further aggravated plastic pollution, with improperly discarded medical waste, such as masks and gloves, increasing the burden of plastic debris in natural environments. MPs exhibit chemical stability and resistance to degradation, enabling their prolonged persistence in terrestrial and aquatic ecosystems (Naidu et al., 2022b). Their hydrophobic nature enhances their capacity to transport pollutants across environmental matrices. In aquatic systems, MPs are disseminated via atmospheric deposition, surface runoff, and hydrodynamic processes, thereby increasing their bioavailability to aquatic organisms (Rukmangada et al., 2023).

MPs exist in various morphological forms, including spheres, fibers, films, and irregular fragments, with their environmental distribution influenced by shape and density variations (Nguyen et al., 2025). Lighter MPs tend to remain suspended or float at the water surface, making them more accessible to pelagic organisms such as phytoplankton and zooplankton, while denser MPs sink and accumulate in sediments, where they are more likely to be encountered by benthic species such as amphipods, polychaetes, mollusks, and echinoderms (Mallik et al., 2021). Both pelagic and benthic fish ingest MPs either directly from their surroundings or indirectly through contaminated prey. Furthermore, aquatic birds and mammals that consume fish or invertebrates are also exposed to MPs, demonstrating their pervasive occurrence across multiple trophic levels (Zeng et al., 2025). The detection of MPs in remote and protected regions underscores their far-reaching ecological impact. Numerous studies have documented MP ingestion, retention, and assimilation in various organisms, highlighting their potential to disrupt biological functions, impair growth, and reduce reproductive success (Wang et al., 2025). Evidence of oxidative stress and neurotoxicity further emphasizes the long-term ecological consequences of MP contamination (Naidu et al., 2022b). Over time, MPs accumulate in aquatic environments, infiltrating food chains and presenting significant risks to human health (Palanisamy et al., 2024). Their ingestion and subsequent retention in biological systems raise concerns about bioaccumulation and trophic transfer (Rukmangada et al., 2023). In vivo studies indicate that nanoplastics, a smaller subset of MPs, can translocate beyond the gastrointestinal tract into critical biological systems such as the circulatory and lymphatic systems (Naidu et al., 2025). MPs have been detected in marine organisms, including digestive tracts, tissues, and excretions, demonstrating their persistence within biological systems (Resmi, 2025). Human exposure to MPs primarily occurs through seafood consumption, with additional pathways including drinking water and inhalation of airborne plastic fibers (Chanikya Naidu et al., 2023). These particles may induce inflammatory responses, oxidative stress, and disruptions in cellular processes (Ezhava et al., 2025). MPs also act as carriers for heavy metals and POPs, further amplifying their toxicity (Kim et al., 2025). Beyond their physical presence, MPs interact with biological systems through mechanisms that influence endocrine function and metabolic regulation. Plastic-derived compounds such as bisphenol A (BPA) and phthalates are known endocrine disruptors capable of interfering with hormonal signaling, with potential carcinogenic and neurotoxic effects (Resmi, 2025). MPs have also been shown to facilitate microbial colonization, potentially serving as vectors for pathogenic bacteria. Their widespread presence in marine and freshwater environments suggests complex interactions with ecological and human health systems. The detection of MPs in various human tissues, including the lungs, liver, placenta, and circulatory system, highlights the potential for systemic exposure (Malik et al., 2017; Megha et al., 2025). Their ability to translocate across biological barriers and accumulate in critical organs raises concerns about their long-term health implications. While excretion mechanisms such as biliary elimination and fecal excretion may mitigate some levels of exposure, the persistence of MPs within the human body remains a significant concern (Zhao et al., 2024). The escalating prevalence of MPs in aquatic environments and their infiltration into food sources necessitate further research to elucidate their toxicokinetics, long-term health effects, and effective mitigation strategies. Addressing MP contamination requires an interdisciplinary approach, integrating advancements in waste management, environmental monitoring, and regulatory policies to minimize human and ecological exposure. Given their persistence, bioavailability, and capacity to act as vectors for toxicants, MPs represent a pressing global concern requiring immediate scientific and policy-driven interventions.

2 Marine and Freshwater Systems as Contamination Hotspots

2.1 Lifespan and Transition of Plastics into Waste

Plastic products are designed to serve a wide range of functions, each with a distinct lifespan determined by factors such as material composition, application, and environmental exposure (Venkatesh et al., 2024). The durability and resistance of plastics make them versatile across multiple industries, yet their persistence also contributes to significant waste management challenges. The longevity of plastics varies widely, spanning from disposable items used for a few months to structural materials that endure for several decades (Zhao et al., 2024).

Short-Lived Plastics

Plastics designed for short-term use are primarily found in packaging, disposable consumer goods, and single-use items. These include plastic bags, food wrappers, beverage bottles, and medical supplies such as syringes and gloves. Such products are typically designed for immediate or temporary use and are discarded after fulfilling their purpose (Kim et al., 2025). Due to their high turnover rate, packaging plastics are considered to have an average lifespan of less than a year. While these products are essential for hygiene, food preservation, and convenience, their rapid disposal leads to significant waste accumulation, often exacerbating environmental pollution (Ezhava et al., 2025).

Long-Lived Plastics
In contrast, plastics used in construction, automotive, and infrastructure applications are engineered for durability and extended service life. Materials such as polyvinyl chloride (PVC), high-density polyethylene (HDPE), and reinforced plastic composites are widely employed in building components, pipelines, insulation, and structural reinforcements (Megha et al., 2025). These plastics have lifespans ranging from 30 to 50 years or more, depending on their exposure to environmental conditions and mechanical stress. Due to their longevity, they accumulate more slowly in the waste stream compared to short-lived plastics, but their disposal poses challenges due to the large volume and complex composition of the materials (Resmi, 2025).

Transition to Waste and Classification
Once plastics reach the end of their functional life, they are classified as waste, which falls into two main categories:

Pre-Consumer Waste (Industrial Waste): This includes plastic scraps, trimmings, and defective products generated during the manufacturing and production process. Such waste is often more straightforward to recycle since it is clean, uncontaminated, and consistent in composition (Li et al., 2025).

Post-Consumer Waste (Municipal Waste): This category comprises plastics discarded by end-users after use. It includes packaging waste, household plastics, and worn-out consumer goods. Post-consumer plastics present greater recycling challenges due to contamination, mixed polymer compositions, and improper disposal (Zhao et al., 2024).

2.2 *Pathways of Plastic Waste Disposal*

Upon disposal, plastic waste follows four primary pathways, each with distinct environmental and economic implications. The most sustainable option is recycling, where plastic materials are collected, sorted, processed, and reintroduced into the production cycle. Recycling helps conserve resources, reduce energy consumption, and minimize landfill waste; however, its effectiveness is hindered by various factors, including contamination, inefficient sorting processes, and the complexity of mixed polymer compositions (Zeng et al., 2025). Consequently, recycling rates remain disproportionately low compared to the vast scale of plastic production, leading to a growing accumulation of plastic waste worldwide (Venkatesh et al., 2024). Another method of plastic waste management is incineration for energy recovery, which involves burning plastics to generate electricity or heat. While this process reduces waste volume and recovers some energy value, it also raises concerns regarding greenhouse gas emissions and the release of toxic pollutants, such as dioxins and furans, which pose risks to both human health and ecosystems (Dey et al., 2024). Advanced waste-to-energy technologies have improved emission controls, but incineration remains a controversial solution due to its environmental trade-offs and potential to discourage sustainable waste reduction practices. Landfill deposition is one of the most common disposal routes for plastic waste (Megha et al., 2025). In controlled landfill environments, plastics are buried and compacted to manage waste accumulation. However, plastics degrade slowly, persisting in the environment for centuries and contributing to long-term pollution risks. Moreover, landfills often experience leakage of MPs and chemical additives into surrounding soil and groundwater, threatening biodiversity and water quality (Yek et al., 2024). The most concerning fate of plastic waste is mismanagement, where plastics are discarded without proper treatment or containment. This occurs due to inadequate waste management infrastructure, lack of public awareness, and improper disposal practices. Mismanaged plastic waste is a major driver of environmental pollution, as it accumulates in natural ecosystems and spreads through wind, water currents, and human activities (Dey et al., 2024; Nabi et al., 2024). Once released into the environment, plastic waste pollutes terrestrial, freshwater, marine, and even atmospheric systems, affecting wildlife, contaminating food chains, and exacerbating global environmental challenges. Addressing plastic mismanagement requires coordinated efforts in policy enforcement, waste collection systems, and public engagement to ensure a more sustainable approach to plastic waste disposal (Venkatesh et al., 2024).

2.3 Sources, Transport, and Redistribution of Plastic Pollution

Plastic pollution originates from multiple anthropogenic sources and enters ecosystems through diverse pathways, leading to its widespread distribution across terrestrial, freshwater, and marine environments. Ineffective waste management practices, including unregulated landfill burial, stormwater runoff, and industrial mishandling, contribute significantly to plastic contamination (Mallik et al., 2021). Tourism and coastal activities exacerbate the issue, particularly in marine ecosystems where vessels, cruise ships, oil platforms, and aquaculture operations release substantial quantities of plastic waste. Urban and residential areas serve as additional sources, generating MPs from consumer products such as personal care items, synthetic textiles, and packaging materials (Naidu et al., 2022a). Many of these products contain microparticles smaller than 100 micrometers, which enter sewage systems and ultimately reach aquatic environments due to the inefficiency of wastewater treatment plants (Naidu et al., 2022b; Rukmangada et al., 2023). Furthermore, agricultural activities, including the application of plastic-based fertilizers and wind-driven dispersion of synthetic fibers, intensify plastic pollution across diverse ecological landscapes. Once introduced into the environment, plastics undergo extensive remobilization through natural and anthropogenic processes (Zeng et al., 2025). Wind plays a critical role in redistributing both micro- and macroplastics across terrestrial and atmospheric domains, facilitating long-range transport and deposition (Xu et al., 2025). Hydrological events such as storms, flooding, and irrigation accelerate the transfer of plastic waste from land to aquatic systems, contributing to its accumulation in riverine and marine habitats. Airborne MPS, derived from textile fiber shedding, industrial emissions, and vehicular wear, further amplify contamination by infiltrating remote ecological zones (Rede et al., 2025). These MPs, suspended in atmospheric currents, undergo global-scale transport, leading to their deposition in otherwise pristine environments. The persistence of plastics across multiple environmental matrices underscores the complexity of their movement, necessitating a comprehensive understanding of their fate, interactions, and ecological impacts to develop effective mitigation strategies (Naidu et al., 2025).

2.4 MPs Fate, and Ecological Consequences in Aquatic Systems

Freshwater and marine ecosystems serve as critical reservoirs and transport pathways for MPs, with lakes, rivers, and estuaries acting as conduits that channel plastic pollutants from terrestrial sources into oceanic environments (Dey et al., 2024). These contaminants originate from diverse anthropogenic activities, including industrial discharge, urban runoff, wastewater treatment effluents, and atmospheric deposition, leading to their accumulation in aquatic sediments, shorelines, and deep-sea environments (Nguyen et al., 2025). Hydrodynamic forces such as ocean currents, tidal action, and storm events further exacerbate MPs redistribution, facilitating their dispersal across various ecological compartments (Rukmangada et al., 2023). Freshwater systems play a pivotal role in this transport cycle, with pollutants settling in calmer zones where biofouling, aggregation, and gravitational settling contribute to their persistence. Industrial processes, vehicular emissions, and synthetic textile fibers released during laundry activities introduce significant MPs loads into aquatic systems, where wastewater treatment plants remain largely ineffective in capturing all MPs fractions before effluent discharge (Ghosh et al., 2025). MPs originate from both primary sources, such as microbeads in cosmetics and industrial abrasives, and secondary sources, including the fragmentation of macroplastic debris through photodegradation, mechanical abrasion, and microbial breakdown (Rede et al., 2025). Their physical and chemical properties, including hydrophobicity, buoyancy, and size distribution, influence their environmental fate, with MPs undergoing prolonged residence times in sedimentary sinks, bioaccumulating in aquatic organisms, and integrating into biogeochemical cycles. While marine environments are often regarded as the ultimate sinks for plastic pollution, a considerable fraction remains entrapped in freshwater ecosystems, where sediment dynamics, organic matter interactions, and anthropogenic modifications determine their spatial distribution (Dey et al., 2024).

The ecological ramifications of MPs contamination extend across trophic levels, affecting microbial communities, planktonic organisms, invertebrates, and higher vertebrates through ingestion, bioaccumulation, and trophic transfer (Naidu et al.,

2022b). MPs ingestion disrupts physiological processes, alters feeding behaviors, and enhances the bioavailability of hydrophobic pollutants that adsorb onto plastic surfaces, exacerbating toxicological risks. Urbanized watersheds exhibit heightened MPs concentrations due to elevated anthropogenic pressures, emphasizing the necessity for stringent pollution mitigation measures (Qiu et al., 2020). In addition to direct ingestion, MPs serve as vectors for persistent organic pollutants (POPs), heavy metals, and pathogenic microorganisms, further complicating their ecological impacts. The combined effects of temperature fluctuations, chemical stressors, and other anthropogenic disturbances intensify the risks associated with MPs exposure, influencing ecosystem stability and biodiversity (Venkatesh et al., 2024). Despite their minuscule size, MPs exhibit remarkable environmental persistence, with long-range atmospheric and hydrological transport mechanisms facilitating their deposition in remote and pristine habitats. Their ubiquitous presence underscores the urgent need for advanced monitoring frameworks, comprehensive risk assessments, and stringent regulatory interventions to curb their environmental influx (Mallik et al., 2021). Developing a holistic understanding of MPs transport dynamics, bioavailability, and ecological interactions is imperative for devising effective mitigation strategies and ensuring long-term environmental sustainability (Ghosh et al., 2025).

3 MPs Bioaccumulation in Seafood

MPs contamination in marine ecosystems has become a significant concern due to its bioaccumulation in seafood species and its implications for food safety and human health. Numerous studies have confirmed MP accumulation in commercially relevant marine organisms, including fish, crustaceans, and mollusks, with trophic transfer occurring through direct ingestion of contaminated prey (Smith et al., 2018). Bioaccumulation and potential biomagnification within marine food webs exacerbate the risks associated with MP exposure. Filter-feeding bivalves such as mussels (Mytilus spp.) and oysters (Crassostrea spp.) are particularly susceptible to MP contamination due to their indiscriminate ingestion of suspended particulate matter (Naidu et al., 2022b). MPs are retained in their gastrointestinal tracts, hemolymph, and soft tissues, disrupting key physiological and metabolic processes (Palanisamy et al., 2024). Laboratory experiments have demonstrated that prolonged MP exposure can alter enzymatic activity, impair immune responses, and affect energy metabolism in bivalves, highlighting the potential for ecosystem-wide consequences (Rede et al., 2025). The mechanisms of MP uptake and retention in bivalves vary depending on polymer composition, particle size, surface charge, and environmental factors such as water temperature, salinity, and MP concentration (Vital et al., 2021). The presence of MPs in these species is particularly concerning because they are consumed whole, allowing for direct ingestion of MPs by humans and facilitating their entry into the human gastrointestinal system (Smith et al., 2018).

Fish species demonstrate varying degrees of MP contamination, largely influenced by feeding strategies, habitat preferences, and trophic levels. Carnivorous species are more vulnerable to secondary ingestion due to their predation on MP-contaminated organisms, whereas planktivorous fish ingest MPs directly from the water column (Qiu et al., 2020). MPs have been detected in multiple tissues, including the gastrointestinal tracts, gills, and liver, raising concerns regarding their systemic distribution and potential toxicity to higher trophic levels, including humans. Studies have shown that MPs can induce oxidative stress, lipid peroxidation, and inflammatory responses in fish, impairing physiological functions such as respiration, digestion, and immune regulation (Mallik et al., 2021). Crustaceans, including shrimp and crabs, also exhibit MP accumulation through dietary exposure and environmental interactions, with evidence of MPs in commercially important shrimp species (Naidu et al., 2025). The extent of MPs retention in seafood varies based on polymer type, particle morphology, and organism-specific physiology. In both wild-caught and farmed seafood, MPs enter through dietary intake and aquatic exposure, underscoring the global scale of this issue. Even in aquaculture settings, where controlled feeding practices are employed, MPs can infiltrate through contaminated feed and water sources, further contributing to human exposure. The risk is particularly pronounced in seafood consumed whole, such as small fish and bivalves, in which MPs remain embedded within their digestive systems (Vital et al., 2021). Beyond their physical presence, MPs pose additional toxicological risks due to their ability to adsorb and transport harmful environmental contaminants, including persistent organic pollutants (POPs), heavy metals, and plastic-derived additives such as bisphenol A (BPA) and phthalates. Once ingested, MPs can act as vectors for these toxicants, facilitating their bioavailability and accumulation in human tissues. The potential health effects of MP ingestion remain under investigation, but emerging evidence suggests associations with gastrointestinal inflammation, oxidative stress, dysbiosis of gut microbiota, endocrine disruption, and metabolic dysfunction (Kim et al., 2025). Additionally, MPs have been linked to immune system dysregulation, potentially increasing susceptibility to infections and inflammatory diseases. Despite growing awareness of MP contamination in seafood, significant research and regulatory gaps persist (Li et al., 2025). Challenges include standardizing MP detection methodologies, assessing the long-term health implications of chronic exposure, and implementing stringent policies to mitigate contamination (Ghosh et al., 2025). Addressing MP pollution in marine environments necessitates an integrated approach involving improved waste management, reduced plastic production, enhanced filtration technologies, and stricter regulatory measures to limit plastic discharge into aquatic ecosystems. Sustainable interventions targeting plastic pollution at its source are crucial for protecting marine biodiversity, preserving ecosystem stability, and safeguarding human health from the pervasive and potentially hazardous effects of MP exposure (Qiu et al., 2020; Dey et al., 2024).

4 Bioaccumulation and Systemic Distribution of MPs in the Human Body

MPs have emerged as a critical environmental and public health concern due to their ability to bioaccumulate and distribute systemically within the human body (Fig. 1). These synthetic polymeric particles, enter biological systems primarily through ingestion, inhalation, and, to a lesser extent, dermal absorption (Kim et al., 2025). Their widespread presence in food, water, and air, coupled with their ability to interact with biological tissues at the cellular and molecular levels, underscores their potential toxicity and long-term health implications (Blackburn & Green, 2022). The ingestion of MPs is one of the most significant exposure pathways, as these particles are commonly detected in dietary sources such as seafood, table salt, honey, sugar, and drinking water (Naidu et al., 2025). Studies have demonstrated that bottled water contains significantly higher MP concentrations than tap water, primarily due to contamination from plastic packaging materials (Rukmangada et al., 2023). Additionally, MPs can accumulate in processed and packaged foods due to plastic leaching and airborne deposition. An often-overlooked source of ingestion is household dust, which contributes considerably to total MP intake, sometimes surpassing dietary exposure from seafood (Vital et al., 2021). The continuous deposition of airborne MPs onto food surfaces and direct hand-to-mouth behavior, particularly in infants and toddlers, further amplifies the risk of ingestion. Once consumed, MPs interact with the gastrointestinal mucosa, where they may undergo transcellular and paracellular transport mechanisms, allowing systemic translocation (Mittal et al., 2024). Inhalation is another critical route of MP exposure, particularly given their prevalence in indoor air, where they are released from synthetic textiles, carpets, and plastic-containing household products. Airborne MPs can be deposited in the respiratory tract, with particle size determining their penetration depth. Larger MPs (>5 μm) tend to be trapped in the upper airways and cleared via mucociliary

Fig. 1 MPs fate and ecological consequences: environmental transport, bioaccumulation in aquatic systems, and human health impacts

transport, while smaller MPs, especially nanoplastics (<1 μm), can reach the alveolar regions, where they may undergo endocytosis by alveolar macrophages or translocate into systemic circulation via the pulmonary capillary network. Detection of MPs in lung tissues, bronchoalveolar lavage fluid, and even fetal and placental tissues suggests that respiratory exposure is a significant yet underappreciated concern (Blackburn & Green, 2022). Prolonged inhalation of MPs has been associated with lung inflammation, fibrosis, and oxidative stress, increasing susceptibility to respiratory diseases such as asthma and chronic obstructive pulmonary disease (COPD) (Jenner et al., 2022). Though less significant than ingestion and inhalation, dermal exposure remains a potential concern, particularly for MPs smaller than 100 nm. These nanoparticles exhibit increased skin permeability due to their ability to penetrate hair follicles and intercellular lipid layers (Winiarska et al., 2024). MPs used in personal care products such as exfoliating agents in cosmetics and microbeads in cleansers can interact with skin cells, leading to oxidative stress and potential cytotoxic effects. While the skin serves as a robust barrier against external insults, compromised skin integrity-due to conditions such as eczema, wounds, or prolonged exposure could facilitate MP penetration and systemic absorption (Amato-Lourenço et al., 2021).

Once MPs breach biological barriers, they enter the circulatory and lymphatic systems, allowing their distribution to various organs. Smaller MPs, particularly nanoplastics, exhibit enhanced translocation efficiency due to their high surface-area-to-volume ratio and increased reactivity (Blackburn & Green, 2022). Studies have demonstrated the presence of MPs in critical organs such as the lungs, liver, kidneys, intestines, and brain, raising concerns about their ability to elicit localized and systemic toxicity (Winiarska et al., 2024). In the gastrointestinal tract, MPs can disrupt gut microbiota homeostasis, compromise intestinal barrier integrity, and induce inflammatory responses (Mittal et al., 2024). Studies have linked MP exposure to dysbiosis, characterized by alterations in microbial composition that can influence host metabolism and immune function. This disruption has been implicated in metabolic disorders such as insulin resistance and nonalcoholic fatty liver disease (NAFLD) (Hunt et al., 2024; Sun et al., 2024). In hepatic tissues, MPs interfere with lipid metabolism, promoting oxidative stress, endoplasmic reticulum stress, and inflammatory cytokine release, all of which contribute to hepatic dysfunction. Inhaled MPs not only cause direct damage to lung tissues but can also enter the bloodstream via alveolar-capillary translocation (Jenner et al., 2022). The accumulation of MPs in the respiratory system has been associated with oxidative stress-induced DNA damage, mitochondrial dysfunction, and inflammatory responses that may exacerbate pulmonary conditions such as asthma, fibrosis, and even carcinogenesis. The cardiovascular system is similarly vulnerable, as MPs have been detected in cardiac tissues and blood vessels, where they may promote vascular inflammation, endothelial dysfunction, and lipid accumulation in macrophages (Zhao et al., 2024). These mechanisms suggest a potential role for MPs in the pathogenesis of atherosclerosis and other cardiovascular diseases. MPs also pose a considerable threat to the nervous system, given their ability to cross the blood-brain barrier (BBB). The BBB is a highly selective endothelial interface designed to protect the brain from neurotoxic compounds; however, studies have shown that nanosized MPs can penetrate this barrier, accumulating in brain tissues and inducing neuroinflammation, oxidative stress, and neurotransmitter dysregulation (Winiarska et al., 2024). Experimental models suggest that MP exposure may be linked to neurodegenerative conditions such as Parkinson's and Alzheimer's diseases due to their role in promoting neurotoxicity and synaptic dysfunction. The presence of MPs in gonadal tissues and reproductive organs raises concerns about their potential impact on fertility and reproductive health. MPs are known to carry endocrine-disrupting chemicals (EDCs) such as bisphenols, phthalates, and persistent organic pollutants (POPs), which can interfere with hormone signaling pathways (Kim et al., 2025). These contaminants have been linked to decreased sperm quality, impaired ovarian function, and developmental abnormalities in offspring (Sun et al., 2024). Studies suggest that MP exposure during critical reproductive windows, such as foetal development and puberty, can lead to permanent physiological alterations, increasing the risk of infertility and adverse pregnancy outcomes. The detection of MPs in placental tissues highlights the potential for prenatal exposure, as these particles can traverse the maternal-foetal interface (Hunt et al., 2024; Ragusa et al., 2021). Prenatal MP exposure has been associated with impaired embryonic development, increased oxidative stress in foetal tissues, and disruptions in foetal programming that may predispose offspring to metabolic and neurological disorders later in life. Given the sensitivity of the developing foetus to environmental contaminants, the long-term consequences of MP-induced endocrine disruption warrant further investigation (Ezhava et al., 2025; Ragusa et al., 2021).

5 Toxicological Risks and Potential Health Impacts

MPs present significant toxicological risks due to their intrinsic chemical composition, capacity to sorb environmental pollutants, and ability to interact with biological systems at the molecular and cellular levels. The effects of MPs and nanoplastics (NPs) on various biological systems is presented in Table 1. Beyond their physical presence, MPs serve as vectors for persistent organic pollutants (POPs), heavy metals, and endocrine-disrupting chemicals (EDCs), exacerbating their toxicity upon human exposure (Taher et al., 2024; Gurjar et al., 2023). Plastics are manufactured with chemical additives such as plasticizers, flame retardants, and stabilizers, which can leach under physiological conditions. Notable among these are bisphenol A (BPA) and phthalates, which interfere with endocrine signaling and have been implicated in reproductive dysfunction, metabolic disorders, and neurological impairments. Additionally, MPs adsorb environmental contaminants, including dioxins, polychlorinated biphenyls (PCBs), polycyclic aromatic hydrocarbons (PAHs), and toxic metals like lead, mercury, and cadmium, all of which can induce oxidative stress, mitochondrial dysfunction, DNA damage, and epigenetic alterations (Sruthisree et al., 2015; Amin et al., 2016, 2017; Prasad et al., 2023; Luo

Table 1 Overview of the major toxic effects of MPs/NPs across different biological systems

System	Key effects of MPs/NPs	References
Digestive system	Accumulation in intestines, gut barrier dysfunction, altered gut microbiota, metabolic disorders (NAFLD, diabetes), inflammation, oxidative stress, DNA damage	Yan et al. (2023) and Fournier et al. (2023)
Liver	Accumulation leading to insulin resistance, lipid metabolism disorders, oxidative stress, histological lesions (inflammation, necrosis)	Wen et al. (2022)
Respiratory system	Airway inflammation, oxidative stress, lung fibrosis, granulomas, disrupted lung microbiota, epithelial damage	Zha et al. (2023)
Cardiovascular system	Pericardial accumulation, endothelial dysfunction, oxidative stress, myocardial injury, vascular inflammation, atherosclerosis	Wang et al. (2025)
Nervous system	Blood-brain barrier penetration, neuroinflammation, cognitive decline, neurotransmitter disruption, Parkinson's-like degeneration	Kwon et al. (2022) and Liang et al. (2022)
Reproductive system	Sperm abnormalities, testicular and ovarian damage, endocrine disruption, transgenerational toxicity	Xie et al. (2022)
Immune system	Immune cell damage, inflammation, altered cytokine levels, compromised gut immune barrier	Kwon et al. (2022)
Endocrine system	Disrupted hormone levels (thyroid, testosterone, estradiol), endocrine axis dysfunction	Torres-Ruiz et al. (2023) and Haddadi et al. (2022)

et al., 2024; Ezhava et al., 2025). These pollutants, once internalized, exert cumulative toxicity that affects multiple organ systems, compounding the risks of metabolic syndromes, carcinogenesis, and neurodegenerative disorders. The ingestion of MPs can lead to gastrointestinal toxicity by mechanically irritating the gut lining, altering microbial composition, increasing gut permeability, and impairing nutrient absorption. Microbiota dysbiosis and gut barrier dysfunction are strongly linked to inflammatory disorders, insulin resistance, nonalcoholic fatty liver disease (NAFLD), and other metabolic abnormalities, making chronic MP exposure a potential driver of systemic inflammation and metabolic disruption (Zhao et al., 2024).

Beyond gastrointestinal toxicity, MPs pose significant immunological and neurological risks, as they can penetrate biological barriers such as the blood-brain barrier (BBB) and accumulate in neural tissues, inducing neuro inflammation, oxidative stress, and neurotransmitter imbalances. MPs have been detected in the brain, where they activate microglial cells and disrupt neural signaling, raising concerns about their potential role in neurodegenerative diseases such as Parkinson's and Alzheimer's (Mittal et al., 2024). Immunotoxicity is another critical concern, as MPs modulate immune responses, impairing antigen-presenting cell (APC) function, suppressing lymphocyte proliferation, and increasing susceptibility to infections and autoimmune conditions. Chronic inflammation triggered by MP exposure has been implicated in oncogenesis through the release of pro-inflammatory cytokines (e.g., TNF-α, IL-6, IL-1β), promoting tumor initiation, genomic instability, and epigenetic dysregulation. Experimental evidence suggests that MPs contribute to carcinogenesis via oxidative DNA damage, chronic inflammation, and epigenetic modifications, but large-scale epidemiological studies are needed to establish definitive causal relationships (Lei et al., 2024). Moreover, MPs have been detected in reproductive tissues, where they disrupt hormone regulation, impair ovarian and testicular function, and potentially contribute to reduced fertility and adverse developmental outcomes in offspring. Given the ability of MPs to bioaccumulate and induce long-term biological stress, their potential role in chronic diseases, including cancer, cardiovascular disorders, and neurodegeneration, is an area of growing concern (Luo et al., 2024; Mittal et al., 2024).

6 Plastic Waste Management and Reduction Strategies

Plastic waste is a significant component of municipal solid waste in most developed countries, with millions of tonnes entering the waste stream annually. Despite efforts to reduce landfill deposits, approximately 25% of plastic waste still ends up in landfills. A primary challenge in mitigating plastic waste and MPs pollution is the establishment of strong, unified legislation, alongside strict enforcement and penalties for non-compliance (Mehran et al., 2025). Effective monitoring systems for marine debris are crucial for understanding spatial and temporal distribution patterns. Additionally, raising public awareness through educational campaigns and events can drive more sustainable behaviour (Ganguly & Chakraborty, 2024). Biodegradable plastics offer promise, but their effectiveness depends on various factors, including the advancement of waste sorting and organic waste processing technologies. The most effective strategy for minimizing MPs is reducing overall plastic production, which can be achieved by promoting alternative materials such as glass and cardboard, as well as encouraging the reuse of discarded plastics through creative repurposing (Suvitha et al., 2024). One particularly successful approach has been container deposit return schemes, which incentivize recycling through financial rewards. Recycling also plays a vital role in plastic waste management, as it conserves energy, reduces resource consumption, and mitigates emissions. When recycling or reuse is not feasible, energy recovery processes such as mechanical, biological, and thermal conversion technologies can transform plastic waste into usable energy or raw materials (Liu et al., 2024).

At the industrial level, the plastics sector must align with the principles of a circular economy, integrating waste reduction into production processes. Key strategies for achieving this include redesigning plastics to enhance recyclability, minimizing the loss of preproduction plastic pellets, extending producer responsibility, banning certain single-use plastics, and supporting markets for recycled plastics

(Doniavi et al., 2024). Ecodesign, guided by environmental standards such as ISO 14006, can play a pivotal role in reducing the environmental impact of plastics throughout their lifecycle. Effective measures include eliminating toxic additives, developing durable and reusable plastic products, and shifting towards alternative materials with lower environmental footprints. Extended Producer Responsibility (EPR) schemes further enhance circularity by holding manufacturers accountable for post-consumer waste management (Mousavi et al., 2024). However, challenges remain in standardizing these schemes across different regions, necessitating harmonized regulations and financial mechanisms that reflect real waste management costs. Measures to prevent plastic pellet loss, such as improved containment, enhanced cleaning techniques, and stringent handling protocols, are essential in reducing environmental contamination (Suvitha et al., 2024; Liu et al., 2024; Debnath et al., 2024). Many governments have implemented bans on single-use plastics to curb marine pollution, yet progress has been hindered in some regions due to setbacks from the COVID-19 pandemic (Naidu et al., 2022b; Prasad et al., 2024; Tabassum et al., 2025). The demand for sustainable materials has driven advancements in bioplastics, which offer environmental benefits such as reduced reliance on fossil fuels and lower carbon footprints. However, confusion persists around the definitions and classifications of bioplastics, as biodegradability depends on chemical structure rather than raw material origin (Mallik et al., 2021). While bioplastics are increasingly used in packaging, pharmaceuticals, and agriculture, challenges such as food safety concerns, collection inefficiencies, and compost quality control must be addressed. Strengthened legislation and improved infrastructure for bioplastic waste management are essential to prevent further environmental harm. MPs pollution from wastewater treatment plants presents another major environmental challenge, as these facilities, despite removing up to 98% of MPs, still release significant quantities into aquatic ecosystems (Naidu et al., 2025; Rukmangada et al., 2023). Mps originate primarily from cosmetics, cleaning agents, and synthetic textiles, with wastewater treatment processes capturing most particles in sludge. However, sludge disposal practices, such as its use as fertilizer, inadvertently reintroduce MPs into the environment. Studies indicate that billions of MPs particles pass through wastewater treatment plants annually, with a substantial fraction ending up in natural water bodies (Venkatesh et al., 2024). Enhancing tertiary treatment processes, such as disc filtration, rapid sand filtration, dissolved air flotation, and membrane bioreactors, is critical for minimizing MPs contamination. Additional solutions include the application of ozone treatment, reverse osmosis, and acoustic wave filtration systems in washing machines to reduce textile-derived MPs emissions (Ganguly & Chakraborty, 2024). As society continues to confront the environmental threats posed by MPs, investing in innovative wastewater treatment technologies and refining sludge management strategies will be essential in mitigating plastic pollution at a global scale (Liu et al., 2024).

7 Conclusion

MPs have emerged as a formidable environmental and public health challenge, with their pervasive presence in marine and freshwater ecosystems, atmospheric compartments, and terrestrial environments raising significant concerns. The rapid escalation of global plastic production, coupled with inadequate waste management practices, has exacerbated the accumulation of MPs in natural habitats, leading to widespread contamination across trophic levels. The small size of MPs facilitates their bioavailability, allowing them to infiltrate aquatic food chains and ultimately enter the human diet through seafood consumption, drinking water, and atmospheric deposition. The persistence of MPs in the environment, their ability to adsorb and transport toxic pollutants, and their capacity to induce adverse biological effects underscore the urgency of addressing this global issue. Despite increasing awareness and scientific research, substantial knowledge gaps remain regarding the long-term ecological and health consequences of chronic MP exposure. The bioaccumulation and trophic transfer of MPs have profound implications for ecosystem stability and biodiversity, as ingestion by aquatic organisms disrupts physiological processes, reduces reproductive success, and increases mortality rates. Additionally, the adsorption of persistent organic pollutants (POPs), heavy metals, and endocrine-disrupting compounds onto MP surfaces heightens their toxicity, exacerbating risks for both wildlife and human populations. Human exposure to MPs is not confined to dietary intake but also extends to inhalation and dermal absorption, raising concerns about systemic distribution and potential toxicokinetic effects. The detection of MPs in human tissues, including the lungs, placenta, liver, and bloodstream, highlights their capacity to traverse biological barriers and induce inflammatory responses, oxidative stress, and endocrine disruption. The interplay between MPs and environmental pollutants further complicates risk assessments, necessitating advanced toxicological studies to elucidate their role in chronic diseases such as metabolic disorders, cardiovascular diseases, neurodegeneration, and cancer. Addressing the growing crisis of MP pollution requires a multifaceted and interdisciplinary approach encompassing policy reforms, technological innovations, and public awareness initiatives. Strengthening global plastic waste management frameworks, enforcing stringent regulations on plastic production and disposal, and fostering a transition toward sustainable alternatives are critical steps in mitigating MP contamination. Scientific advancements in wastewater treatment, MPs filtration technologies, and bioremediation strategies offer promising avenues for reducing MP discharge into natural ecosystems. Additionally, public engagement through education and behavioral interventions can drive systemic change, promoting responsible consumption, waste reduction, and the adoption of circular economy principles. As MPs continue to pose unprecedented challenges to environmental and human health, a concerted effort from policymakers, researchers, industries, and communities is imperative to develop and implement effective mitigation strategies. Future research must focus on refining MP detection methodologies, assessing long-term exposure risks, and exploring innovative solutions to curb their environmental persistence.

Only through sustained global collaboration can we mitigate the detrimental impacts of MPs, safeguard ecological integrity, and ensure the health and well-being of future generations.

References

Amato-Lourenço, L. F., Carvalho-Oliveira, R., Júnior, G. R., dos Santos Galvão, L., Ando, R. A., & Mauad, T. (2021). Presence of airborne microplastics in human lung tissue. *Journal of Hazardous Materials, 416*, 126124.

Amin, A., Indulkar, S. T., & Pai, R. (2016). Sublethal effect of Buprofezin pesticide on carcass composition of Cyprinus carpio communis fingerlings. *Journal of Experimental Zoology India, 20*(1), 269–272.

Amin, A., Ramachandra Naik, A. T., Priyadarshini, N., Nayak, H., & Sree, C. S. (2017). Toxic effect of heavy metal lead on oxygen consumption of rohu (*Labeo rohita*) fingerlings. *Biochemical and Cellular Archives, 17*(1), 225–228.

Blackburn, K., & Green, D. (2022). The potential effects of microplastics on human health: What is known and what is unknown. *Ambio, 51*(3), 518–530.

Chanikya Naidu, B., Sahana, M. D., Hoque, M., & Abuthagir Iburahim, S. (2023). Scientometric analysis of ecotoxicological investigations of xenobiotics in aquatic animals. In *Xenobiotics in aquatic animals: Reproductive and developmental impacts* (pp. 303–323). Springer Nature Singapore.

Debnath, R., Prasad, G. S., Amin, A., Malik, M. M., Ahmad, I., Abubakr, A., Borah, S., Rather, M. A., Impellitteri, F., Tabassum, I., & Piccione, G. (2024). Understanding and addressing microplastic pollution: Impacts, mitigation, and future perspectives. *Journal of Contaminant Hydrology, 266*, 104399.

Dey, S., Veerendra, G. T. N., Babu, P. A., Manoj, A. P., & Nagarjuna, K. (2024). Degradation of plastics waste and its effects on biological ecosystems: A scientific analysis and comprehensive review. *Biomedical Materials & Devices, 2*(1), 70–112.

Doniavi, E., Babazadeh, R., & Hasanzadeh, R. (2024). Polyethylene gasification for sustainable plastic waste management with respect to energy, exergy, and environmental considerations: A non-linear programming optimization. *Process Safety and Environmental Protection, 182*, 86–97.

Ezhava, S. R., Lobo, V. H., & Rodge, S. P. (2025). Hazards of microplastics on human health and its mitigation strategies. *Toxicology and Environmental Health Sciences*, 1–14. https://doi.org/10.1007/s13530-025-00249-9

Fournier, E., Ratel, J., Denis, S., Leveque, M., Ruiz, P., Mazal, C., Amiard, F., Edely, M., Bezirard, V., Gaultier, E., & Lamas, B. (2023). Exposure to polyethylene microplastics alters immature gut microbiome in an infant in vitro gut model. *Journal of Hazardous Materials, 443*, 130383.

Ganguly, R. K., & Chakraborty, S. K. (2024). Plastic waste management during and post Covid19 pandemic: Challenges and strategies towards circular economy. *Heliyon, 10*, e25613.

Ghosh, S., Dey, S., Mandal, A. H., Sadhu, A., Saha, N. C., Barceló, D., Pastorino, P., & Saha, S. (2025). Exploring the ecotoxicological impacts of microplastics on freshwater fish: A critical review. *Journal of Contaminant Hydrology, 269*, 104514.

Gurjar, U. R., Takar, S., Amin, A., & Martin Xavier, K. A. (2023). Microplastic contamination in aquatic organisms: An ecotoxicological perspective. In *Xenobiotics in aquatic animals: Reproductive and developmental impacts* (pp. 353–367). Springer Nature Singapore.

Haddadi, A., Kessabi, K., Boughammoura, S., Rhouma, M. B., Mlouka, R., Banni, M., & Messaoudi, I. (2022). Exposure to microplastics leads to a defective ovarian function and change in cytoskeleton protein expression in rat. *Environmental Science and Pollution Research, 29*(23), 34594–34606.

Hunt, K., Davies, A., Fraser, A., Burden, C., Howell, A., Buckley, K., Harding, S., & Bakhbakhi, D. (2024). Exposure to microplastics and human reproductive outcomes: A systematic review. *BJOG: An International Journal of Obstetrics & Gynaecology, 131*(5), 675–683.

Jenner, L. C., Rotchell, J. M., Bennett, R. T., Cowen, M., Tentzeris, V., & Sadofsky, L. R. (2022). Detection of microplastics in human lung tissue using μFTIR spectroscopy. *Science of the Total Environment, 831*, 154907.

Kim, N., Lee, J. H., Lee, I., Park, J. H., Jung, G. S., Lee, M. J., Im, W., Cho, S., & Choi, Y. S. (2025). Investigation of potential toxic effects of nano-and microplastics on human endometrial stromal cells. *Reproductive Toxicology, 132*, 108848.

Kwon, W., Kim, D., Kim, H. Y., Jeong, S. W., Lee, S. G., Kim, H. C., Lee, Y. J., Kwon, M. K., Hwang, J. S., Han, J. E., & Park, J. K. (2022). Microglial phagocytosis of polystyrene microplastics results in immune alteration and apoptosis in vitro and in vivo. *Science of the Total Environment, 807*, 150817.

Lei, J., Ma, Q., Ding, X., Pang, Y., Liu, Q., Wu, J., Zhang, H., & Zhang, T. (2024). Microplastic environmental behavior and health risk assessment: A review. *Environmental Chemistry Letters, 22*, 2913–2941.

Li, W., Zou, H., Zheng, Y., Zhang, G., Xiang, Y., Zhi, D., & Zhou, Y. (2025). Microplastics in aquatic environments: Detection, abundance, characteristics, and toxicological studies. *Environmental Monitoring and Assessment, 197*(2), 1–22.

Liang, H., Ji, Y., Ge, W., Wu, J., Song, N., Yin, Z., & Chai, C. (2022). Release kinetics of microplastics from disposable face masks into the aqueous environment. *Science of the Total Environment, 816*, 151650.

Liu, L., Barlaz, M. A., & Johnson, J. X. (2024). Economic and environmental comparison of emerging plastic waste management technologies. *Resources, Conservation and Recycling, 205*, 107531.

Luo, D., Chu, X., Wu, Y., Wang, Z., Liao, Z., Ji, X., Ju, J., Yang, B., Chen, Z., Dahlgren, R., & Zhang, M. (2024). Micro-and nano-plastics in the atmosphere: A review of occurrence, properties and human health risks. *Journal of Hazardous Materials, 465*, 133412.

Malik, M., Balkhi, M. H., & Abubakr, A. (2017). Comparative study on the contamination level of Dal Lake, Kashmir based on Total Viable Bacterial Counts over a period of time. *Ecology, Environment and Conservation, 23*(4), 2145–2147.

Mallik, A., Xavier, K. M., Naidu, B. C., & Nayak, B. B. (2021). Ecotoxicological and physiological risks of microplastics on fish and their possible mitigation measures. *Science of the Total Environment, 779*, 146433.

Megha, K. B., Anvitha, D., Parvathi, S., Neeraj, A., Sonia, J., & Mohanan, P. V. (2025). Environmental impact of microplastics and potential health hazards. *Critical Reviews in Biotechnology, 45*(1), 97–127.

Mehran, M. T., Raza Naqvi, S., Ali Haider, M., Saeed, M., Shahbaz, M., & Al-Ansari, T. (2025). Global plastic waste management strategies (technical and behavioral) during and after COVID-19 pandemic for cleaner global urban life. *Energy Sources, Part A: Recovery, Utilization, and Environmental Effects, 47*(1), 4472–4481.

Mittal, N., Tiwari, N., Singh, D., Tripathi, P., & Sharma, S. (2024). Toxicological impacts of microplastics on human health: A bibliometric analysis. *Environmental Science and Pollution Research, 31*(46), 57417–57429.

Mousavi, M., Kowsari, E., Gheibi, M., Cheshmeh, Z. A., Teymoorian, T., & Ramakrishna, S. (2024). Assessing bioplastics' economic, commercial, political, and energy potential with circular economy modeling: A sustainable solution to plastic waste management. *Materials Circular Economy, 6*(1), 6.

Nabi, N., Ahmad, I., Amin, A., Rather, M. A., Ahmed, I., Hajam, Y. A., Khursheed, S., Malik, M. M., & Abubakr, A. (2024). Understanding the sources, fate and effects of microplastics in aquatic environments with a focus on risk profiling in aquaculture systems. *Reviews in Aquaculture, 16*(4), 1947–1980.

Naidu, B. C., Xavier, K. M., Shukla, S. P., Jaiswar, A. K., & Nayak, B. B. (2022a). Microplastics in the foreshore coastal waters, sediment, and coastal fauna of a highly populated megacity-A study on the effect of anthropogenic discharge on clams. *Marine Pollution Bulletin, 185*, 114262.

Naidu, B. C., Xavier, K. M., Shukla, S. P., Jaiswar, A. K., & Nayak, B. B. (2022b). Comparative study on the microplastics abundance, characteristics, and possible sources in yellow clams of different demographic regions of the northwest coast of India. *Journal of Hazardous Materials Letters, 3*, 100051.

Naidu, B. C., Xavier, K. M., Sahana, M. D., Landge, A. T., Jaiswar, A. K., Shukla, S. P., Ranjeet, K., & Nayak, B. B. (2025). Temporal variability of microplastics in shrimp (Litopenaeus vannamei), feed, water and sediments of coastal and inland culture ponds. *Science of the Total Environment, 959*, 178173.

Nguyen, D., Nguyen, M. K., Truong, Q. M., Thai, V. A., Pham, M. T., Chang, S. W., & Nguyen, D. D. (2025). Microplastics and pharmaceuticals from water and wastewater: Occurrence, impacts, and membrane bioreactor-based removal. *Separation and Purification Technology, 361*, 131489.

Palanisamy, S., Naidu, B. C., Venkatesh, S., Porayil, L., Balange, A. K., Nayak, B. B., Shaju, S., & Xavier, M. (2024). Microplastic contamination in salt-cured fish and commercial sea salts: An emerging food safety threat in relation to UN Sustainable Development Goals (SDGs). *Environmental Science and Pollution Research, 31*(21), 30688–30702.

Prasad, G. S., Rout, S. K., Malik, M. M., Karmakar, S., Amin, A., & Ahmad, I. (2023). Occurrence of xenoestrogen alkylphenols (Octylphenols and Nonylphenol) and its impact on the aquatic ecosystem. In *Xenobiotics in aquatic animals: Reproductive and developmental impacts* (pp. 275–284). Springer.

Prasad, G. S., Rout, S. K., Amin, A., Malik, M. M., Ahmad, I., Rather, M. A., Karmakar, S., Khalil, H. S., & Abubakar, A. (2024). Impact of COVID-19 on the aquatic environment and implications for fisheries with special emphasis to Indian perspective. *Proceedings of the Indian National Science Academy, 90*(4), 821–839.

Qiu, R., Song, Y., Zhang, X., Xie, B., & He, D. (2020). Microplastics in urban environments: Sources, pathways, and distribution. In *Microplastics in terrestrial environments: Emerging contaminants and major challenges* (pp. 41–61). Springer International Publishing.

Ragusa, A., Svelato, A., Santacroce, C., Catalano, P., Notarstefano, V., Carnevali, O., Papa, F., Rongioletti, M. C. A., Baiocco, F., Draghi, S., & D'Amore, E. (2021). Plasticenta: First evidence of microplastics in human placenta. *Environment International, 146*, 106274.

Rede, D., Vilarinho, R., Moreira, J. A., Nizzetto, L., Delerue-Matos, C., & Fernandes, V. C. (2025). Screening for microplastics in agricultural soils: Applying green chemistry principles in extraction and analysis. *Environmental Pollution, 367*, 125550.

Resmi, M. R., & Vaishnavi. (2025). Microplastic pollution in Indian aquatic ecosystems: Insights and implications. *Water, Air, & Soil Pollution, 236*(1), 62.

Rukmangada, R., Naidu, B. C., Nayak, B. B., Balange, A., Chouksey, M. K., & Xavier, K. M. (2023). Microplastic contamination in salted and sun dried fish and implications for food security–A study on the effect of location, style and constituents of dried fish on microplastics load. *Marine Pollution Bulletin, 191*, 114909.

Smith, M., Love, D. C., Rochman, C. M., & Neff, R. A. (2018). Microplastics in seafood and the implications for human health. *Current Environmental Health Reports, 5*, 375–386.

Sruthisree, C., Gowda, G., Nayak, H., Harish, B. D., & Amin, A. (2015). Effect of heavy metal (lead) on oxygen consumption and ammonia excretion of Iridescent Shark (*Pangusius hypohthalmus*). *Journal of Experimental Zoology India, 18*(2), 637.

Sun, H., Su, X., Mao, J., Liu, Y., Li, G., & Du, Q. (2024). Microplastics in maternal blood, fetal appendages, and umbilical vein blood. *Ecotoxicology and Environmental Safety, 287*, 117300.

Suvitha, K., Narayanamoorthy, S., Pamucar, D., & Kang, D. (2024). An ideal plastic waste management system based on an enhanced MCDM technique. *Artificial Intelligence Review, 57*(4), 96.

Tabassum, I., Hassan, N., Amin, A., Malik, M. M., Abubakr, A., Prasad, G. S., Latief, F., & Rather, S. A. (2025). COVID-19 and global environmental change: A threat to food Security and nutrition. In *Food security, nutrition and sustainability through aquaculture technologies* (pp. 437–445). Springer Nature Switzerland.

Taher, M. K., Momoli, F., Go, J., Hagiwara, S., Ramoju, S., Hu, X., Jensen, N., Terrell, R., Hemmerich, A., & Krewski, D. (2024). Systematic review of epidemiological and toxicological evidence on health effects of fluoride in drinking water. *Critical Reviews in Toxicology, 54*(1), 2–34.

Torres-Ruiz, M., de Alba González, M., Morales, M., Martin-Folgar, R., González, M. C., Cañas-Portilla, A. I., & De la Vieja, A. (2023). Neurotoxicity and endocrine disruption caused by polystyrene nanoparticles in zebrafish embryo. *Science of the Total Environment, 874*, 162406.

Venkatesh, S., Naidu, B. C., Palanisamy, S., Ngasotter, S., Shaju, S. S., Porayil, L., Balange, A., Nayak, B. B., & Xavier, K. M. (2024). Microplastic accumulation dynamics and risk assessment in dried fish processed with sea salt at different salting ratios. *Journal of Hazardous Materials Advances, 14*, 100415.

Vital, S. A., Cardoso, C., Avio, C., Pittura, L., Regoli, F., & Bebianno, M. J. (2021). Do microplastic contaminated seafood consumption pose a potential risk to human health? *Marine Pollution Bulletin, 171*, 112769.

Wang, S., Yan, K., Dong, Y., Chen, Y., Song, J., Chen, Y., Liu, X., Qi, R., Zhou, X., Zhong, J., & Li, J. (2025). The influence of microplastics on hypertension-associated cardiovascular injury via the modulation of gut microbiota. *Environmental Pollution, 368*, 125760.

Wen, S., Zhao, Y., Liu, S., Chen, Y., Yuan, H., & Xu, H. (2022). Polystyrene microplastics exacerbated liver injury from cyclophosphamide in mice: Insight into gut microbiota. *Science of the Total Environment, 840*, 156668.

Winiarska, E., Jutel, M., & Zemelka-Wiacek, M. (2024). The potential impact of nano-and microplastics on human health: Understanding human health risks. *Environmental Research, 251*, 118535.

Xie, Y., Li, Y., Feng, Y., Cheng, W., & Wang, Y. (2022). Inhalable microplastics prevails in air: Exploring the size detection limit. *Environment International, 162*, 107151.

Xu, Z., Deng, X., Lin, Z., Wang, L., Lin, L., Wu, X., Wang, Y., Li, H., Shen, J., & Sun, W. (2025). Microplastics in agricultural soil: Unveiling their role in shaping soil properties and driving greenhouse gas emissions. *Science of the Total Environment, 958*, 177875.

Yan, J., Pan, Y., He, J., Pang, X., Shao, W., Wang, C., Wang, R., He, Y., Zhang, M., Ye, J., & Lin, C. (2023). Toxic vascular effects of polystyrene microplastic exposure. *Science of the Total Environment, 905*, 167215.

Yek, P. N. Y., Chan, Y. H., Foong, S. Y., Mahari, W. A. W., Chen, X., Liew, R. K., Ma, N. L., Tsang, Y. F., Sonne, C., Cheng, Y. W., & Tan, Y. H. (2024). Co-processing plastics waste and biomass by pyrolysis–gasification: A review. *Environmental Chemistry Letters, 22*(1), 171–188.

Zeng, Y., Cai, J., Zhu, Y., Wang, J., Guo, R., Jian, L., Zheng, X., & Mai, B. X. (2025). Species-specific accumulation of microplastics in different bird species from South China: A comprehensive analysis. *Journal of Hazardous Materials, 483*, 136607.

Zha, H., Xia, J., Li, S., Lv, J., Zhuge, A., Tang, R., Wang, S., Wang, K., Chang, K., & Li, L. (2023). Airborne polystyrene microplastics and nanoplastics induce nasal and lung microbial dysbiosis in mice. *Chemosphere, 310*, 136764.

Zhao, B., Rehati, P., Yang, Z., Cai, Z., Guo, C., & Li, Y. (2024). The potential toxicity of microplastics on human health. *Science of the Total Environment, 912*, 168946.

Plastics in the Food Chain: A Scoping Review of Their Environmental, Health, and Food Security Implications

Joseph Joy Cindrella and K. R. Renjith

1 Introduction

Plastics have become ubiquitous in modern society due to their versatility, durability, and cost-effectiveness (Evode et al., 2021). Due to their widespread use and persistence, plastic pollution has become a defining feature of the Anthropocene era. Since their discovery in the 1900s, plastic production has surged, particularly since the 1950s. In 2022, global plastic production reached about 400 million metric tons, with projections suggesting it could double by 2050 (Stegmann et al., 2022; Nabi et al., 2024; Debnath et al., 2024). Plastics are chains of repeated units called polymers. They are formed by combining monomers like ethylene, propylene, and styrene with a catalyst to create polymers that possess specific physical and chemical properties, such as hydrophobicity, flexibility, and resistance to chemicals, temperature, and UV light (Andrady & Rajapakse, 2019). Chemical additives like plasticizers and flame retardants are included to enhance these properties. The low production cost and versatile characteristics of plastics have made them indispensable in daily life. The food industry extensively uses polyethylene (PE) and polypropylene (PP) for packaging, with these two polymers comprising 40–45% of global plastic production (Geyer, 2020).

Other significant polymers include polyvinyl chloride (PVC), polyethylene terephthalate (PET), polystyrene (PS), polyurethane (PUR), and smaller amounts of polycarbonate (PC) and polyamide (PA). The most commonly encountered plastics are polyethylene (polythene; PE) in low density form (LDPE; bin bags) and high-density construct (HDPE; bottle caps) or as tetraphthalate (PET; bottles), polypropylene (PPL; straws), polyvinyl chloride (PVC; pipes), and polystyrene, both

J. J. Cindrella · K. R. Renjith (✉)
KSCSTE – Centre for Water Resources Development and Management (CWRDM), Kozhikode, Kerala, India
e-mail: renjith@cwrdm.org

A. Abubakr et al. (eds.), *Microplastics: Ecological and Food Security Implications*,
https://doi.org/10.1007/978-3-031-94480-2_2

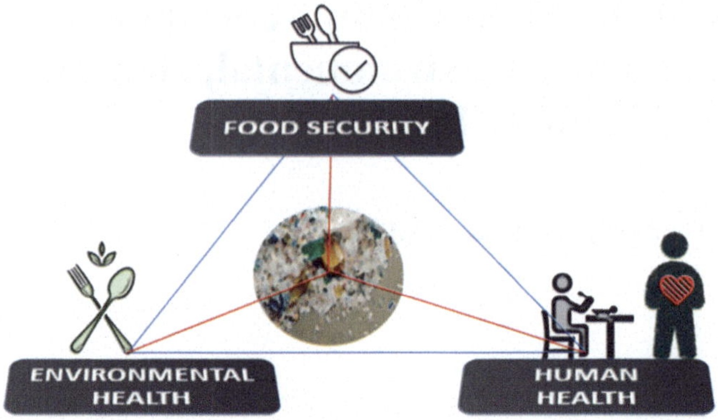

Fig. 1 Interconnection of Food Security, Environmental Health, Human Health with Microplastics

expanded (EPS; packaging, insulation) and rigid (PS; toys). Degradation material of this is subdivided by size into macro (>25 mm), meso (5–25 mm), large micro (1–5 mm), small micro (20 µm—1 mm), and nano (1–1000 nm) plastics, all are considered potential hazards to human health (Chen et al., 2021). These small plastic fragments are now found in soils, freshwater bodies, oceans, and even the atmosphere, leading to their infiltration into the food chain. Food security, environmental health, and human health are interconnected, as evidenced by the contamination of various food products with microplastics, which raises concerns about the potential health risks to humans and highlights the urgent need for research to assess these risks and ensure food safety (Fig. 1).

This chapter highlights the emerge of plastic pollution as a serious environmental challenge, with extensive consequences for ecosystems, human health, and food security.

1.1 Source and Pathways into Environment

The world generates a significant amount of plastic waste due to ineffective management and disposal methods. Plastics persist for centuries due to their stability, durability, and resistance to biodegradation, attributed to their high molecular weight, complex structures, and hydrophobic nature. Despite undergoing weathering processes like photodegradation and biodegradation, plastics are found in various ecosystems, including deserts, farms, mountaintops, and oceans.

1.2 Terrestrial Pathways

The majority of plastic contamination comes from human activity on land. All common plastics are entirely human inventions, with the exception of a few natural plastic polymers like latex which is a natural rubber component. Uncovered landfills, untreated sewage, wind-blown debris, and trash left by people such as bottles, plastic bags and boxes, road marking paints, shoes, and tires for cars, buses, and trucks are the main sources of plastic litter on land (Hurley et al., 2020). Illegal dumping and direct input of waste on land increases the plastic pollution in terrestrial environment.

Usually beginning as macroplastics, these undergo gradual environmental deterioration, releasing smaller particles that are capable to circulate throughout the ecosystem. They also have the ability to remain in the soil for up to 15 years (Pilapitiya & Ratnayake, 2024). These microplastics can influence the soil functionality and the microbial community, impacting both chemical processes and the cycling of water, nutrients, and geochemical elements (Yu et al., 2022). The presence of microplastics alters the size distribution of soil aggregates, indicating potential changes in soil stability. Studies have shown that (Lei et al., 2018) it also affected soil invertebrates, penetrating the intestinal walls of nematodes, causing oxidative stress, and altering gene expression.

1.3 Aquatic Pathways

According to estimates, 70–80% of plastics that end up in the oceans are carried by rivers (Hurley et al., 2020), with the majority of these plastics coming from manufacturing, agriculture, and waste water treatment facilities that release their wastewater into aquatic environments. It has been predicted that a typical 5 kg wash of polyester fabrics can release up to 6,000,000 microfibres (De Falco et al., 2018). Microorganisms that create a biofilm and decompose organic materials can colonize plastics in water. Light transmission will be restricted by the growth of these biofilms (Malik et al., 2017; Sooriyakumar et al., 2022). Microplastics take a very long time to break down. They can therefore infiltrate and thrive in natural environments. Over 700 fish species shown the presence of microplastics (Hossain & Olden, 2022) and is associated with detrimental impacts on fish development, reproduction, and digestion. Fish liver histopathological damage can result from microplastics.

1.4 Atmospheric Pathways

During transport, plastic waste can encounter mechanisms that sort particles based on density, size, and other properties, influencing their subsequent movement, physical and chemical characteristics, and ultimate environmental fate. Similar to waterways, the atmosphere acts as both transport medium and reservoir for environmental plastics. Plastic waste, ranging from microplastics and everyday litter (e.g., bags and wrappers) to large debris mobilized by severe windstorms, can become suspended in the atmosphere. This transport depends on factors such as the size, density, and aerodynamic shape of the plastic, along with wind strength, turbulence, duration, and obstacles along the pathway.

Microplastics from soil, roads, and ocean surfaces, if large enough to be lifted into the atmosphere yet small enough to remain airborne, can undergo long-range transport, with residence times of up to one week (Brahney et al., 2021). The cycles of suspension, deposition, and resuspension create significant atmospheric plastic reservoirs. For instance, the estimated average atmospheric load of microplastics (4–250 μm) over the western United States' land regions is approximately 0.001 MMT (Brahney et al., 2021). Research on particle pathways shows that microplastic transport can occur over regional distances (>100 km) or be influenced by large-scale atmospheric patterns spanning 1000 km, resulting in deposition far from the source (Brahney et al., 2021).

2 Applications of Plastics Across Various Food Sectors: Benefits and Challenges

Plastics play a crucial role in numerous industries due to their adaptability, strength, and affordability. Their applications range from everyday items to advanced technological uses. The versatility of plastics supports innovation and efficiency, though it also raises concerns about environmental impact and sustainability, prompting ongoing research into more eco-friendly alternatives.

2.1 Use of Plastics in Agriculture

Plastic products have become indispensable in modern agriculture, offering a sustainable approach to food production and water-saving irrigation. Their key advantages like affordability, low weight, ease of use, low maintenance, and exceptional durability have driven their widespread adoption. This has given rise to "plasticulture," a term describing the use of plastics in agriculture (Mormile et al., 2017). It aims to enhance crop yield and quality while optimizing water usage. Over the years, traditional materials like glass and paper have been increasingly replaced by plastic products due to their superior performance.

This surge in use of plastic in agriculture now estimated at around 12.5 million tonnes annually (FAO, 2021). These plastics are employed in various applications, from soil fumigation and irrigation systems to packaging and protecting crops. Plastic mulch films, for instance, offer multiple agronomic benefits, such as weed and pest control, moisture conservation, and temperature regulation, significantly boosting food production. High-efficiency irrigation systems, water storage reservoirs, and drainage networks constructed from plastic are more durable, corrosion-resistant, and cost-effective compared to their metal or aluminum counterparts.

Plastic films and rigid panels used for controlled environment farming provide cost-effective alternatives to glass, creating environments that maintain ideal temperature, humidity, and light levels for year-round cultivation. These structures use plastic components for water supply, air distribution, and crop training, among other functions, ensuring a controlled growth environment that mitigates the effects of adverse weather conditions. The rise in agricultural use of plastic products has been accompanied by an increase in plastic waste, leading to contamination of farmlands. The presence of plastics in the soil has been shown to impact soil microbial communities, which are essential for organic matter decomposition and nutrient cycling, thereby impairing soil fertility and crop productivity.

2.2 *Use of Plastics in Food Packaging*

As mentioned earlier, Plastic products have become popular due to their low production costs, light weight, ease of use, and durability. Approximately 39.6% of these materials are used for packaging, including food (Aslam et al., 2023). The most popular thermoplastic polymers used in the food industry for packaging are polyethylene (PE), polypropylene (PP), polyethylene terephthalate (PET), and polystyrene (PS). Sadly, the majority of plastic packaging is thrown away after use, which results plastic pollution in environment. Different additives used in the production of plastic may also be released during degradation (Table 1). Microplastics can also enter the food during its preparation, with an estimated 100 to 300 MPs/mm generated on cutting boards during preparation (Narula & Negi, 2025).

Some of the studies identified the presence of MPs in various food products are listed (Table 1) below:

Table 1 Presence of MPs in various food products: Study Examples

Category	Items	References
Seafood	Fish, shellfish	Wang et al. (2021); Daniel et al. (2021); Ding et al. (2020); Chinfak et al. (2021); Magesh and Ajith (2024)
Poultry	Poultry meat, eggs	Kedzierski et al. (2020); Liu et al. (2022)
Seasonings	Salt, sugar	Kapukotuwa et al. (2022); Lee et al. (2021)
Produce	Fruits, vegetables	Oliveri Conti et al. (2020); Tympa et al. (2021)
Beverages	Water, milk, beer, wine, tea, energy drinks, soft drinks	Koelmans et al. (2019); Kutralam-Muniasamy et al. (2020); Diaz-Basantes et al. (2020); al Naggar et al. (2021)
Sweeteners	Honey	Balzani et al. (2022); Prata et al. (2020); Hernandez et al. (2019); Afrin et al. (2022a, b); Shruti et al. (2020)
Infant products	Infant formula	(Liu et al. (2023)

3 Degradation Pathways

Plastics exhibit limited natural degradation in the environment due to their high stability and durability. Plastic degradation is influenced by environmental factors such as sunlight, heat, chemical/biological activities, and physical properties like polymer size, density, and molecular weight. The degradation of plastics can be categorized into four types: physical/mechanical degradation, photodegradation/photo-oxidative degradation, chemical and thermal degradation, and biological degradation (Balangao, 2023). Physical/mechanical degradation involves structural alterations, such as cracking and embrittlement, due to environmental forces like wave action in coastal areas. This process facilitates chemical changes at the molecular level, including polymer oxidation and the formation of short-chain molecules. Photodegradation/photo-oxidative degradation occurs when plastics are exposed to ultraviolet (UV) radiation, leading to the breakdown of polymer chains, reduction in molecular weight, and generation of free radicals (Gunawardhana et al., 2023). Chemical degradation involves the breakdown of polymers due to exposure to corrosive chemicals, such as nitric, sulfuric, and hydrochloric acids, as well as atmospheric pollutants. Thermal degradation, driven by heat or temperature changes, results in chemical alterations that further degrade polymer chains. Biological degradation occurs when microorganisms like bacteria, fungi, and yeasts metabolize the polymer carbon into carbon dioxide or incorporate biomolecules, altering the physical and chemical properties of plastics. This natural degradation process is an essential pathway for the breakdown of plastics in the environment.

4 Environmental and Health Impacts of Plastic and Associated Chemicals

4.1 Plastic as Pollutants

Turtles, which mistake plastic bags and sheets for jellyfish, are the main victims of macroplastic contamination in aquatic habitats. Intestinal obstructions may result from this. Furthermore, many aquatic animals become stuck and tangled in plastic waste because they are unable to escape. There are several exceptions, even though large plastic objects typically present little risk to people. For instance, accidentally ingesting a plastic sauce package has resulted in cases of intestinal perforation and chronic inflammation (Waring et al., 2018). There have been cases of small bowel blockage and perforation from inadvertently ingesting plastic bag clips, as well as oesophageal obstruction and laceration from swallowed food packaging. Children inadvertently eating or breathing in plastic is still a serious risk (Sripada et al., 2022; Tabassum et al., 2025).

Table 2 Recent findings of the Microplastic associated health issues on Humans

Associated health issues	Reference
Disruption of gut microbiome, leading to gastrointestinal disorders, systemic inflammation, and chronic diseases	Hussain et al. (2001)
Potential harm to reproductive, digestive, and respiratory health; suggested link to colon cancer	Bora et al. (2024)
Gastrointestinal injury, liver dysfunction, asthma exacerbation, hypersensitivity pneumonitis, reproductive toxicity, cardiovascular issues	Chartres et al. (2024)

Relatively little information is available on the potential toxicity of microplastics (MPs), most of which are chemically inert. Any adverse effects are likely due to their physical presence rather than chemical toxicity, as the potential for toxicity from adhered chemicals is considered negligible. MPs are typically resistant to degradation within the body and persist until they are eliminated, primarily via the gastrointestinal tract, where they may concentrate in the mucus layer secreted by gut wall cells. Other elimination pathways include urine, bile, cerebrospinal fluid, pulmonary alveoli, and milk in lactating females (Wu et al., 2022).

Despite their inert nature, in vivo toxicity responses to MPs have been observed. Unlike nanoparticles (NPs), which have been reported to affect fish behaviour due to CNS involvement (Sarasamma et al., 2020; Gurjar et al., 2023), MPs do not typically reach the central nervous system. However, they can impair filter feeders such as mussels and clams. For example, brine shrimp exposed to high concentrations of MPs exhibited altered swimming activity, making them more susceptible to predation (Wang et al., 2020). Polystyrene (PS) MPs have also been linked to DNA damage, oxidative stress, and antioxidant activity in clams (Kong et al., 2023). The toxicity of MPs depends on their size and shape; larger particles and fibers have been found to be more toxic to grass shrimp, while variations in surface charges have altered embryotoxic effects in sea urchins (Dos Santos et al., 2024).

In mammals, MPs are thought to cause only localized inflammatory responses as they do not deeply penetrate organs (Table 2). However, smaller particles and those with negative surface charges show increased uptake. In vitro studies with MPs (polyethylene and polystyrene) have demonstrated oxidative stress-induced damage in cerebral and epithelial human cell lines, suggesting that ingestion could lead to localized immune responses and increased gut inflammation (Wang et al., 2023). Further damage would likely depend on the ability of MPs to cross the gastrointestinal barrier, a process not yet fully understood. Potential mechanisms include paracellular uptake across the intestinal epithelium and active endocytosis.

Studies have shown that MPs can translocate into the lymphatic system in humans (particle size 0.2–150 μm) (Peixoto et al., 2019), rodents (30–40 μm), rabbits (0.1–10 μm), and dogs (3–100 μm), likely via Peyer's patches in the intestine [29]. In dogs, PVC particles (5–110 μm) were detected in the portal vein, indicating potential liver exposure. These absorption mechanisms have been thoroughly reviewed, though they are only applicable to individuals with intact gut walls. Those with inflammatory bowel disease may have increased particle transport into systemic circulation. Additionally, as nanoparticles in aerosols can penetrate the brain, a compromised blood-brain barrier could enhance MP uptake.

4.2 Plastics as Vectors

Plastics can act as vectors for various hazardous chemicals, including persistent organic pollutants (POPs), heavy metals (Otorkpa & Otorkpa, 2024; Prasad et al., 2023; Prasad et al., 2024), and additives such as bisphenol A (BPA), phthalates, and flame retardants, which are commonly used in plastic production to enhance their properties (Hahladakis et al., 2018). These chemicals, often leached into the environment as plastics degrade, have been associated with numerous adverse effects on human health and wildlife. Some of these chemicals, such as BPA and phthalates, are endocrine-disrupting compounds (EDCs) that can interfere with hormonal systems, leading to developmental, reproductive, and metabolic disruptions in both humans and animals (Tumwesigye et al., 2023). Additionally, heavy metals, including lead, mercury, and cadmium, can be absorbed by plastic materials from the surrounding environment and may accumulate in organisms through ingestion, leading to toxic effects (Sruthisree et al., 2015; Amin et al., 2016; Amin et al., 2017; Ghosh et al., 2022).

5 Food Security Challenges

Plastic contamination in the food chain is unlikely to cause significant toxicity under normal conditions, it could pose a risk in cases of high contamination levels in human tissues. Special circumstances, such as a leaky gut, a permeable blood-brain barrier, or prolonged ingestion of contaminated food, might increase the likelihood of adverse effects. Food security faces significant challenges due to the pervasive use of plastics in food systems. The agricultural sector's extensive use of plastic for mulch and packaging has implications for food security by reducing post-harvest losses and extending shelf-life. However, this reliance on plastics also contributes to increased global food waste. Knowledge gaps persist, particularly concerning the health impacts of microplastics and chemicals such as bisphenols and phthalates, which can leach into food and pose potential risks. The fragmented nature of research across disciplines like public health, agriculture, and environmental sciences further complicates the development of coherent policies. Addressing these challenges requires a comprehensive synthesis of evidence to support informed policy-making, balancing the benefits of plastics in food security against their environmental and health risks.

6 Conclusion

Plastics in the food chain present a multifaceted challenge with profound environmental, health, and food security implications. Addressing this issue requires interdisciplinary collaboration to reduce plastic production, enhance waste management, and promote sustainable alternatives. Policymakers, researchers, and stakeholders must prioritize mitigating plastic pollution to safeguard ecosystem health and ensure food security for future generations. Regional and global policy frameworks, such as the UN Sustainable Development Goals (SDGs), provide a critical platform for coordinated action against plastic pollution.

References

Afrin, S., Rahman, M., Akbor, A., Siddique, A. B., Uddin, K., & Malafaia, G. (2022a). Is there tea complemented with the appealing flavor of microplastics? A pioneering study on plastic pollution in commercially available tea bags in Bangladesh. *Science of the Total Environment, 837*, 155833.

Afrin, S., Rahman, M. M., Hossain, M. N., Uddin, M. K., & Malafaia, G. (2022b). Are there plastic particles in my sugar? A pioneering study on the characterization of microplastics in commercial sugars and risk assessment. *Science of the Total Environment, 837*, 155849.

Al Naggar, Y., Brinkmann, M., Sayes, C. M., Al-Kahtani, S. N., Dar, S. A., El-Seedi, H. R., Grünewald, B., & Giesy, J. P. (2021). Are honey bees at risk from microplastics? *Toxics, 9*, 109.

Amin, A., Indulkar, S. T., & Pai, R. (2016). Sublethal effect of Buprofezin pesticide on carcass composition of *Cyprinus carpio* communis fingerlings. *Journal of Experimental Zoology India, 20*(1), 269–272.

Amin, A., Ramachandra Naik, A. T., Priyadarshini, N., Nayak, H., & Sree, C. S. (2017). Toxic effect of heavy metal lead on oxygen consumption of rohu (*Labeo rohita*) fingerlings. *Biochemical & Cellular Archives, 17*(1).

Andrady, A. L., & Rajapakse, N. (2019). Additives and chemicals in plastics. *Hazardous Chemicals Associated with Plastics in the Marine Environment*, 1–17.

Aslam, S., Khurram, A., Hussain, R., Qadir, A., & Ahmad, S. R. (2023). Sources, distribution, and incipient threats of polymeric microplastic released from food storage plastic materials. *Environmental Monitoring and Assessment, 195*(6), 638.

Balangao, J. K. (2023). Degradation of plastics-A brief review. *Journal of Chemical Health Risks, 13*(6), 1413–1420.

Balzani, P., Galeotti, G., Scheggi, S., Masoni, A., Santini, G., & Baracchi, D. (2022). Acute and chronic ingestion of polyethylene (PE) microplastics has mild effects on honey bee health and cognition. *Environmental Pollution, 305*, 119318.

Bora, S. S., Gogoi, R., Sharma, M. R., Anshu, Borah, M. P., Deka, P., et al. (2024). Microplastics and human health: Unveiling the gut microbiome disruption and chronic disease risks. *Frontiers in Cellular and Infection Microbiology, 14*, 1492759.

Brahney, J., Mahowald, N., Prank, M., Cornwell, G., Klimont, Z., Matsui, H., & Prather, K. A. (2021). Constraining the atmospheric limb of the plastic cycle. *Proceedings of the National Academy of Sciences of the United States of America, 118*(16). https://doi.org/10.1073/pnas.2020719118

Chartres, N., Cooper, C. B., Bland, G., Pelch, K. E., Gandhi, S. A., BakenRa, A., & Woodruff, T. J. (2024). *Effects of microplastic exposure on human digestive, reproductive, and respiratory health: A rapid systematic review* (Vol. 58, pp. 22843–22864). Environmental Science & Technology.

Chen, H. L., Selvam, S. B., Ting, K. N., & Gibbins, C. N. (2021). Microplastic pollution in freshwater systems in Southeast Asia: Contamination levels, sources, and ecological impacts. *Environmental Science and Pollution Research, 28*(39), 54222–54237.

Chinfak, N., Sompongchaiyakul, P., Charoenpong, C., Shi, H., Yeemin, T., & Zhang, J. (2021). Abundance, composition, and fate of microplastics in water, sediment, and shellfish in the Tapi-Phumduang River system and Bandon Bay, Thailand. *Science of the Total Environment, 781*, 146700.

Daniel, D. B., Ashraf, P. M., Thomas, S. N., & Thomson, K. T. (2021). Microplastics in the Edible tissues of shellfishes sold for human consumption. *Chemosphere, 264*, 128554.

De Falco, F., Gullo, M. P., Gentile, G., Di Pace, E., Cocca, M., Gelabert, L., et al. (2018). Evaluation of microplastic release caused by textile washing processes of synthetic fabrics. *Environmental Pollution, 236*, 916–925.

Debnath, R., Prasad, G. S., Amin, A., Malik, M. M., Ahmad, I., Abubakr, A., Borah, S., Rather, M. A., Impellitteri, F., Tabassum, I., & Piccione, G. (2024). Understanding and addressing microplastic pollution: Impacts, mitigation, and future perspectives. *Journal of Contaminant Hydrology, 266*, 104399.

Diaz-Basantes, M. F., Conesa, J. A., & Fullana, A. (2020). Microplastics in honey, beer, milk and refreshments in Ecuador as emerging contaminants. *Sustainability, 12*, 5514.

Ding, J., Li, J., Sun, C., Jiang, F., He, C., Zhang, M., Ju, P., & Ding, N. X. (2020). An examination of the occurrence and potential risks of microplastics across various shellfish. *Science of the Total Environment, 739*, 139887.

Dos Santos, J. B., Choueri, R. B., Dos Santos, F. E. M., Santos, L. A. D. O., da Silva, L. F., Nobre, C. R., et al. (2024). Are microfibers a threat to marine invertebrates? A sea urchin toxicity assessment. *Toxics, 12*(10), 753.

Evode, N., Qamar, S. A., Bilal, M., Barceló, D., & Iqbal, H. M. (2021). Plastic waste and its management strategies for environmental sustainability. *Case Studies in Chemical and Environmental Engineering, 4*, 100142.

FAO. (2021). *Assessment of agricultural Plastics and their sustainability—A call for action*. FAO.

Geyer, R. (2020). Production, use, and fate of synthetic polymers. In *Plastic waste and recycling* (pp. 13–32). Academic.

Ghosh, A., Tripathy, A., & Ghosh, D. (2022, March). Impact of endocrine disrupting chemicals (EDCs) on reproductive health of human. In *Proceedings of the zoological society* (Vol. 75, No. 1, pp. 16–30). Springer India.

Gunawardhana, G. M. S. S., Perera, U. L. H. P., & Ratnayake, A. S. (2023). Sources and fate of Plastics into microplastics: Degradation and remediation methods. In *Maritime accidents and environmental pollution-the X-press pearl disaster* (pp. 155–172). CRC Press.

Gurjar, U. R., Takar, S., Amin, A., & Martin Xavier, K. A. (2023). Microplastic contamination in aquatic organisms: An ecotoxicological perspective. In *Xenobiotics in aquatic animals: Reproductive and developmental impacts* (pp. 353–367). Springer Nature Singapore.

Hahladakis, N. J., Costas, A. V., Weber, R., Iacovidou, E., & Purnell, P. (2018). An overview of chemical additives present in plastics: Migration, release, fate and environmental impact during their use, disposal and recycling. *Journal of Hazardous Materials, 344*, 179–199. https://doi.org/10.1016/j.jhazmat.2017.10.014

Hernandez, L. M., Xu, E. G., Larsson, H. C. E., Tahara, R., Maisuria, V. B., & Tufenkji, N. (2019). Plastic teabags release billions of microparticles and nanoparticles into tea. *Environmental Science & Technology, 53*, 12300–12310.

Hossain, M. A., & Olden, J. D. (2022). Global meta-analysis reveals diverse effects of microplastics on freshwater and marine fishes. *Fish and Fisheries, 23*(6), 1439–1454.

Hurley, R., Horton, A., Lusher, A., & Nizzetto, L. (2020). Plastic waste in the terrestrial environment. In *Plastic waste and recycling* (pp. 163–193). Academic.

Hussain, N., Jaitley, V., & Florence, A. T. (2001). Recent advances in the understanding of uptake of microparticulates across the gastrointestinal lymphatics. *Advanced Drug Delivery Reviews, 50*(1–2), 107–142.

Kapukotuwa, R. W. M. G. K., Jayasena, N., Weerakoon, K. C., Abayasekara, C. L., & Rajakaruna, R. S. (2022). High levels of microplastics in commercial salt and industrial Salterns in Sri Lanka. *Marine Pollution Bulletin, 174*, 113239.

Kedzierski, M., Lechat, B., Sire, O., le Maguer, G., le Tilly, V., & Bruzaud, S. (2020). Microplastic contamination of packaged meat: Occurrence and associated risks. *Food Packaging and Shelf Life, 24*, 100489.

Koelmans, A. A., Mohamed Nor, N. H., Hermsen, E., Kooi, M., Mintenig, S. M., & de France, J. (2019). Microplastics in freshwaters and drinking water: Critical review and assessment of data quality. *Water Research, 155*, 410–422.

Kong, C., Pan, T., Chen, X., Junaid, M., Liao, H., Gao, D., et al. (2023). Exposure to polystyrene nanoplastics and PCB77 induced oxidative stress, histopathological damage and intestinal microbiota disruption in white hard clam Meretrix lyrata. *Science of the Total Environment, 905*, 167125.

Kutralam-Muniasamy, G., Pérez-Guevara, F., Elizalde-Martínez, I., & Shruti, V. C. (2020). Branded milks—Are they immune from microplastics contamination? *Science of the Total Environment, 714*, 136823.

Lee, H. J., Song, N. S., Kim, J. S., & Kim, S. K. (2021). Variation and uncertainty of microplastics in commercial table salts: Critical review and validation. *Journal of Hazardous Materials, 402*, 123743.

Lei, L., Wu, S., Lu, S., Liu, M., Song, Y., Fu, Z., et al. (2018). Microplastic particles cause intestinal damage and other adverse effects in zebrafish Danio rerio and nematode Caenorhabditis elegans. *Science of the Total Environment, 619*, 1–8.

Liu, Q., Chen, Z., Chen, Y., Yang, F., Yao, W., & Xie, Y. (2022). Microplastics contamination in eggs: Detection, occurrence and status. *Food Chemistry, 397*, 133771.

Liu, S., Guo, J., Liu, X., Yang, R., Wang, H., Sun, Y., Chen, B., & Dong, R. (2023). Detection of various microplastics in placentas, meconium, infant feces, breastmilk and infant formula: A pilot prospective study. *Science of the Total Environment, 854*, 158699.

Malik, M., Balkhi, M. H., & Abubakr, A. (2017). Comparative study on the contamination level of Dal Lake, Kashmir based on total viable bacterial counts over a period of time. *Ecology, Environment and Conservation, 23*(4), 2145–2147.

Mormile, P., Stahl, N., & Malinconico, M. (2017). The world of plasticulture. In *Soil degradable bioplastics for a sustainable modern agriculture* (pp. 1–21). Springer Berlin Heidelberg.

Nabi, N., Ahmad, I., Amin, A., Rather, M. A., Ahmed, I., Hajam, Y. A., Khursheed, S., Malik, M. M., & Abubakr, A. (2024). Understanding the sources, fate and effects of microplastics in aquatic environments with a focus on risk profiling in aquaculture systems. *Reviews in Aquaculture, 16*(4), 1947–1980.

Narula, D., & Negi, S. (2025). Household microplastic pollutants: A potential contributor to indoor microplastics pollution. In *Microplastics in environment* (pp. 149–155). CRC Press.

Oliveri Conti, G., Ferrante, M., Banni, M., Favara, C., Nicolosi, I., Cristaldi, A., Fiore, M., & Zuccarello, P. (2020). Micro- and nano-plastics in edible fruit and vegetables. The first diet risks assessment for the general population. *Environmental Research, 187*, 109677.

Otorkpa, O. J., & Otorkpa, C. O. (2024). Health effects of microplastics and nanoplastics: Review of published case reports. *Environmental Analysis, Health and Toxicology, 39*, e2024020.

Peixoto, D., Pinheiro, C., Amorim, J., Oliva-Teles, L., Guilhermino, L., & Vieira, M. N. (2019). Microplastic pollution in commercial salt for human consumption: A review. *Estuarine, Coastal and Shelf Science, 219*, 161–168.

Pilapitiya, P. N. T., & Ratnayake, A. S. (2024). The world of plastic waste: A review. *Cleaner Materials, 11*, 100220.

Prasad, G. S., Rout, S. K., Malik, M. M., Karmakar, S., Amin, A., & Ahmad, I. (2023). Occurrence of xenoestrogen alkylphenols (Octylphenols and Nonylphenol) and its impact on the aquatic ecosystem. In *Xenobiotics in aquatic animals: Reproductive and developmental impacts* (pp. 275–284). Springer.

Prasad, G. S., Rout, S. K., Amin, A., Malik, M. M., Ahmad, I., Rather, M. A., Karmakar, S., Khalil, H. S., & Abubakar, A. (2024). Impact of COVID-19 on the aquatic environment and implications for fisheries with special emphasis to Indian perspective. *Proceedings of the Indian National Science Academy, 90*(4), 821–839.

Prata, J. C., Paço, A., Reis, V., da Costa, J. P., Fernandes, A. J. S., da Costa, F. M., Duarte, A. C., & Rocha-Santos, T. (2020). Identification of microplastics in white wines capped with polyethylene stoppers using micro-Raman spectroscopy. *Food Chemistry, 331*, 127323.

Sarasamma, S., Audira, G., Siregar, P., Malhotra, N., Lai, Y. H., Liang, S. T., et al. (2020). Nanoplastics cause neurobehavioral impairments, reproductive and oxidative damages, and biomarker responses in zebrafish: Throwing up alarms of wide spread health risk of exposure. *International Journal of Molecular Sciences, 21*(4), 1410.

Shruti, V. C., Pérez-Guevara, F., Elizalde-Martínez, I., & Kutralam-Muniasamy, G. (2020). First study of its kind on the microplastic contamination of soft drinks, cold tea and energy drinks—Future research and environmental considerations. *Science of the Total Environment, 726*, 138580.

Sooriyakumar, P., Bolan, N., Kumar, M., Singh, L., Yu, Y., Li, Y., et al. (2022). Biofilm formation and its implications on the properties and fate of microplastics in aquatic environments: A review. *Journal of Hazardous Materials Advances, 6*, 100077.

Sripada, K., Wierzbicka, A., Abass, K., Grimalt, J. O., Erbe, A., Röllin, H. B., et al. (2022). A children's health perspective on nano-and microplastics. *Environmental Health Perspectives, 130*(1), 015001.

Sruthisree, C., Gowda, G., Nayak, H., Harish, B. D., & Amin, A. (2015). Effect of heavy metal (lead) on oxygen consumption and ammonia excretion of Iridescent Shark (*Pangusius hypohthalmus*). *Journal of Experimental Zoology India, 18*(2).

Stegmann, P., Daioglou, V., Londo, M., van Vuuren, D. P., & Junginger, M. (2022). Plastic futures and their CO2 emissions. *Nature, 612*(7939), 272–276.

Tabassum, I., Hassan, N., Amin, A., Malik, M. M., Abubakr, A., Prasad, G. S., Latief, F., & Rather, S. A. (2025). COVID-19 and global environmental change: A threat to food Security and nutrition. In *Food security, nutrition and sustainability through aquaculture technologies* (pp. 437–445). Springer Nature Switzerland.

Tumwesigye, E., Nnadozie, C. F., Akamagwuna, F. C., Noundou, X. S., Nyakairu, G. W., & Odume, O. N. (2023). Microplastics as vectors of chemical contaminants and biological agents in freshwater ecosystems: Current knowledge status and future perspectives. *Environmental Pollution, 330*, 121829.

Tympa, L.-E., Katsara, K., Moschou, P. N., Kenanakis, G., & Papadakis, V. M. (2021). Do microplastics enter our food chain via root Vegetables? A Raman based spectroscopic study on Raphanus Sativus. *Materials, 14*, 2329.

Wang, X., Liu, L., Zheng, H., Wang, M., Fu, Y., Luo, X., et al. (2020). Polystyrene microplastics impaired the feeding and swimming behavior of mysid shrimp Neomysis japonica. *Marine Pollution Bulletin, 150*, 110660.

Wang, Q., Zhu, X., Hou, C., Wu, Y., Teng, J., Zhang, C., Tan, H., Shan, E., Zhang, W., & Zhao, J. (2021). Microplastic uptake in commercial fishes from the Bohai Sea, China. *Chemosphere, 263*, 127962.

Wang, W., Guan, J., Feng, Y., Nie, L., Xu, Y., Xu, H., & Fu, F. (2023). Polystyrene microplastics induced nephrotoxicity associated with oxidative stress, inflammation, and endoplasmic reticulum stress in juvenile rats. *Frontiers in Nutrition, 9*, 1059660.

Waring, R. H., Harris, R. M., & Mitchell, S. C. (2018). Plastic contamination of the food chain: A threat to human health? *Maturitas, 115*, 64–68.

Wu, P., Lin, S., Cao, G., Wu, J., Jin, H., Wang, C., et al. (2022). Absorption, distribution, metabolism, excretion and toxicity of microplastics in the human body and health implications. *Journal of Hazardous Materials, 437*, 129361.
Yu, H., Liu, M., Gang, D., Peng, J., Hu, C., & Qu, J. (2022). Polyethylene microplastics interfere with the nutrient cycle in water-plant-sediment systems. *Water Research, 214*, 118191.

Microplastics and Climate Change: Unravelling the Interconnected Global Crisis

Rinkesh Nemichand Wanjari, Karankumar Kishorkumar Ramteke, Adnan Amin, Ishtiyaq Ahmad, Pranali Prabhakar Marbade, Hudisa Banoo, Dechen Chuskit, and Zaheer Abass

1 Introduction

Plastic pollution and climate change are rapidly emerging as two of our era's most severe environmental concerns, with significant consequences for the pedosphere, hydrosphere, and atmosphere (Dong & Huang, 2023; Dewika et al., 2024). The interplay between these challenges poses serious threats to ecosystems and human health (Flandroy et al., 2018). As global temperatures continue to rise, the degradation, distribution, and interaction of microplastics (MPs) across various ecosystems are intensified (Li et al., 2024; Parvez et al., 2024; Nabi et al., 2024; Debnath et al., 2024). Increased temperatures accelerate the fragmentation of plastics into smaller

R. N. Wanjari
Department of Fisheries Resource Management, School of Fisheries, Centurion University of Technology and Management, Paralakhemundi, Odisha, India

K. K. Ramteke
ICAR-Central Institute of Fisheries Education (CIFE), Mumbai, Maharashtra, India

A. Amin
Division of Aquatic Environmental Management, Faculty of Fisheries Rangil, Ganderbal, SKUAST-Kashmir, Jammu and Kashmir, India

I. Ahmad
Division of Fish Genetics and Biotechnology, Faculty of Fisheries, SKUAST-K, Rangil, Ganderbal, Jammu and Kashmir, India

P. P. Marbade
Late Shri Punaram Nishad College of Fisheries Kawardha, Sawaikachhar, Chhattisgarh, India

H. Banoo · D. Chuskit · Z. Abass (✉)
Division of Fisheries Resources Management, Faculty of Fisheries, Sher-e-Kashmir University of Agricultural Science and Technology of Kashmir, Shalimar, Srinagar, Jammu and Kashmir, India

A. Abubakr et al. (eds.), *Microplastics: Ecological and Food Security Implications*,
https://doi.org/10.1007/978-3-031-94480-2_3

particles, while extreme weather events—such as storms and floods—facilitate the widespread dispersal of these microplastics (Lahon & Handique, 2023; Belioka & Achilias, 2024; Fred-Ahmadu et al., 2024). Additionally, climate-induced changes in ocean currents and circulation patterns extend the reach of MPs, contaminating regions previously considered untouched (Murgulet, 2016; Hung et al., 2022). This intricate relationship between climate change and plastic pollution creates a multi-faceted web of environmental challenges that demands urgent attention. The repercussions for marine ecosystems are particularly alarming. Rising temperatures hasten the physical breakdown of plastics and alter the behaviour of aquatic species, increasing their exposure to MPs (Yuan et al., 2020; Lin et al., 2022; Luo et al., 2022). Changes in marine food chains, driven by climate shifts, may lead to higher concentrations of MPs in aquatic organisms, further complicating an already critical situation (Borgå et al., 2022; Viitasalo & Bonsdorff, 2022). This synergistic effect underscores the interconnectedness of these global crises, illustrating that addressing one issue necessitates considering the other (Kurniawan et al., 2024; Thacharodi et al., 2024). Although the urgency of plastic pollution has escalated, research exploring the relationship between climate change and MPs remains limited. Significant studies by Welden and Lusher (2017) and Ford et al. (2022) examine how climate change intensifies marine plastic pollution, while other works, such as those by Li et al. (2016a, b) and Zhang et al. (2021), investigate MP distribution in freshwater and eutrophic systems under different environmental conditions. However, much of the existing research has focused on aquatic ecosystems, leaving critical gaps in our understanding of the combined effects of climate change and MPs on terrestrial and atmospheric environments.

The rise of plastics marks a revolutionary shift in modern life, propelled by their versatility, durability, and affordability (Thompson et al., 2009). Since the 1950s, global plastic production has surged, reaching an astonishing 368 million metric tons by 2020, with projections suggesting a further increase to 600 million tons by 2025 and potentially one billion tons by 2050 (Lusher et al., 2017). This rapid growth, however, comes with dire environmental consequences. Plastics constitute approximately 80% of all marine debris, with an estimated eight million tons entering the oceans annually, over half of which comprises single-use items (Napper et al., 2015). As the global production of plastics exceeded 400 million tons in 2022 (Plastics Europe, 2023), the resulting waste generation has become a significant concern. Alarmingly, only 9% of plastic waste is recycled, while 19% is incinerated, and approximately 50% ends up in landfills, with 22% released into the environment (OECD, 2022). This mismanagement of plastic waste has severe implications for marine and terrestrial ecosystems, harming biodiversity and disrupting the blue economy—affecting fishing, aquaculture, tourism, and water transport (Ajith et al., 2020; Mugilarasan et al., 2023; Lincoln et al., 2023). In recent years, microplastics (MPs), defined as plastic particles smaller than 5 mm, have emerged as a significant environmental concern. These p1articles are pervasive, infiltrating various ecosystems, including aquatic bodies, terrestrial environments, and even the atmosphere. Their origins are twofold: some MPs are released directly from personal care products (primary MPs), while others result from the breakdown of larger plastic items (Mbachu et al., 2021; Villegas et al., 2021; Jain et al., 2024). The widespread distribution of MPs poses substantial risks to environmental health and human well-being, necessitating urgent attention and comprehensive mitigation strategies.

The ingestion of MPs by marine organisms is particularly alarming. Many species mistake these plastic fragments for food, leading to serious health risks. The leaching of harmful chemicals from MPs can disrupt aquatic ecosystems and ultimately affect human health through the food chain (Karthik et al., 2022; Protyusha et al., 2024). The increasing prevalence of MPs in populous coastal regions highlights the critical need for a deeper understanding of their ecological implications and effective strategies to combat plastic pollution. In countries like India, the situation is particularly dire. With a population exceeding 1.2 billion, India generates approximately 3.5 million tons of plastic waste annually, ranking 12th globally in plastic mismanagement (Jambeck et al., 2015). Poor waste management practices lead to significant quantities of plastic entering oceans via inland waterways, wastewater runoff, and wind dispersion, further exacerbating marine plastic pollution. Approximately 80% of oceanic plastic and microplastic pollution originates from terrestrial sources, with rivers serving as the primary transport route into marine environments (Li et al., 2016a, b; Lebreton et al., 2017; Strokal et al., 2023). Beyond their ecological effects, the lifecycle of plastic products contributes significantly to greenhouse gas (GHG) emissions, posing another layer of environmental threat. Projections indicate that by 2050, plastic production could account for up to 13% of the planet's total carbon budget. The persistence of GHGs in the environment exacerbates global warming, creating a troubling feedback loop. Each year, an estimated eight million tons of discarded plastics enter the oceans, raising concerns about plastic toxicity for marine life and, subsequently, human health. The mismanagement of plastic waste not only contributes to GHG emissions but also poses a significant threat to fragile ecosystems, which are already vulnerable to climate change (Iroegbu et al., 2020). Plastics are synthetic polymers derived primarily from fossil fuels, characterized by their lightweight, versatility, and durability. These properties have made them preferred materials across various industries, contributing to a global production value of approximately 580 billion USD in 2020. Since 1950, global plastic production has skyrocketed from 2 to 380 megatons, yet only about 9% of plastic debris is recycled, while a staggering 79% accumulates in landfills (Geyer et al., 2017). Polyethylene (PE), polypropylene (PP), and polyvinyl chloride (PVC) are the most widely manufactured plastic polymers, contributing 21%, 24%, and 19% to global plastic production, respectively (Wright et al., 2013).

Plastic pollution has pervasive adverse effects on marine ecosystems, ranging from shallow coastal waters to the deepest ocean trenches, and even reaching Earth's most remote locations (Napper & Thompson, 2020). Over the last two decades, the accumulation of plastic debris in marine environments has escalated dramatically, playing a substantial role in exacerbating climate change. The consequences of climate change manifest in rising sea levels, ocean acidification, and extreme weather events, inflicting socio-economic and ecological harm (Ford et al., 2022). Therefore, intensive research on the relationship between plastic pollution and climate change is urgently needed to inform policy-making and develop effective mitigation strategies. The environmental ramifications of MPs extend beyond their physical presence. For instance, the interaction of microplastics (MPs) with pollutants, their long-term persistence in the environment, and their impacts on ecosystems and human health constitute a significant area of ongoing research. Studies indicate that MPs can introduce toxic chemicals into marine ecosystems, accumulating in the tissues of aquatic species and ultimately affecting the food web. For example, MPs can act as vectors for pollutants such as hydrophobic organic compounds and trace elements, enhancing their bioavailability and toxicity (Khan et al., 2022; Sruthisree et al., 2015; Amin et al., 2016, 2017). Moreover, hazardous compounds associated with MPs may transfer up the food chain, presenting potential risks to human health. The extent of microplastic pollution is staggering, with estimates suggesting that around 5.25 trillion particles, weighing approximately 268,940 tons, are present on ocean surfaces alone (Eriksen et al., 2014). MPs have been detected in diverse ecosystems, from shorelines and water columns to deep-sea sediments and polar ice caps (Crawford & Quinn, 2017). Their ubiquity makes them readily available to marine organisms, with studies documenting their presence even in remote locations such as the Antarctic Ocean and the Mariana Trench (Peng et al., 2018). Sediment samples from the Mariana Trench revealed a shocking concentration of 200 to 2200 MPs per liter, underscoring the far-reaching impact of human activities on some of Earth's most isolated habitats. The ecological effects of MPs on aquatic organisms are multifaceted. They can cause physical blockages in the digestive systems of marine species, reduce feeding efficiency, and impair reproductive systems (Sussarellu et al., 2016). Furthermore, MPs can damage digestive tracts, hinder nutrient absorption, and disrupt buoyancy, impair swimming and floating capabilities. The complex interactions between MPs and environmental pollutants necessitate a comprehensive understanding of their ecological and physiological impacts. Addressing the challenge of microplastic pollution requires a concerted effort to develop global strategies that mitigate these interconnected crises. Policymakers and scientists must recognize the synergy between plastic pollution and climate change, working together to develop sustainable solutions that address both the causes and consequences of these pressing environmental challenges. This chapter explores the interconnected impacts of microplastic pollution and climate change on ecosystems and human health, emphasizing their combined ecological and physiological consequences while advocating for integrated mitigation strategies.

2 Microplastics as a Driver of Climate Change

Microplastics are the primary carriers for the spread of this type of pollution because of their large surface area and hydrophobicity, which enable them to concentrate harmful substances like heavy metals, polycyclic aromatic hydrocarbons, medications, pesticides, nanoparticles, and organic halogens on their surfaces (Bhagat et al., 2021). Microbial biofilms on the surfaces of microplastic pollution frequently gradually colonize it when it is exposed to the natural environment, creating a phenomenon known as the "plastisphere" (Zhai et al., 2023). The majority of microplastic pollution occurs in soil and aquatic environments, and biological processes are the main means by which it degrades due to their biodegradable qualities, biodegradable pl astics are increasingly utilized in agriculture and domestic settings to reduce plastic pollution (Pratt et al., 2020; Jeyavani et al., 2021).

Organic fertilizers are thought to be one of the newer causes of the buildup of microplastic pollution in soil (O'Connor et al., 2022). The suspension of microplastics in sediments due to the increase in global average temperature intensifies microplastic pollution in the environment, and microplastic pollution exacerbates global climate warming both are mutually reinforcing (Li et al., 2024). Membrane bioreactors and ultrafiltration removed 90% of MPs, whereas magnetic separation successfully extracted 88% of the target MPs from wastewater. About 80 g of plastic could be consumed daily by one kilogram of wax worms in a biological process. This indicates that 2.9 tons of plastic per year, or about 8 kg of plastic per day, can be consumed by 100 kg of wax worms. To address the issues raised by MPs, technology, policy instruments, and legislation must be integrated (Kurniawan et al., 2023).

3 Impact of Climate Change on Microplastic's Fate

Rising atmospheric temperatures drive extreme weather events, including glacier melting, intense winds, increased thunderstorms, severe droughts, and heightened rainfall (IPCC, 2019). One significant consequence of glacier melting is the release of microplastics previously trapped in ice into the ocean (Obbard et al., 2014). Once released, microplastics that have a density lower than that of surrounding water are likely to be carried away by ocean currents, allowing them to be redistributed over long distances. This process transports microplastics from melting glaciers into various aquatic habitats. As microplastics enter freshwater systems, their fate changes due to the lower density of freshwater compared to saline environments, leading to a higher likelihood of sinking to the bottom. Increased rainfall exacerbates this situation by causing coastal flooding and rising sea levels associated with glacier thawing (IPCC, 2019). Consequently, plastic debris and microplastics are more likely to be washed from shorelines into water bodies. Once in these environments, ocean currents contribute to mechanical weathering, breaking down larger plastic debris into microplastics (Aragaw, 2020). These microplastics may either degrade further or accumulate in aquatic habitats. Climate change significantly affects not only the fate of microplastics but also their degradation in the environment. While most studies focus on the impact of microplastics on aquatic and soil biota, this discussion extends to the ecotoxicological effects of microplastics under climate change conditions. Ice melting increases ocean current velocities, while gusty winds and heightened thunderstorms, driven by rising temperatures, also influence microplastic dynamics. Changes in wind patterns can alter surface water movement, redistributing microplastics away from their sources, and complicating the understanding of their ultimate distribution. Moreover, strong wind events can cause vertical mixing of water layers (Kukulka et al., 2012; Reisser et al., 2014), resuspending settled microplastics from the seabed back into the water column. Floods, storms, and extreme weather can transport microplastics to previously unaffected areas, exposing pollution-free zones to contamination. Microplastics are not only moved from rivers to oceans (Napper et al., 2021) but can also migrate from aquatic environments to terrestrial soils (Allen et al., 2020). Furthermore, increased global temperatures can lead to droughts that may trap microplastics in soil (Lozano et al., 2021). Table 1, illustrates the pathways of plastic degradation and their links to climate change factors. The fate of microplastics under various climate change parameters is depicted in (Fig. 1).

4 Sources of Microplastics and Their Environmental Persistence

When large fish populations are present, the pollutants attached to sediment particles may be resuspended and released into the water. A few of the contaminants can also attach themselves to plastic particles and move both inside and outside of the aquaculture setting. In salmon aquaculture, for instance, antifoulants, antibiotics, parasiticides, anesthetics, and disinfectants are used (Prasad et al., 2023; Fred-Ahmadu et al., 2024). The environmental compartments—ditches, streams, rivers, estuaries, temporary and permanent wetlands, ponds, dams, and lakes—all have unique hydrological, chemistry, flora, and fauna characteristics, as well as surrounding watershed and land-use patterns, making microplastic pollution in aquatic freshwater ecosystems extremely complex. By improperly disposing of wrapping and bale twine or by applying sewage sludge to agricultural lands, agricultural activities can also release plastics into the terrestrial environment (Mahon et al., 2017; Corradini et al., 2019).

Microplastics are commonly divided into three main categories: microfibers, which originate from synthetic textiles; plastic pellets, which originate from industrial manufacturing; and microbeads, which are present in skin care products. With synthetic microfiber pollutants finding their way into our waterways and eventually contaminating the air and soil, microfiber pollution is a growing problem (Tyagi, 2024). The important mechanism that can move microplastics from the ocean's surface layer to deep pelagic and mesopelagic zones is the idea of marine snow, which is the continuous settling of primarily organic particles from the upper regions of the water column. It may also increase the bioavailability of microplastics to biota that inhabit benthic habitats (Porter et al., 2018). Microplastics' transmission from cities to rivers and lakes, runoff and subsequent ocean transportation, and marine dispersal throughout ocean basins and deep ocean layers (Peng et al., 2018; Hale et al., 2020).

Table 1 Pathways of plastic degradation and their links to climate change factors

Degradation pathway	Climate change factor	Degradation mechanism	Connection with climate change factor	Reference
Mechanical degradation	Extreme weather events (storms, floods)	Physical fragmentation due to mechanical forces	Increased storm intensity and flooding amplify wave action and abrasion, accelerating the breakdown of plastic into microplastics.	Belioka and Achilias (2024)
	Glacier melting	Fragmentation through ice movement and thawing	Melting glaciers release trapped plastics, and ice movement causes mechanical fragmentation into microplastics.	Haque and Fan (2023)
Thermal degradation	Global warming	Breakdown through elevated temperatures	Rising atmospheric and oceanic temperatures increase the thermal stress on plastics, hastening their degradation.	Saleem et al. (2024)
	Heat waves	Increased thermal stress on polymers	Frequent and intense heat waves raise ambient temperatures, intensifying the thermal breakdown of plastics in terrestrial and marine environments.	Weber et al. (2020) and Wang et al. (2023)
Chemical degradation	Ocean acidification	Hydrolysis and chemical reactions	Increased acidity in oceans alters chemical conditions, enhancing the rate of plastic breakdown through hydrolytic reactions.	Gewert et al. (2015) and Bacha et al. (2021)
	Drought conditions	Increased chemical interactions in soil	Droughts concentrate pollutants and chemicals in soil, accelerating the chemical degradation of microplastics.	Belioka and Achilias (2024) and Li et al. (2024)
Photodegradation	Increased solar radiation	UV-induced breakdown of polymers	Enhanced UV radiation, linked to ozone depletion and climate change, accelerates photochemical degradation of plastics into smaller fragments.	Andrady et al. (2011, 2022, 2023) and Reisser et al. (2014)
	Reduced cloud cover	Enhanced UV exposure	Decreased cloud cover due to climate change increases direct solar	

(continued)

Table 1 (continued)

Degradation pathway	Climate change factor	Degradation mechanism	Connection with climate change factor	Reference
			radiation, accelerating photodegradation of plastics.	
Biological degradation	Climate-induced changes in biota	Microbial breakdown of plastics	Climate change alters microbial communities and enzyme activity, potentially increasing or decreasing biological degradation rates.	Upadhyay (2020), Chowdhury et al. (2022), and Segaran et al. (2023)

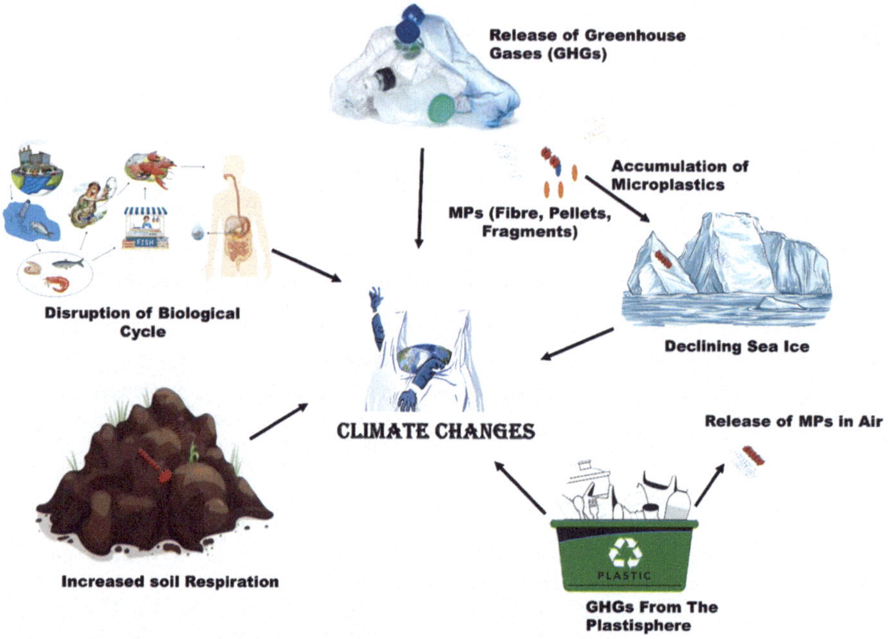

Fig. 1 Fate of microplastics under various climate change parameters

The presence of microplastics in rivers and lakes may have a variety of effects on greenhouse gas emissions. Microplastics can decrease microbial preference for stable carbon during heatwaves and reduce community carbon metabolism capacity. In the river, microplastics can also increase CO_2 release from the microbial community, decrease species interactions and community complexity, and hinder the thermal adaptation of microbiological communities. (Wang et al., 2023). The framework of global environmental change processes, supports the identification of key pathways, mechanisms, and indicators of microplastics, as well as the overall progress of science and efficient policymaking to address this significant environmental issue holistically (Bank & Hansson, 2022). A new artificial ecosystem known as the "plastisphere" has been created by plastic. Because plastic particles, such as microplastics (MPs) and nano-plastics (NPs), are ubiquitous in ecosystems and can interact with biological systems, they have become a global concern (Sridharan et al., 2021). Table 2 depicting the dissemination of microplastics has a distinct origin and affects the biota.

5 Effects of Plastic Pollution on Marine, Terrestrial, and Natural Ecosystems

Plastic pollution has gained significant attention for its detrimental effects on marine ecosystems (Derraik, 2002). A pivotal study by Sarah-Jeanne Royer at the Scripps Institution of Oceanography revealed that plastic debris accumulating on riverbanks, coastlines, and landscapes emits considerable amounts of greenhouse gases (Royer, 2018). As plastic degrades—especially at the ocean's surface—it releases toxic gases like methane, with emissions intensifying as plastics break down into microplastics. This exacerbates the vulnerability of marine species and ecosystems to the combined threats of plastic pollution and climate change (Stubbins et al., 2021). Marine turtles exemplify these impacts, as climate change and plastic pollution increasingly threaten their populations. Rising global temperatures have led to skewed female-biased sex ratios in some turtle hatcheries, further endangering these species (Chatting et al., 2021). For example, green turtles (*Chelonia mydas*) nesting on the northern Great Barrier Reef display alarming sex ratio imbalances: 99.1% of juveniles, 99.8% of sub-adults, and 86.8% of adult-sized turtles are female (Jensen et al., 2018). Microplastics also contribute to altering temperature conditions in marine ecosystems, further amplifying the challenges faced by marine species. Experimental studies have shown that marine turtles are particularly vulnerable to the adverse effects of plastic pollution, which can cause internal injuries, intestinal blockages, and even death due to ingestion or entanglement (Nelms et al., 2016; Duncan et al., 2017). Moreover, the combined impacts of marine plastic pollution and climate change intensify global warming, elevating ocean temperatures and significantly affecting coral reefs. These temperature changes result in widespread coral bleaching (Hughes et al., 2018), leading to high mortality rates and the potential extinction of certain coral species (Hughes et al., 2017; Bento et al., 2016). Corals are essential for sustaining marine biodiversity, offering habitats and shelter for a wide range of marine organisms, facilitating nutrient cycling, and providing vital nutrients such as nitrogen to marine life (Ortiz et al., 2018; Baumann et al., 2019). However, research has shown that plastic fragment ingestion poses serious risks to coral health, impairing photosynthetic performance, slowing growth rates, and compromising overall vitality (Berry et al., 2019; Reichert et al., 2019). Ecosystems are continually adapting to environmental changes such as climate

Table 2 The dissemination of microplastics has a distinct origin and affects the biota

Microplastic produce medium	Mode of microplastic detect	Microplastic	Microplastic impacts	References
Environmental	Fertile soil	Polyethylene (PE) Polypropylene (PP), PE, Polystyrene (PS),	Effect on microbes	Ren et al. (2020)
	Soil	Polyvinyl chloride (PVC),	Influences soil microbial community and structure	Chen et al. (2023)
		Low density polyethylene (LDPE)	Significant inhibition of microorganisms and C and N cycling functional genes in earthworm guts is caused by microplastic.	Gao et al. (2022)
		Polylactic acid (PA),	PA-Microplastic increases the abundance of the accA and pccA genes, which improves soil carbon fixation.	Sun et al. (2023)
	Freshwater sediments	Polyethylene terephthalate (PET)	Microplastics influence N_2O emissions by altering microbial diversity and the community structure involved in nitrogen cycling.	Zhang et al. (2022)
	Mud	Polystyrene (PS)	Microplastic control carbon metabolism by preventing the microbial community from adapting to heat.	Wang et al. (2022)
Aquaculture farms	Fishmeal	PE, PP, PA, PET	Microplastic indirectly entered the food chain, it has an impact on the aquatic ecosystem.	Gündoğdu et al. (2021)
	Fish	Cellophane	The potential high risk microplastic in human food	Feng et al. (2019)
Greenhouse gases	Photochemical ozone	PE and PS	Microplastics produce photo-chemical ozone during the photodegradation process.	Zhao and You (2022)
	CO_2	PE	Bacteria's tricarboxylic acid cycle is used to	Qiang et al. (2023)

(continued)

Table 2 (continued)

Microplastic produce medium	Mode of microplastic detect	Microplastic	Microplastic impacts	References
			enzymatically break down microplastic.	
		LDPE	LDPE-Microplastic is broken down by bacteria that break down polymers.	Pathak and Navneet (2023)

change, temperature variations, water imbalances, rising atmospheric carbon dioxide levels, and extreme weather events. The intricate interactions among organisms make these ecosystems particularly vulnerable to climate change. The combined pressures of plastic pollution and climate change can significantly alter natural ecosystems, jeopardizing biodiversity and adversely affecting global food security. In terrestrial ecosystems, soil microbial communities are critical for regulating biogeochemical cycles. However, plastic pollution linked to climate change can disrupt these communities, leading to biodiversity loss and reduced carbon storage capacity, further exacerbating climate-related challenges (Malhi et al., 2020). Microplastics impede carbon sequestration and disrupt the carbon cycle in soil environments. They also reduce N_2O emissions during nitrogen fertilization and interfere with other nutrient cycles, such as those of nitrogen and phosphorus, often causing oxidative stress. Microplastics interact with soil fauna, altering their biophysical environments and impairing their fitness and functioning. Studies show that organisms like earthworms and springtails can facilitate the horizontal and vertical movement of microplastics within soil systems, exacerbating the disruption of soil processes (De Souza Machado et al., 2018). As a result, soil productivity declines, leading to nutrient cycle imbalances that can significantly reduce crop yields, contributing to the global food crisis (Kumar et al., 2021). The composition of plastics and their link to human activities can profoundly affect ecosystem functioning in terrestrial environments. In wastewater treatment facilities, microplastic surfaces often host pathogenic microorganisms. These treatment plants frequently fail to capture such small plastic particles, allowing them to enter freshwater systems and spread pathogens. Despite their growing significance, the impacts of microplastics on land-based ecosystems remain largely unexplored, marking a critical area for future research.

6 Impacts of Microplastics on Ecosystems and Aquatic Biodiversity

Microplastics are increasingly recognized as a major environmental threat to various aquatic organisms in both freshwater and marine ecosystems. A wide array of species, including zooplankton, corals, lobsters, sea urchins, mussels, aquatic birds, and mammals, are exposed to these persistent pollutants (Browne et al., 2008). Microplastic particles can be ingested selectively, mistaken for prey, or unintentionally consumed during respiration (Derraik, 2002). Studies show that many aquatic animals face difficulty distinguishing between plastic and their natural food sources (Crawford & Quinn, 2017). The physicochemical properties of microplastics—such as their size, density, and particularly their color—play a critical role in deceiving aquatic organisms (Wright et al., 2013; Alomar & Deudero, 2017). Once ingested, microplastics can traverse the food web, reaching higher trophic levels (Lusher et al., 2016). Studies have documented the presence of microplastics in the digestive systems of various aquatic species, including zooplankton (Setälä et al., 2014), macroinvertebrates (Van Cauwenberghe et al., 2015), fish (Foley et al., 2018), aquatic birds (Nelms et al., 2023), and marine mammals (Lusher, 2015). Evidence suggests that microplastic particles can cross biological barriers through endocytotic mechanisms such as phagocytosis and persorption. Persorption is a mechanical process in which particles penetrate underlying tissues via the epithelial layer (Wright & Kelly, 2017). The ability of microplastics to enter cells depends on their size; smaller particles can passively diffuse through cell membranes, while larger particles require endocytosis for uptake (Kettiger et al., 2013). Ingested microplastics can lead to severe consequences for aquatic organisms. They may obstruct digestive tracts, potentially resulting in death, behavioral changes, or a false sense of satiety, weakening the animal and increasing its vulnerability to predation (Derraik, 2002; Wright et al., 2013). Additionally, microplastics can induce oxidative stress, cellular damage, DNA harm, inflammation, and various immune responses (Yong et al., 2020; Gurjar et al., 2023). The prevalence of microplastics in aquatic environments has become a pressing global concern, threatening aquatic life on multiple fronts (Junaid et al., 2023).

It is essential to recognize that the impacts of plastic pollution extend beyond microplastics; all forms of plastic debris adversely affect aquatic ecosystems (Aydin et al., 2023). An estimated 4076 marine species are currently threatened by marine litter, with plastics comprising a significant portion of this waste (Tekman et al., 2024). Data from the LITTERBASE database highlight that fish (23.7%), seabirds (14.3%), crustaceans and arthropods (11.2%), and mollusks (7.9%) are among the most affected organisms. The primary impacts of marine litter on these species include ingestion (40.4%), colonization (34.3%), and entanglement (17.4%). Numerous studies have demonstrated the detrimental effects of microplastics on various aquatic organisms, including fulmars, oysters, mussels, and fish. Even small organisms, such as zooplankton and fish larvae, are susceptible to direct ingestion of microplastics during feeding (Lusher, 2015). As awareness of this issue grows, urgent action is needed to mitigate the impacts of microplastics and protect aquatic ecosystems from further degradation. Ecological and environmental groups are now aware of the connection between biodiversity and micro (nano) plastics. Nevertheless, more work is required to evaluate the effects of micro (nano) plastics and how organisms interact with them (Hu et al., 2019). The actual hazards that microplastic pollution constitutes to people include its sources, spread, health effects, and contamination of the air, water, food, and soil (Manosuthi, 2024). Microplastics harm fish at the molecular level, causing genetic damage through the absorption of polycyclic aromatic hydrocarbons, which cause immunotoxicity, neurotoxicity, and genotoxicity to marine species. They also cause decreased food intake, delayed growth, oxidative damage, abnormal behavior, and a barrier to lipid metabolism (Bubu-Davies & Anwuri, 2022). Because of its pervasiveness, environmental persistence, and diverse interactions with continental biota, microplastic pollution may be a new threat to terrestrial ecosystems as a result of global change. It highlights the physical and chemical characteristics of the corresponding observed effects, discusses the widespread toxicity of nano-plastics derived from plastic breakdown, and examines the widespread microplastic contamination as a possible agent of global change in terrestrial systems (De Souza Machado et al., 2018). Microplastics have a significant impact on the environment due to their tiny sizes, ubiquity, sheer volume, composite chemicals, and growing anthropogenic activities. Their effects may be felt directly by the food ingredients or indirectly by the ecosystems that sustain the food chain's main producers (Okeke et al., 2022). These particles transfer pollutants to other organisms and have negative consequences, such as decreased growth and reproductive function, behavioral changes, oxidative stress, and decreased feeding capacity. They also cause biomagnification as they advance in the trophic chain. The main issue here is that these microplastic and mesoplastics that organisms consume can negatively impact humans through seafood like fish, clams, and oysters (Devi & Kumar, 2024). The two primary goals of the symposium on the development and application of an environmental risk assessment (ERA) framework sponsored by the International Council of Chemical Associations were the following: to examine the scientific benefits and drawbacks of using a conceptual ERA framework proposed for microplastic particles, and to inquire stakeholders about the possibility of such an approach in determining environmental risks. This symposium also aimed to identify key knowledge gaps to boost confidence in using ERA tools for microplastic particles and prioritize research activities to reinforce framework implementation (Gouin et al., 2019). Continued research and concerted global efforts are essential to mitigate the ecological and biological impacts of microplastics on aquatic biodiversity and ecosystems. Table 3 depicting the effect of microplastic on aquatic organisms. Impact of microplastics on ecosystems and aquatic biodiversity, illustrating their pathway in organisms (Fig. 2).

Table 3 Effect of microplastic on aquatic organisms

Organisms	Influence and endpoint	Reference
Fish		
Korean bullhead fish (*Pseudobagrus fulvidraco*)	MP existence causes disturbances in haematological parameters	Lee et al. (2023)
Girella fish (*Girella laevifrons*)	Inflammation, circulatory disturbances, and regressive changes in intestinal tissue, including the loss of crypt/villi cells.	Ahrendt et al. (2020)
Common carp (*Cyprinus carpio*)	Reduced growth rates and immune response alterations	Roch et al. (2020)
Nile tilapia (*Oreochromis niloticus*)	MPs caused anaemia and perturbations leading to mortality of early juvenile tilapia	Hamed et al. (2019)
Japanese medaka (*Oryzias latipes*)	Oxidative stress, bioaccumulation of MPs, and reproductive disruptions	Choi et al. (2018)
Zebrafish (*Danio rerio*)	MPs caused neurotoxicity, immune response alterations, and developmental toxicity in embryos	Mattsson et al. (2017)
European seabass (*Dicentrarchus labrax*)	MPs ingestion led to gut inflammation, reduced feeding efficiency, and lower growth rates	Pedà et al. (2016)
Atlantic cod (*Gadus morhua*)	MPs accumulation caused reduced foraging efficiency and potential trophic transfer	Wieczorek et al. (2018)
Atlantic salmon (*Salmo salar*)	Decreased feeding behavior and altered energy metabolism	Dempsey et al. (2020)
Chondrichthyes		
Galeocerdo cuvier	Impaired feeding behavior and potential bioaccumulation effects	Munno et al. (2024)
Carcharhinus leucas, Carcharodon carcharias, Galeocerdo cuvier, Sphyrna lewini	Disruption in feeding efficiency and health metrics	Lu et al. (2024)
Rhizoprionodon acutus	Inflammatory responses and potential for reduced reproductive success	Janardhanam et al. (2022)
Scyliorhinus canicular	Chronic exposure leading to physiological stress and impaired growth	Impellitteri et al. (2023)
Mammals		
Halichoerus grypus	Reduced foraging efficiency and health complications	Hernandez-Milian et al. (2019) and Desclos-Dukes et al. (2022)
Phoca largha	Ingestion of MPs leading to immune response alterations	Wang et al. (2021b)
Delphinus delphis	Impaired health and reproductive outcomes due to MP ingestion	Stockin et al. (2021)
Delphinus delphis, Kogia breviceps, Mesoplodon ginkgodens, Sousa chinensis	Bioaccumulation effects and potential trophic transfer issues	Aierken et al. (2024)

(continued)

Table 3 (continued)

Organisms	Influence and endpoint	Reference
Rhincodon typus	Reduced feeding efficiency and health indicators	Yong et al. (2021)
Baleen whales	Disruption in feeding and potential health impacts	Zantis et al. (2022)
Sea turtle		
Green sea turtle *Chelonia mydas*	Physiological stress and potential mortality	Aranda et al. (2024)
Green and loggerhead sea turtles	Impaired health and reproductive success due to MP ingestion	Rodríguez et al. (2022)
Caretta caretta and Linnaeus	Ingestion leading to gastrointestinal blockages and health decline	Solomando et al. (2022)
Arthropoda		
Metopograpsus quadridentata (mangrove crab)	Inhibition of growth and reproductive success	Patria et al. (2020)
Lepas (Lepas) *anatifera*	Altered energy balance and reproductive outcomes	Scotti et al. (2023)
Horseshoe crab (*Tachypleus tridentatus*)	Impaired immune responses and reproductive success	Wang et al. (2022)
Echinodermata		
Sea cucumber *Holothuria tubulosa*	MPs were mostly in the form of fibers, and the most commonly detected polymers were PE, PP, and PET	Lombardo et al. (2022)
Sea urchins *Echinometra mathaei* and *Diadema setosum*	PES and PP were reported to have the highest rates	Rahmawati et al. (2023)
Mollusks (Gastropods, Bivalves, Cephalopods)		
Blue mussel (*Mytilus edulis*)	Decreased filtration rate, altered energy balance, and immune response	Browne et al. (2008) and Von Moos et al. (2012)
Pacific oyster (*Crassostrea gigas*)	MP ingestion altered energy allocation, leading to reduced reproductive output	Sussarellu et al. (2016) and Hahladakis et al. (2018)
Common periwinkle (*Littorina littorea*)	MPs induced oxidative stress and tissue inflammation	Cole et al. (2013)
Scallop (*Pecten maximus*)	MPs caused disruption in filter feeding and bioaccumulation in digestive tissues	Al-Sid-Cheikh et al. (2018)
Cuttlefish (*Sepia officinalis*)	MPs exposure led to reduced embryonic development rates and physiological stress	Kaposi et al. (2014)
Common octopus *(Octopus vulgaris)*	Impaired growth and reproductive success	Pedà et al. (2022)
Crustaceans		
Fiddler crab (*Uca arcuata*)	Inhibited burrowing behavior and reduced energy reserves	Wright et al. (2013)
		Yin et al. (2021)

(continued)

Table 3 (continued)

Organisms	Influence and endpoint	Reference
Whiteleg shrimp (*Litopenaeus vannamei*)	MPs ingestion caused histopathological changes in hepatopancreas and oxidative stress	
European lobster (*Homarus gammarus*)	MPs bioaccumulation in gills and hepatopancreas led to impaired metabolic activity	Watts et al. (2014)
Blue crab (*Callinectes sapidus*)	Physiological stress, reduced growth, and increased mortality rates	Schwabl et al. (2019)
Shore crab (*Carcinus maenas*)	MPs accumulation disrupted gill functioning and reduced energy for reproductive processes	Farrell and Nelson (2013)
Brown shrimp (*Crangon crangon*)	MPs ingestion caused gut blockage and lowered fitness	Devriese et al. (2015)
Zooplankton		
Copepods (*Calanus* spp.)	MPs ingestion reduced fecundity and energy transfer efficiency in food webs	Cole et al. (2015)
Daphnia (*Daphnia magna*)	MPs caused reduced growth, impaired reproduction, and mortality in juveniles	Besseling et al. (2014)
Euphausiids (*Euphausia superba*), *Euphasia* spp.	MPs ingestion disrupted nutrient cycling and contributed to energy deficits	Dawson et al. (2018) and Zheng et al. (2020)
Rotifers (*Brachionus plicatilis*)	MPs caused reduced swimming activity and impaired reproduction	Jeong et al. (2016)

7 Effects of Climate Change on Microplastics and Associated Chemical Contaminants

Climate change, as defined by the Intergovernmental Panel on Climate Change (IPCC), encompasses long-term alterations in climate due to natural variability and human activities. The significant rise in greenhouse gas (GHG) emissions has increased global temperatures and global warming. Key GHGs, including water vapor, CO_2, CH_4, O_3, chlorofluorocarbons (CFCs), and N2O, trap infrared radiation, making them crucial for regulating atmospheric temperatures (Borduas & Donahue, 2018). This greenhouse effect is essential for maintaining a habitable climate on Earth (National Geographic Society, 2023). However, GHG concentrations have surged since the industrial revolution, leading to heightened heat retention and greater ultraviolet radiation exposure, accelerating climate change (Borduas & Donahue, 2018). Human activities such as fossil fuel-based power generation, industrial processes, and transportation contribute significantly to the emission of CO_2 and N_2O. Additionally, the clearing of mangroves for aquaculture has diminished carbon sinks, further exacerbating GHG persistence (Grellier et al., 2017;

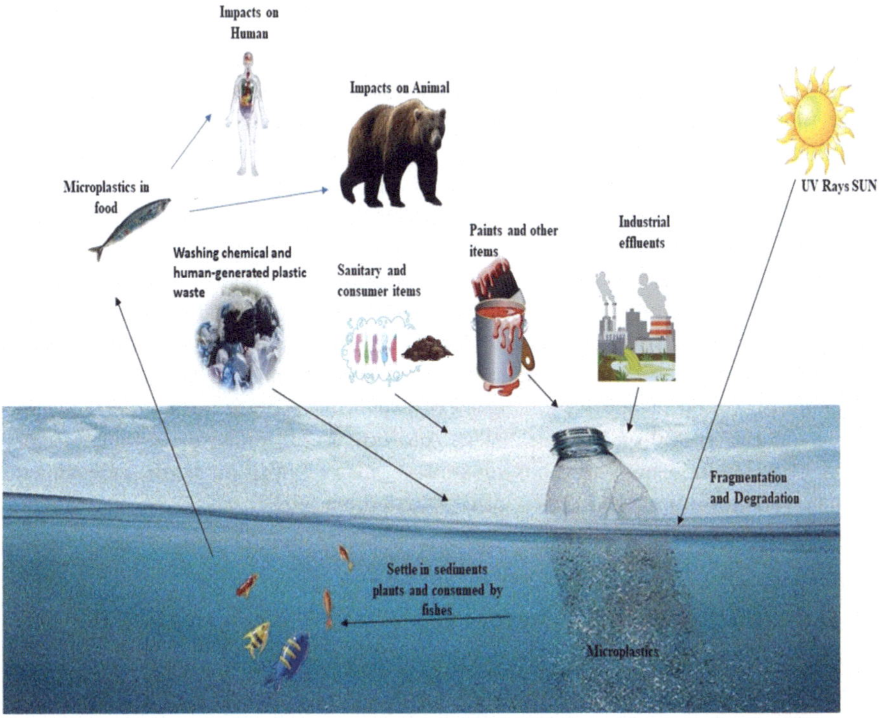

Fig. 2 Microplastic impact on ecosystems and aquatic biodiversity pathways

Sabdaningsih et al., 2023). The consequences of these changes are evident in rising average temperatures, increasingly unpredictable weather patterns, rising sea levels, warmer oceans, frequent wildfires, and extreme weather events like floods, storms, cyclones, landslides, and droughts (Benson et al., 2019). Notably, the adverse effects of GHG emissions and climate change also extend to aquaculture, impacting both production and consumption (Ngarava et al., 2023).

8 Contribution of Plastic to Global Climate Change

8.1 Lifecycle of Plastic and Greenhouse Gas Emissions

The lifecycle of plastic products significantly contributes to greenhouse gas (GHG) emissions, posing a serious environmental threat and exacerbating global climate change. Plastics are primarily derived from fossil fuels such as crude oil, natural gas, and coal, and GHGs are emitted at each stage of their lifecycle (Shen et al., 2020; Jiao et al., 2024).

8.2 Extraction and Transport of Fossil Fuels

The initial phase of plastic production involves the extraction and transportation of fossil fuels, which are the raw materials for plastic manufacturing. This process contributes significantly to GHG emissions through several mechanisms. For instance, natural gas and oil are extracted via drilling and hydraulic fracturing (fracking), which involves injecting pressurized liquids to crack underground rock formations (Alhazmi et al., 2021). These activities are highly carbon-intensive and emit methane and carbon dioxide directly into the atmosphere. Transportation of these fossil fuels through pipelines, trucks, and trains further adds to emissions. A report by the Center for International Environmental Law (CIEL) estimated that in the United States alone, the extraction and transportation of fossil fuels for plastic production resulted in approximately 12.5–13.5 million metric tons of CO_2 emissions annually (Hamilton & Feit, 2019). Additionally, the land disturbance associated with extraction activities has cleared about 19.2 million acres, releasing an estimated 1.686 billion metric tons of CO_2 (Bauman, 2019).

8.3 Refining and Manufacturing

Once extracted, fossil fuels undergo refining to convert crude oil into various petroleum products, including naphtha, which is crucial for plastic production (Wang et al., 2021a, b, 2022). This refining process is energy-intensive, requiring substantial heat and resulting in significant CO_2 emissions. In 2015, emissions from the manufacturing of polyethylene (PE) plastics alone reached between 184.3 and 213 million metric tons of CO_2. Projections indicate that emissions could rise by 34% by 2030 (Bienkowski, 2019; Prasad et al., 2024; Tabassum et al., 2025). The expansion of the petrochemical and plastic sectors has further exacerbated climate challenges. For example, a newly constructed natural gas processing plant in Pennsylvania is expected to emit around 2.25 million tons of GHG annually, while an ethylene plant in Texas is projected to release 1.4 million tons per year (Hamilton & Feit, 2019). Such developments could lead to a cumulative emission of over 125 gigatons of CO_2 by the end of the century, consuming approximately 10% of the Earth's remaining carbon resources.

8.4 Plastic Waste and Global Climate Change

The issue of plastic waste is escalating, with projections indicating that by 2040, between 23 and 37 million metric tons of plastic waste could enter marine ecosystems annually. This alarming trend is underscored by the discovery of plastic debris in the Mariana Trench, the deepest part of the ocean (Morelle, 2019). The persistence of plastic in marine environments poses a severe risk, leading to the release of greenhouse gases and disrupting the carbon sink cycle. Inadequate management of plastic waste has both direct and indirect effects on global climate. Plastics that are not recycled degrade into microplastics, which infiltrate rivers and oceans (Sharma & Chatterjee, 2017). As plastics decompose, they emit greenhouse gases such as CO_2, methane, and ethylene. Research indicates that microplastics can hinder marine organisms' ability to absorb CO_2 and release oxygen, potentially accelerating oceanic oxygen depletion (Edmond, 2022). The limitations of current waste management systems are stark, with only about 25% of produced plastics being recycled. Methods such as incineration, while intended for waste reduction, are significant sources of GHG emissions. Studies show that the incineration of one metric ton of plastic waste can produce nearly one ton of CO_2 (Verma et al., 2016). In the U.S., plastic waste incineration emitted approximately 5.9 million metric tons of CO_2 in 2015, and by 2019, emissions from plastic production and incineration reached about 850 million metric tons, equivalent to the output of 189 coal-fired power plants. The World Energy Council projects that greenhouse gas (GHG) emissions from plastic production and incineration could increase to 49 million metric tons by 2030 and 91 million metric tons by 2050. The plastic industry is estimated to account for approximately 10% of the global warming potential (Hamilton & Feit, 2019).

8.5 Mitigating Climate Impact Through Biodegradable Plastics

The negative impact of plastic waste on climate extends beyond production phases. The degradation of plastics releases greenhouse gases, primarily methane and ethylene, which have a substantial warming potential. One potential solution lies in developing biodegradable plastics, which can mitigate the adverse environmental effects of traditional plastics (Moshood et al., 2022). Biodegradable plastics, made from biological materials rather than fossil fuels, are designed to decompose more easily in the environment. They can be broken down by microorganisms into water, CO_2, and compost, posing minimal risk to ecosystems. Life Cycle Assessment (LCA) studies have indicated that biodegradable plastics generally have a lower environmental impact compared to conventional plastics (Razza & Cerutti, 2017). Future research should focus on optimizing biodegradable plastic production in terms of economic viability, environmental protection, and social responsibility. Regulatory frameworks could support the commercialization of biodegradable plastics, promoting a shift towards more sustainable materials in global markets.

8.6 Mitigation Approaches of Microplastic in Aquatic Ecosystem

Microplastic pollution has emerged as a significant environmental challenge within aquatic ecosystems, posing serious threats to marine organisms and the overall health of these environments. The detrimental effects of microplastics on marine life necessitate the development and implementation of effective mitigation strategies aimed at reducing their presence and impact (Yang et al., 2021). As microplastics accumulate in water bodies, they are ingested by various organisms, leading to adverse health effects and disrupting entire ecosystems (Khan et al., 2023). To effectively combat this issue, a comprehensive approach is essential, focusing on prevention, cleanup, and public awareness. A key strategy for mitigating microplastic pollution is to prevent these materials from entering aquatic environments in the first place (Zhang et al., 2021). This can be achieved through the implementation of policies and regulations that minimize plastic production and consumption, promote the use of biodegradable alternatives, and enhance waste management systems to keep plastic waste out of water bodies. Furthermore, educating consumers through outreach programs can significantly reduce the amount of plastic that ultimately ends up in aquatic ecosystems. Cleanup of existing microplastic pollution is another critical aspect of mitigation. This involves developing innovative technologies for removing microplastics from water bodies and organizing community cleanup efforts that engage non-governmental organizations and government agencies. Research into effective and environmentally sustainable removal methods is also vital for successful mitigation (Lau et al., 2020). Raising public awareness about the harmful effects of microplastic pollution is crucial for garnering support for mitigation efforts. Education and communication campaigns can inform the public about the impacts of microplastics, encouraging active participation in initiatives aimed at reducing their presence in aquatic ecosystems. Addressing microplastic pollution in aquatic ecosystems requires a multifaceted approach encompassing prevention, cleanup, and awareness. By thoroughly exploring these aspects, we can develop and implement effective strategies to alleviate the pervasive threat posed by microplastics in our water bodies. Moreover, mitigating microplastic pollution involves reducing the sources of plastic waste through stricter regulations on single-use plastics and incentivizing manufacturers to create biodegradable alternatives. Public awareness campaigns are essential, as educating individuals about the origins and impacts of microplastics can lead to more responsible consumption and disposal practices. Enhancing waste management systems is another vital component of this effort. Improved recycling programs and increased community participation can significantly reduce the amount of plastic waste entering waterways. Regular waste collection initiatives, particularly in urban and rural areas, are crucial in preventing plastics from reaching rivers and oceans. Engaging communities in cleanup efforts fosters a sense of responsibility and stewardship for local environments. Innovative technologies can also play a role in addressing this challenge. Advanced filtration systems in wastewater treatment plants can capture microplastics before they enter aquatic ecosystems. Additionally, research into bioremediation techniques, which utilize microorganisms capable of degrading plastics, is promising and deserves further exploration to ensure safety and effectiveness. Legislation and policy are pivotal in combating microplastic pollution. Implementing bans on specific items, such as microbeads in cosmetics and certain single-use plastics, can significantly reduce the influx of microplastics into water bodies. Furthermore, fostering international cooperation through agreements similar to the Paris Agreement can enhance global efforts to tackle this pervasive issue. Research and monitoring are essential for understanding the scope of microplastic pollution and its impacts on marine life. Establishing monitoring programs to regularly assess microplastic levels in various aquatic ecosystems can provide valuable data for informed policymaking. Increased funding for research into the sources, distribution, and ecological effects of microplastics will contribute to the development of effective mitigation strategies. Restoration and rehabilitation of aquatic habitats can further aid in mitigating microplastic pollution. By restoring degraded ecosystems, such as wetlands and mangroves, we can enhance the natural filtration of pollutants before they reach larger water bodies. Involving local communities in these restoration projects not only raises awareness but also fosters a deeper connection to the environment.

9 Technological Interventions and Mitigation Strategies

The fishing and aquaculture industries may employ a variety of strategies to address marine litter and microplastic (MP) pollution. Key interventions include reducing net contact with the seabed through gear tagging and tracking technologies, improving management via enhanced deployment procedures, enforcing laws against illicit fishing and dumping, and upgrading port reception facilities. Additionally, stringent towing controls, as outlined in Directive EU 2019, are essential for mitigating marine litter. Recent advancements have highlighted the role of microorganisms as biodegraders, such as the gut microbiota of *Lumbricus terrestris* isolates and wax moth (*Galleria mellonella*) caterpillars, which degrade polyethylene (PE). Similarly, *Linnaeus microbiota* and *Tenebrio molitor* have been shown to induce polystyrene (PS) mineralization (Yang et al., 2015; Huerta et al., 2016; Bombelli et al., 2017). Specific microplastic removal strategies include innovative solutions like the GoJelly prototype, which uses jellyfish mucus to capture MPs, and technologies like Baltimore's "Mr. Trash Wheel." Citizen science initiatives such as the Clean Swell app from Fighting for Trash-Free Seas and Australia's large drain socks for stormwater management further illustrate effective community-driven interventions (McIlwraith et al., 2019; De Poloni, 2020). Upgrading stormwater and wastewater treatment plants (WWTPs) to incorporate advanced filtration technologies, assessing primary MP sources, and enforcing legal tools are critical steps. Promoting microplastic-free formulations in consumer products and implementing certification programs such as Blue Angel and EU Ecolabel can significantly mitigate MP

Table 4 Microplastic pollutants and their biodegraders

Microplastic-degrading biological species	Microplastic pollutant	References
Alcanivorax borkumensis	LDPE	Delacuvellerie et al. (2019)
Zalerion maritimum	PE	Paço et al. (2017)
Ideonella sakaiensis 201-F6	PET	Yoshida et al. (2016)
Rhodococcus spp.	PP	Anjana et al. (2019)
Aspergillus spp. colonizes	HDPE	
Pseudomonas and Alcanivorax	PCL	
Vibrio alginolyticus	PVA-LLDPE	
Muricauda spp.	PET	
Lactobacillus plantarum	PCL	Suzuki et al. (2018)
Pseudomonas and Bacillus	PS, HDPE, PE, LDPE, PVC	Mohan et al. (2016), Muhonja et al. (2018), Rajendran et al. (2019), and Raddadi and Fava (2019)

pollution. Preventive measures, including the development of biodegradable materials, play a central role in current mitigation initiatives. Bioplastics and durable plastic alternatives are pivotal in reducing the fragmentation of plastics into MPs. Public awareness campaigns are equally vital for fostering responsible waste management practices. To address the complexities of MP pollution, interdisciplinary approaches and stakeholder engagement are crucial for developing comprehensive strategies that protect aquatic ecosystems and human health. Future research should focus on advancing detection technologies, understanding the long-term ecological impacts of MPs, and exploring their interactions with pollutants. These efforts will facilitate the creation of holistic plans to combat MP contamination effectively (Akdogan & Guven, 2019). Microplastic pollutants and their biodegraders are presented in Table 4.

10 Integrating Microplastics in Climate Policy Frameworks

A comprehensive timeline of policies addressing microplastics highlights the importance of upstream strategies such as circular economy initiatives, behavioral change campaigns, the development of bio-based polymers, and market-based instruments. Source-specific approaches targeting the apparel industry, tire and road wear particles, antifouling paints, and recreational activities complement these efforts. Downstream strategies, including waste-to-energy technologies, degradation mechanisms, wastewater treatment plant enhancements, and litter clean-up programs, serve as supplemental measures to the primary upstream approaches (Munhoz et al., 2022). Global frameworks like the 2030 Agenda for Sustainable Development underscore the urgency of tackling MP pollution. Pioneering legislative efforts, such as the London Convention (1972), Oslo Convention (1974), and Paris Convention (1974), have laid the groundwork for preventing marine pollution from waste and dumping. The International Convention for the Prevention of Pollution from Ships (MARPOL) further regulates shipborne waste (Agamuthu et al., 2019). The UN General Assembly's resolution A/74/19 on "Oceans and the Law of the Sea" highlights MPs' impact on marine biodiversity, while the G20 Osaka Blue Ocean Vision aims to reduce marine plastic litter to zero by 2050 (UNEP 74/19, 2019). Efforts to mitigate MP pollution include improving waste management, advancing WWTPs, enhancing stormwater systems, and fostering product innovations. Public awareness campaigns, community engagement, and the adoption of certification programs further support these measures. A multifaceted approach—combining upstream prevention and downstream mitigation—is essential for addressing MP contamination in fisheries and aquaculture. Future policies must integrate interdisciplinary research to develop quantitative prediction models and mechanistic insights into MP dynamics, enabling effective regulation and management (Peng et al., 2022).

11 Conclusion and Future Prospects

Microplastic (MP) contamination in aquatic environments represents a critical global issue, threatening ecosystem health, aquatic biodiversity, and human well-being. The pervasive presence of MPs in freshwater, marine, and aquaculture systems necessitates holistic strategies encompassing prevention, monitoring, remediation, and policy reforms. While substantial progress has been made in understanding MPs' distribution, sources, and ecological impacts, significant gaps and emerging challenges persist. Developed nations contribute significantly to MP pollution through the extensive use of plastics, while developing countries face escalating issues due to inadequate waste management systems and weak regulatory enforcement. Addressing this crisis requires enhanced public awareness, effective waste disposal mechanisms, and stringent legal frameworks for plastic management. Strengthening the enforcement of environmental protection laws is vital to curbing MP influx into aquatic systems. Future efforts must prioritize advancing detection technologies, developing sustainable plastic alternatives, and enforcing stringent policy measures. Promising research areas include biodegradable plastics derived from plant-based materials, microbial and enzymatic degradation of MPs, and studies on MPs' interactions with pollutants like heavy metals and antibiotics. Understanding the mechanisms of pollutant adsorption by MPs and the influence of environmental factors on this process remains critical. Stakeholder engagement, interdisciplinary collaborations, and public awareness campaigns are indispensable for mitigating MP pollution. Embracing circular economy principles, integrating low-carbon solutions, and upgrading waste management infrastructure are essential for minimizing MP release. By uniting science, technology, and policy, the global community can address the microplastic crisis, ensuring a cleaner, healthier, and more sustainable aquatic environment for future generations.

References

Agamuthu, P., Mehran, S. B., Norkhairah, A., & Norkhairiyah, A. (2019). Marine debris: A review of impacts and global initiatives. *Waste Management & Research, 37*, 987–1002.

Ahrendt, C., Perez-Venegas, D. J., Urbina, M., Gonzalez, C., Echeveste, P., Aldana, M., & Galbán-Malagón, C. (2020). Microplastic ingestion cause intestinal lesions in the intertidal fish *Girella laevifrons*. *Marine Pollution Bulletin, 151*, 110795.

Aierken, R., Zhang, Y., Zeng, Q., Yong, L., Qu, J., et al. (2024). Microplastics prevalence in different cetaceans stranded along the Western Taiwan Strait. *Animals, 14*(4), 641. https://doi.org/10.3390/ani14040641

Ajith, N., Arumugam, S., Parthasarathy, S., Manupoori, S., & Janakiraman, S. (2020). Global distribution of microplastics and its impact on marine environment—A review. *Environmental Science and Pollution Research, 27*, 25970–25986.

Akdogan, Z., & Guven, B. (2019). Microplastics in the environment: A critical review of current understanding and identification of future research needs. *Environmental Pollution, 254*, 113011.

Alhazmi, H., Almansour, F. H., & Aldhafeeri, Z. (2021). Plastic waste management: A review of existing life cycle assessment studies. *Sustainability, 13*(10), 5340.

Allen, S., Allen, D., Moss, K., Le Roux, G., Phoenix, V. R., & Sonke, J. E. (2020). Examination of the ocean as a source for atmospheric microplastics. *PLoS One, 15*, e0232746.

Alomar, C., & Deudero, S. (2017). Evidence of microplastic ingestion in the shark Galeus melastomus Rafinesque, 1810 in the continental shelf off the western Mediterranean Sea. *Environmental Pollution, 223*, 223–229.

Al-Sid-Cheikh, M., Rowland, S. J., Stevenson, K., Rouleau, C., Henry, T. B., & Thompson, R. C. (2018). Uptake, whole-body distribution, and depuration of nanoplastics by the scallop Pecten maximus at environmentally realistic concentrations. *Environmental Science & Technology, 52*(24), 14480–14486.

Amin, A., Indulkar, S. T., & Pai, R. (2016). Sublethal effect of Buprofezin pesticide on carcass composition of *Cyprinus carpio* communis fingerlings. *Journal of Experimental Zoology India, 20*(1), 269–272.

Amin, A., Ramachandra Naik, A. T., Priyadarshini, N., Nayak, H., & Sree, C. S. (2017). Toxic effect of heavy metal lead on oxygen consumption of rohu (*Labeo rohita*) fingerlings. *Biochemical and Cellular Archives, 17*(1), 225–228.

Andrady, A. L., Hamid, H., & Torikai, A. (2011). Effects of solar UV and climate change on materials. *Photochemical & Photobiological Sciences, 10*, 292–300.

Andrady, A. L., Barnes, P. W., Bornman, J. F., Gouin, T., Madronich, S., White, C. C., Zepp, R. G., & Jansen, M. A. K. (2022). Oxidation and fragmentation of plastics in a changing environment; from UV-radiation to biological degradation. *Science of the Total Environment, 851*, 158022.

Andrady, A. L., Heikkilä, A. M., Pandey, K. K., Bruckman, L. S., White, C. C., Zhu, M., & Zhu, L. (2023). Effects of UV radiation on natural and synthetic materials. *Photochemical & Photobiological Sciences, 22*(5), 1177–1202.

Anjana, K., Hinduja, M., Sujitha, K., & Dharani, G. (2019). Review on plastic wastes in marine environment—Biodegradation and biotechnological solutions. *Marine Pollution Bulletin, 150*, 110733.

Aragaw, T. A. (2020). Surgical face masks as a potential source for microplastic pollution in the COVID-19 scenario. *Marine Pollution Bulletin, 159*, 111517.

Aranda, D. A., Sindou, P., Rodriguez, J. V. C., Saldaña, G. M., Coronado, R. F. V., González, W. D. N., et al. (2024). A non-invasive method of microplastics pollution quantification in green sea turtle Chelonia mydas of the Mexican Caribbean. *Marine Pollution Bulletin, 200*, 116092.

Aydin, I., Terzi, Y., Gündoğdu, S., Aytan, Ü., Öztürk, R., Atamanalp, M., et al. (2023). Microplastic pollution in Turkish aquatic ecosystems: Sources, characteristics, implications, and mitigation strategies. *Turkish Journal of Fisheries and Aquatic Sciences, 23*(12), 24773.

Bacha, A. U. R., Nabi, I., & Zhang, L. (2021). Mechanisms and the engineering approaches for the degradation of microplastics. *ACS ES&T Engineering, 1*(11), 1481–1501.

Bank, M. S., & Hansson, S. V. (2022). The microplastic cycle: An introduction to a complex issue. In *Microplastic in the environment: Pattern and process* (pp. 1–16). Springer.

Bauman, B. (2019). How plastics contribute to climate change. *Yale Climate Connections*, 20.

Baumann, J. H., Ries, J. B., Rippe, J. P., Courtney, T. A., Aichelman, H. E., Westfield, I., & Castillo, K. D. (2019). Nearshore coral growth declining on the Mesoamerican Barrier Reef System. *Global Change Biology, 25*, 3932–3945.

Belioka, M. P., & Achilias, D. S. (2024). The effect of weathering conditions in combination with natural phenomena/disasters on microplastics' transport from aquatic environments to agricultural soils. *Microplastics, 3*(3), 518–538.

Benson, N. U., Nwokike, C., Williams, A. B., Adedapo, A. E., & FredAhmadu, O. H. (2019). Spatial and temporal trends in diurnal temperature and precipitation extremes in North Central Nigeria. *Journal of Physics Conference Series, 1299*(1), 012062. https://doi.org/10.1088/1742-6596/1299/1/012062

Bento, R., Hoey, A. S., Bauman, A. G., Feary, D. A., & Burt, J. A. (2016). The implications of recurrent disturbances within the world's hottest coral reef. *Marine Pollution Bulletin, 105*, 466–472.

Berry, K. L. E., Epstein, H. E., Lewis, P. J., Hall, N. M., & Negri, A. P. (2019). Microplastic contamination has limited effects on coral fertilisation and larvae. *Diversity, 11*, 228.

Besseling, E., Wang, B., Lurling, M., & Koelmans, A. A. (2014). Nanoplastic affects growth of S. obliquus and reproduction of D. magna. *Environmental Science & Technology, 48*(20), 12336–12343.

Bhagat, J., Nishimura, N., & Shimada, Y. (2021). Toxicological interactions of micro-plastics/nanoplastics and environmental contaminants: Current knowledge and future perspectives. *Journal of Hazardous Materials, 405*, 123913.

Bienkowski, B. (2019). From making it to managing it, plastic is a major contributor to climate change. *Environmental Health News, 15*.

Bombelli, P., Howe, C., & Bertocchini, F. (2017). Polyethylene bio-degradation by caterpillars of the wax moth Galleria Mellonella. *Current Biology, 27*, R292–R293.

Borduas, N., & Donahue, N. M. (2018). Chapter 3.1—The natural atmosphere. In B. Török & T. Dransfeld (Eds.), *Green chemistry* (pp. 131–150). Elsevier.

Borgå, K., McKinney, M. A., Routti, H., Fernie, K. J., Giebichenstein, J., Hallanger, I., & Muir, D. C. G. (2022). The influence of global climate change on accumulation and toxicity of persistent organic pollutants and chemicals of emerging concern in Arctic food webs. *Environmental Science: Processes & Impacts, 24*(10), 1544–1576.

Browne, M. A., Dissanayake, A., Galloway, T. S., Lowe, D. M., & Thompson, R. C. (2008). Ingested microscopic plastic translocates to the circulatory system of the mussel, Mytilus edulis (L.). *Environmental Science & Technology, 42*(13), 5026–5031.

Bubu-Davies, O. A., & Anwuri, P. A. (2022). Microplastics: Potential impacts on aquatic biodiversity. *Tropical Freshwater Biology, 31*(1), 45–60.

Chatting, M., Hamza, S., Al-Khayat, J., Smyth, D., Husrevoglu, S., & Marshall, C. (2021). Feminization of hawksbill turtle hatchlings in the twenty-first century at an important regional nesting aggregation. *Endangered Species Research, 44*, 149–158.

Chen, X., Xie, Y., Wang, J., Shi, Z., Zhang, J., Wei, H., & Ma, Y. (2023). Presence of different microplastics promotes greenhouse gas emissions and alters the microbial community composition of farmland soil. *Science of the Total Environment, 879*, 162967.

Choi, J. S., Jung, Y. J., Hong, N. H., Hong, S. H., & Park, J. W. (2018). Toxicological effects of irregularly shaped and spherical microplastics in a marine teleost, the sheepshead minnow (*Cyprinodon variegatus*). *Marine Pollution Bulletin, 129*(1), 231–240.

Chowdhury, G. W., Koldewey, H. J., Niloy, M. N. H., & Sarker, S. (2022). The ecological impact of plastic pollution in a changing climate. *Emerging Topics in Life Sciences, 6*(4), 389–402.

Cole, M., Lindeque, P., Fileman, E., Halsband, C., Goodhead, R., Moger, J., & Galloway, T. S. (2013). Microplastic ingestion by zooplankton. *Environmental Science & Technology, 47*(12), 6646–6655.

Cole, M., Lindeque, P., Fileman, E., Halsband, C., & Galloway, T. S. (2015). The impact of polystyrene microplastics on feeding, function and fecundity in the marine copepod Calanus helgolandicus. *Environmental Science & Technology, 49*(2), 1130–1137.

Corradini, F., Meza, P., Eguiluz, R., Casado, F., Huerta-Lwanga, E., & Geissen, V. (2019). Evidence of microplastic accumulation in agricultural soils from sewage sludge disposal. *Science of the Total Environment, 671*, 411–420.

Crawford, C. B., & Quinn, B. (2017). Plastic production, waste and legislation. In *Microplastic pollutants* (Vol. 30, pp. 39–56). Elsevier.

Dawson, A., Huston, W., Kawaguchi, S., King, C., Cropp, R., Wild, S., Eisenmann, P., Townsend, K., & Bengtson Nash, S. (2018). Uptake and depuration kinetics infuence microplastic bioaccumulation and toxicity in Antarctic Krill (Euphausia superba). *Environmental Science & Technology, 52*, 3195–3201.

De Poloni, G. (2020). *Meet the Drain Sock—A simple pollution solution taking the world by storm 2019*. Available online: https://wattsupwiththat.com/2019/06/10/meet-the-drain-sock-a-simple-pollution-solution-taking-the-world-by-storm/. Accessed 6 December 2020.

De Souza Machado, A. A., Kloas, W., Zarfl, C., Hempel, S., & Rillig, M. C. (2018). Microplastics as an emerging threat to terrestrial ecosystems. *Global Change Biology, 24*(4), 1405–1416.

Debnath, R., Prasad, G. S., Amin, A., Malik, M. M., Ahmad, I., Abubakr, A., Borah, S., Rather, M. A., Impellitteri, F., Tabassum, I., & Piccione, G. (2024). Understanding and addressing microplastic pollution: Impacts, mitigation, and future perspectives. *Journal of Contaminant Hydrology, 266*, 104399.

Delacuvellerie, A., Cyriaque, V., Gobert, S., Benali, S., & Ruddy, W. (2019). The plastisphere in marine ecosystem hosts potential specific microbial degraders including Alcanivorax borkumensis as a key player for the low-density polyethylene degradation. *Journal of Hazardous Materials, 380*, 120899.

Dempsey, J., Martin, T. G., & Sumaila, U. R. (2020). Subsidizing extinction? *Conservation Letters, 13*(1), e12705.

Derraik, J. G. (2002). The pollution of the marine environment by plastic debris: A review. *Marine Pollution Bulletin, 44*(9), 842–852.

Desclos-Dukes, L., Butterworth, A., & Cogan, T. (2022). Using a non- invasive technique to identify suspected microplastics in grey seals (Halichoerus grypus) living in the western North Sea. *Veterinary Record, 190*(11), e1484. https://doi.org/10.1002/vetr.1484

Devi, S. S., & Kumar, A. B. (2024). Impacts of mesoplastics and microplastics on marine biodiversity. In *Meso-and microplastic risk assessment in marine environments* (pp. 309–340). Elsevier.

Devriese, L. I., Van der Meulen, M. D., Maes, T., Bekaert, K., Paul-Pont, I., Frère, L., & Vethaak, A. D. (2015). Microplastic contamination in brown shrimp (Crangon crangon, Linnaeus 1758) from coastal waters of the Southern North Sea and Channel area. *Marine Pollution Bulletin, 98*(1–2), 179–187.

Dewika, M., Markandan, K., Ruwaida, J. N., Sara, Y. Y., Deb, A., Irfan, N. A., & Khalid, M. (2024). Integrating the quintuple helix approach into atmospheric microplastics management policies for planetary health preservation. *Science of the Total Environment, 954*, 176063.

Directive (EU) 2019/904. *On the reduction of the impact of certain plastic products on the environment 2019.*

Dong, L., & Huang, Z. (2023). Some evidence and new insights for feedback loops of human-nature interactions from a holistic earth perspective. *Journal of Cleaner Production, 432*, 139667.

Duncan, E., Botterell, Z., Broderick, A., Galloway, T., Lindeque, P., Nuno, A., & Godley, B. (2017). A global review of marine turtle entanglement in anthropogenic debris: A baseline for further action. *Endangered Species Research, 34*, 431–448.

Edmond, C. (2022). We know plastic pollution is bad—But how exactly is it linked to climate change. In *World economic forum.*

Eriksen, M., Lebreton, L. C., Carson, H. S., Thiel, M., Moore, C. J., Borerro, J. C., & Reisser, J. (2014). Plastic pollution in the world's oceans: More than 5 trillion plastic pieces weighing over 250,000 tons afloat at sea. *PLoS One, 9*(12), e111913.

Farrell, P., & Nelson, K. (2013). Trophic level transfer of microplastic: *Mytilus edulis* (L.) to Carcinus maenas (L.). *Environmental Pollution, 177*, 1–3.

Feng, Z., Zhang, T., Li, Y., He, X., Wang, R., Xu, J., & Gao, G. (2019). The accumulation of microplastics in fish from an important fish farm and mariculture area, Haizhou Bay, China. *Science of the Total Environment, 696*, 133948.

Flandroy, L., Poutahidis, T., Berg, G., Clarke, G., Dao, M. C., Decaestecker, E., Furman, E., Haahtela, T., Massart, S., Plovier, H., & Sanz, Y. (2018). The impact of human activities and lifestyles on the interlinked microbiota and health of humans and of ecosystems. *Science of the Total Environment, 627*, 1018–1038.

Foley, C. J., Feiner, Z. S., Malinich, T. D., & Höök, T. O. (2018). A meta-analysis of the effects of exposure to microplastics on fish and aquatic invertebrates. *Science of the Total Environment, 631*, 550–559.

Ford, H. V., Jones, N. H., Davies, A. J., Godley, B. J., Jambeck, J. R., Napper, I. E., Suckling, C. C., Williams, G. J., Woodall, L. C., & Koldewey, H. J. (2022). The fundamental links between climate change and marine plastic pollution. *Science of the Total Environment, 806*, 150392.

Fred-Ahmadu, O. H., Ahmadu, F. O., Adedapo, A. E., Oghenovo, I., Ogunmodede, O. T., & Benson, N. U. (2024). Microplastics and chemical contamination in aquaculture ecosystems: The role of climate change and implications for food safety—A review. *Environmental Sciences Europe, 36*(1), 181.

Gao, B., Li, Y., Zheng, N., Liu, C., Ren, H., & Yao, H. (2022). Interactive effects of microplastics, biochar, and earthworms on CO_2 and N_2O emissions and microbial functional genes in vegetable-growing soil. *Environmental Research, 213*, 113728.

Gewert, B., Plassmann, M. M., & MacLeod, M. (2015). Pathways for degradation of plastic polymers floating in the marine environment. *Environmental Science: Processes & Impacts, 17*(9), 1513–1521.

Geyer, R., Jambeck, J. R., & Law, K. L. (2017). Production, use, and fate of all plastics ever made. *Science Advances, 3*(7), e1700782.

Gouin, T., Becker, R. A., Collot, A. G., Davis, J. W., Howard, B., Inawaka, K., Lampi, M., Ramon, B. S., Shi, J., & Hopp, P. W. (2019). Toward the development and application of an environmental risk assessment framework for microplastic. *Environmental Toxicology and Chemistry, 38*(10), 2087–2100.

Grellier, S., Janeau, J. L., Hoai, N. D., Kim, C. N. T., Phuong, Q. L. T., Thu, T. P. T., TranThi, N. T., & Marchand, C. (2017). Changes in soil characteristics and C dynamics after mangrove clearing (Vietnam). *Science of the Total Environment, 593–594*, 654–663. https://doi.org/10.1016/j.scitotenv.2017.03.204

Gündoğdu, S., Eroldoğan, O. T., Evliyaoğlu, E., Turchini, G. M., & Wu, X. G. (2021). Fish out, plastic in: Global pattern of plastics in commercial fishmeal. *Aquaculture, 534*, 736316.

Gurjar, U. R., Takar, S., Amin, A., & Martin Xavier, K. A. (2023). Microplastic contamination in aquatic organisms: An ecotoxicological perspective. In *Xenobiotics in aquatic animals: Reproductive and developmental impacts* (pp. 353–367). Springer Nature Singapore.

Hahladakis, J. N., Velis, C. A., Weber, R., Iacovidou, E., & Purnell, P. (2018). An overview of chemical additives present in plastics: Migration, release, fate and environmental impact during their use, disposal and recycling. *Journal of Hazardous Materials, 344*, 179–199.

Hale, R. C., Seeley, M. E., La Guardia, M. J., Mai, L., & Zeng, E. Y. (2020). A global perspective on microplastics. *Journal of Geophysical Research, Oceans, 125*, e2018JC014719.

Hamed, M., Soliman, H. A., Osman, A. G., & Sayed, A. E. D. H. (2019). Assessment the effect of exposure to microplastics in Nile Tilapia (Oreochromis niloticus) early juvenile: I. blood biomarkers. *Chemosphere, 228*, 345–350.

Hamilton, L. A., & Feit, S. (2019). *Plastic & climate: The hidden costs of a plastic planet.*

Haque, F., & Fan, C. (2023). Fate of microplastics under the influence of climate change. *Iscience, 26*, 107649–107649.

Hernandez-Milian, G., Lusher, A., MacGabban, S., & Rogan, E. (2019). Microplastics in grey seal (*Halichoerus grypus*) intestines: Are they associated with parasite aggregations? *Marine Pollution Bulletin, 146*, 349–354. https://doi.org/10.1016/j.marpolbul.2019.06.014

Hu, D., Shen, M., Zhang, Y., Li, H., & Zeng, G. (2019). Microplastics and nanoplastics: Would they affect global biodiversity change? *Environmental Science and Pollution Research, 26*, 19997–20002.

Huerta, E., Gertsen, H., Gooren, H. P. A., Peters, P., Salánki, T., Ploeg, M., Besseling, E., Koelmans, A., & Geissen, V. (2016). Microplastics in the terrestrial ecosystem: Implications for Lumbricus Terrestris (Oligochaeta, Lumbricidae). *Environmental Science & Technology, 50*, 2685–2691.

Hughes, T. P., Kerry, J. T., Álvarez-Noriega, M., Álvarez-Romero, J. G., Anderson, K. D., Baird, A. H., Babcock, R. C., Beger, M., Bellwood, D. R., Berkelmans, R., & Bridge, T. C. (2017). Global warming and recurrent mass bleaching of corals. *Nature, 543*, 373–377.

Hughes, T. P., Anderson, K. D., Connolly, S. R., Heron, S. F., Kerry, J. T., Lough, J. M., Baird, A. H., Baum, J. K., Berumen, M. L., Bridge, T. C., & Claar, D. C. (2018). Spatial and temporal patterns of mass bleaching of corals in the anthropocene. *Science, 359*, 80–83.

Hung, H., Halsall, C., Ball, H., Bidleman, T., Dachs, J., De Silva, A., Hermanson, M., Kallenborn, R., Muir, D., Sühring, R., & Wang, X. (2022). Climate change influence on the levels and trends of persistent organic pollutants (POPs) and chemicals of emerging Arctic concern (CEACs) in the Arctic physical environment–a review. *Environmental Science: Processes & Impacts, 24*(10), 1577–1615.

Impellitteri, F., Multisanti, C. R., Rusanova, P., Piccione, G., Falco, F., & Faggio, C. (2023). Exploring the impact of contaminants of emerging concern on fish and invertebrates physiology in the Mediterranean Sea. *Biology, 12*(6), 767.

IPCC. (2019). In H.-O. Po¨rtner, D. C. Roberts, V. Masson-Delmotte, P. Zhai, M. Tignor, E. Poloczanska, K. Mintenbeck, A. Alegrı´a, M. Nicolai, A. Okem, et al. (Eds.), *In IPCC special report on the ocean and cryosphere in a changing climate*. In press.

Iroegbu, A. O., Sadiku, R. E., Ray, S. S., & Hamam, Y. (2020). Plastics in municipal drinking water and wastewater treatment plant effluents: Challenges and opportunities for South Africa—A review. *Environmental Science and Pollution Research, 27*, 12953–12966.

Jain, Y., Govindasamy, H., Kaur, G., Ajith, N., Ramasamy, K., Robin, R. S., & Ramachandran, P. (2024). Microplastic pollution in high-altitude Nainital Lake, Uttarakhand, India. *Environmental Pollution, 346*, 123598.

Jambeck, J. R., Geyer, R., Wilcox, C., Siegler, T. R., Perryman, M., Andrady, A., Narayan, R., & Law, K. L. (2015). Plastic waste inputs from land into the ocean. *Science, 347*(6223), 768–771.

Janardhanam, M., Sivakumar, P., Srinivasan, G., Sivakumar, R., Marcus, P. N., Balasubramaniam, S., & Harikrishnan, T. (2022). Microplastics in demersal sharks from the southeast Indian coastal region. *Frontiers in Marine Science, 9*, 914391.

Jensen, M. P., Allen, C. D., Eguchi, T., Bell, I. P., LaCasella, E. L., Hilton, W. A., Hof, C. A. M., & Dutton, P. H. (2018). Environmental warming and feminization of one of the largest sea turtle populations in the world. *Current Biology, 28*, 154–159.

Jeong, C. B., Won, E. J., Kang, H. M., Lee, M. C., Hwang, D. S., Hwang, U. K., et al. (2016). Microplastic size-dependent toxicity, oxidative stress induction, and p-JNK and p-p38 activation in the monogonont rotifer (*Brachionus koreanus*). *Environmental Science & Technology, 50*(16), 8849–8857.

Jeyavani, J., Sibiya, A., Shanthini, S., Ravi, C., Vijayakumar, S., Rajan, D. K., & Vaseeharan, B. (2021). A review on aquatic impacts of microplastics and its bioremediation aspects. *Current Pollution Reports, 7*(3), 286–229.

Jiao, H., Ali, S. S., Alsharbaty, M. H. M., Elsamahy, T., Abdelkarim, E., Schagerl, M., Al-Tohamy, R., & Sun, J. (2024). A critical review on plastic waste life cycle assessment and management: Challenges, research gaps, and future perspectives. *Ecotoxicology and Environmental Safety, 271*, 115942.

Junaid, M., Siddiqui, J. A., Liu, S., Lan, R., Abbas, Z., Chen, G., & Wang, J. (2023). Adverse multigeneration combined impacts of micro (nano) plastics and emerging pollutants in the aquatic environment. *Science of the Total Environment, 882*, 163679.

Kaposi, K. L., Mos, B., Kelaher, B. P., & Dworjanyn, S. A. (2014). Ingestion of microplastic has limited impact on a marine larva. *Environmental Science & Technology, 48*(3), 1638–1645.

Karthik, R., Robin, R. S., Purvaja, R., Karthikeyan, V., Subbareddy, B., Balachandar, K., & Ramesh, R. (2022). Microplastic pollution in fragile coastal ecosystems with special reference to the X-Press Pearl maritime disaster, southeast coast of India. *Environmental Pollution, 305*, 119297.

Kettiger, H., Schipanski, A., Wick, P., & Huwyler, J. (2013). Engineered nanomaterial uptake and tissue distribution: From cell to organism. *International Journal of Nanomedicine, 8*, 3255–3269.

Khan, L., Ghias, S., Zafar, M. I., Alhodaib, A., Fatima, H., Ur-Rehman, T., & Howari, H. (2022). Exploration of microplastic pollution with a particular focus on source identification and spatial patterns in riverine water, sediment and fish of the Swat River, Pakistan. *RSC Advances, 12*(16), 9556–9566.

Khan, A., Jie, Z., Wang, J., Nepal, J., Ullah, N., Zhao, Z. Y., & Xiong, Y. C. (2023). Ecological risks of microplastics contamination with green solutions and future perspectives. *Science of the Total Environment, 899*, 165688.

Kukulka, T., Proskurowski, G., Morét-Ferguson, S., Meyer, D. W., & Law, K. L. (2012). The effect of wind mixing on the vertical distribution of buoyant plastic debris. *Geophysical Research Letters, 39*, 7601.

Kumar, R., Verma, A., Shome, A., Sinha, R., Sinha, S., Jha, P. K., Kumar, R., Kumar, P., Das, S., Sharma, P., & Vara Prasad, P. V. (2021). Impacts of plastic pollution on ecosystem services, sustainable development goals, and need to focus on circular economy and policy interventions. *Sustainability, 13*(17), 9963.

Kurniawan, T. A., Haider, A., Mohyuddin, A., Fatima, R., Salman, M., Shaheen, A., Ahmad, H. M., Al-Hazmi, H. E., Othman, M. H. D., Aziz, F., & Anouzla, A. (2023). Tackling microplastics pollution in global environment through integration of applied technology, policy instruments, and legislation. *Journal of Environmental Management, 346*, 118971.

Kurniawan, T. A., Mohyuddin, A., Othman, M. H. D., Goh, H. H., Zhang, D., Anouzla, A., Aziz, F., Casila, J. C., Ali, I., & Pasaribu, B. (2024). Beyond surface: Unveiling ecological and economic ramifications of microplastic pollution in the oceans. *Water Environment Research, 96*(7), e11070.

Lahon, J., & Handique, S. (2023). Impact of flooding on microplastic abundance and distribution in freshwater environment: A review. *Environmental Science and Pollution Research, 30*(56), 118175–118191.

Lau, W. W., Shiran, Y., Bailey, R. M., Cook, E., Stuchtey, M. R., Koskella, J., et al. (2020). Evaluating scenarios toward zero plastic pollution. *Science, 369*(6510), 1455–1461.

Lebreton, L. C. M., Van Der Zwet, J., Damsteeg, J.-W., Slat, B., Andrady, A., & Reisser, J. (2017). River plastic emissions to the world's oceans. *Nature Communications, 8*(1), 15611.

Lee, J.-H., Kang, J.-C., & Kim, J.-H. (2023). Toxic effects of microplastic (Polyethylene) on fish: Accumulation, hematological parameters and antioxidant responses in Korean Bullhead, *Pseudobagrus fulvidraco*. *Science of the Total Environment, 877*, 162874. https://doi.org/10.1016/j.scitotenv.2023.162874

Li, J., Qu, X., Su, L., Zhang, W., Yang, D., Kolandhasamy, P., & Shi, H. (2016a). Microplastics in mussels along the coastal waters of China. *Environmental Pollution, 214*, 177–184.

Li, W. C., Tse, H. F., & Fok, L. (2016b). Plastic waste in the marine environment: A review of sources, occurrence and effects. *Science of the Total Environment, 566–567*, 333–349.

Li, K., Du, L., Qin, C., Bolan, N., Wang, H., & Wang, H. (2024). Microplastic pollution as an environmental risk exacerbating the greenhouse effect and climate change: A review. *Carbon Research, 3*(1), 9.

Lin, Z., Jin, T., Zou, T., Xu, L., Xi, B., Xu, D., He, J., Xiong, L., Tang, C., Peng, J., & Zhou, Y. (2022). Current progress on plastic/microplastic degradation: Fact influences and mechanism. *Environmental Pollution, 304*, 119159.

Lincoln, S., Chowdhury, P., Posen, P. E., Robin, R. S., Ramachandran, P., Ajith, N., et al. (2023). Interaction of climate change and marine pollution in Southern India: Implications for coastal zone management practices and policies. *Science of the Total Environment, 902*, 166061.

Lombardo, J., Solomando, A., Cohen-Sánchez, A., Pinya, S., Tejada, S., et al. (2022). Effects of human activity on markers of oxidative stress in the intestine of *Holothuria tubulosa*, with special reference to the presence of microplastics. *International Journal of Molecular Science, 23*(16), 9018.

Lozano, Y. M., Aguilar-Trigueros, C. A., Onandia, G., Maaß, S., Zhao, T., & Rillig, M. C. (2021). Effects of microplastics and drought on soil ecosystem functions and multifunctionality. *Journal of Applied Ecology, 58*, 988–996.

Lu, H. C., Smith, J. L., Ziajahromi, S., & Leusch, F. D. (2024). Microplastics and other anthropogenic fibres in large apex shark species: Abundance, characteristics, and recommendations for future research. *Chemosphere, 349*, 140957.

Luo, H., Liu, C., He, D., Sun, J., Li, J., & Pan, X. (2022). Effects of aging on environmental behavior of plastic additives: Migration, leaching, and ecotoxicity. *Science of the Total Environment, 849*, 157951.

Lusher, A. (2015). Microplastics in the marine environment: Distribution, interactions and effects. In M. Bergmann, L. Gutow, & M. Klages (Eds.), *Marine anthropogenic litter* (pp. 245–307). Springer International Publishing.

Lusher, A. L., O'Donnell, C., Officer, R., & O'Connor, I. (2016). Microplastic interactions with North Atlantic mesopelagic fish. *ICES Journal of Marine Science, 73*(4), 1214–1225.

Lusher, A., Hollman, P., & Mendoza-Hill, J. (2017). *Microplastics in fisheries and aquaculture: Status of knowledge on their occurrence and implications for aquatic organisms and food safety*. FAO.

Mahon, A. M., O'Connell, B., Healy, M. G., O'Connor, I., Officer, R., Nash, R., & Morrison, L. (2017). Microplastics in sewage sludge: Effects of treatment. *Environmental Science & Technology, 51*, 810–818.

Malhi, Y., Franklin, J., Seddon, N., Solan, M., Turner, M. G., Field, C. B., & Knowlton, N. (2020). Climate change and ecosystems: Threats, opportunities and solutions. *Philosophical Transactions of the Royal Society B, 375*(1794), 20190104.

Manosuthi, J. (2024). Microplastic pollution: The real potential threats to human being. *Science Insights, 44*(2), 1263–1272.

Mattsson, K., Johnson, E. V., Malmendal, A., Linse, S., Hansson, L. A., & Cedervall, T. (2017). Brain damage and behavioural disorders in fish induced by plastic nanoparticles delivered through the food chain. *Scientific Reports, 7*(1), 11452.

Mbachu, O., Jenkins, G., Kaparaju, P., & Pratt, C. (2021). The rise of artificial soil carbon inputs: Reviewing microplastic pollution effects in the soil environment. *Science of the Total Environment, 780*, 146569.

McIlwraith, H. K., Lin, J., Erdle, L. M., Mallos, N., Diamond, M. L., & Rochman, C. M. (2019). Capturing microfibers—Marketed technologies reduce microfiber emissions from washing machines. *Marine Pollution Bulletin, 139*, 40–45.

Mohan, A., Sekhar, V. C., Bhaskar, T., & Nampoothiri, K. M. (2016). Microbial assisted high impact polystyrene (HIPS) degradation. *Bioresource Technology, 213*, 204–207.

Morelle, R. (2019). Mariana trench: Deepest-ever sub dive finds plastic bag. *BBC News*, 13.

Moshood, T. D., Nawanir, G., Mahmud, F., Mohamad, F., Ahmad, M. H., & AbdulGhani, A. (2022). Sustainability of biodegradable plastics: New problem or solution to solve the global plastic pollution? *Current Research in Green and Sustainable Chemistry, 5*, 100273.

Mugilarasan, M., Karthik, R., Robin, R. S., Subbareddy, B., Hariharan, G., Anandavelu, I., et al. (2023). Anthropogenic marine litter: An approach to environmental quality for India's south-eastern Arabian Sea coast. *Science of the Total Environment, 866*, 161363.

Muhonja, C. N., Makonde, H., Magoma, G., & Imbuga, M. (2018). Biodegradability of polyethylene by bacteria and fungi from Dandora dumpsite Nairobi-Kenya. *PLoS One, 13*, e0198446.

Munhoz, D. R., Harkes, P., Beriot, N., Larreta, J., & Basurko, O. C. (2022). Microplastics: A review of policies and responses. *Microplastics, 2*(1), 1–26.

Munno, K., Hoopes, L., Lyons, K., Drymon, M., Frazier, B., & Rochman, C. M. (2024). High microplastic and anthropogenic particle contamination in the gastrointestinal tracts of tiger sharks (*Galeocerdo cuvier*) caught in the western North Atlantic Ocean. *Environmental Pollution, 344*, 123185.

Murgulet, D. (2016). Effects of climate change and sea level rise on coastal water resources. In *Emerging issues in groundwater resources* (pp. 1–30). Springer.

Nabi, N., Ahmad, I., Amin, A., Rather, M. A., Ahmed, I., Hajam, Y. A., Khursheed, S., Malik, M. M., & Abubakr, A. (2024). Understanding the sources, fate and effects of microplastics in aquatic environments with a focus on risk profiling in aquaculture systems. *Reviews in Aquaculture, 16*(4), 1947–1980.

Napper, I. E., & Thompson, R. C. (2020). Plastic debris in the marine environment: History and future challenges. *Global Challenges, 4*(6), 1900081.

Napper, I. E., Bakir, A., Rowland, S. J., & Thompson, R. C. (2015). Characterisation, quantity and sorptive properties of microplastics extracted from cosmetics. *Marine Pollution Bulletin, 99*(1–2), 178–185.

Napper, I. E., Baroth, A., Barrett, A. C., Bhola, S., Chowdhury, G. W., Davies, B. F. R., Duncan, E. M., Kumar, S., Nelms, S. E., Hasan Niloy, M. N., et al. (2021). The abundance and characteristics of microplastics in surface water in the transboundary Ganges River. *Environmental Pollution, 274*, 116348.

National Geographic Society. (2023). *Greenhouse effect*. https://education.nationalgeographic.org/resource/greenhouseeffect

Nelms, S. E., Duncan, E. M., Broderick, A. C., Galloway, T. S., Godfrey, M. H., Hamann, M., Lindeque, P. K., & Godley, B. J. (2016). Plastic and marine turtles: A review and call for research. *ICES Journal of Marine Science, 73*, 165–181.

Nelms, S. E., Clark, B. L., Duncan, E. M., Germanov, E., Godley, B. J., Parton, K. J., et al. (2023). Plastic pollution and marine megafauna: Recent advances and future directions. In *Plastic pollution in the global ocean* (pp. 97–138). World Scientific.

Ngarava, S., Zhou, L., Nyambo, P., Chari, M. M., & Bhungeni, O. (2023). Aquaculture production, GHG emission and economic growth in Sub Sahara Africa. *Environmental Challenges, 12*, 100737. https://doi.org/10.1016/J.ENVC.2023.100737

O'Connor, J., Mickan, B. S., Siddique, K. H. M., Rinklebe, J., Kirkham, M. B., & Bolan, N. S. (2022). Physical, chemical, and microbial contaminants in food waste management for soil application: A review. *Environmental Pollution, 300*, 118860.

Obbard, R. W., Sadri, S., Wong, Y. Q., Khitun, A. A., Baker, I., & Thompson, R. C. (2014). Global warming releases microplastic legacy frozen in Arctic Sea ice. *Earth's Future, 2*, 315–320.

OECD. (2022). *Global plastics outlook: Policy scenarios to 2060.*

Okeke, E. S., Okoye, C. O., Atakpa, E. O., Ita, R. E., Nyaruaba, R., Mgbechidinma, C. L., & Akan, O. D. (2022). Microplastics in agroecosystems-impacts on ecosystem functions and food chain. *Resources, Conservation and Recycling, 177*, 105961.

Ortiz, J. C., Wolff, N. H., Anthony, K. R., Devlin, M., Lewis, S., & Mumby, P. J. (2018). Impaired recovery of the great barrier reef under cumulative stress. *Science Advances, 4*, 6127.

Paço, A., Duarte, K., Da Costa, J., Santos, P., Pereira, R., Pereira, M. E., Freitas, A., Duarte, A., & Rocha-Santos, T. (2017). Biodegradation of polyethylene microplastics by the marine fungus Zalerion maritimum. *Science of the Total Environment, 586*, 10–15.

Parvez, M. S., Ullah, H., Faruk, O., Simon, E., & Czédli, H. (2024). Role of microplastics in global warming and climate change: A review. *Water, Air, & Soil Pollution, 235*(3), 201.

Pathak, V. M., & Navneet. (2023). Exploitation of bacterial strains for microplastics (LDPE) biodegradation. *Chemosphere, 316*, 137845.

Patria, M. P., Santoso, C. A., & Tsabita, N. (2020). Pengingesan mikroplastik daripada siput periwinkel littoraria scabra dan ketam nipah metopograpsus quadridentate di Pulau Pramuka, Teluk Jakarta, Indonesia. *Sains Malaysiana, 49*(9), 2151–2158.

Pedà, C., Caccamo, L., Fossi, M. C., Gai, F., Andaloro, F., Genovese, L., Perdichizzi, A., Romeo, T., & Maricchiolo, G. (2016). Intestinal alterations in European sea bass Dicentrarchus labrax (Linnaeus, 1758) exposed to microplastics: Preliminary results. *Environmental Pollution, 212*, 251–256.

Pedà, C., Longo, F., Berti, C., Laface, F., De Domenico, F., Consoli, P., & Romeo, T. (2022). The waste collector: Information from a pilot study on the interaction between the common octopus (Octopus vulgaris, Cuvier, 1797) and marine litter in bottom traps fishing and first evidence of plastic ingestion. *Marine Pollution Bulletin, 174*, 113185.

Peng, X., Chen, M., Chen, S., Dasgupta, S., Xu, H., Ta, K., Du, M., Li, J., Guo, Z., & Bai, S. (2018). Microplastics contaminate the deepest part of the world's ocean. *Geochemical Perspectives Letters, 9*, 1–5.

Peng, G., Lin, Y., van Bavel, B., Li, D., Ni, J., & Song, Y. (2022). Aggregate exposure pathways for microplastics (mpAEP): An evidence-based framework to identify research and regulatory needs. *Water Research, 209*, 117873.

Plastics Europe. (2023). *Plastics – the facts 2022*. https://plasticseurope.org/knowledge-hub/plastics-the-facts-2022/

Porter, A., Lyons, B. P., Galloway, T. S., & Lewis, C. (2018). Role of marine snows in microplastic fate and bioavailability. *Environmental Science & Technology, 52*, 7111–7119.

Prasad, G. S., Rout, S. K., Malik, M. M., Karmakar, S., Amin, A., & Ahmad, I. (2023). Occurrence of xenoestrogen alkylphenols (Octylphenols and Nonylphenol) and its impact on the aquatic ecosystem. In *Xenobiotics in aquatic animals: Reproductive and developmental impacts* (pp. 275–284). Springer.

Prasad, G. S., Rout, S. K., Amin, A., Malik, M. M., Ahmad, I., Rather, M. A., Karmakar, S., Khalil, H. S., & Abubakar, A. (2024). Impact of COVID-19 on the aquatic environment and implications for fisheries with special emphasis to Indian perspective. *Proceedings of the Indian National Science Academy, 90*(4), 821–839.

Pratt, S., Bolan, N. S., Laycock, B., Lant, P., Bryson, E., & Dilkes-Hoffman, L. (2020). *Particulate plastics in terrestrial and aquatic environments* (pp. 425–434). CRC Press.

Protyusha, G. B., Kavitha, B., Robin, R. S., Nithin, A., Ineyathendral, T. R., Shivani, S. S., et al. (2024). Microplastics in oral healthcare products (OHPs) and their environmental health risks and mitigation measures. *Environmental Pollution, 343*, 123118.

Qiang, L., Hu, H., Li, G., Xu, J., Cheng, J., Wang, J., & Zhang, R. (2023). Plastic mulching, and occurrence, incorporation, degradation, and impacts of polyethylene microplastics in agroecosystems. *Ecotoxicology and Environmental Safety, 263*, 115274–115274.

Raddadi, N., & Fava, F. (2019). Biodegradation of oil-based plastics in the environment: Existing knowledge and needs of research and innovation. *Science of the Total Environment, 679*, 148–158.

Rahmawati, K. M., Riani, E., & Cordova, M. R. (2023). Microplastic contamination in the digestive tract of sea urchins (Echinodermata: Echinoidea) in Kepulauan Seribu, Indonesia. *Environmental Monitoring and Assessment, 19*, 1103.

Rajendran, S., Ramya, R., Kannan, K., Arokiaswamy, R. A., & Rajesh Kannan, V. (2019). Investigation of biodegradation potentials of high density polyethylene degrading marine bacteria isolated from the coastal regions of Tamil Nadu, India. *Marine Pollution Bulletin, 138*, 549–560.

Razza, F., & Cerutti, A. K. (2017). Life cycle and environmental cycle assessment of biodegradable plastics for agriculture. In *Soil degradable bioplastics for a sustainable modern agriculture* (pp. 169–185). Springer Berlin Heidelberg.

Reichert, J., Arnold, A. L., Hoogenboom, M. O., Schubert, P., & Wilke, T. (2019). Impacts of microplastics on growth and health of hermatypic corals are species-specific. *Environmental Pollution, 254*, 113074.

Reisser, J., Slat, B., Noble, K., Du Plessis, K., Epp, M., Proietti, M., de Sonneville, J., Becker, T., & Pattiaratchi, C. (2014). The vertical distribution of buoyant plastics at sea. *Biogeosciences Discuss, 11*, 16207–16226.

Ren, X. W., Tang, J. C., Liu, X. M., & Liu, Q. L. (2020). Effects of microplastics on green-house gas emissions and the microbial community in fertilized soil. *Environmental Pollution, 256*, 113347.

Roch, S., Friedrich, C., & Brinker, A. (2020). Uptake routes of microplastics in fishes: Practical and theoretical approaches to test existing theories. *Scientific Reports, 10*(1), 3896.

Rodríguez, Y., Vandeperre, F., Santos, M. R., Herrera, L., Parra, H., et al. (2022). Litter ingestion and entanglement in green turtles: An analysis of two decades of stranding events in the NE Atlantic. *Environmental Pollution, 298*, 118796. https://doi.org/10.1016/j.envpol.2022.118796

Royer, S. J. (2018). *Yes, everyday plastics are emitting greenhouse*. https://www.greenpeace.org/usa/yes-everyday-plastics-are-emitting-greenhouse-gases/

Sabdaningsih, A., Adyasari, D., Suryanti, S., Febrianto, S., & Eshananda, Y. (2023). Environmental legacy of aquaculture and industrial activities in mangrove ecosystems. *Journal of Sea Research, 196*, 102454. https://doi.org/10.1016/j.seares.2023.102454

Saleem, M. S., Abas, N., Umer, M., & Haider, A. (2024). Migration and plasticity effects on climate change. In *Advances and technology development in greenhouse gases: Emission, capture and conversion* (pp. 211–230). Elsevier.

Schwabl, P., Köppel, S., Königshofer, P., Bucsics, T., Trauner, M., Reiberger, T., & Liebmann, B. (2019). Detection of various microplastics in human stool: A prospective case series. *Annals of Internal Medicine, 171*(7), 453–457.

Scotti, G., D'Alessandro, M., Esposito, V., Vivona, P., & Panti, C. (2023). Anthropogenic fibers and microplastics in the pelagic gooseneck barnacle *Lepas* (Lepas) *anatifera* in Capo Milazzo Marine Protected Area (Tyrrhenian Sea): A first characterization. *Ecological Indicators, 152*, 110368. https://doi.org/10.1016/j.ecolind.2023.110368

Segaran, T. C., Azra, M. N., Lananan, F., & Wang, Y. (2023). Microbe, climate change and marine environment: Linking trends and research hotspots. *Marine Environmental Research, 189*, 106015.

Setälä, O., Fleming-Lehtinen, V., & Lehtiniemi, M. (2014). Ingestion and transfer of microplastics in the planktonic food web. *Environmental Pollution, 185*, 77–83.

Sharma, S., & Chatterjee, S. (2017). Microplastic pollution, a threat to marine ecosystem and human health: A short review. *Environmental Science and Pollution Research, 24*, 21530–21547.

Shen, M., Huang, W., Chen, M., Song, B., Zeng, G., & Zhang, Y. (2020). (Micro) plastic crisis: Un-ignorable contribution to global greenhouse gas emissions and climate change. *Journal of Cleaner Production, 254*, 120138.

Solomando, A., Pujol, F., Sureda, A., & Pinya, S. (2022). Ingestion and characterization of plastic debris by loggerhead sea turtle, Caretta caretta, in the Balearic Islands. *Science of The Total Environment, 826*, 154–159.

Sridharan, S., Kumar, M., Bolan, N. S., Singh, L., Kumar, S., Kumar, R., & You, S. (2021). Are microplastics destabilizing the global network of terrestrial and aquatic ecosystem services? *Environmental Research, 198*, 111243.

Sruthisree, C., Gowda, G., Nayak, H., Harish, B. D., & Amin, A. (2015). Effect of heavy metal (lead) on oxygen consumption and ammonia excretion of Iridescent Shark (*Pangusius hypohthalmus*). *Journal of Experimental Zoology India, 18*(2), 637–640.

Stockin, K. A., Pantos, O., Betty, E. L., Pawley, M. D. M., Doake, F., et al. (2021). Fourier transform infrared (FTIR) analysis identifies microplastics in stranded common dolphins (*Delphinus delphis*) from New Zealand waters. *Marine Pollution Bulletin, 173*(Part B), 113084. https://doi.org/10.1016/j.marpolbul.2021.113084

Strokal, M., Vriend, P., Bak, M. P., Kroeze, C., van Wijnen, J., & van Emmerik, T. (2023). River export of macro-and microplastics to seas by sources worldwide. *Nature Communications, 14*(1), 4842.

Stubbins, A., Law, K. L., Muñoz, S. E., Bianchi, T. S., & Zhu, L. (2021). Plastics in the earth system. *Science, 373*(6550), 51–55.

Sun, X., Tao, R., Xu, D., Qu, M., Zheng, M., Zhang, M., & Mei, Y. (2023). Role of polyamide microplastic in altering microbial consortium and carbon and nitrogen cycles in a simulated agricultural soil microcosm. *Chemosphere, 312*, 137155.

Sussarellu, R., Suquet, M., Thomas, Y., Lambert, C., Fabioux, C., Pernet, M. E. J., & Huvet, A. (2016). Oyster reproduction is affected by exposure to polystyrene microplastics. *Proceedings of the National Academy of Sciences, 113*(9), 2430–2435.

Suzuki, M., Tachibana, Y., Oba, K., Takizawa, R., & Kasuya, K.-I. (2018). Microbial degradation of poly (ε-Caprolactone) in a coastal environment. *Polymer Degradation and Stability, 149*, 1–8.

Tabassum, I., Hassan, N., Amin, A., Malik, M. M., Abubakr, A., Prasad, G. S., Latief, F., & Rather, S. A. (2025). COVID-19 and global environmental change: A threat to food security and nutrition. In *Food security, nutrition and sustainability through aquaculture technologies* (pp. 437–445). Springer Nature Switzerland.

Tekman, M. B., Gutow, L., Bergmann, M., & Peter, C. (2024). *LITTERBASE*. Online portal for marine litter [online]. Website: https://litterbase.awi.de. Accessed 30 May 2024.

Thacharodi, A., Meenatchi, R., Hassan, S., Hussain, N., Bhat, M. A., Arockiaraj, J., Ngo, H. H., Le, Q. H., & Pugazhendhi, A. (2024). Microplastics in the environment: A critical overview on its fate, toxicity, implications, management, and bioremediation strategies. *Journal of Environmental Management, 349*, 119433.

Thompson, R. C., Moore, C., vom Saal, F. S., & Swan, S. H. (2009). Plastics, the environment and human health: Current consensus and future trends. *Philosophical Transactions of the Royal Society B, 364*, 2153–2166.

Tyagi, M. (2024). Water contamination and impacts of synthetic microfibers pollutants to the global ecosystem. In *Sustainable microbial technology for synthetic and cellulosic microfiber bioremediation* (pp. 157–181). Springer Nature Switzerland.

UNEP 74/19. (2019). *Oceans and the Law of the Sea*. Available online: https://www.unep.org/news-and-stories/news/un-environment-programme-unep-during-high-level-week-74th-session-un-general. Accessed 6 December 2020.

Upadhyay, R. K. (2020). Markers for global climate change and its impact on social, biological and ecological systems: A review. *American Journal of Climate Change, 9*(03), 159–203.

Van Cauwenberghe, L., Devriese, L., Galgani, F., Robben, S. J., & Janssen, C. R. (2015). Microplastics in sediments: A review of techniques, occurrence and effects. *Marine Environmental Research, 111*, 5–17.

Verma, R., Vinoda, K. S., Papireddy, M., & Gowda, A. N. S. (2016). Toxic pollutants from plastic waste-a review. *Procedia Environmental Sciences, 35*, 701–708.

Viitasalo, M., & Bonsdorff, E. (2022). Global climate change and the Baltic Sea ecosystem: Direct and indirect effects on species, communities and ecosystem functioning. *Earth System Dynamics, 13*(2), 711–747.

Villegas, L., Cabrera, M., & Capparelli, M. V. (2021). Assessment of microplastic and organophosphate pesticides contamination in fiddler crabs from a Ramsar site in the estuary of Guayas River, Ecuador. *Bulletin of Environmental Contamination and Toxicology, 107*(1), 20–28.

Von Moos, N., Burkhardt-Holm, P., & Köhler, A. (2012). Uptake and effects of microplastics on cells and tissue of the blue mussel Mytilus edulis L. after an experimental exposure. *Environmental Science & Technology, 46*(20), 11327–11335.

Wang, F., Wu, H., Wu, W., Wang, L., Liu, J., An, L., & Xu, Q. (2021a). Microplastic characteristics in organisms of different trophic levels from Liaohe Estuary, China. *Science of the Total Environment, 789*, 148027.

Wang, F., Yu, Y., Wu, H., Wu, W., Wang, L., An, L., & Cai, W. (2021b). Microplastics in spotted seal cubs (*Phoca largha*): Digestion after ingestion? *Science of the Total Environment, 785*, 147426.

Wang, C., Wei, W., Chen, Z., Wang, Y., Chen, X., & Ni, B.-J. (2022). Polystyrene microplastics and nano-plastics distinctively affect anaerobic sludge treatment for hydrogen and methane production. *Science of the Total Environment, 850*, 158085.

Wang, Z., Hu, X., Qu, Q., Hao, W., Deng, P., Kang, W., & Feng, R. (2023). Dual regulatory effects of microplastics and heat waves on river microbial carbon metabolism. *Journal of Hazardous Materials, 441*, 129879.

Watts, A. J., Lewis, C., Goodhead, R. M., Beckett, S. J., Moger, J., Tyler, C. R., & Galloway, T. S. (2014). Uptake and retention of microplastics by the shore crab *Carcinus maenas*. *Environmental Science & Technology, 48*(15), 8823–8830.

Weber, A., Jeckel, N., & Wagner, M. (2020). Combined effects of polystyrene microplastics and thermal stress on the freshwater mussel *Dreissena polymorpha*. *Science of the Total Environment, 718*, 137253.

Welden, N. A., & Lusher, A. L. (2017). Impacts of changing ocean circulation on the distribution of marine microplastic litter. *Integrated Environmental Assessment and Management, 13*, 483–487.

Wieczorek, A. M., Morrison, L., & Croot, P. L. (2018). Frequency of microplastics in mesopelagic fishes from the Northwest Atlantic. *Frontiers in Marine Science, 5*, 39.

Wright, S. L., & Kelly, F. J. (2017). Plastic and human health: A micro issue? *Environmental Science & Technology, 51*(12), 6634–6647.

Wright, S. L., Thompson, R. C., & Galloway, T. S. (2013). The physical impacts of microplastics on marine organisms: A review. *Environmental Pollution, 178*, 483–492.

Yang, Y., Yang, J., Wu, W.-M., Zhao, J., Song, Y., Gao, L., Yang, R., & Jiang, L. (2015). Biodegradation and mineralization of polystyrene by plastic-eating mealworms: Part 2. Role of gut microorganisms. *Environmental Science & Technology, 49*, 12087–12093.

Yang, H., Chen, G., & Wang, J. (2021). Microplastics in the marine environment: Sources, fates, impacts and microbial degradation. *Toxics, 9*(2), 41.

Yin, L., Wen, X., Huang, D., Du, C., Deng, R., Zhou, Z., & Chen, H. (2021). Interactions between microplastics/nanoplastics and vascular plants. *Environmental Pollution, 290*, 117999.

Yong, C. Q. Y., Valiyaveettil, S., & Tang, B. L. (2020). Toxicity of microplastics and nanoplastics in mammalian systems. *International Journal of Environmental Research and Public Health, 17*(05), 1509.

Yong, M. M. H., Leistenschneider, C., Miranda, J. A., Paler, M. K., Legaspi, C., et al. (2021). Microplastics in fecal samples of whale sharks (*Rhincodon typus*) and from surface water in The Philippines. *Microplastics and Nanoplastics, 1*(1), 17. https://doi.org/10.1186/s43591-021-00017-9

Yoshida, S., Hiraga, K., Takehana, T., Taniguchi, I., Yamaji, H., Maeda, Y., Toyohara, K., Miyamoto, K., Kimura, Y., & Oda, K. (2016). A bacterium that degrades and assimilates poly (ethylene terephthalate). *Science, 351*, 1196–1199.

Yuan, J., Ma, J., Sun, Y., Zhou, T., Zhao, Y., & Yu, F. (2020). Microbial degradation and other environmental aspects of microplastics/plastics. *Science of the Total Environment, 715*, 136968.

Zantis, L. J., Bosker, T., Lawler, F., Nelms, S. E., O'Rorke, R., Constantine, R., Sewell, M., & Carroll, E. L. (2022). Assessing microplastic exposure of large marine filter-feeders. *Science of the Total Environment, 818*, 151815.

Zhai, X., Zhang, X. H., & Yu, M. (2023). Microbial colonization and degradation of marine microplastics in the plastisphere: A review. *Frontiers in Microbiology, 14*, 1127308.

Zhang, Y., Jiang, H., Bian, K., Wang, H., & Wang, C. (2021). A critical review of control and removal strategies for microplastics from aquatic environments. *Journal of Environmental Chemical Engineering, 9*(4), 105463.

Zhang, W., Liu, X., Liu, L., Lu, H., Wang, L., & Tang, J. (2022). Effects of microplastics on greenhouse gas emissions and microbial communities in sediment of freshwater systems. *Journal of Hazardous Materials, 435*, 12903.

Zhao, X., & You, F. (2022). Life cycle assessment of microplastics reveals their greater environmental hazards than mismanaged polymer waste losses. *Environmental Science & Technology, 56*(16), 11780–11179.

Zheng, S., Zhao, Y., Liangwei, W., Liang, J., Liu, T., Zhu, M., Li, Q., & Sun, X. (2020). Characteristics of microplastics ingested by zooplankton from the Bohai Sea, China. *Science of the Total Environment, 713*, 136357.

Microplastic Menace: How Contaminated Seafood Puts Food Security at Risk

Vikas Kumar, Siddhnath, Nikheel Bhojraj Rathod, Hafsa Maqbool, Gurjeet Kaur, Simran Kaur, and Vijay Kumar Reddy

1 Introduction

Plastic has emerged as a vital element of everyday life in the modern world. It is a synthetic substance with several unique characteristics, including strength, durability, and lightness. The culinary, healthcare, and electrical product sectors use it substantially owing to its numerous unique features (Kosior & Mitchell, 2020). However, inadequate management of plastic has contributed to the buildup of this non-biodegradable synthetic material in water bodies like rivers, ponds, and the sea (Chae & An, 2018). Plastic constitutes about 80–85% of marine litter and 10% of all plastics generated end up in the ocean (Auta et al., 2017). Plastics disposed of in water bodies vary widely in size, form, density, chemical makeup, and other characteristics. In the marine ecosystem, plastics can break up into smaller pieces (Gigault et al., 2018). Microplastics are tiny (<5 mm) plastic particles, accounting for an estimated 92% of plastic particles (5.25 trillion) on the ocean surface (Auta et al., 2017).

V. Kumar (✉) · G. Kaur
ICAR-Central Institute of Post-Harvest Engineering and Technology, Ludhiana, Punjab, India

Siddhnath · S. Kaur · V. K. Reddy
College of Fisheries, Guru Angad Dev Veterinary and Animal Sciences University, Ludhiana, Punjab, India

N. B. Rathod
Department of Post-harvest Management of Meat, Poultry and Fish, PG Institute of Post-Harvest Management (Dr. Balasaheb Sawant Konkan Krishi Vidyapeeth, Dapoli) Killa-Roha, Raigad, Maharashtra, India

H. Maqbool
Division of Fish Post Harvest Technology, Faculty of Fisheries Rangil, Ganderbal, SKUAST-Kashmir, Jammu & Kashmir, India

A. Abubakr et al. (eds.), *Microplastics: Ecological and Food Security Implications*,
https://doi.org/10.1007/978-3-031-94480-2_4

Microplastics are widely distributed in the environment, particularly in the marine due to ocean currents, hydrodynamic processes, and wind. They can be found in large ocean gyres such as Indian Ocean (Reddy et al., 2006), the equator (Do Sul et al., 2009), the Pacific Ocean (Eriksen et al., 2013), open seas (Taylor et al., 2016), Atlantic Ocean (Cózar et al., 2017), polar regions (Waller et al., 2017; Malik et al., 2017) and on coasts (Martin et al., 2017). The ocean ecosystem has been heavily polluted by microplastics in recent years and received increased scientific and public attention in different fields. The prevalence of microplastics in the aquatic environment is triggered by various human activities such as tourism, the treatment of municipal and industrial effluents, and other similar activities. Marine plastic is an anthropogenic material that comes from either terrestrial sources which make up to 80% of marine waste (Da Costa et al., 2016), or marine activities, like accidental spillage or unlawful disposal during offshore drilling or fishing. Ships also discard around 6.5 million tons of plastic annually (Coyle et al., 2020). The increase in the human population and injudicious use of plastic contribute to the rising concentration of microplastic in different water bodies, including ponds, lakes, rivers, and the sea (Dikareva & Simon, 2019).

Several studies report plastic material to be poisonous to aquatic life, just like dye, heavy metals, and organic contaminants (Bhattacharya et al., 2020; Gola et al., 2020; Jain et al., 2020; Gurjar et al., 2023; Prasad et al., 2023). Microplastics pose a greater risk to aquatic organisms because of their microscopic size. As aquatic species serve as seafood, their pollution creates a simple medium for microplastics to reach the human body. All seafood, including fish, sea cucumbers, amphipods, turtles, copepods, and birds that consume these sea organisms, have been shown to accumulate microplastic (Xu et al., 2020). Globally researchers examined the fate of ingested microplastic and concluded that, once consumed, microplastic can accumulate in a specific body tissue, undergo egestion through pseudo-feces, or translocate within the tissues of various organs. Microplastic can have many detrimental effects on an organism's health once it accumulates in bodily tissue, including growth retardation, infertility, internal or external injuries, obstruction of bodily tracts, etc. (Ogonowski et al., 2016; Sussarellu et al., 2016). The current chapter presents the sources and pathways, impacts of microplastic contamination on seafood, implications for food security, and current mitigation and regulatory measures followed by knowledge gaps and research needs to be addressed to curtail the microplastic menace.

2 Sources, Pathways and Bioaccumulation of Microplastic in Seafood

2.1 Sources of Microplastics

Microplastic contamination in aquatic environments arises from two sources: primary and secondary (Table 1). Primary microplastics are intentionally manufactured

Table 1 Broad classification of microplastics based on sources

Types of microplastic	Forms of microplastic	Applications	References
Primary microplastics	Pellets, microbeads, exfoliating microplastics	Cosmetics, pharmaceuticals, industrial, personal hygiene products, children's goods, insect repellent	GESAMP (2015) and Magnusson et al. (2016)
Secondary microplastics	Fragments from larger plastics, tire wear	Road paint, fishing gear, and ropes	Andrady (2011) and Magnusson et al. (2016)

as small plastic fragments. Examples include pellets (nurdles), microbeads, and exfoliating particles used in cosmetics, pharmaceuticals, and industrial applications such as air blasting (Van Cauwenberghe & Janssen, 2014; GESAMP, 2015; Magnusson et al., 2016). These microplastics are typically composed of polymers like polystyrene, polyethylene, polypropylene, polyolefin, polyvinyl chloride, and polyethylene terephthalate (Magnusson et al., 2016; Dikareva & Simon, 2019; Kor & Mehdinia, 2020). While polystyrene, high-density polyethylene, low-density polyethylene, polypropylene, and polyurethane may float primarily on seawater, polyamide, polyethylene terephthalate, and polyvinyl chloride have densities higher than seawater, which increases their rates of settlement in sediments (Coyle et al., 2020; Nabi et al., 2024; Debnath et al., 2024). The major sources of primary plastic are cosmetics, personal hygiene products, children's goods, insect repellent, etc. (Chatterjee & Sharma, 2019).

Fragmented products made from physical, biological, and chemical breakdown of large-sized plastic materials are significant sources of secondary plastics (Sait et al., 2021). Secondary microplastics also result from larger plastic items breaking down into smaller fragments due to environmental degradation processes such as weathering and photodegradation. These fragments are often released during the use and maintenance of plastic products, including tires, road paint, fishing gear, and ropes (Magnusson et al., 2016). Plastic debris exposed to sunlight degrades more rapidly than submerged or floating on the water surface, although degradation remains incomplete, leaving persistent microplastic fragments (Andrady et al., 1998; Andrady, 2011). Both macroplastics and microplastics are harmful to marine ecosystems. When ingested, they disrupt physiological processes and ecological balance (Fendall & Sewell, 2009).

2.2 *Microplastic Contamination Pathways in Aquatic Ecosystems*

Figure 1 presents potential sources and pathways of microplastic contamination in an aquatic environment. Key pathways for these pollutants into aquatic ecosystems include domestic, industrial, and coastal activities (Hurley et al., 2018) (Table 2).

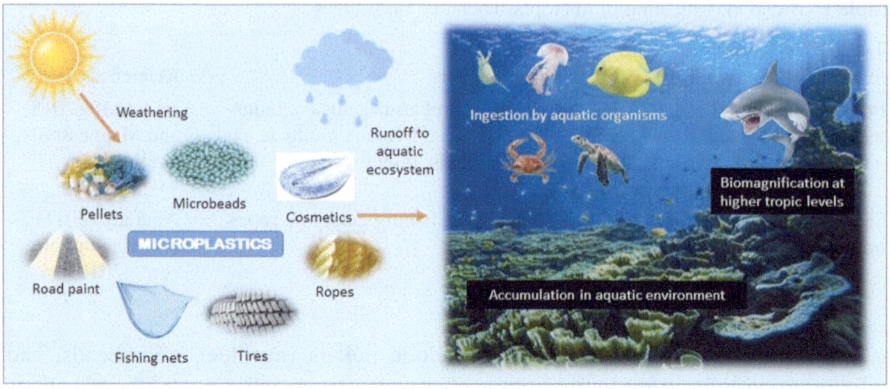

Fig. 1 Microplastic contamination in an aquatic environment: Potential sources and pathways

Table 2 Microplastic contamination pathways in seafood

Pathways	Examples	References
Land-based	Urban runoff, wastewater, industrial discharge	Hurley et al. (2018) and Yang et al. (2021)
Marine-based	Fishing gear, marine debris, ship waste	Liotta et al. (2024) and Andrady et al. (1998)
Atmospheric deposition	Airborne microplastics settling on water surfaces	Hurley et al. (2018)
Bioaccumulation	Transfer through food web to higher trophic levels	Justino et al. (2023) and Akhbarizadeh et al. (2019)

Marine litter including microplastics (approximately 80%), originates from land-based sources. Urban runoff serves as a major conduit, transporting plastics from urban and suburban areas to rivers, lakes, and oceans. Additionally, plastic pellets, resin powders, and waste from coastal tourism, aquaculture, and commercial fishing frequently enter water bodies through runoff, wastewater, and wind currents (Yang et al., 2021). Improper disposal of fishing gear and ship-generated waste further exacerbate the problem (Liotta et al., 2024). Atmospheric deposition also plays a significant role in dispersing microplastics into aquatic systems. Airborne microplastic particles settle on land and water surfaces, contributing to their widespread distribution (Hurley et al., 2018). Marine plastic debris, including discarded fishing nets and plastic packaging, forms another significant pathway, with large plastic debris breaking down into microplastics due to mechanical action and UV exposure.

2.3 *Bioaccumulation of Microplastic in Seafood*

Microplastics persist in aquatic environments due to durability and ubiquity in both the benthic and pelagic zones. Their small size and widespread presence make them bioavailable to a diverse range of marine organisms, including zooplankton, worms, corals, lobsters, sea urchins, and fish (Hall et al., 2015; Okubo et al., 2018; Woods et al., 2020; Mason et al., 2022). Upon ingestion, microplastics accumulate within these organisms and transfer through the food web (Setälä et al., 2014). Higher trophic-level species, including those consumed by humans, often exhibit significant bioaccumulation of microplastics (Justino et al., 2023). This bioaccumulation poses severe ecological and health implications, particularly for human health, as contaminated seafood becomes a direct vector for microplastic exposure (Akhbarizadeh et al., 2019). Furthermore, by binding and transporting different kinds of pollutants in the water, microplastics might function as a vector, causing the bioaccumulation of various hydrocarbons, heavy metals, dyes, and other pollutants (Amelia et al., 2021) which may increase ocean toxicity. The biomagnification process can further enhance the risk to seafood consumers. The environmental consequences of microplastic pollution can indirectly affect food security by disrupting global supply chains. For instance, if microplastic contamination reduces fish stocks or agricultural productivity in certain regions, it can lead to supply shortages, higher prices, and reduced access to nutritious food. Thus, microplastics are becoming a major concern and require addressing issues brought on by microplastics on the environment, health, and food security.

3 Impacts of Microplastic Contamination on Seafood

3.1 *Physiological Effects on Marine Organisms*

The ingestion of microplastics has multiple detrimental effects on marine organisms. These particles can obstruct the intestinal tract, reducing feeding behavior and inhibiting the secretion of gastric enzymes. Such disruptions can lead to imbalances in steroid hormones, causing delayed ovulation and infertility (Zhou et al., 2024). Prolonged reduced food intake often results in mortality among affected species (Eltemsah & Bøhn, 2019). Furthermore, microplastics act as carriers for adsorbed pollutants such as persistent organic pollutants (POPs) and heavy metals. These pollutants, even in low environmental concentrations, adhere to microplastics and magnify their toxic potential as they bioaccumulate through trophic levels (Sruthisree et al., 2015; Amin et al., 2016, 2017; Gao et al., 2019).

Additionally, microplastics have been linked to the proliferation of harmful algae that produce phycotoxins. These toxins can transfer to plankton and benthic organisms such as bivalves and crustaceans, leading to poisoning like paralytic shellfish poisoning (PSP), and diarrhetic shellfish poisoning (DSP). Such toxic outbreaks are associated with microplastic exposure (Sharma & Chatterjee, 2017; Xu et al., 2024). The accumulation of these toxins in shellfish poses significant risks to human health and marine ecosystems, as they pass through the food web, intensifying their impact (Farabegoli et al., 2018; Roopika & Siddhuraju, 2022).

3.2 Microplastic Contamination in Edible Seafood

The prevalence of microplastics in marine environments has resulted in widespread ingestion by marine organisms at various trophic levels. Ingestion occurs through multiple pathways, including feeding, burrowing, and respiration, leading to the accumulation of these anthropogenic particles within the gastrointestinal tracts of numerous commercially important seafood species. The presence of diverse polymer types in marine sediments and biota has been reported in different research studies. Investigations along the Gujarat coast and Vembanad Lake in Kerala, India, have identified polyurethane, nylon, and polystyrene particles, with low-density polyethylene (LDPE) being the dominant polymer type (Reddy et al., 2006; Sruthy & Ramasamy, 2017). This prevalence of LDPE is consistent with its widespread use in packaging and other consumer products, highlighting a direct link to anthropogenic sources. The physical form of microplastic contamination can also vary. In Norway lobster (Nephrops norvegicus) from Scotland's Clyde Sea, microplastic contamination has been observed as tangled filaments, suggesting potential entanglement and physical harm to the organism (Murray & Cowie, 2011).

The microplastic contamination in edible seafood extends across species and geographical locations. Research has consistently demonstrated microplastics in commercially relevant species consumed by humans. Devriese et al. (2015) reported microplastic contamination in brown shrimp (Crangon crangon), and demonstrated the contamination of widely consumed crustaceans in European waters. Studies focusing on bivalve molluscs have also revealed significant levels of microplastic accumulation. Van Cauwenberghe and Janssen (2014) found microplastics in blue mussels and Pacific oysters farmed in the North Sea and Brittany, France, regions known for their intensive shellfish aquaculture. Li et al. (2015) expanded this investigation to include nine different bivalve species sold in Shanghai's fish markets, highlighting the potential for exposure to humans through consumption of these products in major urban centers. This global distribution is further emphasized by studies conducted in other regions. Santana et al. (2016) detected microplastics in brown mussels (Perna perna) from estuaries near São Paulo, Brazil, demonstrating contamination in South American coastal ecosystems. Similarly, Davidson and Dudas (2016) found microplastics in both cultured and wild Manila clams (Venerupis philippinarum) from Canada, highlighting the contamination of both wild-caught and aquacultured seafood in North America.

Furthermore, Rochman et al. (2015) examined marine debris recovered from the gastrointestinal tracts of a range of fish species, providing evidence of microplastic ingestion across different fish taxa. Their study included 11 fish species from

Indonesian waters and 12 species from U.S. waters, further demonstrating the global scale of this issue and its impact on diverse fish populations. These findings collectively underscore the widespread distribution of microplastics in edible seafood and raise concerns regarding potential risks to human health through dietary exposure.

3.3 Health Risks to Consumers

Fishery products contribute approximately 17% of global animal protein consumption, making microplastic contamination a critical concern in seafood (Smith et al., 2018). Microplastics can accumulate in human tissues over time, potentially disrupting neurological functions and contributing to tumor formation (Gaspar et al., 2023; Rafazi et al., 2024; Zhao et al., 2024). The ingestion of microplastics has been associated with cytotoxicity, oxidative stress, and inflammation. These particles can obstruct organs, disrupt immune functions, and translocate to distant tissues, potentially causing neurotoxicity and metabolic disorders. The urgency to address these risks is underscored by the growing ubiquity of microplastics and their persistent nature in the environment (Visalli et al., 2021). Microplastic contaminated seafood, laden with heavy metals, toxins, and pesticides adsorbed onto microplastic surfaces, poses severe health risks. Table 3 presents the impacts of microplastic contamination on seafood and human health.

4 Implication for Food Security

Microplastics pose significant toxicity risks for consumers due to the highly perishable nature of seafood (Siddiqui et al., 2023). This issue negatively affects consumer perceptions of seafood acceptability (El Abed & Özogul, 2023; Tabassum et al., 2025). Many studies reported the occurrence of microplastics and associated toxins

Table 3 Impacts of microplastic contamination on seafood and human health

Category	Examples	References
Physiological Effects	Intestinal blockage, reduced feeding, hormonal disruption	Zhou et al. (2024) and Eltemsah and Bøhn (2019)
Adsorbed pollutants	POPs, heavy metals causing biomagnification	Gao et al. (2019)
Toxic algal blooms	DSP, PSP toxins linked to microplastic exposure	Sharma and Chatterjee (2017) and Xu et al. (2024)
Seafood Contamination	Microplastics in shrimp, mussels, clams, and fish	Murray and Cowie (2011) and Li et al. (2015)
Health Risks to consumers	Inflammation, neurotoxicity, metabolic disruption	Gaspar et al. (2023) and Visalli et al. (2021)

in fresh, dried, and processed seafood (Gündogdu et al., 2023). Pan et al. (2022) analysed the presence of microplastics in various seafood, including bivalves and fish from China, highlighting the implications for human health. Another study focused on microplastic occurrence based on Australian seafood consumption patterns (Dawson et al., 2021), concluded that microplastics found in edible tissues of seafood consumed by the Australian population pose a significant risk. As a consequence of these risks, the consumption of seafood from polluted areas has declined (Bhuyan, 2022). Microplastics and their contaminants in seafood have also led to a reduction in its nutritional value. This is because microplastics interfere with the feeding behavior of fish, ultimately resulting in reduced growth (Smith et al., 2018). The small fragments of microplastics ingested by fish can migrate to surrounding tissues and even the circulatory system (Lusher et al., 2017; Prasad et al., 2024). Additionally, the presence of microplastics in seafood has diminished its market value, directly impacting the livelihoods of fishing communities. Furthermore, the increasing levels of microplastic contamination in the marine environment have led to a decline in export revenue from tuna, lobster, and crab in Mozambique. Plastic pollution in marine environments also results in reduced fishing time, as more effort is needed to clear marine litter from fishing gear. This pollution can damage fishing equipment as well (Raes et al., 2021).

Microplastics can cause significant harm by not only damaging organisms but also by releasing various chemicals and providing surfaces for bacterial colonization and biofilm formation (Puckowski et al., 2021). Pathogenic Vibrio species have also been found on microplastics (Kirstein et al., 2016). Additionally, microplastics are associated with several endocrine disruptors, such as bisphenol A, phthalates, and nonylphenols, which have been linked to various disorders in humans (Rathod et al., 2023). Pan et al. (2022) identified white microplastic fibers less than 1.0 mm in size as the most common contaminant found in seafood. Shellfish are at a higher risk compared to fish for microplastic contamination. Given the accumulation of hazardous chemicals and pathogenic microorganisms, contaminated seafood poses a serious risk to consumers. The ingestion of microplastics by fish can induce oxidative stress and alter their immune response, making the fish more susceptible to infection and spoilage (Bhuyan, 2022). The impact of microplastic pollution on populations that rely on seafood as a staple food is a concerning issue. Coastal communities depend on seafood, whether in fresh or dried forms, as an essential part of their diet (Piyawardhana et al., 2022). The primary risk lies in dried fish, as they are typically dried whole, preserving the intact digestive system, which is often consumed with minimal cleaning (Kutralam-Muniasamy et al., 2023). In contrast, the gut parts of fresh fish are generally removed before consumption, except in the case of smaller fish.

5 Current Mitigation and Regulatory Measures

Several global agencies have adopted policies to combat plastic pollution, which is a leading cause of microplastic contamination. These measures include banning plastics (especially single-use), implementing effective plastic waste policies, and limiting the production of products that contain microplastics (Vuola et al., 2019). Regulatory bodies, such as the OECD, have prohibited single-use plastics and the application of microbeads in rinse-off cosmetics (OECD, 2021). Different countries have established legal regulations regarding single-use plastics and microplastics. For example, Canada has banned the import, the manufacture, and the sale of microbead-containing toiletries. France has prohibited cosmetics that contain microplastics, and India has deemed microbeads unsafe for use in cosmetics (Rathod et al., 2023). The EU has decided to restrict the intentional addition of microplastics to achieve a 30% reduction in microplastic pollution by 2030, defined microplastics as synthetic polymer particles (<5 mm.) which are insoluble, organic, and resistant to degradation in various products (Hrustić, 2022).

China implemented strict regulations against the use of single-use plastics with the Plastic Restriction Order in 2008. It is essential to adopt initiatives aimed at achieving zero plastic waste in landfills for preventing microplastics entering the environment. Abandoned, discarded, and lost fishing gear is a significant source of microplastic in marine environments (Deshpande et al., 2020; Apete et al., 2024). Deshpande et al. (2020) assessed the impact of fishing gear on plastic pollution through a material flow analysis in Norway. The study revealed that commercial fishing alone contributes approximately 380 tonnes of plastic waste annually due to lost fishing gear. Additionally, around 840 tonnes of plastic fishing gear are sent to landfills each year.

Creating public awareness about the dangers associated with the misuse of plastics is essential to reducing plastic pollution. It is important to educate all levels of society about the harmful effects of microplastics and nanoplastics on both human health and the environment. A recent report by Chenier et al. (2024) recommends incorporating plastic waste management topics into school curricula to promote positive behavioural changes. Additionally, it has been suggested that rural communities, which often engage in poor waste management practices, should develop their own policies to address these issues (Mihai et al., 2021).

6 Knowledge Gaps and Research Needs

A comprehensive understanding of the microplastic issue necessitates focused, interdisciplinary investigation. While microplastics in marine environments and seafood are increasingly documented, significant knowledge gaps hinder effective mitigation and risk assessment. Several critical areas require rigorous scientific scrutiny.

6.1 Elucidating Long-Term Human Health Impacts

Determining the chronic health consequences of human microplastic exposure via seafood consumption is paramount. While *in vitro* and *in vivo* studies reveal potential toxicological mechanisms (e.g., inflammation, oxidative stress, cellular disruption), epidemiological studies investigating long-term exposure cohorts and correlating microplastic ingestion with human disease incidence are critically lacking. Key research priorities include:

6.1.1 Bioaccumulation and Translocation

Investigating bioaccumulation patterns of different microplastic polymer types and sizes within human tissues and organs is crucial. Understanding translocation mechanisms, including passage across biological barriers (e.g., intestinal mucosa, blood-brain barrier) and access to sensitive physiological systems, is essential (e.g., Morgan & DeLouise, 2020).

6.1.2 Immunological and Endocrine Effects

Determining microplastic exposure's impact on human immune and endocrine function, particularly in vulnerable populations (infants, pregnant women, elderly), is vital. Research should explore microplastics' potential as endocrine disruptors or allergens, compromising immune function or reproductive health (e.g., Gaspar et al., 2023; Murray & Cowie, 2011).

6.1.3 Toxicological Synergies

Understanding microplastic interactions with other environmental contaminants (e.g., persistent organic pollutants, heavy metals) and potential synergistic health effects requires advanced analytical chemistry to characterize absorbed and desorbed pollutants under physiological conditions (e.g., Rafazi et al., 2024; Zhao et al., 2024).

6.2 Development of Effective Mitigation Technologies

Reducing seafood microplastic contamination depends on developing and implementing effective technologies targeting the entire plastic lifecycle, from source control to wastewater treatment and environmental remediation. Specific research areas include:

6.2.1 Source Reduction Strategies

Evaluating the efficacy of source reduction policies, including single-use plastic reduction and biodegradable alternatives, while considering socioeconomic implications and long-term effects on manufacturing and trade.

6.2.2 Advanced Wastewater Treatment

Researching novel microplastic removal techniques from wastewater, including advanced filtration, biodegradation, and engineered nanomaterials for targeted polymer removal (e.g., Smith et al., 2018).

6.2.3 Remediation of Contaminated Marine Environments

Assessing the feasibility and efficacy of marine microplastic removal technologies (e.g., bioremediation, physical removal), considering cost-effectiveness and environmental impact.

6.3 Establishing Global Monitoring and Reporting Standards

The lack of standardized microplastic sampling, quantification, and reporting methodologies hinders global mitigation and pollution dynamics assessment. Research should focus on:

6.3.1 Standardized Sampling Protocols

Establishing harmonized protocols for microplastic sampling across marine environments and seafood species, considering particle sizes, densities, and polymer types, with interlaboratory validation.

6.3.2 Analytical Techniques

Refining and validating analytical techniques for rapid and accurate microplastic identification and quantification, including micro-spectroscopy (Raman, FTIR) and pyrolysis gas chromatography-mass spectrometry, with standardized data analysis and interpretation.

6.3.3 Data Sharing and Reporting Platforms

Creating a global, publicly accessible database for microplastic pollution data, enabling collaborative research and effective policy development with harmonized metadata reporting.

7 Conclusion

Microplastics are ubiquitous pollutants entering the marine food web. The pervasive microplastic presence in marine environments and seafood poses a significant threat to global food security. The implications of human exposure and effects on marine ecosystems necessitate immediate action. The scale of contamination, uncertain long-term health impacts, and lack of standardized monitoring challenge effective mitigation and seafood supply chain maintenance. A multi-pronged approach encompassing policy, industry, and public behavior is crucial. Governments should enact and enforce robust regulations limiting plastic production, consumption, and disposal, transitioning to a circular economy. They should support international agreements mandating plastic pollution reduction and establishing comprehensive monitoring and reporting, including financial commitment to alternative materials and cleaner production. The seafood industry must adopt sustainable practices, including reducing packaging waste, promoting reusable options, and actively preventing microplastic contamination. Investment in advanced filtration for processing plants and aquaculture is crucial. Public awareness campaigns should educate consumers on microplastic risks and encourage responsible consumption, minimizing plastic use and participating in cleanup efforts, emphasizing circular economy approaches and waste management. Sustainable seafood production and microplastic contamination reduction require a radical shift in our approach to plastics. A future with safe and abundant seafood depends on global cooperation in implementing these recommendations, moving from a linear to a circular economy. Continued, concerted effort involving scientists, policymakers, industry, and the public is essential to secure the future of seafood and marine ecosystems. Continued development and deployment of advanced monitoring and mitigation technologies are critical for long-term success.

References

Akhbarizadeh, R., Moore, F., & Keshavarzi, B. (2019). Investigating microplastics bioaccumulation and biomagnification in seafood from the Persian Gulf: A threat to human health? *Food Additives and Contaminants-Part A Chemistry, Analysis, Control, Exposure and Risk Assessment, 36*(11), 1696–1708. https://doi.org/10.1080/19440049.2019.1649473

Amelia, T. S. M., Khalik, W. M. A. W. M., Ong, M. C., Shao, Y. T., Pan, H. J., & Bhubalan, K. (2021). Marine microplastics as vectors of major ocean pollutants and its hazards to the marine ecosystem and humans. *Progress in Earth and Planetary Science, 8*, 1–26. https://doi.org/10.1186/s40645-020-00405-4

Amin, A., Indulkar, S. T., & Pai, R. (2016). Sublethal effect of Buprofezin pesticide on carcass composition of *Cyprinus carpio* communis fingerlings. *Journal of Experimental Zoology, India, 20*(1), 269–272.

Amin, A., Ramachandra Naik, A. T., Priyadarshini, N., Nayak, H., & Sree, C. S. (2017). Toxic effect of heavy metal lead on oxygen consumption of rohu (*Labeo rohita*) fingerlings. *Biochemical & Cellular Archives, 17*(1), 225–228.

Andrady, A. L. (2011). Microplastics in the marine environment. *Marine Pollution Bulletin, 62*(8), 1596–1605. https://doi.org/10.1016/j.marpolbul.2011.05.030

Andrady, A. L., Hamid, S. H., Hu, X., & Torikai, A. (1998). Effects of increased solar ultraviolet radiation on materials. *Journal of Photochemistry and Photobiology B: Biology, 46*(1–3), 96–103. https://doi.org/10.1016/S1011-1344(98)00188-2

Apete, L., Martin, O. V., & Iacovidou, E. (2024). Fishing plastic waste: Knowns and known unknowns. *Marine Pollution Bulletin, 205*, 116530.

Auta, H. S., Emenike, C. U., & Fauziah, S. H. (2017). Distribution and importance of microplastics in the marine environment: A review of the sources, fate, effects, and potential solutions. *Environment International, 102*, 165–176. https://doi.org/10.1016/j.envint.2017.02.013

Bhattacharya, A., Gola, D., Dey, P., & Malik, A. (2020). Synergistic and antagonistic effects on metal bioremediation with increasing metal complexity in a hexa-metal environment by aspergillus fumigatus. *International Journal of Environmental Research, 14*, 761–770. https://doi.org/10.1007/s41742-020-00295-w

Bhuyan, M. S. (2022). Effects of microplastics on fish and in human health. *Frontiers in Environmental Science, 10*, 827289.

Chae, Y., & An, Y. J. (2018). Current research trends on plastic pollution and ecological impacts on the soil ecosystem: A review. *Environmental Pollution, 240*, 387–395. https://doi.org/10.1016/j.envpol.2018.05.008

Chatterjee, S., & Sharma, S. (2019). Microplastics in our oceans and marine health. Field actions science reports. *The Journal of Field Actions*., (Special Issue, *19*, 54–61.

Chenier, K. A., Englebretson, E., James-Barry, J. A., Rigsby, A. N., Rodolfich, A. E., McQueen, E. P., & Sparks, E. L. (2024). Development of a community-driven waste reduction education and action program. *Sustainability, 16*(11), 4823.

Coyle, R., Hardiman, G., & O'Driscoll, K. (2020). Microplastics in the marine environment: A review of their sources, distribution processes, uptake and exchange in ecosystems. *Case Studies in Chemical and Environmental Engineering, 2*, 100010. https://doi.org/10.1016/j.cscee.2020.100010

Cózar, A., Martí, E., Duarte, C. M., García-de-Lomas, J., Van Sebille, E., Ballatore, T. J., et al. (2017). The Arctic Ocean as a dead end for floating plastics in the North Atlantic branch of the thermohaline circulation. *Science Advances, 3*(4), e1600582. https://doi.org/10.1126/sciadv.1600582

Da Costa, J. P., Santos, P. S., Duarte, A. C., & Rocha-Santos, T. (2016). (Nano) plastics in the environment–sources, fates and effects. *Science of the Total Environment, 566*, 15–26. https://doi.org/10.1016/j.scitotenv.2016.05.041

Davidson, K., & Dudas, S. E. (2016). Microplastic ingestion by wild and cultured Manila clams (*Venerupis philippinarum*) from Baynes Sound, British Columbia. *Archives of Environmental Contamination and Toxicology, 71*(2), 147–156. https://doi.org/10.1007/s00244-016-0286-4

Dawson, A. L., Santana, M. F., Miller, M. E., & Kroon, F. J. (2021). Relevance and reliability of evidence for microplastic contamination in seafood: A critical review using Australian consumption patterns as a case study. *Environmental Pollution, 276*, 116684.

Debnath, R., Prasad, G. S., Amin, A., Malik, M. M., Ahmad, I., Abubakr, A., Borah, S., Rather, M. A., Impellitteri, F., Tabassum, I., & Piccione, G. (2024). Understanding and addressing microplastic pollution: Impacts, mitigation, and future perspectives. *Journal of Contaminant Hydrology, 266*, 104399.

Deshpande, P. C., Philis, G., Brattebø, H., & Fet, A. M. (2020). Using material flow analysis (MFA) to generate the evidence on plastic waste management from commercial fishing gears in Norway. *Resources, Conservation & Recycling: X, 5*, 100024.

Devriese, L. I., van der Meulen, M. D., Maes, T., Bekaert, K., Paul-Pont, I., Frère, L., Robbens, J., & Vethaak, A. D. (2015). Microplastic contamination in brown shrimp (Crangon crangon, Linnaeus 1758) from coastal waters of the Southern North Sea and Channel area. *Marine Pollution Bulletin, 98*(1–2), 179–187. https://doi.org/10.1016/j.marpolbul.2015.06.051

Dikareva, N., & Simon, K. S. (2019). Microplastic pollution in streams spanning an urbanisation gradient. *Environmental Pollution, 250*, 292–299. https://doi.org/10.1016/j.envpol.2019.03.105

Do Sul, J. A. I., Spengler, Â., & Costa, M. F. (2009). Here, there and everywhere. Small plastic fragments and pellets on beaches of Fernando de Noronha (equatorial Western Atlantic). *Marine Pollution Bulletin, 58*(8), 1236–1238.

El Abed, N., & Özogul, F. (2023). The risks of marine micro/nano-plastics on seafood safety and human health. In *Advances in food and nutrition research* (Vol. 103, pp. 229–271). Academic.

Eltemsah, Y. S., & Bøhn, T. (2019). Acute and chronic effects of polystyrene microplastics on juvenile and adult Daphnia magna. *Environmental Pollution, 254*, 112919. https://doi.org/10.1016/j.envpol.2019.07.087

Eriksen, M., Mason, S., Wilson, S., Box, C., Zellers, A., Edwards, W., et al. (2013). Microplastic pollution in the surface waters of the Laurentian Great Lakes. *Marine Pollution Bulletin, 77*(1–2), 177–182. https://doi.org/10.1016/j.marpolbul.2013.10.007

Farabegoli, F., Blanco, L., Rodríguez, L. P., Manuel Vieites, J., & García Cabado, A. (2018). Phycotoxins in marine shellfish: Origin, occurrence and effects on humans. *In Marine Drugs, 16*(6), 188. https://doi.org/10.3390/md16060188

Fendall, L. S., & Sewell, M. A. (2009). Contributing to marine pollution by washing your face: Microplastics in facial cleansers. *Marine Pollution Bulletin, 58*(8), 1225–1228. https://doi.org/10.1016/j.marpolbul.2009.04.025

Gao, F., Li, J., Sun, C., Zhang, L., Jiang, F., Cao, W., & Zheng, L. (2019). Study on the capability and characteristics of heavy metals enriched on microplastics in marine environment. *Marine Pollution Bulletin, 144*, 61–67. https://doi.org/10.1016/j.marpolbul.2019.04.039

Gaspar, L., Bartman, S., Coppotelli, G., & Ross, J. M. (2023). Acute exposure to microplastics induced changes in behavior and inflammation in young and old mice. *International Journal of Molecular Sciences, 24*(15), 12308. https://doi.org/10.3390/ijms241512308

GESAMP. (2015). Sources, fate, and effects of microplastics in the marine environment (part 1). http://www.gesamp.org/publications/reports-and-studies-no-90

Gigault, J., Ter Halle, A., Baudrimont, M., Pascal, P. Y., Gauffre, F., Phi, T. L., El Hadri, H., Grassl, B., & Reynaud, S. (2018). Current opinion: What is a nanoplastic? *Environmental Pollution, 235*, 1030–1034. https://doi.org/10.1016/j.envpol.2018.01.024

Gola, D., Bhattacharya, A., & Dey, P. (2020). Assessment of drain water used for irrigation in the Delhi region. *Journal of Health and Pollution., 10*, 200610. https://doi.org/10.5696/2156-9614-10.26.200610

Gündogdu, S., Rathod, N., Hassoun, A., Jamroz, E., Kulawik, P., Gokbulut, C., et al. (2023). The impact of nano/micro-plastics toxicity on seafood quality and human health: Facts and gaps. *Critical Reviews in Food Science and Nutrition, 63*(23), 6445–6463.

Gurjar, U. R., Takar, S., Amin, A., & Martin Xavier, K. A. (2023). Microplastic contamination in aquatic organisms: An ecotoxicological perspective. In *Xenobiotics in aquatic animals: Reproductive and developmental impacts* (pp. 353–367). Springer.

Hall, N. M., Berry, K. L. E., Rintoul, L., & Hoogenboom, M. O. (2015). Microplastic ingestion by scleractinian corals. *Marine Biology, 162*(3), 725–732. https://doi.org/10.1007/s00227-015-2619-7

Hrustić, A. (2022). EU restrictions on intentionally added microplastics-small plastic, low ambition. *Anali Pravnog fakulteta Univerziteta u Zenici, 15*(31), 7–32.

Hurley, R., Woodward, J., & Rothwell, J. J. (2018). Microplastic contamination of river beds significantly reduced by catchment-wide flooding. *Nature Geoscience, 11*(4), 251–257. https://doi.org/10.1038/s41561-018-0080-1

Jain, A., Ahmad, F., Gola, D., Malik, A., Chauhan, N., Dey, P., & Tyagi, P. K. (2020). Multi dye degradation and antibacterial potential of papaya leaf derived silver nanoparticles. *Environmental Nanotechnology, Monitoring & Management, 14*, 100337. https://doi.org/10.1016/j.enmm.2020.100337

Justino, A. K. S., Ferreira, G. V. B., Fauvelle, V., Schmidt, N., Lenoble, V., Pelage, L., Martins, K., Travassos, P., & Lucena-Frédou, F. (2023). From prey to predators: Evidence of microplastic trophic transfer in tuna and large pelagic species in the southwestern Tropical Atlantic. *Environmental Pollution, 327*, 121532. https://doi.org/10.1016/j.envpol.2023.121532

Kirstein, I. V., Kirmizi, S., Wichels, A., Garin-Fernandez, A., Erler, R., Löder, M., & Gerdts, G. (2016). Dangerous hitchhikers? Evidence for potentially pathogenic Vibrio spp. on microplastic particles. *Marine Environmental Research, 120*, 1–8.

Kor, K., & Mehdinia, A. (2020). Neustonic microplastic pollution in the Persian Gulf. *Marine Pollution Bulletin, 150*, 110665. https://doi.org/10.1016/j.marpolbul.2019.110665

Kosior, E., & Mitchell, J. (2020). Current industry position on plastic production and recycling. In *Plastic waste and recycling* (pp. 133–162). Academic. https://doi.org/10.1016/B978-0-12-817880-5.00006-2

Kutralam-Muniasamy, G., Shruti, V. C., Pérez-Guevara, F., Roy, P. D., & Martínez, I. E. (2023). Consumption of commercially sold dried fish snack "Charales" contaminated with microplastics in Mexico. *Environmental Pollution, 332*, 121961.

Li, J., Yang, D., Li, L., Jabeen, K., & Shi, H. (2015). Microplastics in commercial bivalves from China. *Environmental Pollution, 207*, 190–195. https://doi.org/10.1016/j.envpol.2015.09.018

Liotta, I., Avolio, R., Castaldo, R., Gentile, G., Ambrogi, V., Errico, M. E., & Cocca, M. (2024). Mitigation approach of plastic and microplastic pollution through recycling of fishing nets at the end of life. *Process Safety and Environmental Protection, 182*, 1143–1152. https://doi.org/10.1016/j.psep.2023.12.031

Lusher, A., Hollman, P., & Mendoza-Hill, J. (2017). Microplastics in fisheries and aquaculture: Status of knowledge on their occurrence and implications for aquatic organisms and food safety. FAO fisheries and aquaculture technical paper 615.

Magnusson, K., Eliasson, K., Fråne, A., Haikonen, K., Hultén, J., Olshammar, M., Stadmark, J., & Voisin, A. (2016). *Swedish sources and pathways for microplastics to the marine environment* (p. 183). Swedish Environmental Protection Agency.

Malik, M., Balkhi, M. H., & Abubakr, A. (2017). Comparative study on the contamination level of dal Lake, Kashmir based on Total viable bacterial counts over a period of time. *Ecology, Environment and Conservation, 23*(4), 2145–2147.

Martin, J., Lusher, A., Thompson, R. C., & Morley, A. (2017). The deposition and accumulation of microplastics in marine sediments and bottom water from the Irish continental shelf. *Scientific Reports, 7*(1), 10772. https://doi.org/10.1038/s41598-017-11079-2

Mason, V. G., Skov, M. W., Hiddink, J. G., & Walton, M. (2022). Microplastics alter multiple biological processes of marine benthic fauna. *Science of the Total Environment, 845*, 157362. https://doi.org/10.1016/j.scitotenv.2022.157362

Mihai, F. C., Gündoğdu, S., Markley, L. A., Olivelli, A., Khan, F. R., Gwinnett, C., et al. (2021). Plastic pollution, waste management issues, and circular economy opportunities in rural communities. *Sustainability, 14*(1), 20.

Morgan, S. E., & DeLouise, L. A. (2020). Further studies in translatable model systems are needed to predict the impacts of human microplastic exposure. *Open Access Journal of Toxicology, 4*(3), 79–82.

Murray, F., & Cowie, P. R. (2011). Plastic contamination in the decapod crustacean Nephrops norvegicus (Linnaeus, 1758). *Marine Pollution Bulletin, 62*(6), 1207–1217. https://doi.org/10.1016/j.marpolbul.2011.03.032

Nabi, N., Ahmad, I., Amin, A., Rather, M. A., Ahmed, I., Hajam, Y. A., Khursheed, S., Malik, M. M., & Abubakr, A. (2024). Understanding the sources, fate and effects of microplastics in aquatic environments with a focus on risk profiling in aquaculture systems. *Reviews in Aquaculture, 16*(4), 1947–1980.

OECD. (2021). *Policies to reduce microplastics pollution in water: Focus on textiles and tyres*. OECD. https://www.oecd.org/content/dam/oecd/en/publications/support-materials/2021/10/policies-to-reduce-microplastics-pollution-in-water_3337bc78/policy-highlights-policies-to-reduce-microplastics-pollution-in-water-focus-textiles-and-tyres.pdf

Ogonowski, M., Schür, C., Jarsén, Å., & Gorokhova, E. (2016). The effects of natural and anthropogenic microparticles on individual fitness in Daphnia magna. *PLoS One, 11*(5), e0155063. https://doi.org/10.1371/journal

Okubo, N., Takahashi, S., & Nakano, Y. (2018). Microplastics disturb the anthozoan-algae symbiotic relationship. *Marine Pollution Bulletin, 135*, 83–89. https://doi.org/10.1016/j.marpolbul.2018.07.016

Pan, Z., Liu, Q., Xu, J., Li, W., & Lin, H. (2022). Microplastic contamination in seafood from Dongshan Bay in southeastern China and its health risk implication for human consumption. *Environmental Pollution, 303*, 119163.

Piyawardhana, N., Weerathunga, V., Chen, H. S., Guo, L., Huang, P. J., Ranatunga, R. R. M. K. P., & Hung, C. C. (2022). Occurrence of microplastics in commercial marine dried fish in Asian countries. *Journal of Hazardous Materials, 423*, 127093.

Prasad, G. S., Rout, S. K., Malik, M. M., Karmakar, S., Amin, A., & Ahmad, I. (2023). Occurrence of xenoestrogen alkylphenols (Octylphenols and Nonylphenol) and its impact on the aquatic ecosystem. In M. A. Rather, A. Amin, Y. A. Hajam, A. Jamwal, & I. Ahmad (Eds.), *Xenobiotics in aquatic animals* (pp. 275–284). Springer.

Prasad, G. S., Rout, S. K., Amin, A., Malik, M. M., Ahmad, I., Rather, M. A., Karmakar, S., Khalil, H. S., & Abubakar, A. (2024). Impact of COVID-19 on the aquatic environment and implications for fisheries with special emphasis to Indian perspective. *Proceedings of the Indian National Science Academy, 90*(4), 821–839.

Puckowski, A., Cwięk, W., Mioduszewska, K., Stepnowski, P., & Białk-Bielińska, A. (2021). Sorption of pharmaceuticals on the surface of microplastics. *Chemosphere, 263*, 127976.

Raes, L., Litulo, C., Pereira, M. A. M., Manyara, P., Raes, L., Litulo, C., & Pereira, M. A. (2021). Marine plastics, fisheries, and livelihoods in Mozambique. *Fisheries, 20*, 202004.

Rafazi, P., Bagheri, Z., Haghi-Aminjan, H., Rahimifard, M., & Ahvaraki, A. (2024). Long-term exposure of human U87 glioblastoma cells to polyethylene microplastics: Investigating the potential cancer progression. *Toxicology Reports, 13*, 101757. https://doi.org/10.1016/j.toxrep.2024.101757

Rathod, N. B., Xavier, K. M., Özogul, F., & Phadke, G. G. (2023). Impacts of nano/micro-plastics on safety and quality of aquatic food products. *Advances in Food and Nutrition Research, 103*, 1–40.

Reddy, M. S., Basha, S., Adimurthy, S., & Ramachandraiah, G. (2006). Description of the small plastics fragments in marine sediments along the Alang-Sosiya ship-breaking yard, India. *Estuarine, Coastal and Shelf Science, 68*(3–4), 656–660. https://doi.org/10.1016/j.ecss.2006.03.018

Rochman, C. M., Tahir, A., Williams, S. L., Baxa, D. V., Lam, R., Miller, J. T., Teh, F. C., Werorilangi, S., & Teh, S. J. (2015). Anthropogenic debris in seafood: Plastic debris and fibers from textiles in fish and bivalves sold for human consumption. *Scientific Reports, 5*, 14340. https://doi.org/10.1038/srep14340

Roopika, S., & Siddhuraju, P. (2022). Effect of micro plastics and Nano plastics in terrestrial animals, birds, human and implications on integrated functional ecosystem. *Indian Journal Of Ecology, 49*(6), 2415–2430. https://doi.org/10.55362/ije/2022/3842

Sait, S. T., Sørensen, L., Kubowicz, S., Vike-Jonas, K., Gonzalez, S. V., Asimakopoulos, A. G., & Booth, A. M. (2021). Microplastic fibres from synthetic textiles: Environmental degradation and additive chemical content. *Environmental Pollution, 268*, 115745. https://doi.org/10.1016/j.envpol.2020.115745

Santana, M. F. M., Ascer, L. G., Custódio, M. R., Moreira, F. T., & Turra, A. (2016). Microplastic contamination in natural mussel beds from a Brazilian urbanized coastal region: Rapid evaluation through bioassessment. *Marine Pollution Bulletin, 106*(1–2), 183–189. https://doi.org/10.1016/j.marpolbul.2016.02.074

Setälä, O., Fleming-Lehtinen, V., & Lehtiniemi, M. (2014). Ingestion and transfer of microplastics in the planktonic food web. *Environmental Pollution, 185*, 77–83. https://doi.org/10.1016/j.envpol.2013.10.013

Sharma, S., & Chatterjee, S. (2017). Microplastic pollution, a threat to marine ecosystem and human health: A short review. *Environmental Science and Pollution Research, 24*(27), 21530–21547. https://doi.org/10.1007/s11356-017-9910-8

Siddiqui, S. A., Khan, S., Tariq, T., Sameen, A., Nawaz, A., Walayat, N., & Nagdalian, A. A. (2023). Potential risk assessment and toxicological impacts of nano/micro-plastics on human health through food products. *Advances in Food and Nutrition Research, 103*, 361–395.

Smith, M., Love, D. C., Rochman, C. M., & Neff, R. A. (2018). Microplastics in seafood and the implications for human health. *Current Environmental Health Reports, 5*, 375–386. https://doi.org/10.1007/s40572-018-0206-z

Sruthisree, C., Gowda, G., Nayak, H., Harish, B. D., & Amin, A. (2015). Effect of heavy metal (lead) on oxygen consumption and ammonia excretion of iridescent shark (*Pangusius hypohthalmus*). *Journal of Experimental Zoology, India, 18*(2).

Sruthy, S., & Ramasamy, E. V. (2017). Microplastic pollution in Vembanad Lake, Kerala, India: The first report of microplastics in lake and estuarine sediments in India. *Environmental Pollution, 222*, 315–322. https://doi.org/10.1016/j.envpol.2016.12.038

Sussarellu, R., Suquet, M., Thomas, Y., Lambert, C., Fabioux, C., Pernet, M. E. J., et al. (2016). Oyster reproduction is affected by exposure to polystyrene microplastics. *Proceedings of the National Academy of Sciences, 113*(9), 2430–2435. https://doi.org/10.1073/pnas.1519019113

Tabassum, I., Hassan, N., Amin, A., Malik, M. M., Abubakr, A., Prasad, G. S., Latief, F., & Rather, S. A. (2025). COVID-19 and global environmental change: A threat to food Security and nutrition. In *Food Security, Nutrition and sustainability through aquaculture technologies* (pp. 437–445). Springer.

Taylor, M. L., Gwinnett, C., Robinson, L. F., & Woodall, L. C. (2016). Plastic microfibre ingestion by deep-sea organisms. *Scientific Reports, 6*(1), 33997. https://doi.org/10.1038/srep33997

Van Cauwenberghe, L., & Janssen, C. R. (2014). Microplastics in bivalves cultured for human consumption. *Environmental Pollution, 193*, 65–70. https://doi.org/10.1016/j.envpol.2014.06.010

Visalli, G., Facciolà, A., Ciarello, M. P., De Marco, G., Maisano, M., & Di Pietro, A. (2021). Acute and sub-chronic effects of microplastics (3 and 10 μm) on the human intestinal cells ht-29. *International Journal of Environmental Research and Public Health, 18*(11), 5833. https://doi.org/10.3390/ijerph18115833

Vuola, A., Ruiz, M., & Vianello, A. (2019). Review of existing policies and research related to microplastics–summary for policy makers. *FanpLESStic-sea-2019*, 20.

Waller, C. L., Griffiths, H. J., Waluda, C. M., Thorpe, S. E., Loaiza, I., Moreno, B., et al. (2017). Microplastics in the Antarctic marine system: An emerging area of research. *Science of the Total Environment, 598*, 220–227. https://doi.org/10.1016/j.scitotenv.2017.03.283

Woods, M. N., Hong, T. J., Baughman, D., Andrews, G., Fields, D. M., & Matrai, P. A. (2020). Accumulation and effects of microplastic fibers in American lobster larvae (Homarus americanus). *Marine Pollution Bulletin, 157*, 111280. https://doi.org/10.1016/j.marpolbul.2020.111280

Xu, S., Ma, J., Ji, R., Pan, K., & Miao, A. J. (2020). Microplastics in aquatic environments: Occurrence, accumulation, and biological effects. *Science of the Total Environment, 703*, 134699. https://doi.org/10.1016/j.scitotenv.2019.134699

Xu, H., Dong, C., Yu, Z., Ozaki, Y., Hu, Z., Zhang, B., et al. (2024). Detection and analysis of microplastics in tissues and blood of human cervical cancer patients. *Environmental Research, 259*, 119498.

Yang, H., Chen, G., & Wang, J. (2021). Microplastics in the marine environment: Sources, fates, impacts and microbial degradation. *Toxics, 9*(2), 41. https://doi.org/10.3390/toxics9020041

Zhao, J., Zhang, H., Shi, L., Jia, Y., & Sheng, H. (2024). Detection and quantification of microplastics in various types of human tumor tissues. *Ecotoxicology and Environmental Safety, 283*, 116818. https://doi.org/10.1016/j.ecoenv.2024.116818

Zhou, S., Lin, H., Liu, Z., Lian, X., Pan, C. G., Dong, Z., et al. (2024). The impact of co-exposure to polystyrene microplastics and norethindrone on gill histology, antioxidant capacity, reproductive system, and gut microbiota in zebrafish (Danio rerio). *Aquatic Toxicology, 273*, 107018.

Wastewater as a Source of Microplastic Pollution: Fate and Risk on Biota

G. G. N. Thushari, K. P. G. K. P. Guruge, A. P. Abeygunawardana, and J. D. M. Senevirathna

1 Introduction

Plastic products have become an essential part of the daily life of many people, but their mismanagement poses a serious environment threat. According to recent statistics, more than 6 billion plastics are produced, with a significant portion being single-use plastics (Mazhandu & Muzenda, 2019). A large percentage of both marine and terrestrial plastic waste consists of microplastics (MPs) (>70%), and even small concentrations (10 μg/mL) have been found to be toxic to humans and animals (Priya et al., 2023). Plastic waste exerts immense ecological pressure across all environments due to its persistent nature, lightweight properties, and widespread use (Li et al., 2021; Debnath et al., 2024; Nabi et al., 2024). Therefore, MPs represent an emerging threat to all living organisms, necessitating detailed studies on their sources, distribution, fate, and associated risks.

Global plastic waste generation has been estimated to increase significantly, primarily due to mismanagement, especially in the Asian region. By 2060, plastic waste production is projected to reach 155–265 $Mt. y^{-1}$ (Lebreton & Andrady, 2019). Accordingly, a significant portion of mismanaged plastics from urban areas, household, and industries enters wastewater systems, spreading into terrestrial environments before reaching waterways and aquatic ecosystems. Plastics degrade through various processes, leading to the formation of MPs (particles smaller than 5 mm). These MPs, originating from both primary and secondary sources, are particularly abundant in wastewater. Therefore, wastewater treatment plants are sometimes considered sources of MPs (Okoffo et al., 2019; Wu et al., 2022).

G. G. N. Thushari · K. P. G. K. P. Guruge · A. P. Abeygunawardana · J. D. M. Senevirathna (✉)
Department of Animal Science, Faculty of Animal Science and Export Agriculture, Uva Wellassa University, Badulla, Sri Lanka
e-mail: duminda@uwu.ac.lk

A. Abubakr et al. (eds.), *Microplastics: Ecological and Food Security Implications*,
https://doi.org/10.1007/978-3-031-94480-2_5

The spread of MPs through wastewater is highly complex and unpredictable. Various factors, such as wastewater treatment efficiency, environmental conditions, and the physical and chemical properties of MPs, influence their distribution (Xia et al., 2022). These particles can be transported across different ecosystems, accumulating in soil, water bodies, and even the atmosphere, making their movement difficult to track and control. Therefore, it becomes important to investigate sources and distribution pathways of microplastic in wastewater and their significant impact on biota in the aquatic and terrestrial environments.

Wastewater microplastics (WWMPs) and the accumulated biofilm (plastisphere) pose significant environmental and health risks (Xia et al., 2022). That plastisphere serves as a habitat for diverse microbial communities, including potentially harmful pathogens, which can enhance their survival and transmission (Prata, 2018). Additionally, MPs act as carriers for toxic pollutants and heavy metals, further worsening their harmful effects on soil health, aquatic ecosystems, and even human and animal health through bioaccumulation and trophic transfer.

Studying wastewater MPs is gaining significant attention because wastewater is used or released into various sensitive ecosystems in both aquatic and terrestrial environments, including irrigation systems, agriculture, aquaculture, and marine parks. Numerous studies have examined the abundance, characteristics, fate, distribution, and effects of MPs, as well as the challenges of their removal and potential solutions at a global scale. However, comprehensive research on the fate and risks of wastewater MPs for biota in aquatic and terrestrial ecosystems remains limited. Therefore, this chapter aims to provide an overview of the fate and risks of WWMPs on different organisms in various environments and their ecological interactions. Addressing this emerging threat is crucial for identifying novel technologies to mitigate their impacts on various forms of life.

The chapter consists of background knowledge on WWMPs occurrence, emerging threats, significance, and research gaps (Fig. 1). The second section focuses on the sources of wastewater, providing a comprehensive overview. The third section discusses techniques used for WWMPs identification and treatment methods. The fourth section provides an overall assessment of WWMPs toxicity. The fifth section examines the impact of WWMPs on aquatic biota and its implications for food security. The sixth section explores the effects of WWMPs on terrestrial biota and their ecological consequences. Finally, the chapter concludes with a discussion on WWMPs mitigation methods and future directions.

2 Microplastics in Wastewaters

MPs are easily accumulated in wastewater and it is a significant source of pollution, affecting both aquatic and terrestrial ecosystems. Wastewater treatment plants (WWTPs) act as a major hub where MPs accumulate, get partially removed, and eventually enter the environment through effluent discharge or biosolid application.

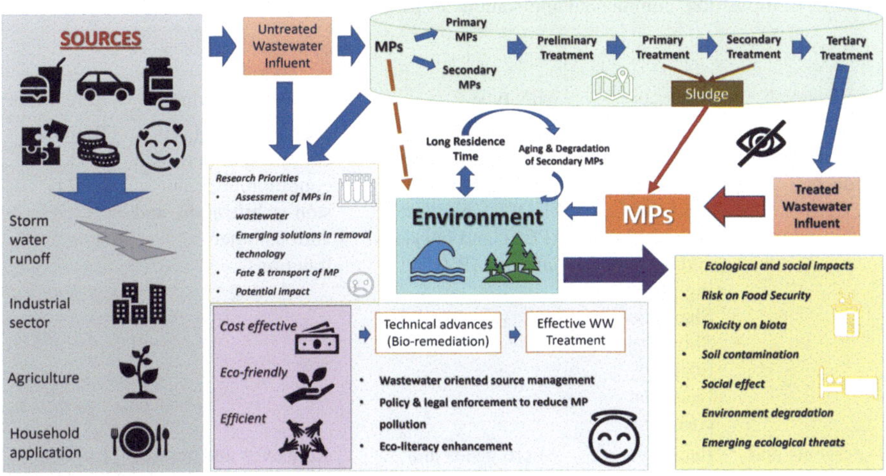

Fig. 1 Overview of MPs transport, implications, feasible management approaches, and research priorities in wastewater

2.1 Primary and Secondary Sources of MPs

Primary sources of MPs are directly released into the environment after the consumption of products. These sources include small-sized (less than 5 mm) microplastic beads, fibers, or particles that are intentionally manufactured and incorporated into these products for various commercial purposes (Reddy & Nair, 2022; Gurjar et al., 2023; Talukdar et al., 2024). Primary source MPs include polyethylene (PE), polypropylene (PP), and polystyrene (PS) particles commonly found in cosmetic and medical products (Horton et al., 2017).

Primary MPs primarily enter the aquatic environment through household sewage, which contains MPs from personal-care and cosmetic products (PCCPs), as well as household products such as detergents and cleaning products (Bikiaris et al., 2024). Mainly, synthetic fibers from textiles are released through washing machine effluent (Talukdar et al., 2024). According to commercial applications, primary sources of MPs can be summarized as follows Table 1.

2.1.1 Secondary Sources of MPs

Secondary sources of MPs refer to accumulated plastic debris or any kind of plastic surfaces that degrade into smaller-sized plastic particles (less than 5 mm) due to physical, chemical, and biological processes (Ashrafy et al., 2023; Reddy & Nair, 2022; Talukdar et al., 2024). Plastic waste and plastic surfaces can be degraded via physicochemical degradation and biodegradation, where the physical forces of the polymers are primarily broken down (Kyrikou & Briassoulis, 2007).

Table 1 Composition, commercial applications of primary sources of MPs

Application	Products	MPs type	Specific characters which hider commercial value	References
Personal-care and cosmetic products (PCCPs)	Hair- and body-bleaching products Hair-coloring and -nourishing products Shampoos Shower gels Soaps Leave-on cosmetics	Micro beads PE, high density polyethylene (HDPE), polyester (PES), and polyvinyl chloride (PVC)	Film formation, viscosity regulation, skin conditioning, emulsion stabilization, and various other functions	Bikiaris et al. (2024)
Detergents and cleaning products	Laundry detergents Toilet bowl cleaners Dishwashing Detergents Scrubbing pads or mops	Micro beads and microfibers (polyamide (PA), silicone, polyurethane (PU), acrylate copolymer (ACR), polyethylene (PE), and polyethylene terephthalate (PET))	Abrasives and act as carriers for active ingredients	Lin et al. (2023)
Textiles	Textiles made of from the synthetic fibers	Microplastic fibers (acrylic(A), PA, PES, recycled PES and PP)	Durability, Cost-effective, versatility, Lightweight, easy care	Akyildiz et al. (2024)
Medical diagnostics and pharmaceuticals	Cardiac invasive catheters Drug carriers	PTFE and PET	Chemical and thermal stability	Dewika et al. (2024)
Inks and paints	Non-aqueous paints Antifouling paints (ships and boats; road markings and buildings are other sources)	Small plastic particles (polyurethanes, PES, polyacrylates and polystyrenes)	Durability, anticorrosive nature, antifouling nature	Gaylarde et al. (2021)
Industrial applications	Polymeric production industries,	Pellets, beads, or nurdles (PE)	Durability, cost effective	Karlsson et al. (2018)

Plastics can potentially undergo degradation and fragmentation processes under various environmental abiotic conditions, such as UV light, pH, salinity, and temperature (Anthony, 2011; David et al., 2009). Naturally, the degradation of plastic wastes begins with photodegradation, which changes the plastic material through the interaction of light with air, leading to oxidation and hydrolysis. These reactions ultimately break down the plastic waste (PWs) into compounds with lower molecular weight (MW), which may then be further degraded through microbial activity (Anthony, 2011; Hayden et al., 2013). Microbial degradation, also known as biodegradation, is a crucial process in the breakdown of (PWs), where microorganisms like bacteria and fungi break down several synthetic polymers such as Low Density Polyethylene (LDPE), PP, Linear Low-Density Polyethylene (LLDPE), and HDPE (Zeenat et al., 2021).

2.2 Sources of Wastewaters and Management

Wastewater originates from various sources, including domestic, industrial, agricultural, and urban runoff. Proper management of wastewater is crucial to minimize environmental contamination, particularly the spread of MPs. Plastics are universally used materials that play a vital role across various industries and in nearly every aspect of daily life, enhancing convenience, durability, and efficiency. All MP types typically enter sewer networks through domestic discharges, rain runoff, and street washing. WWTPs serve as major collection points for MPs acting as the final destination for wastewater that carries above mentioned various MP sources through sewer networks.

2.2.1 Domestic Wastewater

Daily human activities are the primary sources of MPs in wastewater. Personal care products and laundry processes release microbeads and microfibers, with polyester and polyamide components significantly contributing to microplastic pollution in wastewater (Imogen et al., 2016; Stefano et al., 2019). Rinse-off cosmetics are washed down the drain after use, while leave-on cosmetics remain on the body for some time before being removed with a tissue, wipe, or cotton pad, which eventually become waste. These microbeads, have been detected in diverse aquatic environments (Coyle et al., 2020). According to the (Bashir et al., 2021; Prasad et al., 2023), Personal care and cosmetic products (PCCPs) often contain at least one type of microbead, with PE being the most common. It is estimated that the use of PCCPs in a studies Densely Populated Asian Cities may release over 37 billion microbeads per year into the environment via wastewater treatment plants.

Textile materials are one of the primary sources of MPs pollution. The washing process is the most significant pathway through textile products release MP fibers (MPFs), which are made from various textile raw materials (A, PA, PES, Recycled PES, and PP) and are released at different washing stages (pre-washing, soaping/rinsing) (Akyildiz et al., 2024). According to previous studies, household washing has been recorded to emit between 6,000,000 MPFs (Palacios-Mateo et al., 2021) and 18,000,000 MPFs (Galvão et al., 2020) from a 5–6 kg laundry load.

Rainwater runoff serves as a pathway for the entry of MPs (plastics <5 mm) into aquatic ecosystems. The concentrations observed were 0.99 ± 1.10 MPs/L for the 500–1000 μm size range and 0.41 ± 0.30 MPs/L for the 1000–5000 μm size range which recorded the highest percentages of these polymers PE (36.23%), PP (9.66%), and PS (9.66%) (Ochoa et al., 2024). Several factors potentially control microplastic abundance, including land use, watershed characteristics, drainage systems, and precipitation factors such as antecedent dry days, accumulated precipitation, and rainfall intensity (Ross et al., 2023; Ochoa et al., 2024).

Tire and road wear particles have been identified as a major source of MPs incorporated into stormwater, sweep sand, and wash water. Tire and bitumen microplastic particles (TBMP) $\geq$100 μm reached up to 2561 particles/L. Sweep sand and washwater contained high amounts of TBMP $\geq$100 μm, with up to 2170 particles/kg dry weight and 4500 particles/L, respectively (Järlskog et al., 2020). The results suggest that sweepers collect significant amounts of TBMP, indicating that regular sweeping could help prevent the further transport of TBMP to receiving stormwater. In stormwater, the number of particles $\geq$100 μm reached up to 3 particles/L, and those $\geq$20 μm reached up to 5900 particles/L.

2.2.2 Industrial Wastewater

Industrial and residential activities are major contributors to MP pollution. The link between industrialization levels and MP pollution is widely recognized (Townsend et al., 2019). Previous studies indicate that industrial emissions of MPs are generally higher than those from other sources (Wu et al., 2020). But limited studies were conducted in respect to the industrial wastewater.

MPs originating from paints have become a growing concern in both the fisheries and maritime industries. A significant source of these MPs comes from antifouling paints applied to commercial vessels and recreational boats. As these paints degrade, they not only release MPs but also leach harmful substances like heavy metals and biocides, which contribute to pollution (Gaylarde et al., 2021).

In addition to the maritime sector, other common sources of paint-related MPs include road markings and the external surfaces of buildings. Over time, these painted surfaces break down due to weathering, traffic, and environmental conditions, leading to the release of MPs into the surrounding environment.

3 Microplastics in Wastewaters

MPs, defined as plastic particles (fragments, fibers, pellets, foam, and films) smaller than 5 millimeters in size, are vitally distributed throughout environmental systems (Qaiser et al., 2023). Wastewater serves as a critical aggregation point for these particles, collecting MPs from both primary sources (directly manufactured plastic particles) and secondary sources (fragments derived from larger plastic items) (Mathew et al., 2024; Na et al., 2024).

3.1 Detection of MPs

WWTPs represent the largest point source for microplastic accumulation, functioning as convergence points for diverse wastewater streams from residential, commercial, and industrial sources (Ziajahromi et al., 2017; Hajji et al., 2023; Mathew et al., 2024). As the final barrier between anthropogenic waste streams and natural aquatic ecosystems, WWTPs play a crucial role in preventing microplastic pollution (Ziajahromi et al., 2017; Talukdar et al., 2024). These facilities constitute the last line of defense against MP discharge into natural waterways. Consequently, the comprehensive identification and characterization of MPs within WWTP systems is essential for developing effective management strategies for MP pollution in aquatic environments.

3.1.1 Sampling Techniques

Effluent and influent water and sludge samples are commonly collected from different treatment stages (primary, secondary, and tertiary treatment) based on the treatment facility's capacity (Bakaraki Turan et al., 2021; Mathew et al., 2024). Water samples were collected using simple metal meshes or container collection (grab sampling), surface filtration, autosampler collection, separate pumping and filtration combined pumping-filtration approaches (Bakaraki Turan et al., 2021; Jani et al., 2024). The sample volume varies significantly among studies, depending on the analytical method's limit of determination (Bretas et al., 2020). Sludge samples are obtained using grab samplers and undergo pretreatment processes, which vary based on the type of sludge treatment, MPs types, and concentrations (Bakaraki Turan et al., 2021).

3.1.2 Separation and Extraction Techniques

The high organic load present in wastewater and sludge samples poses a significant challenge in the identification and verification of MPs due to strong interferences. However, pretreatment processes, including digestion and separation, must be carefully designed to minimize any potential harm to the MPs in the sample. In particular, the high organic content in sludge samples requires gradual preprocessing before MP identification.

Chemical digestion is the most common preprocessing step for wastewater samples, while sludge samples undergo a series of digestion methods tailored for microplastic analysis. Wet peroxide oxidation (WPO) and the Fenton reaction are widely used digestion processes, with modifications based on sample composition. In these processes, hydrogen peroxide (H_2O_2) serves as the oxidizing agent, accelerating the breakdown of complex organic matter with the aid of catalysts such as iron (Fe) or copper (Cu) at elevated temperatures. This treatment effectively reduces organic interference while preserving MPs.

In addition to oxidative digestion, enzymatic degradation using protease, lipase, and cellulase has been explored. However, conventional oxidative treatment is often preferred, as enzymatic digestion can take up to 13 days (Mintenig et al., 2017). After the digestion process the density separation were compiled to separate the particle from the suspensions (Masura et al., 2015). The chemically inert high density solutions sodium iodide (NaI), sodium chloride (NaCl), sodium polytungstate (SPT) and zinc chloride ($ZnCl_2$) to float the existing MPs in the retain. The selection of a saturated solution is based on the density of the polymer to be extracted (Sadia et al., 2022).

3.2 *Identification of MPs*

The identification of MPs involves assessing both physical characteristics (color, shape, mass) and chemical properties (polymer-type and associated chemicals) (Rochman et al., 2019). The identification methods can be categorized into three main approaches: visual/microscopic methods, spectrometric methods, and thermo-analytical methods.

3.2.1 Visual or Microscopic Observation

The physical characteristics of the MPs were analyzed using visual observations through the microscope. Dried fitter paper with filtrates were observed under microscope by following systematic visual observation technique. When examining a filter under a microscope, it is recommended to move the filter in a zigzag (Z-shaped) pattern rather than scanning randomly (Masura et al., 2015). This movement ensures that every part of the filter is systematically covered and that no MPs are overlooked. The identification of the MP particles from the organic matters are challengeable. Hidalgo-Ruz et al., 2012 suggested a set of rules to proper identification of the MPs. During the observation drag your forceps across the particles to sort out and collect MPs. The hot needle test used as a simple and commonly used verification of the identified MPs. Quality assurance aspects need to be highly concerned to prevent the cross contaminations by cleaning the microscope, forceps with ethanol (Masura et al., 2015).

The primary identification of the MPs based on shape, colour and size were widely analyzed using Stereomicroscope. But Stereomicroscope have limitation with the size range of maps restricted more than 1 mm and nature of the MPs based on the degradation status, and compositions. The advanced microscopic techniques polarized-light optical microscopy (PLOM) and fluorescent microscope used as a kind of solutions for the above-mentioned limitations. A fluorescent microscope is used to identify MPs stained with dyes like Nile red, which emit green fluorescence under blue light (Meyers et al., 2022). Additionally, Polarized Light Optical Microscopy (PLOM) is employed for detecting and characterizing MPs in complex media, particularly for identifying PE, PP, and PET (Sheriff et al., 2024).

Scanning electron microscopy (SEM) is expensive advanced mechanism which required a fully trained personals for the observation and identifications (Kalaronis et al., 2022). The surface morphology of a particle elemental content of MPs, and changes of due to weathering can be identified based on the interaction of the electron beam (Sheriff et al., 2024). Energy dispersive X-ray (EDX) (Zhang et al., 2021) X-ray spectroscopy (FESEM-EDX) (Parashar & Hait, 2022) are the major two techniques used in the SEM. But the encountered human error raged from the 6–8% during the microscopic observations due to that most of the studies suggest to perform verification and characterization through the μ- Fourier transform infrared (FTIR) technique (Ren et al., 2020).

3.2.2 Spectroscopic Methods

Spectroscopic methods such as FTIR and Raman spectroscopy supplied light beam which passes through or reflects off a sample, specific molecular bonds absorb certain wavelengths of the radiation. Each type of chemical bond (e.g., C-H, C=O, C-N) absorbs radiation at a unique frequency, leading to characteristic absorption peaks in a spectrum. These absorption peaks serve as a "fingerprint" of the material, enabling identification based on available reference libraries.

FTIR provides valuable insights into the chemical structure, abundance, and resistance of MPs, contributing to a better understanding of their environmental pathways (Andrea et al., 2016). It considered as a rapid, accurate and immunity to fluorescence interferences, but required a skillful human resource, with resourceful data library making it a reliable tool for MP analysis.

Mode selection is a crucial aspect of FTIR spectroscopy, as it determines the effectiveness of microplastic identification. The most commonly used modes are reflectance, transmission, and attenuated total reflectance (ATR) (Kang et al., 2020). ATR-FTIR has a detection range of >10 μm (Dong et al., 2022), making it particularly suitable for identifying MPs larger than 500 μm (Kang et al., 2020). Micro-Fourier Transform Infrared Spectroscopy (μ-FTIR) is particularly suitable for analyzing small polymer particles and has been reported to detect MPs with diameters of 50 μm (Maurizi et al., 2023), >25 μm (Rathore et al., 2023), 20 μm (Rathore et al., 2023), and even as small as 10 μm (Chen et al., 2020a, b) for directly measuring MPs on filters.

The Raman spectroscopy extends the detection limits with a particle size >1 mm compared to plastics from non-plastics components. But Raman spectroscopy has some limitation to distinguish the plastic particles possible interferences with colour (blue and black) by fluorescence from dyes and pigment (Sheriff et al., 2024).

Laser Direct Infrared (LDIR) spectroscopy is an automated and rapid technique for identifying MPs, providing detailed information on their shape, size, and polymer composition, similar to FTIR spectroscopy (Ghanadi et al., 2024). LDIR can detect MPs as small as 20 µm (Cheng et al., 2022) and offers a fast analysis time, processing particles in as little as 8 s each (Whiting et al., 2022). A key advantage of LDIR is its ability to analyze multiple particles simultaneously, making it ideal for large-scale sample analysis (Ghanadi et al., 2024). Additionally, LDIR efficiently scans samples without analyzing empty spaces or requiring liquid nitrogen. The likelihood of false positives is minimal, as the technique reduces misidentifications of natural particles as MPs (Whiting et al., 2022). However, there remains a potential for misclassification when analyzing weathered plastic particles.

3.2.3 Thermal-Based Analysis for Microplastic Identification

MPs can be identified through various thermal-based analytical techniques that detect their degradation products.Pyrolysis Gas Chromatography-Mass Spectrometry (Py-GC–MS)—This technique enables the identification of MPs' chemical composition and organic additives by analyzing the pyrolysis products of the sample (Fries et al., 2013; Hermabessiere et al., 2018). Thermogravimetry Coupled with Differential Scanning Calorimetry (TGA-DSC)—Used to assess the thermal stability and decomposition patterns of MPs (Szymańska & Obolewski, 2020). Thermal Extraction Desorption Gas Chromatography-Mass Spectrometry (TED-GC–MS)—A method that detects and quantifies MPs based on their gaseous decomposition products (Dümichen et al., 2017).

These thermal analytical methods allow both quantitative and qualitative MP analysis by generating gaseous compounds that are subsequently detected using gas chromatography-mass spectrometry (GC–MS) (Szymańska & Obolewski, 2020). Among them, Py-GC–MS is particularly valuable for identifying MPs' polymer composition and additives. They are destructive, meaning samples cannot be recovered after analysis. They require complex instrumentation and operation, making them less accessible for routine monitoring. They struggle with the analysis of high-molecular-weight compounds, which may affect detection efficiency (Dümichen et al., 2017).

3.3 Microplastic Characterization

Characterizing MPs is essential for understanding their sources, composition, environmental behavior, and potential impacts. The characterization process typically involves morphological, physical, and chemical analyses using the techniques described above Table 2. Color, size, and shape are the primary morphological

Table 2 Summary of the various sampling, pretreatment, extraction, and identification methods used in selected studies across different regions of the world

Region	Country	Sludge/ Water	Sampling	Pretreatment	Extraction	Identification	Reference
North America	Mexico	Wastewater	Mesh filtration	Wet peroxide oxidation	Density separation ($ZnCl_2$)	Optical microscope FTIR (ATR)	Flores-Munguía et al. (2023)
	Canada	Waste water	Container collection	Fenton's reagent	Filtered through a 0.7-μm filter paper (Whatman)	SEM Raman FTIR (ATR)	Prajapati et al. (2021)
Australia/ Oceania	Australia	Waste water and sludge	Mesh screen	Wet peroxide oxidation	Passed through a stack of filters (500,190,100 and 25 mm)	(mFTIR) and attenuated total reflection (ATR) FTIR	Ziajahromi et al. (2021)
South America	Colombia	Waste water and sludge	800 mL wastewater samples were collected at a depth of 5 cm using glass containers	Wet peroxide oxidation	(Whatman™ glass fiber filters, 47 mm, 1.2 μm pore size).	Stereomicroscope FTIR (ATR)	Arregocés-Garcés et al. (2024)
Europe	Finland	Wastewater and sludge	Grab sampling Pumping mesh filtration	Enzymatic purification	Density separation ($ZnCl_2$)	Stereomicroscope Raman	Salmi et al. (2021)
	Greece	Waste water	Pumping mesh filtration	Wet peroxide oxidation	N/A	Stereomicroscope μ-FTIR	Zeri et al. (2021)
	Italy	Wastewater	Autosampler pumping mesh filtration	Wet peroxide oxidation	Visual sorting	Stereomicroscope μ-FTIR and ATR Raman	Galafassi et al. (2022)
Asia	Thailand	Wastewater and sludge	Grab sampling	Fenton's reagent	Density separation (Nal)	Stereomicroscope FTIR (ATR)	Maw et al. (2024)
	South Korea	Wastewater	Grab sampling mesh filtration	Fenton's reagent	Density separation (Li2O13W4–24)	Microscope FTIR	Kim et al. (2022)

and physical characteristics assessed in MP studies, while polymer composition is analyzed as a chemical characteristic to determine the type of plastic present. Table to summarize the above-mentioned morphological, physical, and chemical characters recorded in the WWTPs as the kind end point transportation of MPs in environment.

3.3.1 Colour

Plastic particles of various colors have been detected in WWTPs (Table 3). Some studies have quantified the abundance of these colored particles, while others have only recorded their presence. The presence of colored plastic particles in wastewater can be attributed to the release of large amounts from different sources majorly textile fibers and personal care and cosmetic products through domestic and industrial discharges. Notably, coloured plastic particles are more likely to be accidentally ingested by visual predators due to their resemblance to prey, potentially leading to direct harm such as suffocation or malnutrition, as well as indirect toxic effects like disruptions in organ homeostasis (Ali et al., 2021). Since, Overall, transparent plastic particles accounted for the greatest proportion (40–75%) of the observed MPs in the influent of WWTPs (Ali et al., 2021). Therefore, the future research on transparent MPs would be required to understand their fate and behavior in WWTPs.

3.3.2 Size

Researchers have employed various techniques to estimate the size distribution and abundances, including the use of different sieve sizes and advanced microscopic imaging methods (Tables 1 and 2). The sizes of the plastic particles can be categorized into different size ranges (i.e., >1000 μm, 500–1000 μm, 250–500 μm, 100–250 μm, 15–100 μm, <1 μm, and 1–15 μm), as a single size number may be insufficient due to the asymmetrical shapes of MPs (Bakaraki Turan et al., 2021; Ziajahromi et al., 2021). From the size estimation, it is evident that all detected plastic particles fall within the classification of MPs (<5 mm or < 5000 μm or > 0.1 μm).

3.3.3 Shape

The shape of MPs can be easily recognized due to their distinct appearances (Fiber, fragments, films, microbeads, foam and others). Table 3 illustrates that various shapes of MPs have been observed in different WWTPs. The large presence of Fibers in wastewater indicates the release of MPs from domestic washing and textile industries (Lant et al., 2020). The appearance of film- and foam-shaped particles in sewage is likely due to the erosion of plastic bags and packaging products, while pellets and microbeads primarily originate from personal care and cosmetic

Table 3 Concentration of MPs in influent and effluent of wastewater in different WWTPs (meanings of some abbreviations; High-Density Polyethylene (HDPE), Medium-Density Polyethylene (MDPE), polyphenylene ether (PPE), polymethyl methacrylate (PMMA), Acrylonitrile Butadiene Styrene (ABS), Polysulfone (PSU), polyoxymethylene (POM), Ethylene Vinyl Acetate (EVA), Polydimethylsiloxane (PDMS), Polytetrafluoroethylene (PTFE), polycaprolactone (PCL), polyurethane (PUR), Polyether Urethane (PEUR), Styrene-Butadiene Rubber (SBR), Polyphenylsulfone (PPSU / PPSF), Perfluoroalkoxy (PFA), Polyacrylonitrile (PAN), Bisphenol A-polycarbonate (BPA-PC), Fluorinated ethylene propylene (FEP), Polytrimethylene terephthalate (PTT), polylactic acid (PLA)).

Region	Country	Capacity of water treatment	Identified polymer type	Shape	Color	Size	MP's concentration (influent) (n/l)	MP's concentration (Effluent) (n/l)	Removal percentage (%)	Reference
North America	Mexico	Primary secondary (Municipal wastewater)	PP, PE, PET, PVC	Fragment, filament, Granule, foam, beads, films	Blue, black, red, purple, transparent, green	38–150 μm	20.7–218.8 45.50–194.7	1.0–24.3 0.8–34	82.5–97.5	Flores-Munguía et al. (2023)
	Canada	Territory (Municipal wastewater)	PE, other HDPE, MDPE, LDPE, ABS, Nylon 6, PET, PMMA, PPE, PS, and PSU	Particles and fiber	N/A	> 500 μm	N/A	1.76 ± 0.03	N/A	Prajapati et al. (2021)
Australia/ Oceania	Australia	Advanced secondary treatment (Municipal wastewater)	PET, PE, PP, nylon	Fiber, fragment, granule	N/A	>25 mm	92.0	0.80	69–79% of microplastic	Ziajahromi et al. (2021)
South America,	Colombia	Secondary treatment (Municipal wastewater)	N/A	Fragment, fiber, forms	White	5 mm and 0.2 mm	15 items/l	5 items/l	67%	Arregocés-Garcés et al. (2024)
Europe	Finland				N/A	N/A	61 ± 26	0.8 ± 0.4	99%	

(continued)

Table 3 (continued)

Region	Country	Capacity of water treatment	Identified polymer type	Shape	Color	Size	MP's concentration (influent) (n/l)	MP's concentration (Effluent) (n/l)	Removal percentage (%)	Reference
		Tertiary treatment (Municipal wastewater)	PP, PE, PET, PS, POM, EVA, PDMS, PTFE, PMMA.	Fiber, fragment						Salmi et al. (2021)
	Greece	Secondary tertiary (pilot- scale MBR) (Municipal wastewater)	PP, PE, PET, acrylic, PA-6.	Films, filaments, fragments, foams	N/A	<3.0 mm	38 41	0.4 0.002	>99%	Zeri et al. (2021)
	Italy	Primary and secondary Primary, secondary, and tertiary (Municipal wastewater)	PE, PP, PET, PCL, PUR, PEUR, SBR, ABS, Phenoxy resin, and other	Fiber, fragments	N/A	Smaller than 500 μm	18.1–9.7	0.56–0.06	94.9–99.7	Galafassi et al. (2022)
Asia	Thailand	Primary and secondary (Municipal wastewater)	PE, PP, PET, PVC, PA, PS, alkyd, and other	Fiber, fragments, film, form, fiber bundle, pellet, sphare	n/a	Large (1000 μ m–5000 μ m) Small (50 μ m–1000 m)	2.9–51.9	1.3–5.9	50 to 96.8% in centralized WWTPs 14.2–53.6% in decentralized WWTP	Maw et al. (2024)
	South Korea	Primary, secondary,			N/A	20–200 μm			99.8	Kim et al. (2022)

		and tertiary (Municipal and industrial water)	PP, PE, acrylic, PES/PET, and other	Fiber, ragment, and pellet			114 ± 17–216 ± 65	0.26 ± 0.29–0.48 ± 0.11		
Africa	Mauritius	Wastewater treatment and sludge Municipal water)	Cotton-PA fibers, PE spheres, and EVA foams	Fibers, fragments, flakes, films, spheres, and foams	N/A	Less than 1 mm in size,	N/A	EFF: 117.8–355.6 particles/L^{-1} (Avg. 276.3 ± 137.3 particles/L^{-1})	N/A	Ragoobur et al. (2021)
Middle East/ North Africa	Iran	Primary, secondary (Municipal water)	PPSF or PPSU, PET, ABS, cellulose, PSU, poly (m-phenylene terephthalamide), PFA, PAN, BPA-PC, PMMA, nylon 6,12; FEP, PLA, and PTT	Fiber, fragment, and film	Clear, green, blue and red	25–840 μm	180 ± 4.3	EFF: 5.3 ± 0.31	99.06%	Takdastan et al. (2021)

products, such as masks, soaps, and toothpaste. Some researchers also categorize MPs into spherical and irregular shapes. Irregular particles may result from eroded plastic products or particles originating from personal care items like toothpaste (Ali et al., 2021).

3.3.4 Chemical Characters- Polymer Type

The most widely detected polymers in the influent include PS, PES, PE, PP, PET, Other polymers, listed in Table 3, which summarized detected polymers in WWTPs. These findings suggest that a large portion of MPs originates from daily activities, though the precise pathways by which these particles enter WWTPs are not fully understood. Therefore, further research should focus on comprehensively exploring these pathways to better control MP pollution at the source.

3.4 Mismanagement and Release to the Environment

WWTPs serve as major hotspots for MPs contamination. Although conventional WWTPs are not specifically designed for MPs removal (Carnevale Miino et al., 2024; Dayal et al., 2024). Conventional WWTPs, equipped with primary and secondary treatment processes, can remove up to 99% of MPs from wastewater (Azizi et al., 2022). However, despite their high removal efficiency, WWTPs remain a significant source of MPs due to the large volumes of effluent discharged. As a result, WWTPs function both as a barrier preventing MPs from entering the environment and as a pathway through which MPs are released, particularly into aquatic ecosystems (Talvitie et al., 2017; Azizi et al., 2022). According to the comprehensive meta-analysis was conducted to assess relevant 77 studies on microplastic abundance in wastewater treatment processes. The analysis revealed that the average MP concentrations ranged from 1 to 31,400 items/L in influent, 0.2 to 12,580 items per liter after primary treatment, 0.002 to 7863 items/L after secondary treatment, and 0.003 to 447 items per liter following tertiary treatment (Azizi et al., 2022). Further, recent meta study revealed that considering 38 WWTPs representing 11 countries still discharged 5.00×10^{5}–1.39×10^{10} microplastic particles discharged into the aquatic environment each day after the treatment process. Table 3 summarized MPs concentrations in influent and effluent water. However, the remaining MPs in treated effluent has recorded high discharge volumes of WWTPs and residual MP concentrations can be than those found in natural receiving waters. Another study by Murphy et al., 2016 examined a WWTP in the UK and estimated that despite high MP removal efficiency, the plant discharged approximately 65 million microplastic particles/day into the receiving river. This highlights that even a small percentage of MPs in effluent can contribute significantly to environmental contamination due to the vast amount of wastewater processed daily.

Wastewater is confined to a pool where MPs particles accumulate from both primary and secondary sources. Domestic and industrial wastewater contain a wide range of MPs, as summarized as the MP concentration in influents in Table 3. In most countries, domestic wastewater is collected and treated through municipal wastewater treatment systems. However, in some regions, direct discharge of untreated or improperly treated domestic wastewater remains a major source of non-point pollution which release to the aquatic environment (Kye et al., 2023).

The different treatment processes recorded wide range of removal efficiencies based on the concentration of the influents and effluents associated with capacity of the treatment process (Table 3). The capacity of the treatment depends on the available steps such as preliminary, primary, secondary and tertiary) which were applied in WWTPs. According to the above table the removal efficiency was calculated based on available treatment steps in the WWTP i.e., primary + secondary; secondary + tertiary, and primary + secondary + tertiary.

3.4.1 Preliminary Treatment

The removal of larger solids particles and grits to ensure the protection of the equipment in the upcoming treatment process were completed based on physical process such as filtration, screening and aeration. MPs larger than the screen openings, particularly in fine (6–150 mm) and micro (< 6 mm) screens, are mostly removed during screening (Reddy & Nair, 2022). However, some fibers still pass through, aligning perpendicularly to the screen mesh (Ziajahromi et al., 2017). Further, aerated grit chambers remove over 60% of microfibers primarily through density-based separation (Yang et al., 2019), while rotary grit chambers may increase MPs fragmentation due to mechanical forces (Asadi et al., 2025). According to estimates based on systematic literature reviews preliminary treatment removed approximately 30–60% of MPs (Ali et al., 2021) 40 to>90% (Reddy & Nair, 2022) of removed through the preliminary treatment process. In this treatment step accumulated 40% to 70% of MPs in the sludge (Ali et al., 2021).

3.4.2 Primary Treatment

The primary treatment operations (filtration, screening, skimming and settling) have eradicated significant number of MPs in the WWTPS. These physical treatment operations were implied in both preliminary and primary treatment as a combined operation. High-density MPs settle quickly through gravity, while low-density MPs float on the surface and are removed through skimming before settling (Sun et al., 2019). The pre-treatment had the largest impact on MPs size distribution, as it could effectively remove MPs of larger size. According to Ali et al. (2021) 45–75% of MPs were removed from wastewater and 65–75% were accumulated in the sludge.

3.4.3 Secondary Treatment

A significant amount of MPs can be removed during secondary treatment (Wu et al., 2021). This process is primarily biological, aiming to degrade contaminants in influent water through microbial activity. However, since MPs are resistant to biological degradation, their breakdown and removal in this stage remain challenging. The extracellular secretions of microorganisms can aid in the retention of some MPs, while sludge within the system functions as a natural filter, contributing to their removal (Moyal et al., 2023). Studies indicate that secondary treatment can remove approximately 2% to 55% of MPs (Yang et al., 2019; Parashar & Hait, 2023). More specifically, activated sludge—the most widely used secondary treatment method—has been reported to eliminate around 3.6% to 42.9% of MPs from wastewater (Steve et al., 2016; Mason et al., 2016; Mirka et al., 2018).

3.4.4 Tertiary Treatment

Tertiary treatment serves as the final purification step, enhancing wastewater quality before reuse, recycling, or discharge into the environment. This stage targets the removal of residual inorganic compounds, nitrogen, phosphorus, and additional solid matter following secondary treatment. Common tertiary treatment methods include coagulation (for solid matter and phosphorus removal), membrane filtration (for solid matter removal), and denitrification-nitrification (for nitrogen removal). Ma et al. (2024) reported that coagulation alone could remove over 35% of MPs. While the overall MP removal efficiency in tertiary treatment can reach up to 81.6% (Hidayaturrahman & Lee, 2019).

3.4.5 Sludge Treatment

As a result, a significant number of MPs remains in the WWTPs effluent, leading to their release into the environment. While a large portion of MPs is retained in sewage sludge, their disposal through land application can further contribute to environmental contamination. An estimated maximum application rate of 4.8 g of MPs/m^2/year, or 11,489 MP particles/m^2/year, indicates that the land application of sewage sludge could be one of the largest global sources of MPs pollution (Lofty et al., 2022). Given this potential environmental impact, the practice of recycling raw sewage sludge onto agricultural soils should be carefully reassessed to prevent excessive MP contamination.

All the portion (65%–90%) of removed MP incorporated with influent accumulating in sewage sludge. According to Carnevale Miino et al. (2024), when treated through a structured sludge management system, as assumed, approximately 25–50% of MPs are recirculated from the sludge line back into the wastewater stream as supernatant. Additionally, around 20–40% of MPs—primarily fragments and fibers—remain in the dewatered sludge. A study estimated that approximately 774.25 million microplastic particles were deposited daily in 81.5 tons of sludge produced by a selected WWTP in Turkey (Üstün et al., 2022). The previous meta study estimated abundance of MPs in sewage sludge has been reported to range from 4.40×10^3 to 2.40×10^5 particles per kilogram (Liu et al., 2021).

MPs are highly concentrated in sewage sludge (Sun et al., 2019), raising concerns about their potential impact when sludge is reused as a soil improver in agriculture. The agricultural application of sludge serves as a pathway for MPs dispersion into the environment, potentially contaminating soil, water, and crops (Carnevale Miino et al., 2024). This highlights the need for careful management and further research on the long-term implications of sludge reuse. The report "Progress on wastewater treatment—Global status and acceleration needs for SDG indicator 6.3.1" published by UN habitat and WHO highlighted in 2020, approximately 56% of household wastewater was treated globally, based on data from 128 countries representing 80% of the global population. However, significant regional disparities were observed in the treatment of household wastewater. In contrast, the proportion of industrial wastewater treated was much lower, at 30%, with data for industrial wastewater available from only 14 countries, representing just 4% of the global population under data imitated conditions (UN Habitat and WHO, 2021). This figure highlights the amount of untreated domestic and industrial wastewater being released into the natural environment. Furthermore, it underscores the significant quantity of MPs present in the influent water, which are directly discharged into the environment, contributing to pollution and environmental contamination.

3.5 Distribution and Fate of MPs

Understanding the sources, abundance, and composition of MPs in the environment is a significant challenge, as hundreds of millions of tonnes of plastic are produced annually for societal use, with a portion inevitably entering the environment (Horton et al., 2017). MPs from both treated and untreated effluents accumulate in various critical destinations, including inland waters (both natural and artificial water bodies), soil, and animal bodies, before ultimately reaching the ocean.

MPs have been detected in the freshwaters of Europe, North America, and Asia, with initial baseline studies revealing that freshwater fauna across various feeding guilds ingest these particles (Eerkes-Medrano et al., 2015). Although data related to abundance in freshwater systems are limited, reported MPs concentrations in freshwater ecosystems vary widely, ranging from 0.00031 particles per cubic meter (p/m^3) to 10,200 p/m^3 (Eerkes-Medrano et al., 2015) comparable to contamination levels in marine environments (Peng et al., 2017). Accumulated MPs in freshwater systems ultimately enter the ocean, contributing to marine pollution. Additionally, a considerable number of MPs accumulate at different trophic levels within food webs.

According to the Kılıç et al., 2022, high occurrence of MPs was detected in all examined fresh water fish species: carp and mullet range from 5.1 ± 2 MPs/fish and

4.4 $\pm$ 2 MPs/fish in Gastro intestinal track and gills respectively. MPs are commonly ingested or passively taken up by various freshwater fish species. Ingestion patterns are often influenced by individual traits, such as body size and trophic level, as well as environmental factors, including local urbanization and habitat characteristics (Parker et al., 2021).

Soil also serves as a major sink for MPs. Previous studies recorded the presence of MPs in high concentration of the soil range from 13,000 items/kg of dry soil and 4.5 mg/kg of dry soil in agricultural and horticultural sites which exposed to sewage sludge and mulching film applications (Büks & Kaupenjohann, 2020). MPs significantly alter soil physicochemical properties, including porosity, enzymatic activity, microbial activity, plant growth, and crop yield and play a crucial role in transporting toxic chemicals, such as plasticizers (Sajjad et al., 2022). Again, the pool of MPs in soil provides chance to enter different trophic levels within the food web.

Atmospheric interactions facilitate the dispersion of MPs in soil and dust due to their lightweight nature (Torres-Agullo et al., 2021). They are released into the air from various sources, including material abrasion, synthetic textiles, and the re-suspension of MPs from surfaces (Dris et al., 2016). Atmospheric fallout of MPs was recorded at rates ranging from 2 to 355 particles/m^2/day (Dris et al., 2016). The MPs concentrations of outdoor air varied between less than 1 and more than 1000 MPs per cubic meter of air, and between 0.5 and 1357 MPs/m2/day for deposition (O'Brien et al., 2023).

In other hand, MPs can travel through various layers of soil and ultimately end up in groundwater water sources (Viaroli et al., 2022). Surface runoff, animal disturbance, and water infiltration are the major migratory pathways of MPs to ground water (Dey et al., 2023). A previous systematic review, estimated that globally, the concentration of MPs varied from 0 to 6832 particles/L (Lee et al., 2024). Ultimately, all MPs from various sources; soil, air, freshwater, and the biosphere eventually make their way into the ocean. The marine environment is also impacted by MPs through various pathways. Some MPs are transported to the sea via runoff and sewage systems. Additionally, ports and landfills act as entry points for MPs onto shorelines. Adverse weather conditions, such as strong winds and currents, facilitate the movement of these MPs, which then degrade and break down into MPs.

According to a range of previous studies, the presence of MPs has been proven across all biotic and abiotic components of complex and integrated marine ecosystems. These ecosystems include integrated coastal environments such as mangroves (Talukdar et al., 2023), salt marshes (Lloret et al., 2021), seagrasses (Li et al., 2023, b), coral reefs (Hansani et al., 2023), sandy beaches, sand dunes (Costello & Ebert, 2020), and the deep ocean (Van Cauwenberghe et al., 2013). MPs have been recorded in water, sediments, and within living communities throughout these diverse marine habitats. MPs are unevenly distributed, with the highest concentrations found in oligotrophic subtropical gyres, as well as in biologically productive semi-enclosed seas and coastal waters (Auta et al., 2017; Setälä et al., 2018). MPs can infiltrate marine food webs through four primary routes: ingestion, inhalation, entanglement, and trophic transfer (Setälä et al., 2018).

MPs can be ingested directly through accidental consumption, often due to indiscriminate feeding strategies, such as filter-feeding Oysters (Sussarellu et al., 2016) (2024 Alternatively, MPs can enter organisms indirectly through trophic transfer, where predators consume prey items that are contaminated with MPs such as Zooplanktons (Cole et al., 2013). Previous studies and reports documented the wildlife becoming entangled in discarded bags, fishing nets, and other plastic waste (Mghili et al., 2023; Apete et al., 2024). Furthermore, Ma et al., 2024 documented the process how this entanglement occurs on a micro level, focusing on the smaller particles and their effects in Daphnia magna.

While studies on the transportation of MPs through marine food webs have been extensively conducted, much less research has been focused on this process in freshwater environments. However, the same processes that occur in marine ecosystems such as ingestion, trophic transfer, and entanglement can also happen in freshwater food webs. Aquatic and soil organisms, as well as humans, represent all biological communities that are exposed to ecotoxicological risks from MPs (Wu et al., 2021). This leads to bioaccumulation and biomagnification, where the concentration of MPs increases as they move up the food chain, affecting both ecosystem health and human well-being. The irreversible impacts of MPs on the entire globe are alarming, highlighting the urgent need for further research. Investigating the distribution and fate of MPs in terrestrial and freshwater environments is essential to fully understand their impacts across ecosystems and to develop strategies for mitigating their widespread effects.

4 Toxicity of MPs in Wastewater

Microplastics bearing toxicity risks to organisms and ecosystems due to their physical properties, chemical composition, and ability to adsorb pollutants. Their toxicity can be directly or indirectly effects health of microorganism to human, health of plants, and ecological interactions.

4.1 Entry Pathways of Toxic Chemicals/Pollutants

Toxic chemicals and pollutants enter wastewater from different categories of wastewater. Some categories of wastewater undergo treatment before being discharged into the environment, while others do not. Pollutants from untreated wastewater are released into aquatic environments, though their discharge from treated wastewater is controlled. However, the pollutant removal efficiency in wastewater treatment plants is not at an optimal level worldwide for all pollutants due to technological and financial constraints (e.g., MPs), leading to their release into aquatic environments.

4.1.1 Stormwater Runoff Wastewater

Water that originates from heavy rainfall, storms, or floods and flows on the surface of the Earth is classified as stormwater runoff wastewater. This water travels across various surfaces, such as industrial and construction sites and agricultural farms, and ultimately drains into the aquatic environment, either directly or through channeled drains, without any treatment. Plastics, pesticides, herbicides, heavy metals, oil, and grease, along with other toxic contaminants, are included in this category of wastewater (Ahmed et al., 2021).

4.1.2 Domestic Wastewater

Wastewater produced from laundry, sanitary, and cooking discharges in household activities belongs to this category (Ahmed et al., 2021). Detergents, pharmaceutical active compounds such as antibiotics, tranquilizers, diuretics, and psychiatric drugs (emerging contaminants), personal care products, oil, biowaste, nutrients and plastics are major contaminants in domestic wastewater (Huang et al., 2020). These pollutants are often released into the aquatic environment without treatment.

4.1.3 Agricultural Wastewater

Runoff from agricultural and animal production farms belongs to this category, releasing agrochemicals (e.g., fertilizers, pesticides, herbicides), crop residues, plastics, and animal and crop waste into the environment (Ahmed et al., 2021). Agricultural wastewater from large farms generally undergoes treatment. However, wastewater from small farm sites often does not, likely releasing fertilizers, herbicides, and pesticides into the aquatic environment.

4.1.4 Industrial Wastewater

Wastewater originating from industrial processes (e.g., water released during manufacturing and cleaning) containing dissolved and suspended materials is defined as industrial wastewater. Heavy metals, phenol and its compounds (such as petroleum hydrocarbons, sulfides, aniline, naphthalenic acid, organochlorines, olefins, nitrobenzene, alkanes, and chloroalkanes), nutrients, and plastics are released into the environment through industrial wastewater (Ahmed et al., 2021; Shah, 2022). Industrial wastewater is released into the environment through the wastewater treatment process; therefore, the magnitude of pollutant and contaminant release into the aquatic environment is controlled according to wastewater treatment standards. Moreover, the pollutant entry pathways from different wastewater categories can be further subdivided into point and non-point source pollution.

4.1.5 Point Source Pollution

Pollutants discharge from a single source, such as a domestic sewage outfall or effluent from an oil refinery, belongs to point source pollution. The annual discharge of pollutants from direct sources is approximately 300–400 metric tons (Elgarahy et al., 2021).

4.1.6 Non-point Source Pollution

The combination of multiple wastewater sources releasing into a particular environment (e.g., domestic, agricultural, and industrial wastewater entering a lagoon) is referred to as non-point source pollution (Elgarahy et al., 2021).

4.2 Pollutant Composition in Wastewater

Wastewater from different sources introduce various toxic chemicals and pollutants, including MPs, into aquatic ecosystems.

4.2.1 Trace Elements (TEs)/Heavy Metals

TEs are usually present in low concentrations in the environment and animal tissues. TEs can be categorized as essential and non-essential (Idrissi et al., 2023). Essential TEs, such as Mn, Cu, Zn, Ni, and Cr, play an important role in metabolic activities in biota at small concentrations (Table 4), but they can have toxic impacts when they exceed the prescribed limits (Abba et al., 2024; Idrissi et al., 2023).

Non-essential TEs, such as Pb, Cd, As, and Hg (heavy metals), are not involved in any metabolic activities in organisms and can create toxic impacts on the environment and biota even at very low concentrations:

- **Pb**: Permanent impacts on brain development in children, anemia, hypertension, renal impairment, immunotoxicity, and toxicity to the reproductive organs (WHO, 2024).
- **Cd**: Chronic bronchitis, impaired heart function, and cognitive and behavioral deterioration (Charkiewicz et al., 2023).
- **As**: Cancers, diabetes, and pulmonary and cardiovascular diseases (WHO, 2022).
- **Hg**: Neurological and behavioral disorders (Guzzi et al., 2020).

Industrial activities are likely the major source of toxic heavy metal pollution (Ibrahim & Fakhre, 2018). Unfortunately, the conventional wastewater treatment process is not efficient in completely removing heavy metals (Elgarahy et al., 2021).

Table 4 The roles and toxic effects of TEs on organisms

Essential TE	Role in organisms	Toxic effects
Mn	Coenzyme for biological processes, bone formation, ammonia clearance and neurotoxicity (Erikson & Aschner, 2019)	Neurological impairment (Erikson & Aschner, 2019), cardiotoxicity, hepatotoxicity, and increased mortality in infants (O'Neal & Zheng, 2015)
Cu	an essential cofactor for a wide variety of enzymes (cell growth, differentiation and survival) in organisms from bacteria to mammals, metabolism (Samanovic et al., 2012)	Oxidative stress, genotoxicity, cellular destructions (Oe et al., 2016)
Zn	Stabilization of DNA, gene expression, immune functions (Frassinetti et al., 2006)	Nausea, vomiting, and fatigue (Fosmire, 1990)
Ni	Involved in lipid metabolism, increase hormone activity, seeds germination (Genchi et al., 2020)	Allergy, cardiovascular and kidney diseases, lung and nasal cancer germination (Genchi et al., 2020)
Cr	Carbohydrate and lipid metabolism, insulin-receptor activation, control cholesterol and triglycerides in blood (Terpiłowska & Zaporowska, 2004)	DNA damage, Allergic reactions, immune system disorders (Shin et al., 2023).

A significant number of toxic metals, such as Pb^{2+}, Cr^{6+}, Cd^{2+}, and As^{3+}, is released into the aquatic environment without proper treatment (Dobrowolski et al., 2019).

4.2.2 Persistent Organic Pollutants (POPs)

POPs are hazardous chemicals and priority pollutants that persist in the environment in an intact form for a long period. They are capable of long-range transport, affecting the environment and biota far from their origin. Pesticides (e.g., toxaphene, DDT), industrial chemicals (e.g., polychlorinated biphenyls (PCBs), polybrominated diphenyl ethers (PBDEs), tributyl tin (TBT), industrial by-products (e.g., dioxins, furans, polyaromatic hydrocarbons (PAHs), and crude oil production (PAHs) are major sources of POPs (Alharbi et al., 2018; UNEP, 2002).

POPs have a high potential for bioaccumulation along the food chain, causing sub-lethal and lethal impacts on organisms, including death, cancer, endocrine disruption, hypersensitivity, damage to the nervous system, and adverse effects on the developmental, behavioral, and immune systems of animals, including humans (US EPA, 2024). Therefore, the production, use, and trade of POPs are controlled under the Stockholm Convention (Alharbi et al., 2018; US EPA, 2024). However, POPs are also unintentionally produced as by-products of industrial processes and combustion, such as medical and municipal waste incineration (US EPA, 2024).

4.2.3 Organometals

A compound containing a metal and a carbon in an organic group, bonded through single σ-bonds or multiple π-bonds, is called an organometal (Zheng & Greedan, 2003). Organometallic compounds have been detected in paper mill wastewater, including decane, 2,9-dimethyl eicosane, 2-bromotetradecane, 1,2-dihydro-1,4-diphenylphthalazine, benzene acetic acid, 4-[(trimethylsilyl)oxy]-, trimethylsilyl ester, and docosane (Singh et al., 2021) as well as in sewage sludge, such as organolead compounds (Lu et al., 2012), which are highly toxic to aquatic organisms.

4.2.4 Pharmaceutically Active Compounds (PhAC)

The use of pharmaceutical products (antibiotics, anti-inflammatory drugs, lipid regulators, hormones) in health treatments has increased to improve quality of life and lifespan in animals, including humans. Pharmaceutical active compounds (PhACs) are released into aquatic systems through human excretion (sewage), improper disposal, industrial effluents, and landfill leachate (Majumder et al., 2019). The complete removal of PhACs in conventional wastewater treatment processes is inefficient, as most of them are soluble, highly polar, and resistant to hydrolysis or biodegradation under normal conditions (Gupta et al., 2024). The European Union has reported the presence of PhACs in its waters, including the synthetic hormone 17-α-ethinylestradiol, antibiotics such as azithromycin and erythromycin, and painkillers like diclofenac (Elgarahy et al., 2021). Prolonged exposure to PhACs can adversely impact non-target species; for example, diclofenac can cause DNA damage and lipid inhibition in mussels (*Mytilus* sp.) at environmentally relevant concentrations (Schmidt et al., 2011). Since PhACs are designed to be active in very low dosages, they can severely affect the metabolism of microorganisms in aquatic environments (Majumder et al., 2019).

4.2.5 Dyes

Dyes are extensively used in various industries (paper, textile, leather) due to their cost-effectiveness and resistance to environmental conditions. Dyes can be categorized into two types: natural and synthetic. Natural dyes are extracted from natural materials (minerals, animals, plants) and are cost-effective and environmentally friendly. Synthetic dyes are synthesized from organic and inorganic compounds, as the production of dyes from natural materials does not meet industrial demand. However, synthetic dyes show resistance to oxidation agents due to their complex chemical structure, which creates a high chemical oxygen demand in wastewater (Elgarahy et al., 2021).

4.2.6 Nutrients

Nitrogen (N), phosphorus (P), and potassium (K) fertilizers account for 13% of the global agricultural fertilizer demand (UNEP, 2025). The annual average amounts of N, P, and K in global wastewater production are 16.6, 3, and 6.3 million metric tons, respectively (Malik et al., 2017; Qadir et al., 2020). An excessive amount of nutrients in aquatic environments leads to eutrophication, which reduces dissolved oxygen in the water, creating an anoxic environment that directly affects the health of aquatic plants and animals.

4.2.7 MPs

MPs are present in wastewater and are not completely removed, even during the wastewater treatment process. A significant amount is released back into the environment with treated water, identifying wastewater treatment plants as a source of MPs (Hamidian et al., 2021). MPs released from wastewater into aquatic environments can cause acute and chronic toxic effects on aquatic organisms (Bhuyan, 2022; Félix et al., 2023; Subaramaniyam et al., 2023; Zhang et al., 2019). Additionally, MPs adsorb different pollutants onto their surface, serving as vectors for the transfer of pollutants into organisms' bodies. Therefore, interactions between MPs and various toxic contaminants (adsorbents) in wastewater can amplify their toxic impacts on aquatic and terrestrial ecosystems (Hou et al., 2024; Li et al., 2023, b; O'Donovan et al., 2018; Santos et al., 2022), as MPs with adsorbents are likely not efficiently removed from wastewater. Consequently, MPs can be defined as an emerging pollutant that is directly released into aquatic environments through untreated wastewater (e.g., domestic wastewater) and even from treated wastewater.

4.3 Types of Pollutants and Toxic Components Associated with MPs

The impact of pollutant toxicity on biota depends on their ability to dissolve in or bind with water or lipids. Pollutants that interact with water molecules through hydrogen bonding to form water soluble compounds are called hydrophilic (water loving) pollutants (Wang & Wang, 2018). These pollutants are more easily degradable in the environment, preventing their accumulation along food chains (Kozioł et al., 2017). Nutrients and most dyes used in industrial applications, except for disperse and vat dyes, are hydrophilic (Elgarahy et al., 2021; Halling-Sørensen et al., 1998).

The water-soluble 'Direct dye' category (produced for direct application to substrates such as cotton and blended fibers) and the 'Reactive dye' category (in which only the reactive group of the dye interacts with the fiber polymer) have toxic effects.

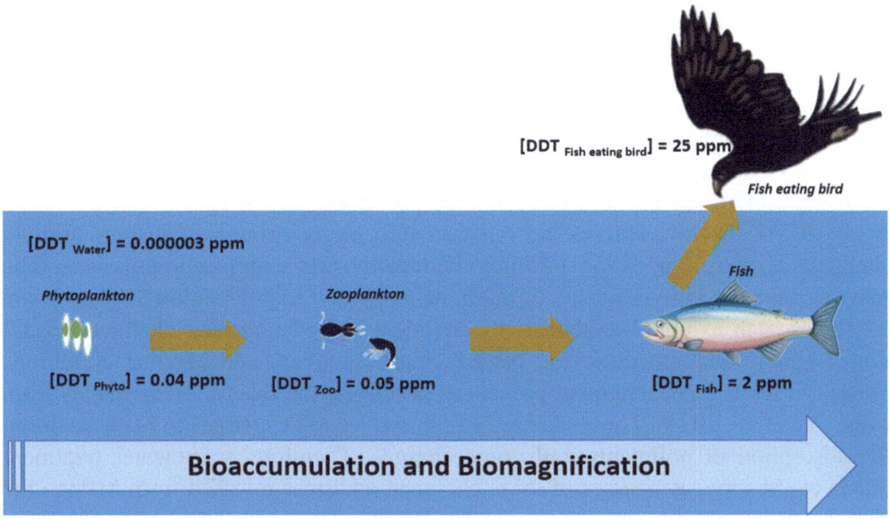

Fig. 2 Bioaccumulation and biomagnification of DDT (a lipophilic pollutant) along the aquatic food chain, showing DDT concentrations in water (0.000003 ppm), phytoplankton (0.04 ppm), zooplankton (0.05 ppm), fish (2 ppm), and fish-eating bird (25 ppm) (Lushchak et al., 2018)

The toxic effects of direct dyes are associated with bladder cancer, while reactive dyes are linked to respiratory issues in organisms (Elgarahy et al., 2021). Pollutants that have a high affinity for lipids and accumulate in the fatty tissues of organisms, such as POPs and organometals, are called lipophilic pollutants (lipid loving) (Gräslund & Bengtsson, 2001). As lipophilic pollutants are stored and persistent in the bodies of organisms, they undergo trophic transfer along the food chain, facilitating bioaccumulation (Fig. 1). Therefore, higher trophic levels in the food chain are more vulnerable to their toxic effects (Kozioł et al., 2017). DDT is one of the POPs, and its commercial use and production have been banned. The bioaccumulation and biomagnification of DDT along the aquatic food chain are illustrated in Fig. 2 (Lushchak et al., 2018). These organic pollutants potentially interact with MPs and alter biological pathways in organisms.

MPs predominantly exhibit hydrophobic characteristics; as a result, they interact with co-occurring hydrophobic pollutants in wastewater, indicating their potential for bioaccumulation along the food chain. Hydrophobic contaminants adsorb onto the surface of MPs through various mechanisms, including hydrophobic interactions (Prajapati et al., 2022) pore blockage, and the availability of binding surfaces, which also depends on the type of plastic (Joo et al., 2021). Moreover, the surface properties of MPs can also change due to plastic degradation and modifications in additives during the wastewater treatment process. These changes can introduce ionic charges to MPs, enabling the adsorption of pollutants with ionic charges. Electrostatic interactions, intermolecular hydrogen bonding, van der Waals forces are the driving forces for the adsorption of Pharmaceuticals and personal care products like hydrophilic contaminants onto the MPs surface (Sun et al., 2019).

Several chemical additives, such as plasticizers (phthalates), flame retardants [brominated flame retardants (BFRs)], and stabilizers (Bisphenol A), are incorporated during the plastic manufacturing process to enhance plastic quality (Andrady & Neal, 2009; Hahladakis et al., 2017). Ingestion of additives by organisms creates toxic effects, impacting growth, development, and offspring. It has been recorded that over 13,000 chemicals are added to plastics as additives, and plastic additives are likely to be used in more than 50% of plastic objects (Megill et al., 2024; Yu et al., 2024). These additives are released into the environment in their original chemical composition, as most do not chemically react with plastic polymers. The release of these additives occurs through the aging and fragmentation of plastics via processes such as photo-oxidation, biochemical degradation, and chemical oxidation. Light irradiation, dissolved organic matter content, and alkaline conditions present in wastewater treatment plants can accelerate the release of chemicals from MPs. (Yu et al., 2024). These changes can introduce ionic charges to MPs, enabling the adsorption of pollutants with ionic charges. Therefore, wastewater treatment plants could serve as sources of toxic chemical additives released from MPs to the aquatic environment.

MPs undergo extensive fragmentation as they are exposed to high shear forces during mixing processes in wastewater treatment. This fragmentation facilitates the formation of secondary MPs and nanoplastics, increasing the surface-to-volume ratio (Monira et al., 2025) and likely enhancing their adsorption properties for both hydrophilic and hydrophobic contaminants and pollutants. The adsorption of metals (e.g., Cd), PhAc, and personal care products onto MPs has been shown to increase significantly after the wastewater treatment process (Hamidian et al., 2021; Li et al., 2019). Additionally, microorganisms present in wastewater aggregate on MPs to form biofilms or plastispheres (Rajcoomar et al., 2023), further increasing the available surface area for the secondary adsorption of pollutants onto MPs. The attachment of pathogenic microorganisms to plastispheres leads to their ingestion by biota, potentially causing health and immune system issues (Zhong et al., 2023).

The sludge treatment process in wastewater treatment acts as a sink for MPs, removing them from the water (Hamidian et al., 2021). However, MPs, along with adsorbed contaminants and additives, accumulate in sludge. This sludge is subsequently used in various applications (e.g., landfilling and fertilizers), posing a potential risk of contamination from wastewater runoff, such as agricultural runoff in to the fresh water sources. This creates a feedback loop that reintroduces MPs and associated pollutants into the wastewater treatment process.

5 Effect of MPs in Wastewater on Aquatic biota and Food Security

Aquatic organisms come into contact with MPs through entanglement, ingestion, and inhalation, which can lead to physical and chemical changes in their bodies, resulting in acute and chronic toxicities.

5.1 *Physical Damages*

The size of MPs is similar to that of prey items consumed by filter-feeding organisms (e.g., bivalves and fish), misleading them into unintentionally ingesting MPs. These ingested MPs create physical blockages in the biological systems of these animals. Digestive tract obstructions and a feeling of gut fullness caused by MPs reduce the feeding appetite of fish, thereby decreasing their growth performance (Bhuyan, 2022). MP entanglement in zooplankton causes disruptions to their swimming and predatory efficiency (Lehtiniemi et al., 2018). *Mytilus edulis* and *Perna viridis* mussel species showed a reduction in their filtration activity after exposure to MPs, likely due to the trapping of MPs in their gills, which blocks the available surface area for water filtration (Zhang et al., 2019).

5.2 *Chemical Damages*

MP accumulation alone or in combination with adsorbed chemicals can alter various chemical processes that influence the metabolism of aquatic organisms, leading to oxidative stress, genotoxicity, neurotoxicity and endocrine disruption in their bodies. Zebrafish (*Danio rerio*) is an ideal animal model to study the human pathogenesis of metabolic diseases and possible therapeutic options due to the similarity of its genome structure to that of humans (Choi et al., 2021). Bivalves are benthic, filter-feeding organisms; therefore, they can concentrate and accumulate floating micro sized particles and in water, making them suitable sentinel organisms for monitoring environmental contaminants and pollutants. They can be used for both MP pollution monitoring and studying the effects of MPs (Baralla et al., 2021; Bråte et al., 2018; Pizzurro et al., 2023). Fish species accumulate MPs in their bodies from their habitats and through trophic transfer (Das et al., 2023; Justino et al., 2023), facilitating MP and associated pollutants accumulation in the human body through fish consumption. Therefore, the toxic effects of MPs on the chemical changes in cellular and molecular functions, leading to lethal and sublethal impacts in zebrafish, bivalves, and food fish species, are discussed to provide a broader overview of their impact on humans, the environment, and food safety.

5.2.1 Oxidative Stress

Free radicals, including reactive oxygen species (ROS) originate from both endogenous sources (e.g., peroxisomes) and exogenous sources (e.g., pollutants/MPs) (Fig. 3). These ROS can create harmful effects on biological molecules (e.g., lipids and proteins) and cellular organelles (e.g., mitochondria). Antioxidants serve as a defense system in organisms, counteracting ROS (Fig. 2). Catalase (CAT), superoxide dismutase (SOD), and glutathione peroxidase (GPx) are major antioxidants that function as intracellular defenders. SOD is a first line antioxidant enzyme that reacts with oxyradicals, accelerating the dismutation of superoxide ($O2^-$) into H_2O_2. CAT is involved in breaking down H_2O_2 produced by SOD into H_2O and O_2, while GPx is responsible for converting H_2O_2 into alcohol, protecting biota from the harmful effects of free radicals (Nair & Perumal, 2024). Oxidative stress occurs due to a physiological imbalance between oxidants and antioxidants within the cell (Carvalho et al., 2012; Ighodaro & Akinloye, 2017; Mandal et al., 2022; Phaniendra et al., 2014) (Fig. 3). Therefore, changes in cellular antioxidant enzyme levels indicate the toxic effects of contaminants, serving as biomarkers to assess their impact on organisms (Carvalho et al., 2012).

Felix et al. 2023, exposed adult zebrafish to naturally weathered MP concentrations of 0.1 and 1 mg/L for 21 days. CAT activity in the brains of adult zebrafish

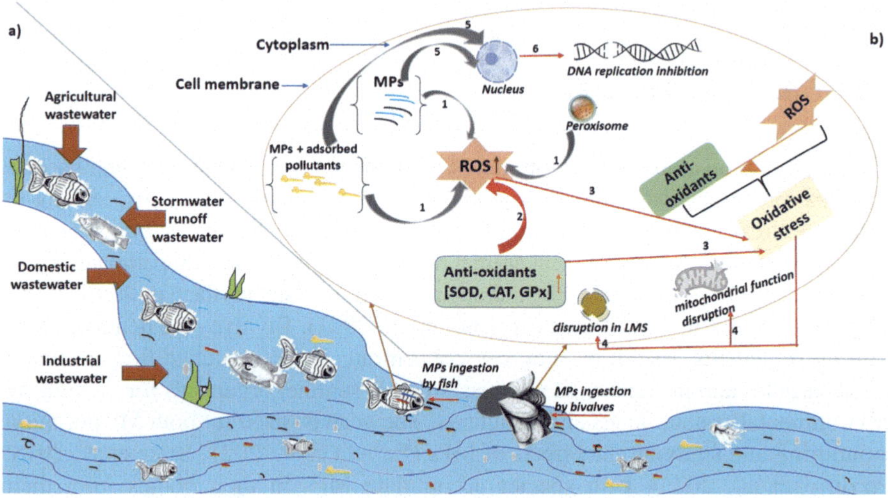

Fig. 3 The effect of MPs alone and with adsorbed pollutants on cellular organelles. The exposure of freshwater and marine biota to MPs and MP-adsorbed pollutants (a). Small, colorful fibers and fragments in the water represent MPs and MP-adsorbed pollutants. The effects of MPs alone and with adsorbed pollutants on oxidative stress and genotoxicity in cell (b). Formation of ROS from endogenous (peroxisomes) and exogenous (MPs) sources (1). Antioxidant defense mechanisms against ROS (2). Imbalance between ROS and antioxidant levels leading to oxidative stress in the cell (3). Impact of oxidative stress on lysosomes and mitochondria (4). Effects of MPs and MP-adsorbed pollutants on the nucleus (5). Inhibition of DNA replication/genotoxicity (6)

significantly increased ($p < 0.05$) in both MP treatment concentrations. On the other hand, liver SOD activity was significantly higher in the 1 mg/L MP treatment group. Intestinal oxidative stress caused by MPs in zebrafish was reported by Qiao et al., 2019. CAT activity significantly increased ($p < 0.05$) by 56% and 124.6% at MP concentrations of 50 μg/L and 500 μg/L, respectively, compared to the control group. Similarly, intestinal SOD activity significantly increased ($p < 0.05$) by 23.8% at a 50 μg/L MP concentration relative to the control (Qiao et al., 2019). GPx activity was significantly increased in zebrafish larvae exposed to a mixture of Cu (125 μg/L) and MPs (2 mg/L) (Santos et al., 2020). Exposure to Cu and MPs mixture ([MPs] = 2 mg/L, [Cu] = 25 μg/L) also increased glutathione transferase enzyme activity in the gills of adult zebrafish (Santos et al., 2022).

Pacific blue mussel *(Mytillus trossulus)* was exposed to MPs treatment with HDPE) fragments (4 fragments/L, each with an area of 0.01 m^2) and PS microspheres (10^5 spheres/L) separately for 72 h. Following exposure, oxidative stress and destructive cellular processes were observed. The lysosomal membrane stability (LMS) of hemocytes decreased 2.7-fold compared to the control in the HDPE treatment (Fig. 2) (Dovzhenko et al., 2024). LMS has been identified as a biomarker for chemical pollution-induced stress (Martínez-Gómez et al., 2015). Additionally, a significant reduction in LMS was observed in juvenile Pacific oysters (*Crassostrea gigas*) exposed to spherical PS (size = 6 μm, concentration = 10^6 particles/L) for 80 days, compared to the control ($p = 0.017$). Moreover, Dovzhenko et al. (2024) demonstrated the impact of HDPE fragments and PS microspheres on lipid peroxidation in the gills and digestive gland of Pacific blue mussel, leading to a significant increase in malondialdehyde (MDA) levels. MDA is an end product of polyunsaturated fatty acid peroxidation. MDA is commonly used as a marker of oxidative stress since high levels of free radicals contribute to its overproduction (Gaweł et al., 2004).

Fingerlings of Nile tilapia (*Oreochromis niloticus*) fed with MP-incorporated feed (4.5 g plastic powder/kg of feed) for 7 weeks showed significantly higher SOD and GPx levels in the liver (Nair & Perumal, 2024). Nylon MP exposure in crucian carp (*Carassius carassius*) caused oxidative stress by disrupting mitochondrial function and increasing ROS levels (Fig. 3) (Subaramaniyam et al., 2023). Therefore, significantly elevated antioxidant levels in various cellular compartments and organelles across different life stages of zebrafish, bivalves and food fishes highlight the role of the antioxidant defense system against MPs and MPs-adsorbed pollutants induced toxicity.

5.2.2 Genotoxicity

It is the ability of a chemical or substance to cause damage to genetic information, resulting in mutations in cells. This can occur through the interaction of a toxic chemical or substance with the DNA sequence and structure (Seukep et al., 2014). Zebrafish exhibited inhibited DNA replication (Fig. 3), and chromosome arrangement and separation after exposure to PS microbeads (0.07 and 7 μm in size) (Gao et al., 2021). Moreover, they showed DNA damage in erythrocytes and erythrocytic nuclear abnormalities after exposure to PE MPs for 15 days (Da Costa Araújo et al., 2022).

Comet assay results confirmed a significant increase in DNA damage in the gills and digestive gland of *Mytillus trossulus* exposed to HDPE fragments and PS microspheres compared to the control ($p < 0.05$) (Dovzhenko et al., 2024). Significant DNA damage was quantified using the Olive Tail Moment (OTM), with higher OTM values observed in peppery furrow shell clam exposed to the LDPE+ BaP compared to the LDPE alone treatment ($p < 0.05$) (O'Donovan et al., 2018).

Jeyavani et al. (2023) exposed *Oreochromis mossambicus* to different concentrations of PP MPs (MP) (Group 1—100 mg/kg; Group 2—500 mg/kg; Group 3—1000 mg/kg) for 14 days. This exposure induced DNA damage in the fish, which was directly proportional to the MP exposure doses. Juvenile *Cyprinus carpio* were exposed to two environmentally relevant MP concentrations (0.75 and 8.25 μg/L). The frequency of micronucleated cells increased with MP concentration, indicating genotoxicity. Although *C. carpio* is considered a highly resistant fish species, exposure to environmentally relevant MPs concentrations induces genotoxicity, which may lead to long term health issues (Menezes et al., 2024), creating a threat to the continuous supply of nutrient rich, affordable protein sources, such as freshwater food fish.

5.2.3 Neurotoxicity

Micro- and nano-sized persistent particles in the water can enter aquatic organisms through feeding and respiration. These particles can penetrate the bloodstream and cross the blood-brain barrier, leading to neurotoxicity (Barboza et al., 2023). Neurotransmitters function by transferring messages from one nerve cell to another nerve, muscle, or gland cell. The levels of inhibitory neurotransmitters, 5-hydroxytryptamine (serotonin, 5-HT) and gamma-aminobutyric acid (GABA), which regulate brain function, were significantly higher in zebrafish larvae exposed to photoaged PS MPs (0.1–100 g/L) compared to the control group. The behavioral patterns of zebrafish larvae (e.g., swimming pattern and speed) showed a significant negative correlation ($p < 0.01$) with neurotransmitter biomarkers (5-HT, GABA, Acetylcholine [Ach]) (Ding et al., 2023). These findings suggest that photoaged PS MPs may induce neurotoxicity in zebrafish larvae by disrupting neurotransmitter function.

Peppery furrow shell clam (*Scrobicularia plana*) was exposed to LDPE MPs (size = 11–13 μm, concentration = 1 mg/L), with and without adsorbed contaminants, at environmentally relevant concentrations of benzo[a]pyrene (BaP) (16.87 ± 0.22 μg/g) for 14 days. A significant increase in acetylcholinesterase (AChE) activity was observed in the peppery furrow shell clam exposed to the LDPE + BaP treatment compared to the control ($p < 0.05$), indicating neurotoxic effects (O'Donovan et al., 2018). Similarly, blood clams (*Tegillarca granosa*) exposed to MPs (1 mg/L) in combination with two different concentrations of bisphenol A (BPA; 10 and 100 ng/L) showed significantly higher levels of GABA, Dopamine and AChE neurotransmitters compared to the control. However, the expression of genes encoding modulatory enzymes and receptors for these neurotransmitters was reduced, leading to neurotoxicity in the blood clams (Tang et al., 2020).

Exposure to PS MPs at concentrations of 1, 10, and 100 μg/L for 14 days in O. *niloticus* caused AchE inhibition in their brain (Ding et al., 2018). AChE is found at postsynaptic neuromuscular junctions, particularly in muscles and nerves, and plays a crucial role in neuronal transmission and signaling. It terminates neuronal transmission and signaling between synapses to prevent excessive AChE activity and the overstimulation of nearby receptors (Pryor et al., 1992). Inhibition of AChE can cause malfunction in mitochondria and impair autophagy, leading to neurotransmitter disruptions and protein-folding abnormalities, which may contribute to neurotoxicity in organisms, including brain oxidative stress and damage (Barboza et al., 2023; Jadhav et al., 2022).

5.2.4 Immunotoxicity

Exposure to toxic substances leads to adverse effects on the immune system, which is referred to as immunotoxicity (Gulati & Ray, 2009). Zebrafish were exposed to a mixture of HDPE (50%) and PS (50%) at concentrations of 100 and 1000 μg/L for 20 days. Immune system gene disruptions were confirmed by the transcriptomic results, and a correlation was observed between the downregulation of genes related to epithelium integrity and lipid metabolism after the exposure (Limonta et al., 2019). A relatively high number of innate immune cells, including macrophages and neutrophils, were observed after MP exposure (Li et al., 2024; Limonta et al., 2019).

Crassostrea gigas oysters were exposed to 50 μg/L MPs alone, as well as 10 or 100 ng/L polychlorinated biphenyls (PCBs) individually or in combination with 50 μg/L MPs for 28 days. Oysters in the PCB + MPs treatments showed drastic suppression of nuclear factor *kappa B* (NF-κB), tumour necrosis factor alpha (TNF-α), and toll like receptor 6 (TLR-6) immune-related genes, while a significant increase in genes related to apoptosis was observed. The decrease in total hemocyte count and inhibition of phagocytosis in hemocytes caused oxidative stress in the animals by disrupting nucleus and lysosome functions (Hou et al., 2024). NF-κB is a regulator of innate immunity (Albensi, 2019); TNF-α has the ability to regulate peripheral and local immune responses (Caldito, 2023); and TLRs are involved in innate immunity recognition and the stimulation of antimicrobial activities (Moraes et al., 2012).

Gilthead seabream (*Sparus aurata*) were fed a diet incorporating PS MPs for 21 days. The expression of pro-inflammatory cytokine genes (e.g., cyclooxygenase-2 (COX-2)) increased, whereas the expression of anti-inflammatory genes (e.g., interleukin-10 (IL-10)) decreased, altering immune system functions in the gut. The mitogen-activated protein kinase signaling pathway was activated, disrupting the intestinal epithelial integrity (Del Piano et al., 2023). Soliman et al. (2023) exposed African catfish (*Clarias gariepinus*) to a mixture of MPs (100 mg/L) and $Pb(NO_3)_2$ (1 mg/L) for 15 days. A significant reduction in red blood cells, thrombocytes, and lymphocytes was observed relative to the control group. Changes in blood cells (e.g., lymphocytes) indicate alterations in the immune system of the fish (Soliman et al., 2023).

MPs may serve as vectors for pathogen transmission, particularly in wastewater treatment plants, where effluent can still contain considerable levels of pathogens (e.g., *Arcobacter* sp., hepatitis E virus, and SARS-CoV-2) despite disinfection. MPs are not efficiently removed from wastewater treatment plants; therefore, once pathogens attach to MPs in these facilities, they can be transported to streams, rivers, and ultimately the ocean, potentially posing health risks to humans as well (Zhong et al., 2023). The presence of pathogens on MPs in the marine environment was confirmed by Viršek et al. (2017). They found *Aeromonas salmonicida*, a pathogenic fish bacterium, on MPs in the northern Adriatic Sea, indicating its effect on fish health and, ultimately, on humans through fish consumption.

5.2.5 Endocrine Disruptors

Endocrine disrupting chemicals (EDCs) are natural or human made substances that may mimic, block, or interfere with the endocrine systems (hormones) of humans and animals. Since hormone balance regulates all body systems in animals, EDCs induced hormonal imbalances can create adverse health effects, such as abnormalities in reproductive cells. The impact of EDCs on insulin regulation may also increase the risk of type 2 diabetes. Pesticides, polycyclic aromatic hydrocarbons, dioxins, brominated flame retardants, steroid hormones, and alkylphenols are known EDCs, and they are not completely removed by wastewater treatment plants (Chen et al., 2022; Ullah et al., 2023; Tabassum et al., 2025). There are three types of EDCs:

1. Androgenic compounds (EDsA): Mimic or block natural testosterone.
2. Thyroid compounds (EDsT): Directly or indirectly influence the thyroid gland.
3. Estrogenic compounds (EDsE): Mimic or block natural estrogens (Chen et al., 2022).

Plastic additives can act as EDCs, and plastics can also adsorb EDCs (e.g., POPs) from the environment. Therefore, the role of MPs in transferring EDCs into organisms' body is significant. Zebrafish individuals were exposed to different concentrations of the plasticizer, mono-(2-ethylhexyl) phthalate (1.6, 8, 40, and 200 mg/L) from 2 to 168 hrs post-fertilization. A significant reduction in whole body T4 (thyroxine) levels, along with an increase in whole body T3 (triiodothyronine) levels, led to thyroid endocrine disruption (Zhai et al., 2014). T3 and T4 are important components of the endocrine system and are involved in regulating metabolism (e.g., weight and energy level regulation). *Mytilus galloprovincialis* were subjected to tetrabromobisphenol A (POP) for 30 days. Following exposure, there was an increase in sulfotransferase and steroid sulfatase levels, due to the influence of tetrabromobisphenol A on steroidogenesis in *Mytilus galloprovincialis*, facilitating steroid sulfonation and the breakdown of sulfate steroids (Wang et al., 2021a, b). Exposure to bisphenol S (plasticizer) in *Oreochromis mossambicus* caused a drastic change in thyroxine hormone levels and a significant alteration in steroid hormones ($p < 0.05$) due to the disruption of the thyroid gland and the hypothalamic-pituitary-gonadal axis in fish, respectively (Anjali et al., 2019).

5.3 Ecological Impacts

There is a high potential for MPs and MP-adsorbed pollutants to accumulate in the sludge of wastewater treatment plants. Globally, the concentration of MPs in sludge from wastewater treatment plants ranges from >1000 particles/kg to 301,400 particles/kg (Hassan et al., 2023). Sludge collected from wastewater treatment plants is commonly used as a fertilizer on agricultural soil due to its high organic matter and nutrient content (Corradini et al., 2019; Hassan et al., 2023). Additionally, sludge is used in landfills and for producing building materials, such as ceramics. However, the application of sludge to agricultural soil has been identified as a source of MPs contamination. Corradini et al. (2019) observed an increasing trend in MP abundance in topsoil with the number of sludge applications. Soils that received 1, 2, 3, 4, and 5 sludge applications had median MPs abundance of 1.1, 1.6, 1.7, 2.3, and 3.5 particles/g of dry soil, respectively. Similarly, Zubris and Richards (2005) reported 1.21 ± 0.25 synthetic fibers/g of soil 5 years after sludge application.

MPs in agricultural soil have the potential to influence soil properties (such as soil structure and porosity) and affect interactions between soil and plants. Degradable poly (butylene adipate-co-terephthalate) micro particles have been shown to cause significantly greater oxidative damage, reduce the photosynthetic rate, alter soil aggregation and soil ammonium levels, and disrupt microbial activity, ultimately leading to growth reduction in lettuce (Wang et al., 2023a, b, c). The co-occurrence of Cd and polyethylene MPs in the soil affects the chloroplast structure of maize (*Zea mays*), inhibiting its photosynthetic capacity (Li, et al., 2023a, b). Moreover, these pollutants may lead to biomagnification through food chains. These findings present an alarming signal regarding the impact of MPs on agricultural food production. Han et al. (2024) revealed that MPs alter the soil bacterial community structure and keystone taxa, creating cascading effects on soil biodiversity and organic matter decomposition processes.

Phthalates and BPA plasticizers have affected reproduction in annelids, mollusks, crustaceans, amphibians, and fish (Oehlmann et al., 2009). Therefore, the leaching of plasticizers from sludge-treated soil into terrestrial and aquatic environments is likely to influence the health and vitality of future wildlife generations, leading to long term impacts on community structure and biodiversity.

6 Impacts of MPs on Terrestrial and Soil Biota

Terrestrial and soil biota can be affected by various types of environment pollution activities and sources, with wastewater being a significant contributor. Wastewater typically contains a cocktail of pollutants, toxins, and other contaminants that can harm organisms within these ecosystems. Additionally, it may carry pathogens that pose a threat to terrestrial life. Furthermore, wastewater often contains high levels of nutrients, which can promote excessive growth of certain organisms, potentially disrupting the natural ecological balance. Activities such as landfilling, crop fertilizing, and irrigation are common ways through which wastewater enters terrestrial environments. However, with proper treatment before its application in terrestrial systems, wastewater can significantly reduce its negative impact on soil organisms and terrestrial ecosystems. In many cases, the impact of wastewater on aquatic environments has been widely discussed; however, its impact on terrestrial environments has rarely been addressed.

Wastewater often contains a variety of contaminants, including human waste, soaps, solvents, pharmaceuticals, detergents, fertilizers, germs, pesticides, radioactive materials, heavy metals, and plastics. When released into the terrestrial environment, wastewater with these pollutants can become a significant source of pollution. Some of these contaminants may disrupt plant-animal interactions and negatively impact biodiversity (Talukdar et al., 2024; Prasad et al., 2024). Discharges from WWTPs can also impact the overall structure and functions of ecosystems. In particular, nanomaterials present in wastewater can accumulate in terrestrial animals, leading to negative consequences (Batley et al., 2013). For example, the impact of pharmaceutical contaminants in wastewater on terrestrial wildlife has recently been identified and extensively discussed (Taggart et al., 2015). Wastewater is a main source of MPs in the terrestrial environment. According to previous research, MPs in the terrestrial environment can be transported by advection, dispersion, settling, adsorption, diffusion, degradation, and aggregation (Qaiser et al., 2023).

6.1 *Exposure of Terrestrial Life to WWMPs*

The wastewater generated in homes, shops, and industries are usually contaminated with MPs (Talukdar et al., 2024). It includes water from sinks, showers, toilets, industrial processes, and stormwater runoff. This water can contain microbeads primarily from cosmetics, toothpaste, exfoliants, resin pellets, microbeads form synthetic textiles, and etc. Secondarily, MPs come to wastewater through breaking down of plastic litter in urban runoff, releasing during laundry processes, and from tire ware by stormwater. Hence, MPs in wastewater are a significant environmental concern due to their widespread presence and potential harmful effects on ecosystem, biota and finally human health (Qaiser et al., 2023). These small plastic

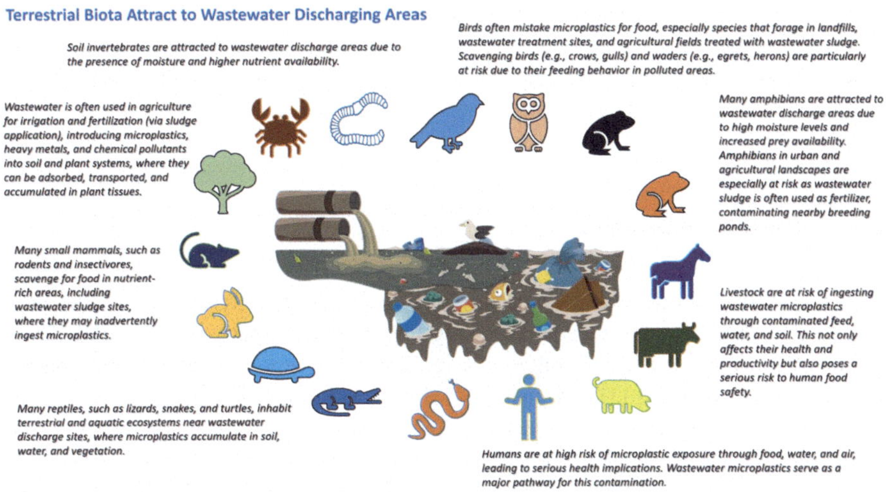

Fig. 4 Exposure and attraction of terrestrial biota such as plants, invertebrates, and vertebrates on wastewater discharging areas

particles, typically less than 5 mm in size, are difficult to remove in conventional wastewater treatment systems. Therefore, these particles are accumulated in water bodies, soils, and food chains through wastewater removal processes (Fig. 4).

Wastewater is normally treated at the treatment plants before release to the environment. However, MPs in wastewater are difficult to capture using traditional screening and sedimentation techniques due to their small size and lightweight. Also, these MPs are non-degradable, hence accumulate in sludge or effluent. Biosolids from wastewater treatment plants are generally used as fertilizers in agriculture or horticulture purposes, hence can release a significant number of MPs to terrestrial environment and potentially contaminate the soil (Batley et al., 2013). Finally, these MPs can be ingested by organisms and bioaccumulate up to the food chain. Later, these accumulated MPs can harm the biota directly (physical harm) and indirectly (toxic chemical exposure).

In the terrestrial environment, MPs derived from wastewater can impact terrestrial animals, earthworms, insects, soil microorganisms, plants and human (Qaiser et al., 2023; Rillig & Lehmann, 2020). Moreover, these MPs can also contribute to changes of soil characteristics, nutrient storage, and cycling processes (Kumar et al., 2023). These changes in the soil environment can directly impact soil bacteria, fungal communities, and decomposers. Alterations in microbial communities may, in turn, affect plants, particularly crop cultivation. Accordingly, a study has found a significant impact of MPs derived from sludge on plant biomass, tissue elemental composition, root traits, and soil microbial activities (de Souza Machado et al., 2019). MPs can significantly alter the abundance and diversity of soil microorganisms. Beneficial microorganisms may decrease, while harmful microbial communities could thrive (Khalid et al., 2020). These changes can negatively impact crucial processes like nitrogen fixation and the decomposition of organic matter in the soil. Ultimately, MPs from wastewater discharge in the terrestrial environment can impact terrestrial biota through physical, chemical, and combined effects (de Souza Machado et al., 2018).

6.2 Physical Damages

MPs originating from wastewater can physically impact terrestrial biota, both directly and indirectly. However, assessing the specific impact of MPs solely from wastewater in terrestrial environments is challenging due to multiple contamination sources (M. Wu et al., 2021). A possible solution is to identify terrestrial organisms that are particularly attracted to or dependent on wastewater discharge areas. Monitoring these organisms can provide critical insights into the ecological consequences of MPs in terrestrial systems. Earthworms, for instance, may mistakenly ingest these MPs, mistaking them for organic particles (Huerta Lwanga et al., 2016). This can lead to gut blockages, impair nutrient absorption, and adversely affect their overall health and soil conditioning functions. Moreover, other invertebrates such as protists, nematodes, snails like gastropods, and insects can also be affected in a same way (Büks et al., 2020; Daghighi et al., 2023). Sharp or irregular-shaped MPs can physically damage the internal tissues of soil invertebrates. Furthermore, soil compaction caused by MPs hinders the burrowing activity of earthworms and other soil invertebrates, adversely affecting soil aeration and fertility. The altered burrowing behavior resulting from soil compaction can disrupt critical soil processes. Studies have also reported reduced cocoon production and slower growth rates in earthworms exposed to MPs, highlighting the long-term ecological implications of such contamination (Huerta Lwanga et al., 2016). Further, sharp or irregular-shaped MPs can physically damage the internal tissues of soil invertebrates (Sanchez-Hernandez, 2024). These particles can particularly harm the gut epithelium of earthworms and may translocate to various internal organs and tissues. Such physical damages ultimately lead to changes in cellular responses, alterations in microbiota, increased mortality rates, and disruptions in development, behavior, and reproduction among soil invertebrates (Möhrke et al., 2022). Moreover, it has been noted that these particles can disrupt the cellular environment, induce DNA alterations, and ultimately lead to changes in cellular responses. A large number of soil invertebrates are attracted to wastewater discharges in the terrestrial environment due to the high nutrient load. However, the presence of MPs in the wastewater sludge poses a significant threat to their survival, leading to increased mortality rates.

Wastewater sludge is often applied to agricultural land as a fertilizer, inadvertently introducing MPs into the soil. These MPs directly affect plant roots and their interactions with the soil by physically obstructing root pathways, thereby limiting root elongation and branching (Khalid et al., 2020). Compacted soil due to MPs further reduces the essential contact between roots and soil particles, hindering nutrient and water absorption. MPs can alter soil characteristics, affecting its texture, aeration, and water retention capacity. Notably, MPs can directly impact plant seeds by clogging seed pores, which may hinder water uptake and gas exchange, ultimately affecting seed germination and plant development (Khalid et al., 2020; Zantis et al., 2023). Studies have found that MPs can significantly alter root traits, including total root length and diameter, as well as plant leaf traits and overall biomass (de Souza Machado et al., 2019; Xu et al., 2024). Additionally, MPs can enter plants through the roots and accumulate in various cells and tissues. This accumulation may affect plant physiology and impair critical physical processes such as transpiration, ultimately compromising plant health and productivity (Tang, 2020). Therefore, although wastewater sludge is often used to fertilize agricultural lands and crop plantations, it can have negative consequences, potentially leading to a total loss of production. This is due to the physical damage caused by wastewater MPs, affecting the entire growth cycle from seed germination to plant development. These impacts include blocking pores or light, causing mechanical damage to roots, hindering genes expression, and releasing additives (J. Li et al., 2022).

Birds have been identified as a model to monitor terrestrial microplastic level as they are feed on several fields including agriculture lands (Sherlock et al., 2022). Wastewater sludge accumulation sites in the terrestrial environment are highly attractive to birds such as egrets, crows, eagles and waders. These sites often harbor a rich presence of soil invertebrates and flying insects, which serve as a food source for these birds. As a result, the bioaccumulation of WWMPs in birds becomes inevitable, potentially causing adverse effects on their physiology. These impacts may include digestive blockages, tissue damage, and disruptions in metabolic processes. A recent study conducted in Florida, USA, revealed a significant presence of MPs in the gastrointestinal tracts of various prey birds (Carlin et al., 2020). In Pakistan, a study on cattle egrets reported a high accumulation of MPs in their gastrointestinal tracts, which was identified as a potential cause of mortality (Khaliq et al., 2024). This finding underscores the pervasive nature of WWMPs pollution and its potential threat to terrestrial avian health and ecosystems. This also highlights the severe impact of microplastic pollution on terrestrial wildlife, particularly in species that forage in contaminated environments.

Another group of terrestrial animals affected by wastewater MPs is small mammals, as they are also attracted to wastewater sludge-filled sites in the terrestrial environment. These areas often provide food sources such as insects and plants, but exposure to MPs can negatively impact their health and well-being. A study in the UK identified the presence of MPs in the feces of various small mammals, including herbivorous, insectivorous, and omnivorous species, highlighting the widespread environmental contamination and its potential impact on terrestrial wildlife (Thrift et al., 2022). Other mammals, such as grazing animals, may congregate in wastewater sludge landfill areas due to the abundant growth of grasses fueled by high nutrient levels. These animals may inadvertently consume microplastic particles along with forage, leading to gastrointestinal blockages and potential health complications. MPs with jagged edges can cause abrasions and internal injuries in the digestive systems of these animals. A study has found that MPs can cause significant damage to the reproductive system, including oxidative stress, apoptosis, inflammatory responses, and dysregulation of the endocrine system (Urli et al., 2023). Furthermore, these particles may accumulate in various organs, exacerbating their harmful effects, and ultimately pose a risk to humans if consumed through the food chain. A study conducted on a livestock farm in Spain (Beriot et al., 2021) revealed microplastic contamination in soil (~2×10^3 particles/kg) of agriculture/grassland and faces of sheep (~10^3 particles/kg). This study clearly demonstrated that agricultural and grassland areas are contaminated with MPs, with potential sources being wastewater sludge application as fertilizer, landfill activities, and atmospheric distribution. Wind-blown MPs from biosolids can become airborne, potentially causing respiratory damage when inhaled by terrestrial animals, including humans. Further, the presence of MPs in the gut reduces the surface area available for nutrient absorption in these animals, potentially leading to poor nutrition and production loss. However, there are still very limited studies on how wastewater MPs impact small mammals, predatory mammals, and their bioaccumulation within the food chain. The physical impacts on mammalian bodies remain largely unexplored. Recently, large mammals like elephants have been observed visiting wastewater sludge landfill areas due to the availability of food materials. However, the potential exposure of these animals to MPs has never been investigated.

6.3 Chemical Damages and Metabolic Changes

WWMPs not only cause physical harm but also act as carriers and sources of chemical pollution. Their chemical composition, additives, and ability to adsorb toxic pollutants from the environment make them hazardous to soil, plants, animals, and humans in terrestrial ecosystems (Hassan et al., 2023). These MPs contain harmful chemical additives such as phthalates (plasticizers), BPA, flame retardants (PBDEs), and heavy metals (e.g., lead, cadmium). These substances can disrupt endocrine functions, interfere with hormonal balance, and impair reproduction and development in organisms. The leaching of these chemicals alters the pH and nutrient balance of the soil, negatively affecting plant growth and microbial diversity. Persistent chemicals such as PBDEs accumulate in plants and animals, entering and propagating through the terrestrial food web, leading to long-term ecological and health consequences.

Wastewater contains a diverse range of chemicals and MPs originating from household activities, industrial processes, and various businesses in urban environments. MPs, in particular, act as sponges, absorbing toxic chemicals from the wastewater. Upon discharge into terrestrial environments, such as soil, these toxic substances and MPs accumulate, creating prolonged exposure risks for organisms. This contamination can have harmful consequences for soil health, plant growth, and terrestrial wildlife, leading to bioaccumulation and ecological imbalances.

MPs disrupt microbial communication and enzymatic activities essential for nutrient cycling. Adsorbed toxins and chemical leachates from wastewater MPs inhibit critical enzymatic functions. Sensitive microbial populations decline due to chemical toxicity. Notably, BPA produced by PVC can break down the cell walls of anaerobic digesting bacteria, further impairing microbial processes in the soil (Arab et al., 2024). Also, another study has witnessed that these MPs can be a platform for microorganism colonization and chemical adsorption (Mammo et al., 2020). Hence, these MPs can directly cause biotoxicity and indirectly change the microhabitats and nutrient availability (Aralappanavar et al., 2024). Accordingly, they mentioned that the biotoxicity generally cause by oxidative stress by reactive oxygen species (ROS) or the toxic effects of the attached chemicals. Soil toxicity caused by chemicals attached to wastewater MPs can lead to changes in soil structure and properties. These changes may result in a reduction in microbial diversity and abundance, disrupting essential soil functions such as nutrient cycling and organic matter decomposition. Changes in soil characteristics, such as alterations in soil pH, can become toxic to soil microbial communities. These changes may inhibit enzyme activities, disrupt cell metabolism, and reduce nutrient availability, ultimately affecting soil health and fertility (Aralappanavar et al., 2024; Sajjad et al., 2022).

Several soil invertebrates, including earthworms, nematodes, isopods, and snails, have been investigated for the toxic effects of wastewater MPs (Guo et al., 2020; Sajjad et al., 2022). The observed impacts on these organisms include bioaccumulation and transformation through the food chain, reduced growth and increased mortality, immune system damage, harm to cholinergic and GABAergic neurons, impaired movement, oxidative stress, and oxidative damage. Another study found that polystyrene MPs (PSMP), polytetrafluoroethylene (PTFE), and arsenic can reduce the abundance of Proteobacteria, while increasing the abundance of Chloroflexi and Acidobacteria. These MPs and contaminants also inhibit key soil enzyme activities, such as urease, acid phosphatase, protease, dehydrogenase, and peroxidase, by affecting the tertiary structure of the enzymes. This can disrupt essential biochemical processes in the soil ecosystem (Dong et al., 2021).

A study found that high concentrations of PBAT (polybutylene adipate terephthalate) MPs (PBAT-MPs) severely inhibited plant growth (Liu et al., 2024). These MPs hindered root development, reduced overall plant biomass, and affected the plant's ability to absorb nutrients and water, potentially due to physical blockages or interactions with soil microorganisms. Compounds and by-products from the degradation of biodegradable plastics, such as PBAT and PLA (polylactic acid), have been found to have direct toxic effects on plant growth (Liu et al., 2024). As these plastics degrade in the soil, they release harmful substances, including organic acids, oligomers, and other by-products, which can alter soil pH and nutrient balance. These changes can disrupt plant root functions, inhibit germination, impair nutrient uptake, oxidative stress, hindering nutrient absorption, metal absorption, physical plant damage, toxicity mechanisms, and synergy with other stressors (Qiu et al., 2024). Additionally, the accumulation of degradation products can lead to a decline in soil microbial diversity, further affecting plant health and growth. Recent discussions have highlighted the influence of MPs and their associated factors on phytotoxicity levels, with a focus on food safety and security (Roy et al., 2023). MPs in soil can alter the chemical environment, affecting plant growth and productivity.

These particles can release harmful chemicals and disrupt essential microbial processes that plants rely on for nutrient absorption and growth. Differentially charged MPs can have varying impacts on plant health, particularly on the structure and function of photosystem II (PSII), which plays a key role in photosynthesis (Z. Xu et al., 2022). The damage caused by MPs is influenced by their surface charge, which determines how they interact with plant tissues. MPs can cause abnormal lignification in plant roots, making them stiffer and less able to absorb water and nutrients effectively. This can affect overall plant growth and development. Polyethylene terephthalate (PET) MPs containing phenyl groups have been found to exhibit the most significant adverse effects on plant health, particularly on fresh weight, chlorophyll content (both a and b), and hydrogen peroxide (H_2O_2) levels (Wang et al., 2023a, b, c). A recent study highlighted the ecotoxicity and genotoxicity effects of PS-MPs on the higher plant *Vicia faba* (broad bean) (Jiang et al., 2019). The findings reveal critical insights into how these MPs affect plant health and genetic stability. Polystyrene caused reduced seed germination rates, diminished root and shoot lengths, and overall impaired plant development. DNA fragmentation and other genotoxic markers were detected, highlighting the risks posed by microplastic exposure.

WWMPs can have toxic impacts on birds in various ways, such as causing delayed ovulation, reduced reproductive output, delayed sexual maturation, impaired digestive functions, activation of inflammatory responses, and weakened immune functions (Wang et al., 2021a, b). Exposure to MPs and their chemical additives can interfere with hormonal regulation, leading to delayed ovulation. Hormonal disruption may result in slower development of sexual maturity in juvenile birds. Persistent exposure to MPs and their toxic additives can weaken birds' immune systems, making them more susceptible to diseases. Also, numerous environmental contaminants absorbed by MPs in wastewater can cause malnutrition, endocrine disruption, and reproductive issues in various types of birds, according to previous studies (Wang et al., 2021a, b). According to three studies on birds using PS-MPs, MPs were found to affect intestinal necrosis & pyroptosis in the intestine, and oxidative stress, intracellular toxicity, and lipid metabolism in bird livers (Yin et al., 2023; Zhang et al., 2024). Specifically, exposure to MPs significantly increased reactive oxygen species (ROS), alanine aminotransferase (ALT), aspartate aminotransferase (AST), and hepatic triglyceride (TG) levels.

Many amphibians are attracted to wastewater discharge areas in terrestrial environments. As a result, animals such as land frogs can be exposed to high levels of MPs. Studies have confirmed the cytotoxic effects of PE MPs on tadpoles, evidenced by smaller erythrocyte size and reduced nucleus area, perimeter, width, length, and radius, as well as a lower nucleus/cytoplasm ratio (da Costa Araújo et al., 2020), and on intestine & skin (Burgos-Aceves et al., 2022). Three species of terrestrial herpetofauna *Rhinella diptycha*, *Hemidactylus mabouia*, and *Tropidurus torquatus* were studied in Paragua and microplastic ingestion was observed (Mackenzie & Vladimirova, 2023). However, there are only a few studies on how wastewater MPs impact land amphibians, highlighting a significant potential for future research in this area.

Lizards of the *Podarcis* genus have been of interest as a good model system in ecotoxicological and cytotoxicological research (Scudiero et al., 2025). They provide an accurate description of the effects of pollutants such as WWMPs, clarify the defense mechanisms activated in response to different exposure routes. As a semi-aquatic animal, Asian water monitor (*Varanus salvator*) is usually found in polluted areas such as wastewater discharge or sludge land filling areas in the terrestrial environment. However, there are currently no studies focusing on these potential animals to investigate the toxicity of MPs or their ability to overcome the toxic effects in their habitats.

Wastewater discharge areas often attract small mammals due to nutrient-rich vegetation and water availability. Hence, small mammals are a vulnerable group of terrestrial biotas exposed to microplastic toxicity in wastewater discharge areas within the terrestrial environment. These animals may ingest MPs accidentally while foraging on contaminated vegetation or soil. Otherwise, consumption of prey such as insects or invertebrates that have accumulated MPs can expose them to secondary contamination. MPs can bioaccumulate in small mammals, and their position in the food chain increases the risk of passing toxins to higher predators. The leaching of toxic chemicals can interfere with hormonal balance, reproduction, and immune functions of these animals. The concentrations of heavy metals in the liver and kidney of meadow voles (*Microtus pennsylvanicus*) and white-footed mice (*Peromyscus leucopus*) from wastewater-irrigated areas are of significant concern (Anthony & Kozlowski, 1982). These organs are critical sites for metal accumulation due to their role in detoxification and metabolic processes. The liver is a primary detoxification organ and accumulates metals such as lead (Pb), cadmium (Cd), and mercury (Hg). Prolonged exposure can impair liver functions, leading to oxidative stress and enzyme disruption. The kidney plays a key role in filtering blood and excreting toxins, making it highly susceptible to heavy metal accumulation. Cadmium accumulation in particular has been linked to nephrotoxicity and tissue damage. In a forest land where wastewater sewage sludge was introduced as fertilizer, liver and kidney tissue samples from Trowbridge's shrews (*Sorex trowbridgii*) were examined for evidence of heavy metal-induced lesions (Hegstrom & West, 1989). Irrigation with wastewater introduces metals such as arsenic (As), chromium (Cr), and zinc (Zn) into the ecosystem. These metals are also absorbed by vegetation, which is consumed by small mammals, leading to bioaccumulation. Heavy metal exposure in small mammals has been associated with immune suppression, reproductive impairments, and altered behavioral patterns. Chronic exposure may affect population dynamics due to lower survival and reproduction rates. A study has highlighted livestock as a potential missing link in microplastic exposure to humans (Corte Pause et al., 2024). These particles can induce inflammation, cytotoxicity, genotoxicity, and immune toxicity in livestock. As animals can ingest or inhale MPs, these particles may accumulate in their tissues and transfer to humans through the food chain. However, comprehensive research on this critical pathway of microplastic exposure remains limited and requires further investigation.

6.4 Ecological Impacts

Wastewater MPs can cause significant ecological risks to terrestrial ecosystems by altering soil structure, soil properties, microbial communities, plant health, and other animals in food webs. Without proper mitigation, these WWMPs contamination in soils could lead to long-term environmental and agricultural challenges in the terrestrial environment. Unlike aquatic environments, where MPs accumulate in sediments, their effects in soil are more complex due to interactions with soil organisms, nutrient cycles, and pollutant transport.

There is an interesting perspective on the ecological effects of wastewater biosolids application in the terrestrial environment. The interaction between indigenous and introduced microbial populations, protozoa, and bacteriophages is crucial for understanding the fate of pathogens and overall soil health (Rogers & Smith, 2007). Plastic particles in wastewater affect soil health, plant growth, microbial communities, and terrestrial food webs, causing long-term ecological disturbances. Accordingly, MPs in wastewater biosolids serve as novel habitats for microbial communities, providing surfaces that promote biofilm formation and enhance pathogen survival. This has significant implications for agricultural soil health, microbial ecology, and the spread of potentially harmful microorganisms (Surendran et al., 2023). The surface properties of MPs, such as hydrophobicity, surface roughness, and electrostatic interactions, may regulate these processes. Most MPs (e.g., PE, PP) are hydrophobic, encouraging microbial adhesion. Further, irregularities on MP surfaces provide anchoring sites for microbial attachment and biofilm development. Some MPs carry charges that attract bacteria, viruses, and fungi, promoting biofilm formation. MPs in biosolids adsorb organic compounds, creating nutrient-rich microenvironments that sustain microbial communities. A recent study found that microplastic-rich biosolids increased the persistence of *E. coli* and antibiotic-resistant bacteria in agricultural soil (Grobelak et al., 2024). Horizontal gene transfer also increases due to microbial interactions on plastic surfaces. Other experiments also showed that pathogenic *Salmonella* formed robust biofilms on PE MPs recovered from biosolids (Quilliam et al., 2023). Biofilm-coated MPs in biosolids can be transported by wind, runoff, and soil movement, spreading pathogens beyond treated fields in the terrestrial environment, making more ecological risks. This also increases the potential of contamination in nearby water bodies, crops, and even livestock.

A recent study in Kenya also showed that MPs can act as carriers for fungal pathogen transmission among organisms in the terrestrial environment, including important animal and plant pathogens. These pathogens form the plastisphere core mycobiome, leading to significant ecological impacts (Gkoutselis et al., 2021). Moreover, in the soil environment, disruptions in detritivores and decomposer activity slow down organic matter breakdown and reducing these activities affect soil fertility. These particles can persist in the soil environment, impacting soil functioning and biodiversity, ultimately affecting terrestrial ecosystems (Rillig, 2012). Since WWMPs pose a potential threat in urban environments as well, this issue warrants increased attention from policymakers and regulatory bodies to reduce the ecological damage.

Herbivores consuming contaminated plants transfer MPs up the food chain. Grazing animals consuming soil or forage with MPs may ingest biofilm-embedded pathogens. The introduction of wastewater sewage sludge as fertilizer in forest lands has raised ecological concerns due to heavy metal contamination in small mammals, such as Trowbridge's shrews (*Sorex trowbridgii*), shrew-moles (*Neurotrichus gibbsii*), and granivorous deer mice (*Peromyscus maniculatus*) (Hegstrom & West, 1989). Trowbridge's shrews and shrew-moles are insectivores that consume soil-dwelling invertebrates, making them highly susceptible to heavy metal contamination through trophic transfer. Deer mice are granivorous, ingesting contaminated seeds and plant material directly exposed to heavy metals. The decline or health impairment of these small mammals can disrupt forest ecosystem functions, as they play essential roles in seed dispersal, soil aeration, and trophic interactions.

7 Mitigation and Management of MPs in Wastewater

As per the literature review, MPs in wastewater even after subjecting to the treatments end up from the water bodies and terrestrial systems by causing severe ecological and social consequences. This chapter will review and suggest feasible solutions including effective removal technologies for mitigation and management of microplastic loading in wastewater.

Scientific community can play a major role in addressing the research gaps and emerging priorities related to MPs contamination in wastewater. Toxicity, implications, fate, sources, transport, and behavior of wastewater-based MPs are yet to be explored (Raju et al., 2018). Advanced scientific tools (satellite images, remote sensing) and modeling approaches can be used for interpretation of the transport pathway of discharged MPs from wastewater (Reddy & Nair, 2022). Further, the differences MPs and removal rate depend on the sample type, location, analytical protocol, nature & type of treatment process in treated wastewater or effluents (Tagg et al., 2020). When implementing effective wastewater treatment techniques, developing countries face financial issues which is one of the significant challenges of reducing wastewater-based microplastic pollution. The future research implications can be focused on emerging low-cost, ecofriendly treatment techniques by addressing the existing challenges and drawbacks in wastewater treatment plants.

Conventional techniques in wastewater treatment plants generally reduce the discharge of large-sized MPs into the environmental setting (Masiá et al., 2020), Conversely, tiny MPs particles can remain in the final effluent. Accordingly, existing technological approaches do not effectively remove the MPs since those treatments are not particularly targeted on remediation of MPs particles (Van Do et al., 2022). Authors of this chapter also confirmed this finding during literature review process. Considering existing removal efficiencies of MPs in wastewater treatment plants, resultant MPs reduction would be 90% to the global aquatic ecosystems, If the wastewater treatment technologies could be implemented and improved adequately (Uddin et al., 2020). Accordingly, improvement of MPs removal efficiency is important in ecological and social point of view at the international, regional, and national level. As per the previous recommendations, new technological approaches, modifications in the treatment plants, improvements of the current systems or further treatments and managements to the treated water are most feasible solutions to reduce or make zero microplastic input from wastewater to the environment (Conley et al., 2019; Lv et al., 2019; Ngo et al., 2019; Reddy & Nair, 2022; Talvitie et al., 2017).

In general, wastewater-associated MPs removal technology corresponded with degradation and removal-based processing techniques (biological, chemical, thermal, and physical techniques). Thermal and biological approaches make degradation of MPs, while chemical and physical methods remove the MPs from wastewater in the treatment plants (Reddy & Nair, 2022). Scientific records emphasize the potential applications of bacteria, fungi and tiny pelagic Euphausiids for biological degradation of MPs. For instance, *Aspergillus niger* is capable in removal of Polymethyl methacrylate (5 μm) and PS (200 nm) with 0.01% and 0.005% of solid loading (Wang et al., 2023a, b, c). However, biological degradation could generate greenhouse gases (e.g.: Methane, and Carbon dioxide) under anaerobic circumstances, while aerobic condition could result in water and carbon dioxide. Accordingly introducing these microorganisms with an advanced bioengineering process makes a promising result in MPs remediation in the wastewater treatment plants (Reddy & Nair, 2022). Further, microalgae are a potential candidate for the treatment of MPs from wastewater through formation of algae-MPs aggregation (Sousa et al., 2025).

Advanced wastewater treatment processes could effectively remove the tiny MPs without discharging into the environment. As per the previous records, the lowest MPs concentration resulted by different tertiary treatment techniques (Kittipongvises et al., 2022). Advanced oxidation process, filtration (membrane filtration: ultrafiltration, microfiltration, reverse osmosis, nanofiltration systems, disk type filtration, and sand filters), coagulation, air flotation, disinfection, membrane bioreactors, and air flotation (Liu et al., 2021; Reddy & Nair, 2022). As an advanced remediation method, membrane filtration is one of the efficient treatment methods for removal of MPs (Hamidian et al., 2021). The major drawback of this method is clogging the pores on the membrane surface by emerging fouling issues (particularly ultrafiltration, and microfiltration) which is a critical challenge to implement membrane filtration at a commercial level for a longer period (Kumar & Ismail, 2015; Reddy & Nair, 2022). For large scale stable applications, the issue of membrane fouling is required to control effectively with in-depth research. The literature noted ozone technology, chlorination, and UV disinfection are also applicable for microplastic removal from wastewater treatment plants. Dissolved air floatation allows to enhance microplastic removal efficiency up to 95% as plastics have lower density (Talvitie et al., 2017). In a modern wastewater treatment plant, MPs in the final treated water contained 0.25 particles/L with a substantial reduction of >98% during the grease removal phase, showing the promising applications of advanced approaches to reduce discharge of plastics into the environment (Murphy et al., 2016). Also, membrane bioreactors have been recorded as an excellent technology to removal of MPs (removal efficiency ≈ 99%) compared to other traditional and advanced treatment processes (Talvitie et al., 2017). In tertiary treatment systems, advanced oxidation process break down the carbon polymer chain and induce the oxidation with the presence of photocatalysts particularly for high density and low density PE (Reddy & Nair, 2022). The same literature suggests that the electrooxidation is an advanced, promising wastewater treatment technology which forms gas without discharging toxic derivatives. Effective MPs removal technology should ensure the reduced or zero microplastic accumulation in sludge and effluent (Cydzik et al., 2022) by preventing the discharge of this tiny pollutants into the environment.

Accordingly, sludge management and MPs removal is strictly suggested since a proportion of MPs shift into sludge in addition to treated effluents causing ecological toxicity (Liu et al., 2021). MPs removal is suggested prior entering the biological treatment system in which sludge formation happens (Cydzik et al., 2022). Anaerobic digestion is recorded lower microplastic abundance in sludge than thermal drying and lime stabilization by signifying potential application for MPs removal from sludge (Mahon et al., 2017). Composting is one management approach of the sludge. Moreover, hyper-thermophilic composting is another effective option to remove MPs (approximately 43.7%) from the sludge and removal rate of hyper-thermophilic composting is nine times higher compared to conventional compositing method (Chen et al., 2020a, b). Incineration is the most successful method to remove plastics from the environment under thermal treatments. Number of EU countries, and Japan practice the incineration as a thermal technology to management of plastic pollution. This will ensure the prevention of soil contamination by MPs. However, bottom ash could be a source of environmental microplastic after incineration (Yang et al., 2021). Mono incineration (incineration of sewage sludge separately) removes MPs effectively compared with co incineration: application of wastes as a fuel or thermal treatment of wastes (Salva et al., 2025) since co incineration would still remain microplastic in bottom ash (Cydzik et al., 2022). However, incineration is a costly option and developing countries with financial issues face difficulties during implementation of this method.

For minimize MPs discharge, sludge management and wastewater treatment can be effectively followed using the advanced infrastructure and equipment. For most developing countries, establishment of modern, advanced technological approaches could be a critical challenge due to high initial operational cost. In that scenario, international and regional level collaboration and cooperation is required in multiple ways such as provision of financial incentives (e.g.: loan system with no interest), technology transfer programs and capacity building campaigns for upgrading existing knowledge of stakeholders (e.g.: technical officers).

Moreover, MPs discharge control at the source is one of most effective feasible solutions to reduce the MPs load in wastewater. Existing household, agricultural and industrial system can be upgraded and modified to prevent microplastic discharge into sewage or near-by environment. For instance, synthetic textile fibers (secondary microplastic) discharge during laundering in washing machines. We also confirmed the abundance of fiber particles in wastewater samples during literature review. As revealed by past literature, the source of nearly 35% of marine microfibers are laundering textiles (Prata, 2018) possibly through wastewater effluents. As a mitigation measure, filtering devices are suggested to reduce the load of microfibers originating from laundering and prevent discharge of MPs into wastewater treatment plants (Reddy & Nair, 2022). A community scale study conducted to assess the efficiency of washing machine filters had recorded substantial reduction of microfibers in treated effluents of municipal wastewater treatment systems after installing the filters. Accordingly, microfiber filters in washing machines are suggested to reduce microplastic emission from household wastewater (Erdle et al., 2021). However, this kind of novel filter devices are expensive, and installation of filter is a time-consuming process. Hence, the cost of the filtration device may not be affordable for all household consumers. At this point, cost-effective, and user-friendly filtering innovations are suggested to reduce the microfiber discharge, specially focusing on developing nations.

Policy formulation and effective legal enforcement is a viable option in reduction of microplastic input via source management and treatment within the municipal and industrial wastewater treatment plants. In Canada, microfiber pollution control based legislation has been prioritized to prevent microfiber discharge via different ways such as community level house-hold applications (Erdle et al., 2021). This kind of legal approaches can be adopted by other nations also considering country-specific ecological, social, and financial factors. The full life cycle of plastics could be considered for national level environmental policy governance focusing on wastewater based microplastic input. For instance, ecofriendly biodegradable designs, green raw materials, and eco-labelling could be promoted during the manufacturing phase in the upstream of product life cycle by ensuring the reduction of microplastic discharge into sewer line or wastewater (e.g.: prevent or minimize microfibers shredding from ecofriendly garments). The entry of primary microplastic input can be restricted by policies which focus on incorporation of microbeads (as raw materials) in personal care products, and cosmetics.

Environmental policy could be introduced at the national level as aligning with Polluter Pays Principle (PPP) considering waste producers such as consumers, industrial sector, and households depending on cost for plastic waste collection, disposal, and treatments through taxes and fees. This kind of legislations could reduce the generation of plastic waste at the source and minimize the plastic (microplastic) entry into wastewater treatment. Further, the financial charges could be used for wastewater treatment operations. However, these policies are to be carefully adopted by considering the views of all stakeholder parties at the national level.

Eco-literacy and community awareness also act as an effective strategy to inform about the plastic/microplastic pollution in wastewater and its emerging impacts. The enhanced evo-literacy level is a powerful tool to make changes in attitudes and behaviors of all stakeholders (e.g.: general public and businesspersons) to minimize wastewater based microplastic pollution compared to legal enforcement. For instance, the consumers will tend to purchase biodegradable products rather than conventional plastic products with the attitudes on plastic free environment. It will reduce the microplastic input at the source and entry into the wastewater treatment system and/or ecosystems. As the environmental-conscious industries, they will also ensure minimum/zero microplastic discharge by final effluent and sludge through implementation of effective microplastic remediation techniques.

8 Conclusion

In this chapter, WWMPs abundance, qualitative & quantitative characters, distribution, fate in the environment, emerging risks and management measures were discussed in detail emphasizing critical challenges for effectively mitigate the microplastic emission from wastewater. From various sources, microplastic contamination in wastewater causes significant environmental toxicity, affecting both aquatic and terrestrial biological communities. Despite existing wastewater treatment methods, discharged microplastics and associated toxins persist, and accumulate in biota through the food chain. In general wastewater receiving and treatment plants serve as MPs hotspots. We further highlight the urgent need for advanced technologies to enhance MPs removal and management of sources of MPs. Future research implications should focus on innovative and cost-effective sustainable wastewater management strategies to mitigate impacts of MPs effectively towards a zero-plastic environment.

References

Abba, S., Yassin, M. A., Shah, S. M. H., Egbueri, J. C., Elzain, H. E., Agbasi, J. C., & Aljundi, I. H. (2024). Trace element pollution tracking in the complex multi-aquifer groundwater system of Al-Hassa oasis (Saudi Arabia) using spatial, chemometric and index-based techniques. *Environmental Research, 249*, 118320. https://doi.org/10.1016/j.envres.2024.118320

Ahmed, J., Thakur, A., & Goyal, A. (2021). Industrial wastewater and its toxic effects. In *The royal society of chemistry eBooks* (pp. 1–14). https://doi.org/10.1039/9781839165399-00001

Akyildiz, S. H., Fiore, S., Bruno, M., Sezgin, H., Yalcin-Enis, I., Yalcin, B., & Bellopede, R. (2024). Release of microplastic fibers from synthetic textiles during household washing. *Environmental Pollution, 357*, 124455. https://doi.org/10.1016/j.envpol.2024.124455

Albensi, B. C. (2019). What is nuclear factor Kappa B (NF-KB) doing in and to the mitochondrion? *Frontiers in Cell and Developmental Biology, 7*. https://doi.org/10.3389/fcell.2019.00154

Alharbi, O. M., Basheer, A. A., Khattab, R. A., & Ali, I. (2018). Health and environmental effects of persistent organic pollutants. *Journal of Molecular Liquids, 263*, 442–453. https://doi.org/10.1016/j.molliq.2018.05.029

Ali, I., Ding, T., Peng, C., Naz, I., Sun, H., Li, J., & Liu, J. (2021). Micro- and nanoplastics in wastewater treatment plants: Occurrence, removal, fate, impacts and remediation technologies – A critical review. *Chemical Engineering Journal, 423*, 130205. https://doi.org/10.1016/j.cej.2021.130205

Andrady, A. L., & Neal, M. A. (2009). Applications and societal benefits of plastics. *Philosophical Transactions of the Royal Society B Biological Sciences, 364*(1526), 1977–1984. https://doi.org/10.1098/rstb.2008.0304

Andrea, Käppler Dieter, Fischer Sonja, Oberbeckmann Gerald, Schernewski Matthias, Labrenz Klaus-Jochen, Eichhorn Brigitte, Voit (2016) Analysis of environmental microplastics by vibrational microspectroscopy: FTIR Raman or both? Analytical and Bioanalytical Chemistry 408(29) 8377–8391 https://doi.org/10.1007/s00216-016-9956-3

Anjali, V. R., Remya, V. S., Reshmi, S., Mahim, S. S., & Devi, C. A. (2019). Impact of Bisphenol S as an endocrine disruptor in a freshwater fish, *Oreochromis mossambicus*. *Journal of Endocrinology & Reproduction, 23*(2), 49–63. https://doi.org/10.18311/jer/2019/27184

Anthony L., Andrady (2011) Microplastics in the marine environment Marine Pollution Bulletin 62(8), 1596–1605. https://doi.org/10.1016/j.marpolbul.2011.05.030

Anthony, R. G., & Kozlowski, R. (1982). Heavy metals in tissues of small mammals inhabiting waste-water-irrigated habitats. *Journal of Environmental Quality, 11*(1), 20–22. https://doi.org/10.2134/jeq1982.00472425001100010005x

Apete, L., Martin, O. V., & Iacovidou, E. (2024). Fishing plastic waste: Knowns and known unknowns. *Marine Pollution Bulletin, 205*, 116530. https://doi.org/10.1016/j.marpolbul.2024.116530

Arab, M., Yu, J., & Nayebi, B. (2024). Microplastics in Sludges and soils: A comprehensive review on distribution, characteristics, and effects. *ChemEngineering, 8*(5), 86. https://doi.org/10.3390/chemengineering8050086

Aralappanavar, V. K., Mukhopadhyay, R., Yu, Y., Liu, J., Bhatnagar, A., Praveena, S. M., Li, Y., Paller, M., Adyel, T. M., Rinklebe, J., Bolan, N. S., & Sarkar, B. (2024). Effects of microplastics on soil microorganisms and microbial functions in nutrients and carbon cycling – A review. *Science of the Total Environment, 924*, 171435. https://doi.org/10.1016/j.scitotenv.2024.171435

Arregocés-Garcés, R., Garcés-Ordóñez, O., Vivas-Aguas, L.-J., & Canals, M. (2024). Microplastics transfer from a malfunctioning municipal wastewater oxidation pond into a marine protected area in the Colombian Caribbean. *Regional Studies in Marine Science, 69*, 103361. https://doi.org/10.1016/j.rsma.2023.103361

Asadi, A., Khodadoost, F., Daglioglu, N., & Eris, S. (2025). A systematic review on the occurrence and removal of microplastics during municipal wastewater treatment plants. *Environmental Engineering Research, 30*(2), 240366–240360. https://doi.org/10.4491/eer.2024.366

Ashrafy, A., Liza, A. A., Islam, M. N., Billah, M. M., Arafat, S. T., Rahman, M. M., & Rahman, S. M. (2023). Microplastics pollution: A brief review of its source and abundance in different aquatic ecosystems. *Journal of Hazardous Materials Advances, 9*, 100215. https://doi.org/10.1016/j.hazadv.2022.100215

Auta, H. S., Emenike, C. U., & Fauziah, S. H. (2017). Distribution and importance of microplastics in the marine environment: A review of the sources, fate, effects, and potential solutions. *Environment International, 102*, 165–176. https://doi.org/10.1016/j.envint.2017.02.013

Azizi, N., Nasseri, S., Nodehi, R. N., Jaafarzadeh, N., & Pirsaheb, M. (2022). Evaluation of conventional wastewater treatment plants efficiency to remove microplastics in terms of abundance, size, shape, and type: A systematic review and meta-analysis. *Marine Pollution Bulletin, 177*, 113462. https://doi.org/10.1016/j.marpolbul.2022.113462

Bakaraki Turan, N., Sari Erkan, H., & Onkal Engin, G. (2021). Microplastics in wastewater treatment plants: Occurrence, fate and identification. *Process Safety and Environmental Protection, 146*, 77–84. https://doi.org/10.1016/j.psep.2020.08.039

Baralla, E., Pasciu, V., Varoni, M. V., Nieddu, M., Demuro, R., & Demontis, M. P. (2021). Bisphenols' occurrence in bivalves as sentinel of environmental contamination. *The Science of the Total Environment, 785*, 147263. https://doi.org/10.1016/j.scitotenv.2021.147263

Barboza, L. G. A., Otero, X. L., Fernández, E. V., Vieira, L. R., Fernandes, J. O., Cunha, S. C., & Guilhermino, L. (2023). Are microplastics contributing to pollution-induced neurotoxicity? A pilot study with wild fish in a real scenario. *Heliyon, 9*(1), e13070. https://doi.org/10.1016/j.heliyon.2023.e13070

Bashir, S. M., Kimiko, S., Mak, C.-W., Fang, J. K.-H., & Gonçalves, D. (2021). Personal care and cosmetic products as a potential source of environmental contamination by microplastics in a densely populated Asian City. *Frontiers in Marine Science, 8*, 683482. https://doi.org/10.3389/fmars.2021.683482

Batley, G. E., Kirby, J. K., & McLaughlin, M. J. (2013). Fate and risks of nanomaterials in aquatic and terrestrial environments. *Accounts of Chemical Research, 46*(3), 854–862. https://doi.org/10.1021/ar2003368

Beriot, N., Peek, J., Zornoza, R., Geissen, V., & Huerta Lwanga, E. (2021). Low density-microplastics detected in sheep faeces and soil: A case study from the intensive vegetable farming in Southeast Spain. *Science of the Total Environment, 755*, 142653. https://doi.org/10.1016/j.scitotenv.2020.142653

Bhuyan, M. S. (2022). Effects of microplastics on fish and in human health. *Frontiers in Environmental Science, 10*. https://doi.org/10.3389/fenvs.2022.827289

Bikiaris, N., Nikolaidis, N. F., & Barmpalexis, P. (2024). Microplastics (MPs) in cosmetics: A review on their presence in personal-care, cosmetic, and cleaning products (PCCPs) and sustainable alternatives from biobased and biodegradable polymers. *Cosmetics, 11*(5), 145. https://doi.org/10.3390/cosmetics11050145

Bråte, I. L. N., Hurley, R., Iversen, K., Beyer, J., Thomas, K. V., Steindal, C. C., et al. (2018). Mytilus spp. as sentinels for monitoring microplastic pollution in Norwegian coastal waters: A qualitative and quantitative study. *Environmental Pollution, 243*, 383–393. https://doi.org/10.1016/j.envpol.2018.08.077

Bretas Alvim, C., M.A., Bes-Piá J.A., Mendoza-Roca (2020) Separation and identification of microplastics from primary and secondary effluents and activated sludge from wastewater treatment plants. Chemical Engineering Journal 402126293-https://doi.org/10.1016/j.cej.2020.126293

Büks, F., & Kaupenjohann, M. (2020). Global concentrations of microplastics in soils – A review. *SOIL, 6*(2), 649–662. https://doi.org/10.5194/soil-6-649-2020

Büks, F., Loes van Schaik, N., & Kaupenjohann, M. (2020). What do we know about how the terrestrial multicellular soil fauna reacts to microplastic? *The Soil, 6*(2), 245–267. https://doi.org/10.5194/soil-6-245-2020

Burgos-Aceves, M. A., Faggio, C., Betancourt-Lozano, M., González-Mille, D. J., & Ilizaliturri-Hernández, C. A. (2022). Ecotoxicological perspectives of microplastic pollution in amphibians. *Journal of Toxicology and Environmental Health, Part B, 25*(8), 405–421. https://doi.org/10.1080/10937404.2022.2140372

Caldito, N. G. (2023). Role of tumor necrosis factor-alpha in the central nervous system: A focus on autoimmune disorders. *Frontiers in Immunology, 14*, 1213448. https://doi.org/10.3389/fimmu.2023.1213448

Carlin, J., Craig, C., Little, S., Donnelly, M., Fox, D., Zhai, L., & Walters, L. (2020). Microplastic accumulation in the gastrointestinal tracts in birds of prey in Central Florida, USA. *Environmental Pollution, 264*, 114633. https://doi.org/10.1016/j.envpol.2020.114633

Carnevale Miino, M., Galafassi, S., Zullo, R., Torretta, V., & Rada, E. C. (2024). Microplastics removal in wastewater treatment plants: A review of the different approaches to limit their release in the environment. *Science of the Total Environment, 930*, 172675. https://doi.org/10.1016/j.scitotenv.2024.172675

Steve A., Carr Jin, Liu Arnold G., Tesoro (2016) Transport and fate of microplastic particles in wastewater treatment plants Water Research 91174–182. https://doi.org/10.1016/j.watres.2016.01.002

Carvalho, C. D. S., Bernusso, V. A., De Araújo, H. S. S., Espíndola, E. L. G., & Fernandes, M. N. (2012). Biomarker responses as indication of contaminant effects in *Oreochromis niloticus*. *Chemosphere, 89*(1), 60–69. https://doi.org/10.1016/j.chemosphere.2012.04.013

Charkiewicz, A. E., Omeljaniuk, W. J., Nowak, K., Garley, M., & Nikliński, J. (2023). Cadmium toxicity and health effects—A brief summary. *Molecules, 28*(18), 6620. https://doi.org/10.3390/molecules28186620

Chen, Y., Wen, D., Pei, J., Fei, Y., Ouyang, D., Zhang, H., & Luo, Y. (2020a). Identification and quantification of microplastics using Fourier-transform infrared spectroscopy: Current status and future prospects. *Current Opinion in Environmental Science & Health, 18*, 14–19. https://doi.org/10.1016/j.coesh.2020.05.004

Chen, Z., Zhao, W., Xing, R., Xie, S., Yang, X., Cui, P., Lü, J., Liao, H., Yu, Z., Wang, S., & Zhou, S. (2020b). Enhanced in situ biodegradation of microplastics in sewage sludge using hyperthermophilic composting technology. *Journal of Hazardous Materials, 384*, 121271. https://doi.org/10.1016/j.jhazmat.2019.121271

Chen, Y., Yang, J., Yao, B., Zhi, D., Luo, L., & Zhou, Y. (2022). Endocrine disrupting chemicals in the environment: Environmental sources, biological effects, remediation techniques, and perspective. *Environmental Pollution, 310*, 119918. https://doi.org/10.1016/j.envpol.2022.119918

Cheng, Y.-L., Zhang, R., Tisinger, L., Cali, S., Yu, Z., Chen, H. Y., & Li, A. (2022). Characterization of microplastics in sediment using stereomicroscopy and laser direct infrared (LDIR) spectroscopy. *Gondwana Research, 108*, 22–30. https://doi.org/10.1016/j.gr.2021.10.002

Choi, T., Choi, T., Lee, Y., Choe, S., & Kim, C. (2021). Zebrafish as an animal model for biomedical research. *Experimental & Molecular Medicine, 53*(3), 310–317. https://doi.org/10.1038/s12276-021-00571-5

Cole, M., Lindeque, P., Fileman, E., Halsband, C., Goodhead, R., Moger, J., & Galloway, T. S. (2013). Microplastic ingestion by zooplankton. *Environmental Science & Technology, 47*(12), 6646–6655. https://doi.org/10.1021/es400663f

Conley, K., Clum, A., Deepe, J., Lane, H., & Beckingham, B. (2019). Wastewater treatment plants as a source of microplastics to an urban estuary: Removal efficiencies and loading per capita over one year. *Water Research X, 3*, 100030. https://doi.org/10.1016/j.wroa.2019.100030

Corradini, F., Meza, P., Eguiluz, R., Casado, F., Huerta-Lwanga, E., & Geissen, V. (2019). Evidence of microplastic accumulation in agricultural soils from sewage sludge disposal. *The Science of the Total Environment, 671*, 411–420. https://doi.org/10.1016/j.scitotenv.2019.03.368

Corte Pause, F., Urli, S., Crociati, M., Stradaioli, G., & Baufeld, A. (2024). Connecting the dots: Livestock animals as missing links in the chain of microplastic contamination and human health. *Animals, 14*(2), 350. https://doi.org/10.3390/ani14020350

Costello, J. D., & Ebert, J. R. (2020). Microplastic pollutants in the coastal dunes of Lake Erie and Lake Ontario. *Journal of Great Lakes Research, 46*(6), 1754–1760. https://doi.org/10.1016/j.jglr.2020.08.001

Coyle, R., Hardiman, G., & Driscoll, K. O. (2020). Microplastics in the marine environment: A review of their sources, distribution processes, uptake and exchange in ecosystems. *Case Studies in Chemical and Environmental Engineering, 2*, 100010. https://doi.org/10.1016/j.cscee.2020.100010

Cydzik, K., Agnieszka, N. M., & Piotr, J. (2022). The fate of microplastic in sludge management systems. *Science of the Total Environment, 848*(June), 157466. https://doi.org/10.1016/j.scitotenv.2022.157466

da Costa Araújo, A. P., de Melo, N. F. S., de Oliveira Junior, A. G., Rodrigues, F. P., Fernandes, T., de Andrade Vieira, J. E., Rocha, T. L., & Malafaia, G. (2020). How much are microplastics harmful to the health of amphibians? A study with pristine polyethylene microplastics and Physalaemus cuvieri. *Journal of Hazardous Materials, 382*, 121066. https://doi.org/10.1016/j.jhazmat.2019.121066

Da Costa Araújo, A. P., Da Luz, T. M., Rocha, T. L., Ibrahim Ahmed, M. A., Melo E Silva, D. d., Rahman, M. M., & Malafaia, G. (2022). Toxicity evaluation of the combination of emerging pollutants with polyethylene microplastics in zebrafish: Perspective study of genotoxicity, mutagenicity, and redox unbalance. *Journal of Hazardous Materials, 432*, 128691. https://doi.org/10.1016/j.jhazmat.2022.128691

Daghighi, E., Shah, T., Chia, R., Lee, J.-Y., Shang, J., & Rodríguez-Seijo, A. (2023). The forgotten impacts of plastic contamination on terrestrial micro- and mesofauna: A call for research. *Environmental Research, 231*, 116227. https://doi.org/10.1016/j.envres.2023.116227

Das, B. C., Ramanan, P. A., Gorakh, S. S., Pillai, D., & Jayadradhan, R. K. V. (2023). Sub-chronic exposure of Oreochromis niloticus to environmentally relevant concentrations of smaller microplastics: Accumulation and toxico-physiological responses. *Journal of Hazardous Materials, 458*, 131916. https://doi.org/10.1016/j.jhazmat.2023.131916

David K. A., Barnes Francois, Galgani Richard C., Thompson Morton, Barlaz (2009). Accumulation and fragmentation of plastic debris in global environments Philosophical Transactions of the Royal Society B: Biological Sciences 364(1526), 1985–1998 https://doi.org/10.1098/rstb.2008.0205

Dayal, L., Yadav, K., Dey, U., Das, K., Kumari, P., Raj, D., & Mandal, R. R. (2024). Recent advancement in microplastic removal process from wastewater – A critical review. *Journal of Hazardous Materials Advances, 16*, 100460. https://doi.org/10.1016/j.hazadv.2024.100460

de Souza Machado, A. A., Kloas, W., Zarfl, C., Hempel, S., & Rillig, M. C. (2018). Microplastics as an emerging threat to terrestrial ecosystems. *Global Change Biology, 24*(4), 1405–1416. https://doi.org/10.1111/gcb.14020

de Souza Machado, A. A., Lau, C. W., Kloas, W., Bergmann, J., Bachelier, J. B., Faltin, E., Becker, R., Görlich, A. S., & Rillig, M. C. (2019). Microplastics Can change soil properties and affect plant performance. *Environmental Science & Technology, 53*(10), 6044–6052. https://doi.org/10.1021/acs.est.9b01339

Debnath, R., Prasad, G. S., Amin, A., Malik, M. M., Ahmad, I., Abubakr, A., Borah, S., Rather, M. A., Impellitteri, F., Tabassum, I., & Piccione, G. (2024). Understanding and addressing microplastic pollution: Impacts, mitigation, and future perspectives. *Journal of Contaminant Hydrology, 266*, 104399. https://doi.org/10.1016/j.jconhyd.2024.104399

Del Piano, F., Lama, A., Piccolo, G., Addeo, N. F., Iaccarino, D., Fusco, G., et al. (2023). Impact of polystyrene microplastic exposure on gilthead seabream (Sparus aurata Linnaeus, 1758): Differential inflammatory and immune response between anterior and posterior intestine. *The Science of the Total Environment, 879*, 163201. https://doi.org/10.1016/j.scitotenv.2023.163201

Dewika, M., Kantha, N., Markandan, K., Nagaratnam, S., Irfan, N. A., & Khalid, M. (2024). Microplastics release from coronary catheters: Insights from catheter analysis. *Chemosphere, 366*, 143428. https://doi.org/10.1016/j.chemosphere.2024.143428

Dey, U., Raj, D., Mondal, M., Roy, P., Mukherjee, A., Mondal, N. K., & Das, K. (2023). Microplastics in groundwater: An overview of source, distribution, mobility constraints and potential health impacts during the anthropocene. *Groundwater for Sustainable Development, 23*, 101036. https://doi.org/10.1016/j.gsd.2023.101036

Ding, J., Zhang, S., Razanajatovo, R. M., Zou, H., & Zhu, W. (2018). Accumulation, tissue distribution, and biochemical effects of polystyrene microplastics in the freshwater fish red tilapia (Oreochromis niloticus). *Environmental Pollution, 238*, 1–9. https://doi.org/10.1016/j.envpol.2018.03.001

Ding, P., Xiang, C., Li, X., Chen, H., Shi, X., Li, X., et al. (2023). Photoaged microplastics induce neurotoxicity via oxidative stress and abnormal neurotransmission in zebrafish larvae (Danio rerio). *The Science of the Total Environment, 881*, 163480. https://doi.org/10.1016/j.scitotenv.2023.163480

Dobrowolski, R., Krzyszczak, A., Dobrzyńska, J., Podkościelna, B., Zięba, E., Czemierska, M., et al. (2019). Extracellular polymeric substances immobilized on microspheres for removal of heavy metals from aqueous environment. *Biochemical Engineering Journal, 143*, 202–211. https://doi.org/10.1016/j.bej.2019.01.004

Dong, Y., Gao, M., Qiu, W., & Song, Z. (2021). Effect of microplastics and arsenic on nutrients and microorganisms in rice rhizosphere soil. *Ecotoxicology and Environmental Safety, 211*, 111899. https://doi.org/10.1016/j.ecoenv.2021.111899

Dong, S., Gao, P., Li, B., Feng, L., Liu, Y., Du, Z., & Zhang, L. (2022). Occurrence and migration of microplastics and plasticizers in different wastewater and sludge treatment units in municipal wastewater treatment plant. *Frontiers of Environmental Science & Engineering, 16*(11), 142. https://doi.org/10.1007/s11783-022-1577-9

Dovzhenko, N. V., Slobodskova, V. V., Mazur, A. A., Kukla, S. P., Istomina, A. A., Chelomin, V. P., & Beskhmelnov, D. D. (2024). Oxidative stress in mussel Mytilus trossulus induced by different-sized plastics. *Journal of Xenobiotics, 14*(4), 1826–1835. https://doi.org/10.3390/jox14040097

Dris, R., Gasperi, J., Saad, M., Mirande, C., & Tassin, B. (2016). Synthetic fibers in atmospheric fallout: A source of microplastics in the environment? *Marine Pollution Bulletin, 104*(1), 290–293. https://doi.org/10.1016/j.marpolbul.2016.01.006

Dümichen, E., Eisentraut, P., Bannick, C. G., Barthel, A.-K., Senz, R., & Braun, U. (2017). Fast identification of microplastics in complex environmental samples by a thermal degradation method. *Chemosphere, 174*, 572–584. https://doi.org/10.1016/j.chemosphere.2017.02.010

Eerkes-Medrano, D., Thompson, R. C., & Aldridge, D. C. (2015). Microplastics in freshwater systems: A review of the emerging threats, identification of knowledge gaps and prioritisation of research needs. *Water Research*, [online] 75,, *75*, 63–82. https://doi.org/10.1016/j.watres.2015.02.012

Elgarahy, A., Elwakeel, K., Mohammad, S., & Elshoubaky, G. (2021). A critical review of biosorption of dyes, heavy metals and metalloids from wastewater as an efficient and green process. *Cleaner Engineering and Technology, 4*, 100209. https://doi.org/10.1016/j.clet.2021.100209

Erdle, L. M., Parto, D. N., Sweetnam, D., & Rochman, C. M. (2021). Washing machine filters reduce microfiber emissions: Evidence from a community-scale pilot in Parry Sound, Ontario. *Frontiers in Marine Science, 8*(November), 1–9.

Erikson, K. M., & Aschner, M. (2019). *Manganese: Its role in disease and health* (pp. 253–266). De Gruyter eBooks. https://doi.org/10.1515/9783110527872-016

Félix, L., Carreira, P., & Peixoto, F. (2023). Effects of chronic exposure of naturally weathered microplastics on oxidative stress level, behaviour, and mitochondrial function of adult zebrafish (Danio rerio). *Chemosphere, 310*, 136895. https://doi.org/10.1016/j.chemosphere.2022.136895

Flores-Munguía, E. J., Rosas-Acevedo, J. L., Ramírez-Hernández, A., Aparicio-Saguilan, A., Brito-Carmona, R. M., & Violante-González, J. (2023). Release of microplastics from urban wastewater treatment plants to aquatic ecosystems in Acapulco, Mexico. *Water, 15*(20), 3643. https://doi.org/10.3390/w15203643

Fosmire, G. (1990). Zinc toxicity. *American Journal of Clinical Nutrition, 51*(2), 225–227. https://doi.org/10.1093/ajcn/51.2.225

Frassinetti, S., Bronzetti, G. L., Caltavuturo, L., Cini, M., & Della Croce, C. (2006). The role of zinc in life: A review. *Journal of Environmental Pathology Toxicology and Oncology, 25*(3), 597–610. https://doi.org/10.1615/jenvironpatholtoxicoloncol.v25.i3.40

Fries, E., Dekiff, J. H., Willmeyer, J., Nuelle, M.-T., Ebert, M., & Remy, D. (2013). Identification of polymer types and additives in marine microplastic particles using pyrolysis-GC/MS and scanning electron microscopy. *Environmental Science: Processes & Impacts, 15*(10), 1949–1956. https://doi.org/10.1039/C3EM00214D

Galafassi, S., Di Cesare, A., Di Nardo, L., Sabatino, R., Valsesia, A., Fumagalli, F. S., Corno, G., & Volta, P. (2022). Microplastic retention in small and medium municipal wastewater treatment plants and the role of the disinfection. *Environmental Science and Pollution Research, 29*(7), 10535–10546. https://doi.org/10.1007/s11356-021-16453-2

Galvão, A., Aleixo, M., De Pablo, H., Lopes, C., & Raimundo, J. (2020). Microplastics in wastewater: Microfiber emissions from common household laundry. *Environmental Science and Pollution Research, 27*(21), 26643–26649. https://doi.org/10.1007/s11356-020-08765-6

Gao, N., Huang, Z., Xing, J., Zhang, S., & Hou, J. (2021). Impact and molecular mechanism of microplastics on zebrafish in the presence and absence of copper nanoparticles. *Frontiers in Marine Science, 8*. https://doi.org/10.3389/fmars.2021.762530

Gaweł, S., Wardas, M., Niedworok, E., & Wardas, P. (2004). Malondialdehyde (MDA) as a lipid peroxidation marker. *PubMed, 57*(9–10), 453–455. Retrieved from https://pubmed.ncbi.nlm.nih.gov/15765761

Gaylarde, C. C., Neto, J. A. B., & da Fonseca, E. M. (2021). Paint fragments as polluting microplastics: A brief review. *Marine Pollution Bulletin, 162*, 111847. https://doi.org/10.1016/j.marpolbul.2020.111847

Genchi, G., Carocci, A., Lauria, G., Sinicropi, M. S., & Catalano, A. (2020). Nickel: Human health and environmental toxicology. *International Journal of Environmental Research and Public Health, 17*(3), 679. https://doi.org/10.3390/ijerph17030679

Ghanadi, M., Joshi, I., Dharmasiri, N., Jaeger, J. E., Burke, M., Bebelman, C., Symons, B., & Padhye, L. P. (2024). Quantification and characterization of microplastics in coastal environments: Insights from laser direct infrared imaging. *Science of The Total Environment, 912*, 168835. https://doi.org/10.1016/j.scitotenv.2023.168835

Gkoutselis, G., Rohrbach, S., Harjes, J., Obst, M., Brachmann, A., Horn, M. A., & Rambold, G. (2021). Microplastics accumulate fungal pathogens in terrestrial ecosystems. *Scientific Reports, 11*(1), 13214. https://doi.org/10.1038/s41598-021-92405-7

Gräslund, S., & Bengtsson, B. (2001). Chemicals and biological products used in south-east Asian shrimp farming, and their potential impact on the environment — a review. *The Science of the Total Environment, 280*(1–3), 93–131. https://doi.org/10.1016/s0048-9697(01)00818-x

Grobelak, A., Całus-Makowska, K., Jasińska, A., Klimasz, M., Wypart-Pawul, A., Augustajtys, D., Baor, E., Sławczyk, D., & Kowalska, A. (2024). Environmental impacts and contaminants management in sewage sludge-to-energy and fertilizer technologies: Current trends and future directions. *Energies, 17*(19), 4983. https://doi.org/10.3390/en17194983

Gulati, K., & Ray, A. (2009). Immunotoxicity. In *Elsevier eBooks* (pp. 595–609). https://doi.org/10.1016/b978-012374484-5.00040-7

Guo, J.-J., Huang, X.-P., Xiang, L., Wang, Y.-Z., Li, Y.-W., Li, H., Cai, Q.-Y., Mo, C.-H., & Wong, M.-H. (2020). Source, migration and toxicology of microplastics in soil. *Environment International, 137*, 105263. https://doi.org/10.1016/j.envint.2019.105263

Gupta, A., Kumar, S., Bajpai, Y., Chaturvedi, K., Johri, P., Tiwari, R. K., et al. (2024). Pharmaceutically active micropollutants: Origin, hazards and removal. *Frontiers in Microbiology, 15*. https://doi.org/10.3389/fmicb.2024.1339469

Gurjar, U. R., Takar, S., Amin, A., & Martin Xavier, K. A. (2023). Microplastic contamination in aquatic organisms: An ecotoxicological perspective. In *Xenobiotics in aquatic animals: Reproductive and developmental impacts* (pp. 353–367). Springer.

Peng, Peng Bangshang, Zhu Dongqi, Yang Lei, Su Huahong, Shi Daoji, Li (2017) Microplastics in sediments of the Changjiang Estuary China Environmental Pollution 225283-290. https://doi.org/10.1016/j.envpol.2016.12.064

Guzzi, G., Ronchi, A., & Pigatto, P. (2020). Toxic effects of mercury in humans and mammals. *Chemosphere, 263*, 127990. https://doi.org/10.1016/j.chemosphere.2020.127990

Hahladakis, J. N., Velis, C. A., Weber, R., Iacovidou, E., & Purnell, P. (2017). An overview of chemical additives present in plastics: Migration, release, fate and environmental impact during their use, disposal and recycling. *Journal of Hazardous Materials, 344*, 179–199. https://doi.org/10.1016/j.jhazmat.2017.10.014

Hajji, S., Ben-Haddad, M., Abelouah, M. R., De-la-Torre, G. E., & Alla, A. A. (2023). Occurrence, characteristics, and removal of microplastics in wastewater treatment plants located on the Moroccan Atlantic: The case of Agadir metropolis. *Science of The Total Environment, 862*, 160815. https://doi.org/10.1016/j.scitotenv.2022.160815

Halling-Sørensen, B. S., Nors Nielsen P.F., Lanzky F., Ingerslev H.C., Holten Lützhøft S.E., Jørgensen (1998) Occurrence fate and effects of pharmaceutical substances in the environment-A review Chemosphere 36(2) 357–393 https://doi.org/10.1016/S0045-6535(97)00354-8

Hamidian, A. H., Ozumchelouei, E. J., Feizi, F., Wu, C., Zhang, Y., & Yang, M. (2021). A review on the characteristics of microplastics in wastewater treatment plants: A source for toxic chemicals. *Journal of Cleaner Production, 295*, 126480. https://doi.org/10.1016/j.jclepro.2021.126480

Han, L., Chen, L., Feng, Y., Kuzyakov, Y., Chen, Q., Zhang, S., et al. (2024). Microplastics alter soil structure and microbial community composition. *Environment International, 185*, 108508. https://doi.org/10.1016/j.envint.2024.108508

Hansani, K. U. D. N., Thilakarathne, E. P. D. N., Koongolla, J. B., Gunathilaka, W. G. I. T., Perera, B. G. D. O., Weerasingha, W. M. P. U., & Egodauyana, K. P. U. T. (2023). Contamination of microplastics in tropical coral reef ecosystems of Sri Lanka. *Marine Pollution Bulletin, 194*, 115299. https://doi.org/10.1016/j.marpolbul.2023.115299

Hassan, F., Prasetya, K. D., Hanun, J. N., Bui, H. M., Rajendran, S., Kataria, N., Khoo, K. S., Wang, Y.-F., You, S.-J., & Jiang, J.-J. (2023). Microplastic contamination in sewage sludge: Abundance, characteristics, and impacts on the environment and human health. *Environmental Technology & Innovation, 31*, 103176. https://doi.org/10.1016/j.eti.2023.103176

Hayden, Webb Jaimys, Arnott Russell, Crawford Elena, Ivanova (2013) Plastic Degradation and Its Environmental Implications with Special Reference to Poly(ethylene terephthalate) Polymers 5(1) 1–18. https://doi.org/10.3390/polym5010001

Hegstrom, L. J., & West, S. D. (1989). Heavy metal accumulation in small mammals following sewage sludge application to forests. *Journal of Environmental Quality, 18*(3), 345–349. https://doi.org/10.2134/jeq1989.00472425001800030018x

Hermabessiere, L., Himber, C., Boricaud, B., Kazour, M., Amara, R., Cassone, A.-L., Laurentie, M., Paul-Pont, I., Soudant, P., Dehaut, A., & Duflos, G. (2018). Optimization, performance, and application of a pyrolysis-GC/MS method for the identification of microplastics. *Analytical and Bioanalytical Chemistry, 410*(25), 6663–6676. https://doi.org/10.1007/s00216-018-1279-0

Hidalgo-Ruz, V., Gutow, L., Thompson, R. C., & Thiel, M. (2012). Microplastics in the marine environment: A review of the methods used for identification and quantification. *Environmental Science & Technology, 46*(6), 3060–3075. https://doi.org/10.1021/es2031505

Hidayaturrahman, H., & Lee, T.-G. (2019). A study on characteristics of microplastic in wastewater of South Korea: Identification, quantification, and fate of microplastics during treatment process. *Marine Pollution Bulletin, 146*, 696–702. https://doi.org/10.1016/j.marpolbul.2019.06.071

Horton, A. A., Walton, A., Spurgeon, D. J., Lahive, E., & Svendsen, C. (2017). Microplastics in freshwater and terrestrial environments: Evaluating the current understanding to identify the knowledge gaps and future research priorities. *Science of The Total Environment, 586*, 127–141. https://doi.org/10.1016/j.scitotenv.2017.01.190

Hou, T., Yu, J., Li, C., Wang, Z., & Liu, H. (2024). Immunotoxicity of microplastics and polychlorinated biphenyls alone or in combination to Crassostrea gigas. *Marine Pollution Bulletin, 200*, 116161. https://doi.org/10.1016/j.marpolbul.2024.116161

Huang, Y., Dsikowitzky, L., Yang, F., & Schwarzbauer, J. (2020). Emerging contaminants in municipal wastewaters and their relevance for the surface water contamination in the tropical coastal city Haikou, China. *Estuarine Coastal and Shelf Science, 235*, 106611. https://doi.org/10.1016/j.ecss.2020.106611

Huerta Lwanga, E., Gertsen, H., Gooren, H., Peters, P., Salánki, T., van der Ploeg, M., Besseling, E., Koelmans, A. A., & Geissen, V. (2016). Microplastics in the terrestrial ecosystem: Implications for *Lumbricus terrestris* (Oligochaeta, Lumbricidae). *Environmental Science & Technology, 50*(5), 2685–2691. https://doi.org/10.1021/acs.est.5b05478

Ibrahim, B., & Fakhre, N. (2018). Crown ether modification of starch for adsorption of heavy metals from synthetic wastewater. *International Journal of Biological Macromolecules, 123*, 70–80. https://doi.org/10.1016/j.ijbiomac.2018.11.058

Idrissi, O. E., Ternengo, S., Monnier, B., Lepoint, G., Aiello, A., Bastien, R., & Gobert, S. (2023). Assessment of trace element contamination and effects on *Paracentrotus lividus* using several approaches: Pollution indices, accumulation factors and biochemical tools. *The Science of the Total Environment, 869*, 161686. https://doi.org/10.1016/j.scitotenv.2023.161686

Ighodaro, O., & Akinloye, O. (2017). First line defence antioxidants-superoxide dismutase (SOD), catalase (CAT) and glutathione peroxidase (GPX): Their fundamental role in the entire antioxidant defence grid. *Alexandria Journal of Medicine, 54*(4), 287–293. https://doi.org/10.1016/j.ajme.2017.09.001

Imogen E., Napper Richard C., Thompson (2016) Release of synthetic microplastic plastic fibres from domestic washing machines: Effects of fabric type and washing conditions Marine Pollution Bulletin 112(1-2) 39–45. https://doi.org/10.1016/j.marpolbul.2016.09.025

Jadhav, K., Raghuraj, N., Ray, E., Vaghasiya, K., & Verma, R. K. (2022). Neurodegenerative disorders due to inhalation of various small particles. In *Elsevier eBooks* (pp. 41–54). https://doi.org/10.1016/b978-0-323-85544-0.00010-1

Jani, V., Wu, S., & Venkiteshwaran, K. (2024). Advancements and Regulatory Situation in Microplastics Removal from Wastewater and Drinking Water: A Comprehensive Review. Microplastics, 3(1), 98–123. https://doi.org/10.3390/microplastics3010007

Järlskog, I., Strömvall, A.-M., Magnusson, K., Gustafsson, M., Polukarova, M., Galfi, H., Aronsson, M., & Andersson-Sköld, Y. (2020). Occurrence of tire and bitumen wear microplastics on urban streets and in sweepsand and washwater. *Science of the Total Environment, 729*, 138950. https://doi.org/10.1016/j.scitotenv.2020.138950

Jeyavani, J., Sibiya, A., Stalin, T., Vigneshkumar, G., Al-Ghanim, K. A., Riaz, M. N., & Vaseeharan, B. (2023). Biochemical, genotoxic and histological implications of polypropylene microplastics on freshwater fish *Oreochromis mossambicus*: An aquatic eco-toxicological assessment. *Toxics, 11*(3), 282. https://doi.org/10.3390/toxics11030282

Jiang, X., Chen, H., Liao, Y., Ye, Z., Li, M., & Klobučar, G. (2019). Ecotoxicity and genotoxicity of polystyrene microplastics on higher plant Vicia faba. *Environmental Pollution, 250*, 831–838. https://doi.org/10.1016/j.envpol.2019.04.055

Joo, S. H., Liang, Y., Kim, M., Byun, J., & Choi, H. (2021). Microplastics with adsorbed contaminants: Mechanisms and treatment. *Environmental Challenges, 3*, 100042. https://doi.org/10.1016/j.envc.2021.100042

Justino, A. K., Ferreira, G. V., Fauvelle, V., Schmidt, N., Lenoble, V., Pelage, L., et al. (2023). From prey to predators: Evidence of microplastic trophic transfer in tuna and large pelagic species in the southwestern tropical Atlantic. *Environmental Pollution, 327*, 121532. https://doi.org/10.1016/j.envpol.2023.121532

Kalaronis, D., Ainali, N. M., Evgenidou, E., Kyzas, G. Z., Yang, X., Bikiaris, D. N., & Lambropoulou, D. A. (2022). *Microscopic techniques as means for the determination of microplastics and nanoplastics in the aquatic environment: A concise review* (Vol. 3, p. 100036). Green Analytical Chemistry. https://doi.org/10.1016/j.greeac.2022.100036

Kang, P., Ji, B., Zhao, Y., & Wei, T. (2020). How can we trace microplastics in wastewater treatment plants: A review of the current knowledge on their analysis approaches. *Science of The Total Environment, 745*, 140943. https://doi.org/10.1016/j.scitotenv.2020.140943

Karlsson, T. M., Arneborg, L., Broström, G., Almroth, B. C., Gipperth, L., & Hassellöv, M. (2018). The unaccountability case of plastic pellet pollution. *Marine Pollution Bulletin, 129*(1), 52–60. https://doi.org/10.1016/j.marpolbul.2018.01.041

Khalid, N., Aqeel, M., & Noman, A. (2020). Microplastics could be a threat to plants in terrestrial systems directly or indirectly. *Environmental Pollution, 267*, 115653. https://doi.org/10.1016/j.envpol.2020.115653

Khaliq, Z., Ashraf, M. B., Abbasi, N. A., Ahmad, S. R., Shahid, S. U., & Qadir, A. (2024). Assessment of microplastics in gastrointestinal tract of cattle egret (Bubulcus ibis) from a metropolitan city Lahore. *Pakistan. Environmental Science and Pollution Research, 31*(56), 64903–64912. https://doi.org/10.1007/s11356-024-35540-8

Kılıç, E., Yücel, N., & Mübarek Şahutoğlu, S. (2022). First record of microplastic occurence at the commercial fish from Orontes River. *Environmental Pollution, 307*, 119576. https://doi.org/10.1016/j.envpol.2022.119576

Kim, M.-J., Na, S.-H., Batool, R., Byun, I.-S., & Kim, E.-J. (2022). Seasonal variation and spatial distribution of microplastics in tertiary wastewater treatment plant in South Korea. *The Journal of Hazardous Materials, 438*, 129474. https://doi.org/10.2139/ssrn.4110879

Kittipongvises, S., Phetrak, A., Hongprasith, N., & Lohwacharin, J. (2022). Unravelling capability of municipal wastewater treatment plant in Thailand for microplastics: Effects of seasonality on detection, fate and transport. *Journal of Environmental Management, 302*, 113990. https://doi.org/10.1016/j.jenvman.2021.113990

Kozioł, K., Kozak, K., & Polkowska, Ż. (2017). Hydrophobic and hydrophilic properties of pollutants as a factor influencing their redistribution during snowpack melt. *The Science of the Total Environment, 596–597*, 158–168. https://doi.org/10.1016/j.scitotenv.2017.04.061

Kumar, R., & Ismail, A. F. (2015). Fouling control on microfiltration/ultrafiltration membranes: Effects of morphology, hydrophilicity, and charge. *Journal of Applied Polymer Science, 132*, 42042. https://doi.org/10.1002/app.42042

Kumar, A., Mishra, S., Pandey, R., Yu, Z. G., Kumar, M., Khoo, K. S., Thakur, T. K., & Show, P. L. (2023). Microplastics in terrestrial ecosystems: Un-ignorable impacts on soil characterises, nutrient storage and its cycling. *TrAC Trends in Analytical Chemistry, 158*, 116869. https://doi.org/10.1016/j.trac.2022.116869

Kye, H., Kim, J., Ju, S., Lee, J., Lim, C., & Yoon, Y. (2023). Microplastics in water systems: A review of their impacts on the environment and their potential hazards. *Heliyon, 9*(3), e14359. https://doi.org/10.1016/j.heliyon.2023.e14359

Kyrikou, I., Briassoulis, D. (2007). Biodegradation of Agricultural Plastic Films: A Critical Review. J Polym Environ 15, 125–150. https://doi.org/10.1007/s10924-007-0053-8

Lant, N. J., Hayward, A. S., Peththawadu, M. M. D., Sheridan, K. J., & Dean, J. R. (2020). Microfiber release from real soiled consumer laundry and the impact of fabric care products and washing conditions. *Plos One, 15*(6), e0233332. https://doi.org/10.1371/journal.pone.0233332

Lebreton, L., & Andrady, A. (2019). Future scenarios of global plastic waste generation and disposal. *Palgrave Communications, 5*(1), 6. https://doi.org/10.1057/s41599-018-0212-7

Lee, J.-Y., Cha, J., Ha, K., & Viaroli, S. (2024). Microplastic pollution in groundwater: A systematic review. *Environmental Pollutants and Bioavailability, 36*(1), 2299545. https://doi.org/10.1080/26395940.2023.2299545

Lehtiniemi, M., Hartikainen, S., Näkki, P., Engström-Öst, J., Koistinen, A., & Setälä, O. (2018). Size matters more than shape: Ingestion of primary and secondary microplastics by small predators. *Food Webs, 17*, e00097. https://doi.org/10.1016/j.fooweb.2018.e00097

Li, X., Mei, Q., Chen, L., Zhang, H., Dong, B., Dai, X., He, C., & Zhou, J. (2019). Enhancement in adsorption potential of microplastics in sewage sludge for metal pollutants after the wastewater treatment process. *Water Research, 157*, 228–237. https://doi.org/10.1016/j.watres.2019.03.069

Li, P., Wang, X., Su, M., Zou, X., Duan, L., & Zhang, H. (2021). Characteristics of plastic pollution in the environment: A review. *Bulletin of Environmental Contamination and Toxicology, 107*(4), 577–584. https://doi.org/10.1007/s00128-020-02820-1

Li, J., Yu, S., Yu, Y., & Xu, M. (2022). Effects of microplastics on higher plants: A review. *Bulletin of Environmental Contamination and Toxicology, 109*(2), 241–265. https://doi.org/10.1007/s00128-022-03566-8

Li, C., Zhu, L., Li, W.-T., & Li, D. (2023a). Microplastics in the seagrass ecosystems: A critical review. *Science of The Total Environment, 902*, 166152. https://doi.org/10.1016/j.scitotenv.2023.166152

Li, Y., Feng, H., Xian, S., Wang, J., Zheng, X., & Song, X. (2023b). Phytotoxic effects of polyethylene microplastics combined with cadmium on the photosynthetic performance of maize (Zea mays L.). *Plant Physiology and Biochemistry, 203*, 108065. https://doi.org/10.1016/j.plaphy.2023.108065

Li, H., Liu, H., Bi, L., Liu, Y., Jin, L., & Peng, R. (2024). Immunotoxicity of microplastics in fish. *Fish & Shellfish Immunology, 150*, 109619. https://doi.org/10.1016/j.fsi.2024.109619

Limonta, G., Mancia, A., Benkhalqui, A., Bertolucci, C., Abelli, L., Fossi, M. C., & Panti, C. (2019). Microplastics induce transcriptional changes, immune response and behavioral alterations in adult zebrafish. *Scientific Reports, 9*(1), 15775. https://doi.org/10.1038/s41598-019-52292-5

Lin, Q., Pang, L., Ngo, H. H., Guo, W., Zhao, S., Liu, L., Chen, L., & Li, F. (2023). Occurrence of microplastics in three types of household cleaning products and their estimated emissions into the aquatic environment. *Science of The Total Environment, 902*, 165903. https://doi.org/10.1016/j.scitotenv.2023.165903

Liu, W., Zhang, J., Liu, H., Guo, X., Zhang, X., Yao, X., Cao, Z., & Zhang, T. (2021). A review of the removal of microplastics in global wastewater treatment plants: Characteristics and mechanisms. *Environment International, 146*, 106277. https://doi.org/10.1016/j.envint.2020.106277

Liu, J., Han, S., Wang, P., Zhang, X., Zhang, J., Hou, L., Zhang, Y., Wang, Y., Li, L., & Lin, Y. (2024). Soil microorganisms play an important role in the detrimental impact of biodegradable microplastics on plants. *Science of the Total Environment, 933*, 172933. https://doi.org/10.1016/j.scitotenv.2024.172933

Lloret, J., Pedrosa-Pamies, R., Vandal, N., Rorty, R., Ritchie, M., McGuire, C., Chenoweth, K., & Valiela, I. (2021). Salt marsh sediments act as sinks for microplastics and reveal effects of current and historical land use changes. *Environmental Advances, 4*, 100060. https://doi.org/10.1016/j.envadv.2021.100060

Lofty, J., Muhawenimana, V., Wilson, C. A. M. E., & Ouro, P. (2022). Microplastics removal from a primary settler tank in a wastewater treatment plant and estimations of contamination onto European agricultural land via sewage sludge recycling. *Environmental Pollution, 304*, 119198. https://doi.org/10.1016/j.envpol.2022.119198

Lu, M., Wu, X., Zeng, D., & Liao, Y. (2012). Distribution of PCDD/Fs and organometallic compounds in sewage sludge of wastewater treatment plants in China. *Environmental Pollution, 171*, 78–84. https://doi.org/10.1016/j.envpol.2012.07.035

Lushchak, V. I., Matviishyn, T. M., Husak, V. V., Storey, J. M., & Storey, K. B. (2018). Pesticide toxicity: A mechanistic approach. *EXCLI Journal, 17*, 1101–1136. https://doi.org/10.17179/excli2018-1710

Lv, X., Dong, Q., Zuo, Z., Liu, Y., Huang, X., & Wu, W. M. (2019). Microplastics in a municipal wastewater treatment plant: Fate, dynamic distribution, removal efficiencies, and control strategies. *Journal of Cleaner Production, 225*, 579–586. https://doi.org/10.1016/j.jclepro.2019.03.321

Ma, C., Shi, H., & Slaveykova, V. I. (2024). Entanglement of Daphnia magna by fibrous microplastics through "hook and loop" action. *Environmental Science & Technology Letters, 11*(5), 433–437. https://doi.org/10.1021/acs.estlett.4c00279

Masura, Julie et al. (2015). Laboratory methods for the analysis of microplastics in the marine environment : recommendations for quantifying synthetic particles in waters and sediments.

Mackenzie, C. M., & Vladimirova, V. (2023). Preliminary study and first evidence of presence of microplastics in terrestrial herpetofauna from southwestern Paraguay. *Studies on Neotropical Fauna and Environment, 58*(1), 16–24. https://doi.org/10.1080/01650521.2021.1895466

Mahon, A. M., O'Connell, B., Healy, M. G., O'Connor, I., Officer, R., Nash, R., & Morrison, L. (2017). Microplastics in sewage sludge: Effects of treatment. *Environmental Science & Technology, 51*, 810–818. https://doi.org/10.1021/acs.est.6b04048

Majumder, A., Gupta, B., & Gupta, A. K. (2019). Pharmaceutically active compounds in aqueous environment: A status, toxicity and insights of remediation. *Environmental Research, 176*, 108542. https://doi.org/10.1016/j.envres.2019.108542

Malik, M., Balkhi, M. H., & Abubakr, A. (2017). Comparative study on the contamination level of Dal Lake, Kashmir based on Total viable bacterial counts over a period of time. *Ecology, Environment and Conservation, 23*(4), 2145–2147.

Mammo, F. K., Amoah, I. D., Gani, K. M., Pillay, L., Ratha, S. K., Bux, F., & Kumari, S. (2020). Microplastics in the environment: Interactions with microbes and chemical contaminants. *Science of the Total Environment, 743*, 140518. https://doi.org/10.1016/j.scitotenv.2020.140518

Mandal, A., Hossain, U., & Sil, P. C. (2022). Antioxidants and cardiovascular diseases. In *Elsevier eBooks* (pp. 613–640). https://doi.org/10.1016/b978-0-12-819096-8.00072-0

Martínez-Gómez, C., Bignell, J., & Lowe, D. (2015). *Lysosomal membrane stability in mussels*. International Council for the Exploration of the Sea (ICES). Available at https://repository.oceanbestpractices.org/bitstream/handle/11329/683/TIMES%2056.pdf?sequence=1&isAllowed=y

Masiá, P., Sol, D., Ardura, A., Laca, A., Borrell, Y. J., Dopico, E., Laca, A., Machado-Schiaffino, G., Díaz, M., & Garcia-Vazquez, E. (2020). Bioremediation as a promising strategy for microplastics removal in wastewater treatment plants. *Marine Pollution Bulletin, 156*, 111252. https://doi.org/10.1016/j.marpolbul.2020.111252

Mason, S. A., Garneau, D., Sutton, R., Chu, Y., Ehmann, K., Barnes, J., Fink, P., Papazissimos, D., & Rogers, D. L. (2016). Microplastic pollution is widely detected in US municipal wastewater treatment plant effluent. Environmental pollution (Barking, Essex : 1987), 218, 1045–1054. https://doi.org/10.1016/j.envpol.2016.08.056

Mathew, J., Pulicharla, R., Rezai, P., & Brar, S. K. (2024). Microplastics in wastewaters: Pretreatment to detection trail. *Journal of Water Process Engineering, 64*, 105702. https://doi.org/10.1016/j.jwpe.2024.105702

Maurizi, L., Iordachescu, L., Kirstein, I. V., Nielsen, A. H., & Vollertsen, J. (2023). It matters how we measure – Quantification of microplastics in drinking water by μFTIR and μRaman. *Heliyon, 9*(9), e20119. https://doi.org/10.1016/j.heliyon.2023.e20119

Maw, M. M., Boontanon, N., Aung, H. K. Z. Z., Jindal, R., Fujii, S., Visvanathan, C., & Boontanon, S. K. (2024). Microplastics in wastewater and sludge from centralized and decentralized wastewater treatment plants: Effects of treatment systems and microplastic characteristics. *Chemosphere, 361*, 142536. https://doi.org/10.1016/j.chemosphere.2024.142536

Mazhandu, Z. S., & Muzenda, E. (2019). Global plastic waste pollution challenges and management. In: *2019 7th international renewable and sustainable energy conference (IRSEC)*, 1–8. https://doi.org/10.1109/IRSEC48032.2019.9078268.

Megill, C., Shaw, K., Knauer, K., Seeley, M., & Lynch, J. (2024). Plastic additives in the ocean: Use of a comprehensive dataset for meta-analysis and method development. *Chemosphere, 358*, 142172. https://doi.org/10.1016/j.chemosphere.2024.142172

Menezes, M., De Mello, F. T., Ziegler, L., Wanderley, B., Gutiérrez, J. M., & Dias, J. D. (2024). Revealing the hidden threats: Genotoxic effects of microplastics on freshwater fish. *Aquatic Toxicology, 276*, 107089. https://doi.org/10.1016/j.aquatox.2024.107089

Meyers, N., Catarino, A. I., Declercq, A. M., Brenan, A., Devriese, L., Vandegehuchte, M., De Witte, B., Janssen, C., & Everaert, G. (2022). Microplastic detection and identification by Nile red staining: Towards a semi-automated, cost- and time-effective technique. *Science of The Total Environment, 823*, 153441. https://doi.org/10.1016/j.scitotenv.2022.153441

Mghili, B., Keznine, M., Analla, M., & Aksissou, M. (2023). The impacts of abandoned, discarded and lost fishing gear on marine biodiversity in Morocco. *Ocean & Coastal Management*, [online] 239,, *239*, 106593. https://doi.org/10.1016/j.ocecoaman.2023.106593

Mintenig, S. M., Int-Veen, I., Löder, M. G. J., Primpke, S., & Gerdts, G. (2017). Identification of microplastic in effluents of waste water treatment plants using focal plane array-based micro-Fourier-transform infrared imaging. *Water Research, 108*, 365–372. https://doi.org/10.1016/j.watres.2016.11.015

Mirka, Lares Mohamed Chaker, Ncibi Markus, Sillanpää Mika, Sillanpää (2018) Occurrence identification and removal of microplastic particles and fibers in conventional activated sludge process and advanced MBR technology Water Research 133236-246. https://doi.org/10.1016/j.watres.2018.01.049

Möhrke, A. C. F., Haegerbaeumer, A., Traunspurger, W., & Höss, S. (2022). Underestimated and ignored? The impacts of microplastic on soil invertebrates—Current scientific knowledge and research needs. *Frontiers in Environmental Science, 10*. https://doi.org/10.3389/fenvs.2022.975904

Monira, S., Roychand, R., Hai, F. I., Bhuiyan, M., & Pramanik, B. K. (2025). Microplastic fragmentation into nanoplastics by water shear forces during wastewater treatment: Mechanical insights and theoretical analysis. *Environmental Pollution, 364*, 125310. https://doi.org/10.1016/j.envpol.2024.125310

Moraes, T. J., Chow, C., & Downey, G. P. (2012). Pulmonary host defenses. In *Elsevier eBooks* (pp. 275–287). https://doi.org/10.1016/b978-1-4557-0792-8.00022-2

Moyal, J., Dave, P. H., Wu, M., Karimpour, S., Brar, S. K., Zhong, H., & Kwong, R. W. M. (2023). Impacts of biofilm formation on the physicochemical properties and toxicity of microplastics: A concise review. *Reviews of Environmental Contamination and Toxicology, [online], 261*(1), 8. https://doi.org/10.1007/s44169-023-00035-z

Murphy, F., Ewins, C., Carbonnier, F., & Quinn, B. (2016). Wastewater treatment works (WwTW) as a source of microplastics in the aquatic environment. *Environmental Science & Technology, [online], 50*(11), 5800–5808. https://doi.org/10.1021/acs.est.5b05416

Na, S.-H., Kim, M.-J., Kim, J., Batool, R., Cho, K., Chung, J., Lee, S., & Kim, E.-J. (2024). Fate and potential risks of microplastic fibers and fragments in water and wastewater treatment processes. *Journal of Hazardous Materials, 463*, 132938. https://doi.org/10.1016/j.jhazmat.2023.132938

Nabi, N., Ahmad, I., Amin, A., Rather, M. A., Ahmed, I., Hajam, Y. A., Khursheed, S., Malik, M. M., & Abubakr, A. (2024). Understanding the sources, fate and effects of microplastics in aquatic environments with a focus on risk profiling in aquaculture systems. *Reviews in Aquaculture, 16*(4), 1947–1980. https://doi.org/10.1111/raq.12941

Nair, H. T., & Perumal, S. (2024). Growth performance, hematological and oxidative stress responses in Nile tilapia (*Oreochromis niloticus*) exposed to polypropylene microplastics. *Environmental Quality Management, 34*(1). https://doi.org/10.1002/tqem.22179

Ngo, P. L., Pramanik, B. K., Shah, K., & Roychand, R. (2019). Pathway, classification and removal efficiency of microplastics in wastewater treatment plants. *Environmental Pollution, 255*, 113326. https://doi.org/10.1016/j.envpol.2019.113326

O'Brien, S., Rauert, C., Ribeiro, F., Okoffo, E. D., Burrows, S. D., O'Brien, J. W., Wang, X., Wright, S. L., & Thomas, K. V. (2023). There's something in the air: A review of sources, prevalence and behaviour of microplastics in the atmosphere. *Science of The Total Environment, 874*, 162193. https://doi.org/10.1016/j.scitotenv.2023.162193

O'Donovan, S., Mestre, N. C., Abel, S., Fonseca, T. G., Carteny, C. C., Cormier, B., et al. (2018). Ecotoxicological effects of chemical contaminants adsorbed to microplastics in the clam *Scrobicularia plana. Frontiers in Marine Science, 5*, 143. https://doi.org/10.3389/fmars.2018.00143

O'Neal, S. L., & Zheng, W. (2015). Manganese toxicity upon overexposure: A decade in review. *Current Environmental Health Reports, 2*(3), 315–328. https://doi.org/10.1007/s40572-015-0056-x

Ochoa, L., Chan, J., Auguste, C., Arbuckle-Keil, G., & Fahrenfeld, N. L. (2024). Stormwater runoff microplastics: Polymer types, particle size, and factors controlling loading rates. *Science of The Total Environment, 929*, 172485. https://doi.org/10.1016/j.scitotenv.2024.172485

Oe, S., Miyagawa, K., Honma, Y., & Harada, M. (2016). Copper induces hepatocyte injury due to the endoplasmic reticulum stress in cultured cells and patients with Wilson disease. *Experimental Cell Research, 347*(1), 192–200. https://doi.org/10.1016/j.yexcr.2016.08.003

Oehlmann, J., Schulte-Oehlmann, U., Kloas, W., Jagnytsch, O., Lutz, I., Kusk, K. O., & Tyler, C. R. (2009). A critical analysis of the biological impacts of plasticizers on wildlife. *Philosophical Transactions of the Royal Society B Biological Sciences, 364*(1526), 2047–2062. https://doi.org/10.1098/rstb.2008.0242

Okoffo, E. D., O'Brien, S., O'Brien, J. W., Tscharke, B. J., & Thomas, K. V. (2019). Wastewater treatment plants as a source of plastics in the environment: A review of occurrence, methods for identification, quantification and fate. *Environmental Science: Water Research & Technology, 5*(11), 1908–1931. https://doi.org/10.1039/C9EW00428A

Palacios-Mateo, C., van der Meer, Y., & Seide, G. (2021). Analysis of the polyester clothing value chain to identify key intervention points for sustainability. *Environmental Sciences Europe, 33*(1), 2. https://doi.org/10.1186/s12302-020-00447-x

Parashar, N., & Hait, S. (2022). Occurrence and removal of microplastics in a hybrid growth sewage treatment plant from Bihar, India: A preliminary study. *Journal of Cleaner Production, 376*, 134295. https://doi.org/10.1016/j.jclepro.2022.134295

Parashar, N., & Hait, S. (2023). Abundance, characterization, and removal of microplastics in different technology-based sewage treatment plants discharging into the middle stretch of the Ganga River, India. *Science of The Total Environment, 905*, 167099. https://doi.org/10.1016/j.scitotenv.2023.167099

Parker, B., Andreou, D., Green, I. D., & Britton, J. R. (2021). Microplastics in freshwater fishes: Occurrence, impacts and future perspectives. *Fish and Fisheries, 22*(3), 467–488. https://doi.org/10.1111/faf.12528

Phaniendra, A., Jestadi, D. B., & Periyasamy, L. (2014). Free radicals: Properties, sources, targets, and their implication in various diseases. *Indian Journal of Clinical Biochemistry, 30*(1), 11–26. https://doi.org/10.1007/s12291-014-0446-0

Pizzurro, F., Nerone, E., Ancora, M., Di Domenico, M., Mincarelli, L. F., Cammà, C., et al. (2023). Exposure of Mytilus galloprovincialis to microplastics: Accumulation, depuration and evaluation of the expression levels of a selection of molecular biomarkers. *Animals, 14*(1), 4. https://doi.org/10.3390/ani14010004

Prajapati, S., Beal, M., Maley, J., & Brinkmann, M. (2021). Qualitative and quantitative analysis of microplastics and microfiber contamination in effluents of the City of Saskatoon wastewater treatment plant. *Environmental Science and Pollution Research, 28*(25), 32545–32553. https://doi.org/10.1007/s11356-021-12898-7

Prajapati, A., Vaidya, A. N., & Kumar, A. R. (2022). Microplastic properties and their interaction with hydrophobic organic contaminants: a review. *Environmental Science and Pollution Research, 29*(33), 49490–49512. https://doi.org/10.1007/s11356-022-20723-y

Prasad, G. S., Rout, S. K., Malik, M. M., Karmakar, S., Amin, A., & Ahmad, I. (2023). Occurrence of xenoestrogen alkylphenols (Octylphenols and Nonylphenol) and its impact on the aquatic ecosystem. In M. A. Rather, A. Amin, Y. A. Hajam, A. Jamwal, & I. Ahmad (Eds.), *Xenobiotics in aquatic animals: Reproductive and developmental impacts* (pp. 275–284). Springer. https://doi.org/10.1007/978-981-99-1214-8_13

Prasad, G. S., Rout, S. K., Amin, A., Malik, M. M., Ahmad, I., Rather, M. A., Karmakar, S., Khalil, H. S., & Abubakar, A. (2024). Impact of COVID-19 on the aquatic environment and implications for fisheries with special emphasis to Indian perspective. *Proceedings of the Indian National Science Academy, 90*(4), 821–839. https://doi.org/10.1007/s43538-024-00289-8

Prata, J. C. (2018). Microplastics in wastewater: State of the knowledge on sources, fate and solutions. *Marine Pollution Bulletin, 129*(1), 262–265. https://doi.org/10.1016/j.marpolbul.2018.02.046

Priya, A. K., Muruganandam, M., Imran, M., Gill, R., Vasudeva Reddy, M. R., Shkir, M., Sayed, M. A., AlAbdulaal, T. H., Algarni, H., Arif, M., Jha, N. K., & Sehgal, S. S. (2023). A study on managing plastic waste to tackle the worldwide plastic contamination and environmental remediation. *Chemosphere, 341*, 139979. https://doi.org/10.1016/j.chemosphere.2023.139979

Pryor, A. N., Selwood, T., Leu, L. S., Andracki, M. A., Lee, B. H., Rao, M., et al. (1992). Simple general acid-base catalysis of physiological acetylcholinesterase reactions. *Journal of the American Chemical Society, 114*(10), 3896–3900. https://doi.org/10.1021/ja00036a043

Qadir, M., Drechsel, P., Cisneros, B. J., Kim, Y., Pramanik, A., Mehta, P., & Olaniyan, O. (2020). Global and regional potential of wastewater as a water, nutrient and energy source. *Natural Resources Forum, 44*(1), 40–51. https://doi.org/10.1111/1477-8947.12187

Qaiser, Z., Aqeel, M., Sarfraz, W., Fatima Rizvi, Z., Noman, A., Naeem, S., & Khalid, N. (2023). Microplastics in wastewaters and their potential effects on aquatic and terrestrial biota. *Case Studies in Chemical and Environmental Engineering, 8*, 100536. https://doi.org/10.1016/j.cscee.2023.100536

Qiao, R., Sheng, C., Lu, Y., Zhang, Y., Ren, H., & Lemos, B. (2019). Microplastics induce intestinal inflammation, oxidative stress, and disorders of metabolome and microbiome in zebrafish. *The Science of the Total Environment, 662*, 246–253. https://doi.org/10.1016/j.scitotenv.2019.01.245

Qiu, G., Wang, Q., Wang, T., Zhang, S., Song, N., Yang, X., Zeng, Y., Sun, Z., Wu, G., & Yu, H. (2024). Microplastic risk assessment and toxicity in plants: A review. *Environmental Chemistry Letters, 22*(1), 209–226. https://doi.org/10.1007/s10311-023-01665-4

Quilliam, R. S., Pow, C. J., Shilla, D. J., Mwesiga, J. J., Shilla, D. A., & Woodford, L. (2023). Microplastics in agriculture—a potential novel mechanism for the delivery of human pathogens onto crops. *Frontiers in Plant Science, 14*. https://doi.org/10.3389/fpls.2023.1152419

Ragoobur, D., Huerta-Lwanga, E., & Somaroo, G. D. (2021). Microplastics in agricultural soils, wastewater effluents and sewage sludge in Mauritius. *Science of The Total Environment, 798*, 149326. https://doi.org/10.1016/j.scitotenv.2021.149326

Rajcoomar, S., Amoah, I. D., Abunama, T., Mohlomi, N., Bux, F., & Kumari, S. (2023). Biofilm formation on microplastics in wastewater: Insights into factors, diversity and inactivation strategies. *International journal of Environmental Science and Technology, 21*(4), 4429–4444. https://doi.org/10.1007/s13762-023-05266-0

Raju, S., Carbery, M., Kuttykattil, A., et al. (2018). Transport and fate of microplastics in wastewater treatment plants: Implications to environmental health. *Reviews in Environmental Science and Biotechnology, 17*, 637–653. https://doi.org/10.1007/s11157-018-9480-3

Rathore, C., Saha, M., Gupta, P., Kumar, M., Naik, A., & de Boer, J. (2023). Standardization of micro-FTIR methods and applicability for the detection and identification of microplastics in environmental matrices. *Science of The Total Environment, 888*, 164157. https://doi.org/10.1016/j.scitotenv.2023.164157

Reddy, A. S., & Nair, A. T. (2022). *The fate of microplastics in wastewater treatment plants: An overview of source and remediation technologies* (Vol. 28, p. 102815). Environmental Technology & Innovation. https://doi.org/10.1016/j.eti.2022.102815

Ren, P., Dou, M., Wang, C., Li, G., & Jia, R. (2020). Abundance and removal characteristics of microplastics at a wastewater treatment plant in Zhengzhou. *Environmental Science and Pollution Research, 27*(29), 36295–36305. https://doi.org/10.1007/s11356-020-09611-5

Rillig, M. C. (2012). Microplastic in terrestrial ecosystems and the soil? *Environmental Science & Technology, 46*(12), 6453–6454. https://doi.org/10.1021/es302011r

Rillig, M. C., & Lehmann, A. (2020). Microplastic in terrestrial ecosystems. *Science, 368*(6498), 1430–1431. https://doi.org/10.1126/science.abb5979

Rochman, C. M., Brookson, C., Bikker, J., Djuric, N., Earn, A., Bucci, K., Athey, S., Huntington, A., McIlwraith, H., Munno, K., De Frond, H., Kolomijeca, A., Erdle, L., Grbic, J., Bayoumi, M., Borrelle, S. B., Wu, T., Santoro, S., Werbowski, L. M., Zhu, X., Giles, R. K., Hamilton, B. M., Thaysen, C., Kaura, A., Klasios, N., Ead, L., Kim, J., Sherlock, C., Ho, A., & Hung, C. (2019). Rethinking microplastics as a diverse contaminant suite. *Environmental Toxicology and Chemistry, 38*(4), 703–711. https://doi.org/10.1002/etc.4371

Rogers, M., & Smith, S. R. (2007). Ecological impact of application of wastewater biosolids to agricultural soil. *Water and Environment Journal, 21*(1), 34–40. https://doi.org/10.1111/j.1747-6593.2006.00043.x

Ross, M. S., Loutan, A., Groeneveld, T., Molenaar, D., Kroetch, K., Bujaczek, T., Kolter, S., Moon, S., Huynh, A., Khayam, R., Franczak, B. C., Camm, E., Arnold, V. I., & Ruecker, N. J. (2023). Estimated discharge of microplastics via urban stormwater during individual rain events. Frontiers in environmental. *Science, 11*, 1090267. https://doi.org/10.3389/fenvs.2023.1090267

Roy, T., Dey, T. K., & Jamal, M. (2023). Microplastic/nanoplastic toxicity in plants: An imminent concern. *Environmental Monitoring and Assessment, 195*(1), 27. https://doi.org/10.1007/s10661-022-10654-z

Sadia, M., Mahmood, A., Ibrahim, M., Irshad, M. K., Quddusi, A. H. A., Bokhari, A., Mubashir, M., Chuah, L. F., & Show, P. L. (2022). Microplastics pollution from wastewater treatment plants: A critical review on challenges, detection, sustainable removal techniques and circular economy. *Environmental Technology & Innovation, 28*, 102946. https://doi.org/10.1016/j.eti.2022.102946

Sajjad, M., Huang, Q., Khan, S., Khan, M. A., Liu, Y., Wang, J., Lian, F., Wang, Q., & Guo, G. (2022). Microplastics in the soil environment: A critical review. *Environmental Technology & Innovation, 27*, 102408. https://doi.org/10.1016/j.eti.2022.102408

Salmi, P., Ryymin, K., Karjalainen, A. K., Mikola, A., Uurasjärvi, E., & Talvitie, J. (2021). Particle balance and return loops for microplastics in a tertiary-level wastewater treatment plant. *Water Science aned Technology, 84*(1), 89–100. https://doi.org/10.2166/wst.2021.209

Salva, J., Sečkár, M., Schwarz, M. et al. Analysis of the current state of sewage sludge treatment from the perspective of current European directives. Environ Sci Eur 37, 59 (2025). https://doi.org/10.1186/s12302-025-01097-7

Samanovic, M. I., Ding, C., Thiele, D. J., & Darwin, K. H. (2012). Copper in microbial pathogenesis: Meddling with the metal. *Cell Host & Microbe, 11*(2), 106–115. https://doi.org/10.1016/j.chom.2012.01.009

Sanchez-Hernandez, J. C. (2024). *Microplastic contamination and earthworms: Current trends and research needs*. CABI Reviews. https://doi.org/10.1079/cabireviews.2024.0037

Santos, D., Félix, L., Luzio, A., Parra, S., Cabecinha, E., Bellas, J., & Monteiro, S. M. (2020). Toxicological effects induced on early life stages of zebrafish (Danio rerio) after an acute exposure to microplastics alone or co-exposed with copper. *Chemosphere, 261*, 127748. https://doi.org/10.1016/j.chemosphere.2020.127748

Santos, D., Luzio, A., Félix, L., Bellas, J., & Monteiro, S. M. (2022). Oxidative stress, apoptosis and serotonergic system changes in zebrafish (Danio rerio) gills after long-term exposure to microplastics and copper. *Comparative Biochemistry and Physiology Part C Toxicology & Pharmacology, 258*, 109363. https://doi.org/10.1016/j.cbpc.2022.109363

Schmidt, W., O'Rourke, K., Hernan, R., & Quinn, B. (2011). Effects of the pharmaceuticals gemfibrozil and diclofenac on the marine mussel (Mytilus spp.) and their comparison with standardized toxicity tests. *Marine Pollution Bulletin, 62*(7), 1389–1395. https://doi.org/10.1016/j.marpolbul.2011.04.043

Scudiero, R., Chianese, T., Cretì, P., & Rosati, L. (2025). Risk assessment arising from the exposure of terrestrial vertebrates to soil contamination: Learning from field lizards of the Podarcis genus. *Journal of Xenobiotics, 15*(1), 21. https://doi.org/10.3390/jox15010021

Setälä, O., Lehtiniemi, M., Coppock, R., & Cole, M. (2018). Chapter 11 – Microplastics in marine food webs. In E. Y. Zeng (Ed.), *Microplastic contamination in aquatic environments* (pp. 339–363). Elsevier. https://doi.org/10.1016/B978-0-12-813747-5.00011-4

Seukep, A. J., Noumedem, J. A., Djeussi, D. E., & Kuete, V. (2014). Genotoxicity and teratogenicity of African medicinal plants. In *Elsevier eBooks* (pp. 235–275). https://doi.org/10.1016/b978-0-12-800018-2.00009-1

Shah, M. P. (2022). *Biological treatment of industrial wastewater*. Royal Society of Chemistry. https://doi.org/10.1039/9781839165399

Sheriff, I., Awang, N. A., Halim, H. B., Ikechukwu, O. S., & Jusoh, A. F. (2024). Extraction and analytical methods of microplastics in wastewater treatment plants: Isolation patterns, quantification, and size characterization techniques. *Desalination and Water Treatment, 318*, 100399. https://doi.org/10.1016/j.dwt.2024.100399

Sherlock, C., Fernie, K. J., Munno, K., Provencher, J., & Rochman, C. (2022). The potential of aerial insectivores for monitoring microplastics in terrestrial environments. *Science of the Total Environment, 807*, 150453. https://doi.org/10.1016/j.scitotenv.2021.150453

Shin, D. Y., Lee, S. M., Jang, Y., Lee, J., Lee, C. M., Cho, E., & Seo, Y. R. (2023). Adverse human health effects of chromium by exposure route: A comprehensive review based on toxicogenomic approach. *International Journal of Molecular Sciences, 24*(4), 3410. https://doi.org/10.3390/ijms24043410

Singh, A. K., Kumar, A., Bilal, M., & Chandra, R. (2021). Organometallic pollutants of paper mill wastewater and their toxicity assessment on stinging catfish and sludge worm. *Environmental Technology & Innovation, 24*, 101831. https://doi.org/10.1016/j.eti.2021.101831

Soliman, H. A. M., Salaah, S. M., Hamed, M., & Sayed, A. E. H. (2023). Toxicity of co-exposure of microplastics and lead in African catfish (*Clarias gariepinus*). *Frontiers in Veterinary Science, 10*, 1279382. https://doi.org/10.3389/fvets.2023.1279382

Sousa, P. M. S., Kinney, K. A., Sousa, C. A., et al. (2025). Microalgae for microplastic removal from water and wastewater: A review. *Environmental Chemistry Letters, 23*, 611–648. https://doi.org/10.1007/s10311-024-01809-0

Stefano, Magni Andrea, Binelli Lucia, Pittura Carlo Giacomo, Avio Camilla, Della Torre Camilla Carla, Parenti Stefania, Gorbi Francesco, Regoli (2019) The fate of microplastics in an Italian Wastewater Treatment Plant Science of The Total Environment 652602-610. https://doi.org/10.1016/j.scitotenv.2018.10.269

Subaramaniyam, U., Allimuthu, R. S., Vappu, S., Ramalingam, D., Balan, R., Paital, B., & Sahoo, D. K. (2023). Effects of microplastics, pesticides and nano-materials on fish health, oxidative stress and antioxidant defense mechanism. *Frontiers in Physiology, 14*, 1217666. https://doi.org/10.3389/fphys.2023.1217666

Sun, J., Dai, X., Wang, Q., van Loosdrecht, M. C. M., & Ni, B.-J. (2019). Microplastics in wastewater treatment plants: Detection, occurrence and removal. *Water Research, 152*, 21–37. https://doi.org/10.1016/j.watres.2018.12.050

Surendran, U., Jayakumar, M., Raja, P., Gopinath, G., & Chellam, P. V. (2023). Microplastics in terrestrial ecosystem: Sources and migration in soil environment. *Chemosphere, 318*, 137946. https://doi.org/10.1016/j.chemosphere.2023.137946

Sussarellu, R., M. Suquet, Y. Thomas, C. Lambert, C. Fabioux, M.E.J. Pernet, N. Le Goïc, V. Quillien, C. Mingant, Y. Epelboin, C. Corporeau, J. Guyomarch, J. Robbens, I. Paul-Pont, P. Soudant, & A. Huvet (2016) Oyster reproduction is affected by exposure to polystyrene microplastics, Proc. Natl. Acad. Sci. U.S.A. 113(9) 2430–2435. https://doi.org/10.1073/pnas.1519019113

Szymańska, M., & Obolewski, K. (2020). Microplastics as contaminants in freshwater environments: A multidisciplinary review. *Ecohydrology & Hydrobiology, 20*(3), 333–345. https://doi.org/10.1016/j.ecohyd.2020.05.001

Tabassum, I. et al. (2025). COVID-19 and Global Environmental Change: A Threat to Food Security and Nutrition. In: Sundaray, J.K., Rather, M.A., Ahmad, I., Amin, A. (eds) Food Security, Nutrition and Sustainability Through Aquaculture Technologies. Springer, Cham. https://doi.org/10.1007/978-3-031-75830-0_25

Tagg, A. S., Sapp, M., Harrison, J. P., Sinclair, C. J., Bradley, E., Ju-Nam, Y., & Ojeda, J. J. (2020). Microplastic monitoring at different stages in a wastewater treatment plant using reflectance micro-FTIR imaging. *Frontiers in Environmental Science, 8*, 1–9. https://doi.org/10.3389/fenvs.2020.00145

Taggart, M. A., Richards, N., & Kinney, C. A. (2015). Impacts of pharmaceuticals on terrestrial wildlife. In *Pharmaceuticals in the Environment* (pp. 216–254). The Royal Society of Chemistry. https://doi.org/10.1039/9781782622345-00216

Takdastan, A., Niari, M. H., Babaei, A., Dobaradaran, S., Jorfi, S., & Ahmadi, M. (2021). Occurrence and distribution of microplastic particles and the concentration of Di 2-ethyl hexyl phthalate (DEHP) in microplastics and wastewater in the wastewater treatment plant. *Journal of Environmental Management, 280*, 111851. https://doi.org/10.1016/j.jenvman.2020.111851

Talukdar, A., Kundu, P., Bhattacharjee, S., Dey, S., Dey, A., Biswas, J. K., Chaudhuri, P., & Bhattacharya, S. (2023). Microplastics in mangroves with special reference to Asia: Occurrence, distribution, bioaccumulation and remediation options. *Science of the Total Environment, 904*, 166165. https://doi.org/10.1016/j.scitotenv.2023.166165

Talukdar, A., Kundu, P., Bhattacharya, S., & Dutta, N. (2024). Microplastic contamination in wastewater: Sources, distribution, detection and remediation through physical and chemical-biological methods. *Science of the Total Environment, 916*, 170254. https://doi.org/10.1016/j.scitotenv.2024.170254

Talvitie, J., Mikola, A., Koistinen, A., & Setälä, O. (2017). Solutions to microplastic pollution – Removal of microplastics from wastewater effluent with advanced wastewater treatment technologies. *Water Research, 123*, 401–407. https://doi.org/10.1016/j.watres.2017.07.005

Tang, K. H. D. (2020). Effects of microplastics on agriculture: A mini-review. *Asian Journal of Environment & Ecology, 13*, 1–9. https://doi.org/10.9734/ajee/2020/v13i130170

Tang, Y., Zhou, W., Sun, S., Du, X., Han, Y., Shi, W., & Liu, G. (2020). Immunotoxicity and neurotoxicity of bisphenol A and microplastics alone or in combination to a bivalve species. *Tegillarca granosa. Environmental Pollution, 265*, 115115. https://doi.org/10.1016/j.envpol.2020.115115

Terpiłowska, S., & Zaporowska, H. (2004). The role of chromium in cell biology and medicine. *PubMed*, 51–54. Retrieved from https://europepmc.org/article/MED/15682944

Thrift, E., Porter, A., Galloway, T. S., Coomber, F. G., & Mathews, F. (2022). Ingestion of plastics by terrestrial small mammals. *Science of the Total Environment, 842*, 156679. https://doi.org/10.1016/j.scitotenv.2022.156679

Torres-Agullo, A., Karanasiou, A., Moreno, T., & Lacorte, S. (2021). Overview on the occurrence of microplastics in air and implications from the use of face masks during the COVID-19 pandemic. *The Science of the Total Environment, 800*, 149555. https://doi.org/10.1016/j.scitotenv.2021.149555

Townsend, K. R., Lu, H.-C., Sharley, D. J., & Pettigrove, V. (2019). Associations between microplastic pollution and land use in urban wetland sediments. *Environmental Science and Pollution Research, 26*(22), 22551–22561. https://doi.org/10.1007/s11356-019-04885-w

Uddin, S., Fowler, S. W., & Behbehani, M. (2020). An assessment of microplastic inputs into the aquatic environment from wastewater streams. *Marine Pollution Bulletin, 160*, 111538. https://doi.org/10.1016/j.marpolbul.2020.111538

Ullah, S., Ahmad, S., Guo, X., Ullah, S., Ullah, S., Nabi, G., & Wanghe, K. (2023). A review of the endocrine disrupting effects of micro and nano plastic and their associated chemicals in mammals. *Frontiers in Endocrinology, 13*. https://doi.org/10.3389/fendo.2022.1084236

UN Habitat and WHO. (2021). *Progress on wastewater treatment—Global status and acceleration needs for SDG indicator 6.3.1*. United Nations Human Settlements Programme (UN-Habitat) and World Health Organization (WHO). Available at: https://unhabitat.org/sites/default/files/2021/08/sdg6_indicator_report_631_progress_on_wastewater_treatment_2021_english_pages.pdf

UNEP. (2002). *Regionally based assessment of persistent toxic substances*. UNEP Chemicals. Retrieved from https://digitallibrary.un.org/record/487298?ln=en&v=pdf

UNEP. (2025). *Facts about wastewater and nutrient management*. Retrieved from https://www.unep.org/facts-about-wastewater-and-nutrient-management

Urli, S., Corte Pause, F., Crociati, M., Baufeld, A., Monaci, M., & Stradaioli, G. (2023). Impact of microplastics and Nanoplastics on livestock health: An emerging risk for reproductive efficiency. *Animals, 13*(7), 1132. https://doi.org/10.3390/ani13071132

US EPA. (2024). *Persistent organic pollutants: A global issue, a global response*. Retrieved from https://www.epa.gov/international-cooperation/persistent-organic-pollutants-global-issue-global-response

Üstün, G. E., Bozdaş, K., & Can, T. (2022). Abundance and characteristics of microplastics in an urban wastewater treatment plant in Turkey. *Environmental Pollution, 310*, 119890. https://doi.org/10.1016/j.envpol.2022.119890

Van Cauwenberghe, L., Vanreusel, A., Mees, J., & Janssen, C. R. (2013). Microplastic pollution in deep-sea sediments. *Environmental Pollution, 182*, 495–499. https://doi.org/10.1016/j.envpol.2013.08.013

Van Do, M., Le, T. X. T., Vu, N. D., & Dang, T. T. (2022). Distribution and occurrence of microplastics in wastewater treatment plants. *Environmental Technology and Innovation, 26*, 102286. https://doi.org/10.1016/j.eti.2022.102286

Viaroli, S., Lancia, M., & Re, V. (2022). Microplastics contamination of groundwater: Current evidence and future perspectives. A review. *Science of the Total Environment, 824*, 153851. https://doi.org/10.1016/j.scitotenv.2022.153851

Viršek, M. K., Lovšin, M. N., Koren, Š., Kržan, A., & Peterlin, M. (2017). Microplastics as a vector for the transport of the bacterial fish pathogen species *Aeromonas salmonicida*. *Marine Pollution Bulletin, 125*(1–2), 301–309. https://doi.org/10.1016/j.marpolbul.2017.08.024

Vollertsen, J., & Hansen, A. (2017). *Microplastic in Danish wastewater: Sources, occurrences and fate*. The Danish Environmental Protection Agency. Available at https://www2.mst.dk/udgiv/publications/2017/03/978-87-93529-44-1.pdf

Wang, Y., & Wang, R. (2018). *Reverse osmosis membrane separation technology* (pp. 1–45). Elsevier eBooks. https://doi.org/10.1016/b978-0-12-812815-2.00001-6

Wang, L., Nabi, G., Yin, L., Wang, Y., Li, S., Hao, Z., & Li, D. (2021a). Birds and plastic pollution: Recent advances. *Avian Research, 12*(1), 59. https://doi.org/10.1186/s40657-021-00293-2

Wang, S., Ji, C., Li, F., Zhan, J., Sun, T., Tang, J., & Wu, H. (2021b). Tetrabromobisphenol A induced reproductive endocrine-disrupting effects in mussel *Mytilus galloprovincialis*. *Journal of Hazardous Materials, 416*, 126228. https://doi.org/10.1016/j.jhazmat.2021.126228

Wang, C., Luo, Q., Zhang, J., Zhang, X., Yang, N., & Feng, L. (2023a). Toxic effects of microplastics and nanoplastics on plants: A global meta-analysis. *Environmental Pollution, 337*, 122593. https://doi.org/10.1016/j.envpol.2023.122593

Wang, H., Neal, B., White, B., Nelson, B., Lai, J., Long, B., Arreola-Vargas, J., Yu, J., Banik, M. T., & Dai., S. Y. (2023b). Microplastics removal in the aquatic environment via fungal pelletization. *Bioresource Technology Reports., 23*, 101545. https://doi.org/10.1016/j.biteb.2023.101545

Wang, W., Xie, Y., Li, H., Dong, H., Li, B., Guo, Y., & Zhou, W. (2023c). Responses of lettuce (*Lactuca sativa* L.) growth and soil properties to conventional non-biodegradable and new biodegradable microplastics. *Environmental Pollution, 341*, 122897. https://doi.org/10.1016/j.envpol.2023.122897

Whiting, Q. T., O'Connor, K. F., Potter, P. M., & Al-Abed, S. R. (2022). A high-throughput, automated technique for microplastics detection, quantification, and characterization in surface waters using laser direct infrared spectroscopy. *Analytical and Bioanalytical Chemistry, 414*(29), 8353–8364. https://doi.org/10.1007/s00216-022-04371-2

WHO. (2022, December 7). Arsenic. Retrieved from https://www.who.int/news-room/fact-sheets/detail/arsenic

WHO. (2024, September 27). *Lead poisoning*. Retrieved from https://www.who.int/news-room/fact-sheets/detail/lead-poisoning-and-health.

Wu, P., Tang, Y., Dang, M., Wang, S., Jin, H., Liu, Y., Jing, H., Zheng, C., Yi, S., & Cai, Z. (2020). Spatial-temporal distribution of microplastics in surface water and sediments of Maozhou River within Guangdong-Hong Kong-Macao Greater Bay Area. *Science of The Total Environment, 717*, 135187. https://doi.org/10.1016/j.scitotenv.2019.135187

Wu, M., Tang, W., Wu, S., Liu, H., & Yang, C. (2021). Fate and effects of microplastics in wastewater treatment processes. *Science of the Total Environment, 757*, 143902. https://doi.org/10.1016/j.scitotenv.2020.143902

Wu, X., Zhao, X., Chen, R., Liu, P., Liang, W., Wang, J., Teng, M., Wang, X., & Gao, S. (2022). Wastewater treatment plants act as essential sources of microplastic formation in aquatic environments: A critical review. *Water Research, 221*, 118825. https://doi.org/10.1016/j.watres.2022.118825

Xia, Y., Zhang, X., Zhang, M., Chen, L., Tang, X., Sun, Y., & Li, X. (2022). Plastic materials and water sources actively select and shape wastewater plastispheres over time. *Frontiers of Environmental Science & Engineering, 16*(11), 145. https://doi.org/10.1007/s11783-022-1580-1

Xu, Z., Zhang, Y., Lin, L., Wang, L., Sun, W., Liu, C., Yu, G., Yu, J., Lv, Y., Chen, J., Chen, X., Fu, L., & Wang, Y. (2022). Toxic effects of microplastics in plants depend more by their surface functional groups than just accumulation contents. *Science of the Total Environment, 833*, 155097. https://doi.org/10.1016/j.scitotenv.2022.155097

Xu, H., Chen, C., Pang, Z., Zhang, G., Zhang, W., & Kan, H. (2024). Effects of microplastics concentration on plant root traits and biomass: Experiment and meta-analysis. *Ecotoxicology and Environmental Safety, 285*, 117038. https://doi.org/10.1016/j.ecoenv.2024.117038

Yang, L., Li, K., Cui, S., Kang, Y., An, L., & Lei, K. (2019). Removal of microplastics in municipal sewage from China's largest water reclamation plant. *Water Research, 155*, 175–181. https://doi.org/10.1016/j.watres.2019.02.046

Yang, Z., Lü, F., Zhang, H., Wang, W., Shao, L., Ye, J., & He, P. (2021). Is incineration the terminator of plastics and microplastics? *Journal of Hazardous Materials, 401*, 123429. https://doi.org/10.1016/j.jhazmat.2020.123429

Yin, K., Wang, D., Zhang, Y., Lu, H., Wang, Y., & Xing, M. (2023). Dose-effect of polystyrene microplastics on digestive toxicity in chickens (Gallus gallus): Multi-omics reveals critical role of gut-liver axis. *Journal of Advanced Research, 52*, 3–18. https://doi.org/10.1016/j.jare.2022.10.015

Yu, Y., Kumar, M., Bolan, S., Padhye, L. P., Bolan, N., Li, S., Wang, L., Hou, D., & Li, Y. (2024). Various additive release from microplastics and their toxicity in aquatic environments. *Environmental Pollution, 343*, 123219. https://doi.org/10.1016/j.envpol.2023.123219

Zantis, L. J., Borchi, C., Vijver, M. G., Peijnenburg, W., Di Lonardo, S., & Bosker, T. (2023). Nano- and microplastics commonly cause adverse impacts on plants at environmentally relevant levels: A systematic review. *Science of the Total Environment, 867*, 161211. https://doi.org/10.1016/j.scitotenv.2022.161211

Zeenat, Elahi, A., Bukhari, D. A., Shamim, S., & Rehman, A. (2021). Plastics degradation by microbes: A sustainable approach. *Journal of King Saud University - Science, 33*(6), 101538. https://doi.org/10.1016/j.jksus.2021.101538

Zeri, C., Adamopoulou, A., Koi, A., Koutsikos, N., Lytras, E., & Dimitriou, E. (2021). Rivers and wastewater-treatment plants as microplastic pathways to eastern Mediterranean waters: First Records for the Aegean Sea, Greece. *Sustainability, 13*(10), 5328. https://doi.org/10.3390/su13105328

Zhai, W., Huang, Z., Chen, L., Feng, C., Li, B., & Li, T. (2014). Thyroid endocrine disruption in zebrafish larvae after exposure to mono-(2-Ethylhexyl) phthalate (MEHP). *PLoS One, 9*(3), e92465. https://doi.org/10.1371/journal.pone.0092465

Zhang, F., Man, Y. B., Mo, W. Y., Man, K. Y., & Wong, M. H. (2019). Direct and indirect effects of microplastics on bivalves, with a focus on edible species: A mini-review. *Critical Reviews in Environmental Science and Technology, 50*(20), 2109–2143. https://doi.org/10.1080/10643389.2019.1700752

Zhang, L., Liu, J., Xie, Y., Zhong, S., & Gao, P. (2021). Occurrence and removal of microplastics from wastewater treatment plants in a typical tourist city in China. *Journal of Cleaner Production*, [online] 291,, *291*, 125968. https://doi.org/10.1016/j.jclepro.2021.125968

Zhang, H., Gao, Y., Zheng, Y., Zheng, J., He, J., Shi, J., Zhang, K., Song, Y., Zhang, J., Shi, X., Zhang, R., Ding, Y., Jing, Y., Xu, K., & Wang, J. (2024). Potential toxicity of microplastics on vertebrate liver: A systematic review and meta–analysis. *Ecotoxicology and Environmental Safety, 286*, 117166. https://doi.org/10.1016/j.ecoenv.2024.117166

Zheng, Z., & Greedan, J. E. (2003). Rare earth elements and materials. In *Elsevier eBooks* (pp. 1–22). https://doi.org/10.1016/b0-12-227410-5/00653-0

Zhong, H., Wu, M., Sonne, C., Lam, S. S., Kwong, R. W., Jiang, Y., & Ren, H. (2023). The hidden risk of microplastic-associated pathogens in aquatic environments. *Eco-Environment & Health, 2*(3), 142–151. https://doi.org/10.1016/j.eehl.2023.07.004

Ziajahromi, S., Neale, P. A., Rintoul, L., & Leusch, F. D. L. (2017). Wastewater treatment plants as a pathway for microplastics: Development of a new approach to sample wastewater-based microplastics. *Water Research, 112*, 93–99. https://doi.org/10.1016/j.watres.2017.01.042

Ziajahromi, S., Neale, P. A., Telles Silveira, I., Chua, A., & Leusch, F. D. L. (2021). An audit of microplastic abundance throughout three Australian wastewater treatment plants. *Chemosphere, 263*, 128294. https://doi.org/10.1016/j.chemosphere.2020.128294

Zubris, K. A. V., & Richards, B. K. (2005). Synthetic fibers as an indicator of land application of sludge. *Environmental Pollution, 138*(2), 201–211. https://doi.org/10.1016/j.envpol.2005.04.013

Microplastics in Freshwater Ecosystem: Implication for Tropic Transfer, Food Security and Human Health Risk

Fariha Latif, Aiman, Saman Iram, Muhammad Mudassar Shahzad, and Monisa M. Malik

1 Introduction

1.1 Microplastics

Microplastics (MPs) are plastic particles measuring no more than 5 mm and no less than 0.1 mm that includes microbeads, microfibers, fragmented plastics, and Nano plastics. They consist of various plastic types, including polyethylene, polypropylene, polyvinyl chloride (PVC), and polystyrene (Arif et al., 2024).

1.2 Types of Microplastics

Microplastics can be divided into various categories on the basis of the toxicology they produce and the size.

(a) **Primary microplastics** are small fragments, which are the products from manufacturing. These types of microplastics are used in some drugs, toilet utilities, cleansers, and cosmetics.

F. Latif (✉) · Aiman · S. Iram
Institute of Zoology, Bahauddin Zakariya University, Multan, Pakistan

M. M. Shahzad
Department of Zoology, University of Education, Lahore, Pakistan

M. M. Malik
Division of Aquatic Environmental Management, Faculty of Fisheries Rangil, Ganderbal, SKUAST-Kashmir, Jammu and Kashmir, India

A. Abubakr et al. (eds.), *Microplastics: Ecological and Food Security Implications*,
https://doi.org/10.1007/978-3-031-94480-2_6

(b) **Secondary microplastics** are initially formed from the breakdown of larger plastic contents, which are present in the marine systems (Boucher & Friot, 2017).

1.3 Sources of Microplastics

(i) **Cosmetics and Personal Care Products:**

The incorporation of microplastic beads in personal care and cosmetic products is a popular source and pathway for persistent and potentially hazardous primary microplastic constituents to infiltrate aquatic ecosystems. Plastic particles are directly discharged into wastewater during and subsequent to the daily application of personal care and cosmetic products, as most of these items are rinsed off or disposed of via the drain (Megha et al., 2025; Prasad et al., 2024).

(ii) **Textiles and Commercial Cleaning of Synthetic Fibers:**

Recent field research identified fibers as a significant category of microplastics in various habitats, including remote locations and polar glaciers. Textile microfibers represent a significant source of microplastic fibers. Fibers detach from the textile article at various stages of its lifecycle, especially during laundering. Nonfiltered effluents from commercial laundries are discharged into drains, serving as a point source of microfibers, similar to home laundries, where a single garment can release over 1900 fibers in a single wash. The various washing parameters, including detergent usage, washing duration, and fabric type, influence the quantity of fiber released during laundry processes (Akyildiz et al., 2024).

(iii) **From Household/Indoor:**

Numerous polymer-based products, both synthetic and natural, are present in our daily lives and are utilized in indoor and outdoor settings. Common household items, including plastic carpets, furniture, and curtains, release thousands of fibers daily (Friedrich, 2024). Fakes and chips from ancient interior paint on walls, polyurethane particles from mattresses, and plastic dust from electronics serve as sources of microplastics.

(iv) **From Fisheries and Aquaculture:**

Plastics are substantial elements in both fisheries and aquaculture. A variety of instruments, including nets, fishing gear, trawls, dredges, hooks and lines, fish hold insulation, fish crates, and packing materials, are utilized in both industries. As synthetic equipment advances, a significant concern arises: when these devices deteriorate, their plastic components swiftly enter the water, where they decompose into microplastics and nanoparticles (Pataki, 2024; Nabi et al., 2024).

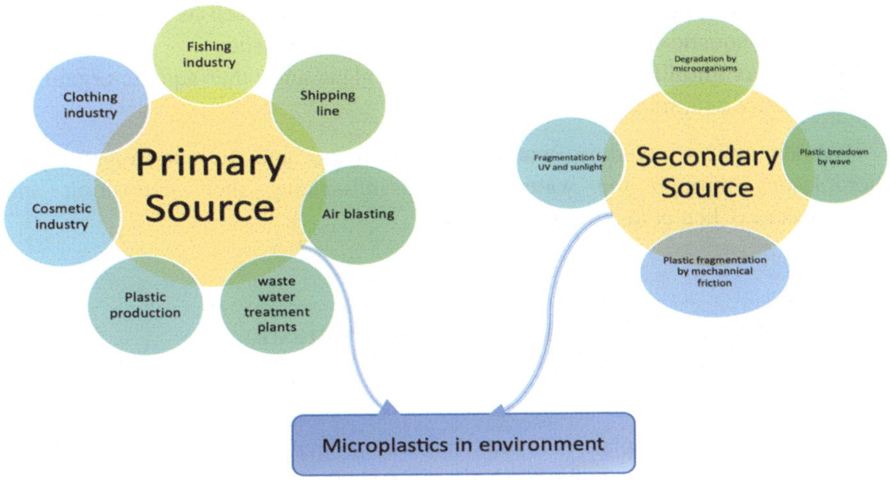

Fig. 1 Main sources of microplastics

(v) **From Wastewater Treatment Plants:**

Wastewater treatment facilities may serve as a significant source of microplastics in aquatic ecosystems. The final destination of those MPs will be wastewater, subsequently entering a wastewater treatment facility. Granulated polyethylene, polypropylene, and polystyrene particles present in personal care products, cleaning agents, and air blasting media are sufficiently diminutive to evade the treatment processes in wastewater treatment facilities (Wu et al., 2022).

(vi) **Spills during Production and Transport:**

Granulated forms of plastic serve as raw materials in the production of various plastic products. The plastics utilized in the recycling process are transformed into plastic granules (Meneses et al., 2022). During the preproduction and recycling phases, the likelihood of accidental discharge of these granules into the surrounding environment is significant. The plastic manufacturing and conversion process is frequently regarded as a point source for microplastic discharge into the environment (Fig. 1).

Environmental Factors

Microplastics in freshwater ecosystems undergo various physical, chemical, and biological processes that influence their distribution, fate, and potential impacts on the environment. Environmental factors, such as photooxidation, temperature, nutrient levels, and hydrodynamics, also play a crucial role in shaping microplastic dynamics.

- **Photooxidation,** driven by sunlight, breaks down microplastics, forming oxygen-containing functional groups (Wang et al., 2023).

- **Temperature** influences the rate of photooxidation, as well as the growth and activity of aquatic organisms that interact with microplastics
- **Nutrient availability** impacts the growth of algae and other aquatic organisms that interact with microplastics.

Hydrodynamics including water flow, currents, and turbulence, transport microplastics through the water column, influencing their deposition and resuspension. (Chen et al., 2019).

2 Distribution and Persistence of Microplastic in Freshwater System

The distribution of microplastics in the riverine ecosystem is significantly influenced by seasonal and climatic conditions like heavy rains resulting in flood-related events that could expedite the transport of terrestrial MPs to riverine systems (Scircle et al., 2020). The type of distribution are:

(a) **Vertical distribution**

The vertical distribution of microplastics (MPs) in freshwater environments is influenced by their aggregation, homoaggregation, and heteroaggregation. Aggregation, heteroaggregation, and trophic transfer are key processes that affect microplastic dynamics. Aggregation occurs when microplastics cluster together, forming larger particles that can settle faster (Besseling et al., 2019). Heteroaggregation, on the other hand, involves microplastics combining with other particles, such as algae, sediment, or organic matter, which can alter their density and distribution. Trophic transfer occurs when microplastics are ingested by aquatic organisms, such as zooplankton, fish, and other invertebrates, and are transferred through the food chain.

(b) **Horizontal distribution**

The Horizontal distribution of microplastics (MPs) in freshwater environments is a complex process influenced by both natural conditions and biorelated factors. Hydrodynamic conditions, such as water flow and currents, play a significant role in shaping the distribution of MPs. Additionally, bio related factors, including the ingestion and adhesion of MPs by freshwater biota, also contribute to their distribution. The type of biota, shape, color, and size of MPs all affect their distribution, highlighting the need for a comprehensive understanding of these factors to assess the potential hazards of MPs in freshwater ecosystems (Zhang et al., 2019).

3 Ecological Impacts on Freshwater Organisms

Microplastics have become a ubiquitous contaminant in marine and freshwater environments, posing serious threats to biota. Due to their small size and large specific surface area, microplastics can adsorb toxic substances and be inhabited by pathogenic bacteria (McCormick et al., 2014).

(i) **Ingestion**

The ingestion of microplastics by aquatic organisms results in significant adverse effects. Microplastics (MPs) inflict physical and histophysiological harm on biota within freshwater ecosystems due to their size and shape. The physical damage associated with MPs encompasses broken cell walls or membranes, altered feeding and locomotion, digestive tract blockage, diminished nutrient absorption, inflammation, oxidative stress, immune system suppression, decreased reproductive output, and reduced growth or survival rates in freshwater organisms, potentially leading to mortality. MPs impact both producers, such as bacteria and algae, and consumers, thereby potentially disrupting the overall structure of freshwater ecosystems (Lei et al., 2018). MPs can bind to the surfaces of producers due to their specific properties (size, shape, zeta potential, etc.), leading to broken cell walls or membranes and reduced growth. This can ultimately alter the population densities of producers in certain locations, having a ripple effect throughout the ecosystem (Cai et al., 2019; Debnath et al., 2024).

(ii) **Bioaccumulation**

microplastics can also facilitate the transfer of toxic substances into the food chain through a process called bioaccumulation. This occurs when microplastics are ingested by small aquatic organisms, which are then consumed by larger animals, leading to the accumulation of toxins in the tissues of these organisms (Sruthisree et al., 2015; Amin et al., 2016; Amin et al., 2017; Vaid et al., 2023; Gurjar et al., 2023).

Bioaccumulation of microplastics and associated contaminants in freshwater fishes is a growing concern. Several factors can influence bioaccumulation, including diet, size and type of microplastics, water quality, and fish species and physiology. Fish that feed on plankton, invertebrates, or smaller fish are more likely to ingest microplastics. (Malik et al., 2017; Taylor et al., 2016; Prasad et al., 2023). Smaller microplastics, typically less than 1 mm, are more easily ingested and bio accumulated. Poor water quality, such as high levels of pollutants or low dissolved oxygen, can also increase bioaccumulation.

(iii) **Biodiversity disruption**

The disruption of biodiversity is another critical consequence of microplastic pollution. Microplastics can alter the composition of aquatic communities, leading to changes in species interactions, nutrient cycling, and ecosystem functioning

(Nguyen et al., 2023). This can have cascading effects on the entire ecosystem, including:

- Changes in food web dynamics
- Loss of species diversity
- Decreased ecosystem resilience
- Impacts on human livelihoods and well-being

4 Tropic Transfer in Freshwater Food Chains

Microplastics pollution has emerged as a new ecological concern, posing a risk to both fish and human health. Fish contaminated with microplastics (MPs) can transfer these pollutants to humans through the food chain. The human diet relies heavily on seafood, and the ingestion of MPs through seafood consumption can lead to intestinal infection, which can spread to other body systems. Two common pathways for MP entry into the human body are endocytosis and persorption (Bilal et al., 2023). Trophic transfer of microplastics (MPs) in freshwater food chains is a complex process that involves the transfer of MPs from one trophic level to another.

4.1 Pathways and Biomagnification Risk

The transfer of microplastics and nanoparticles (MNPs) through aquatic food webs represents a significant focus within the field of ecotoxicology. Microplastic nanoparticles (MNPs) have been identified in various aquatic organisms, encompassing 427 fish species across multiple trophic levels and habitats. Research indicates that MNPs can be transmitted across different levels of the ecosystem through consumption, with investigations employing direct feeding experiments to examine this phenomenon (Habumugisha et al., 2024).

Microplastics (MPs) can enter the food chain via multiple pathways, including ingestion by primary producers like algae and phytoplankton, as well as by filter-feeding organisms. The ingestion of microplastics (MPs) can have substantial physical and biological effects on organisms, including disruptions in feeding patterns, weight reduction, increased mortality, and decreased fertility. The effects of MPs on freshwater organisms have been studied, with findings indicating that MPs can cause physical harm, toxicity, and changes in behavior and physiology.

The implications of microplastic trophic transfer are still unclear and require further study. However, it is potential that microplastic trophic transfer can lead to bioaccumulation and biomagnification in aquatic food chains. Moreover, the effects of microplastic ingestion on aquatic organisms and ecosystems are still unknown. Further research is needed to understand the mechanisms and effects of microplastic trophic transfer in higher trophic levels and multilevel aquatic food chains.

5 Impact on Fisheries and Food Security

Microplastics have been found in various fish species worldwide, including freshwater, estuarine, and marine environments. Fish can ingest microplastics directly from the environment or through secondary ingestion from their prey (Parker et al., 2021). The ingestion of microplastics by fish is influenced by factors such as fish species, life stages, feeding strategy, and plastic properties. Studies have shown that microplastics can accumulate in fish tissues, including the gastrointestinal tract, gills, liver, and even brain.

The ecotoxicological effects of microplastics on fish have been verified in experimental studies, affecting tissue and organ health, behavioral and neurological functions, intestinal permeability, metabolism, and intestinal microbiome diversity. Smaller plastic particles have been found to pose a greater hazard than larger ones, and higher plastic concentrations also play a significant role. However, some field investigations have suggested that microplastics retained in fish are minimal and may not accumulate inside the intestinal tracts for extended periods, having limited effects on wild fish, especially top predatory fish (Chen et al., 2022).

Microplastics in freshwater systems pose a significant threat to fisheries and food security. The ingestion of microplastics by fish and other aquatic organisms can lead to physical harm, toxicity, and altered behavior, ultimately affecting the productivity and sustainability of fisheries. Microplastics have been found in various freshwater fish species, including those that are commercially important and consumed by humans. This raises concerns about the potential transfer of microplastics and associated pollutants to humans through the food chain.

6 Human Health Risk

Microplastics can affect human beings in several ways, including through drinking water, consuming fish products, using cosmetics, and inhalation (Revel et al., 2020). While microplastics have also been found in other products like beer and table salts, the health effects of consuming these are considered negligible. However, the presence of microplastics in food has raised concerns about their potential chronic effects on human health (Liebezeit & Liebezeit, 2014).

6.1 Ingestion of Contaminated Freshwater fish and Sea Food

One of the primary pathways of exposure is dietary intake, particularly through consuming seafood and other contaminated food items. Shellfish, such as mussels, oysters, and clams, are especially vulnerable to absorbing microplastics due to their filter-feeding habits. (Hermsen et al., 2017) These particles can accumulate in their

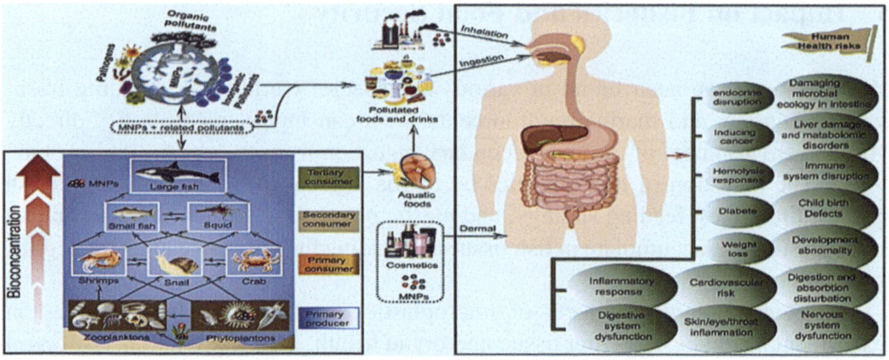

Fig. 2 Relationship between the pollution of plastic particles in aquatic organisms in food web and related potential effects of MNPs on human health

tissues and be transferred to humans through consumption. Microplastics have been found in fish and shellfish, and even removing the gastrointestinal tract from fish does not eliminate the risk of human exposure. Furthermore, microplastics can also be present in fishmeal used in aquaculture and animal husbandry (Santillo et al., 2017).

- **Toxicity concerns**

The ingestion of microplastics can lead to physical harm, oxidative stress, and inflammation in the body. Small particles can penetrate cell membranes, the blood-brain barrier, and the human placenta, potentially causing harm to the developing fetus. Inhaled microplastics can also cause respiratory problems, including inflammation, oxidative stress, and lung dysfunction.

Studies have found microplastics in human tissues, including lung, kidney, and placental tissues. Microplastics have also been detected in breast milk and infant formula, raising concerns about the exposure of vulnerable populations to these pollutants. The health risks associated with microplastic exposure include inflammation, oxidative stress, DNA damage, and potentially even cancer (Fig. 2).

Despite the growing evidence of microplastic contamination in human tissues, the full impact of microplastics on human health is not yet well understood. Further research is needed to understand the mechanisms of 'microplastic toxicity, the extent of human exposure, and the potential health risks associated with microplastic contamination (Liu & You, 2023).

7 Challenges in Assessing Microplastics Impacts

Microplastic pollution in freshwater environments has become a growing concern worldwide. Despite initial assumptions that sewage treatment plants could remove all plastic debris, research has shown that microplastics persist in freshwater ecosystems (Mishra et al., 2023; Tabassum et al., 2025). The main sources of microplastics in freshwater environments include synthetic textiles, personal care products, industrial raw materials, and improper disposal of plastic waste.

Studies have reported varying concentrations of microplastics in freshwater environments globally, ranging from 3.5 × 10^3 microplastic units·L − 1 in Lake Huron to 1.2 × 10–4 units·L − 1 in Mongolia (Haitao & Tangbin, 2023). The primary polymer constituents of microplastics found in freshwaters are polyethylene (PE), polypropylene (PP), polystyrene (PS), and polyethylene terephthalate (PET).

The lack of a harmonized and standardized analyzing framework for microplastics (MPs) in freshwater systems is a significant issue. Key challenges include:

- **No unified methods:** Different studies use varying analysis protocols for sampling, pre-treatment, and detection of MPs.
- **Data incomparability**: The use of different protocols hinders the comparability of data across studies.

To address this issue, a universally accepted standard analysis protocol is needed to harmonize MP analysis practices. This will:

- Ensure data quality and comparability
- Facilitate a comprehensive understanding of MP pollution

A concerted global effort is necessary to standardize future MP analysis practices and promote a more complete understanding of MP pollution in freshwater systems. To accurately monitor freshwater microplastics (MP), harmonized procedures are necessary for:

- **Sampling**: Standardized methods for collecting MP samples.
- **Identification**: Accurate identification of MP to avoid misclassification.
- **Characterization**: Detailed analysis of MP properties, such as size, shape, and material.

Current challenges include:

- **Separation of MP from sample materials**: A time-consuming and biased process.
- **Detection of small MP**: Difficulty detecting MP in the low micrometer range.

To address these challenges, technological innovation is needed to:

- Increase sample throughput: Faster and more efficient processing of samples.
- Improve accuracy: More accurate identification and characterization of MP.
- Enhance detection capabilities: Development of methods to detect very small MP.

8 Conclusion and Future Perspectives

Microplastic pollution is a critical issue that requires further research and attention. Understanding the mechanisms of adsorption, bioaccumulation, and biomagnification, as well as the influence of environmental factors, is crucial for assessing the potential health effects of microplastic exposure. Developing effective methods for identifying microplastics and implementing legal regulations can help mitigate this problem.

Research has shown that microplastics are present in freshwaters and drinking water, highlighting the need for effective removal strategies. While various treatment options are available, further research is necessary to improve existing techniques and develop new methodologies. Key areas for future research include understanding microplastic movement and fate, developing best practices for treatment and removal, and assessing the environmental impacts of treatment technologies. Additionally, there is a need to investigate the effects of microplastic-associated biofilms and the removal of fibrous microplastics by membrane filtering. Overall, addressing the knowledge gaps in microplastic removal will require a multidisciplinary approach, including research, development, and implementation of effective treatment strategies.

References

Akyildiz, S. H., Fiore, S., Bruno, M., Sezgin, H., Yalcin-Enis, I., Yalcin, B., & Bellopede, R. (2024). Release of microplastic fibers from synthetic textiles during household washing. *Environmental Pollution, 357*, 124455.

Amin, A., Indulkar, S. T., & Pai, R. (2016). Sublethal effect of Buprofezin pesticide on carcass composition of *Cyprinus carpio* communis fingerlings. *Journal of Experimental Zoology, India, 20*(1), 269–272.

Amin, A., Ramachandra Naik, A. T., Priyadarshini, N., Nayak, H., & Sree, C. S. (2017). Toxic effect of heavy metal lead on oxygen consumption of rohu (*Labeo rohita*) fingerlings. *Biochemical & Cellular Archives, 17*(1), 225–228.

Arif, Y., Mir, A. R., Zieliński, P., Hayat, S., & Bajguz, A. (2024). Microplastics and nanoplastics: Source, behavior, remediation, and multi-level environmental impact. *Journal of Environmental Management, 356*, 120618.

Besseling, E., Redondo-Hasselerharm, P., Foekema, E. M., & Koelmans, A. A. (2019). Quantifying ecological risks of aquatic micro-and nanoplastic. *Critical Reviews in Environmental Science and Technology, 49*(1), 32–80.

Bilal, M., Qadir, A., Yaqub, A., Hassan, H. U., Irfan, M., & Aslam, M. (2023). Microplastics in water, sediments, and fish at Alpine River, originating from the Hindu Kush Mountain, Pakistan: Implications for conservation. *Environmental Science and Pollution Research, 30*(1), 727–738.

Boucher, J., & Friot, D. (2017). *Primary microplastics in the oceans: A global evaluation of sources* (Vol. 10). Iucn.

Cai, W., Wu, L., Lengaigne, M., Li, T., McGregor, S., Kug, J. S., et al. (2019). Pantropical climate interactions. *Science, 363*(6430), eaav4236.

Chen, Q., Zhang, H., Allgeier, A., Zhou, Q., Ouellet, J. D., Crawford, S. E., ... & Hollert, H. (2019). Marine microplastics bound dioxin-like chemicals: model explanation and risk assessment. *Journal of Hazardous Materials, 364*, 82–90.

Chen, J., Rao, C., Yuan, R., Sun, D., Guo, S., Li, L., et al. (2022). Long-term exposure to polyethylene microplastics and glyphosate interferes with the behavior, intestinal microbial homeostasis, and metabolites of the common carp (Cyprinus carpio L.). *Science of the Total Environment, 814*, 152681.

Debnath, R., Prasad, G. S., Amin, A., Malik, M. M., Ahmad, I., Abubakr, A., Borah, S., Rather, M. A., Impellitteri, F., Tabassum, I., & Piccione, G. (2024). Understanding and addressing microplastic pollution: Impacts, mitigation, and future perspectives. *Journal of Contaminant Hydrology, 266*, 104399.

Friedrich, D. (2024). *Applied building physics and materials science of natural fiber reinforced plastics*. Springer Fachmedien Wiesbaden.

Gurjar, U. R., Takar, S., Amin, A., & Martin Xavier, K. A. (2023). Microplastic contamination in aquatic organisms: An ecotoxicological perspective. In *Xenobiotics in aquatic animals: Reproductive and developmental impacts* (pp. 353–367). Singapore.

Habumugisha, T., Zhang, Z., Uwizewe, C., Yan, C., Ndayishimiye, J. C., Rehman, A., & Zhang, X. (2024). Toxicological review of micro-and nano-plastics in aquatic environments: Risks to ecosystems, food web dynamics and human health. *Ecotoxicology and Environmental Safety, 278*, 116426.

Haitao, W., & Tangbin, H. (2023) Distribution of Mps pollution and ecological risk assessment in the surface water, fish and sediments of Jingpo Lake-the world's second largest High-Mountain barrier Lake. Fish and Sediments of Jingpo Lake-the World's Second Largest High-Mountain Barrier Lake.

Hermsen, E., Pompe, R., Besseling, E., & Koelmans, A. A. (2017). Detection of low numbers of microplastics in North Sea fish using strict quality assurance criteria. *Marine Pollution Bulletin, 122*(1–2), 253–258.

Lei, L., Wu, S., Lu, S., Liu, M., Song, Y., Fu, Z., et al. (2018). Microplastic particles cause intestinal damage and other adverse effects in zebrafish Danio rerio and nematode Caenorhabditis elegans. *Science of the Total Environment, 619*, 1–8.

Liebezeit, G., & Liebezeit, E. (2014). Synthetic particles as contaminants in German beers. *Food Additives & Contaminants: Part A, 31*(9), 1574–1578.

Liu, Z., & You, X. Y. (2023). Recent progress of microplastic toxicity on human exposure base on in vitro and in vivo studies. *Science of the Total Environment, 903*, 166766

Malik, M., Balkhi, M. H., & Abubakr, A. (2017). Comparative study on the contamination level of dal Lake, Kashmir based on Total viable bacterial counts over a period of time. *Ecology, Environment and Conservation, 23*(4), 2145–2147.

McCormick, A., Hoellein, T. J., Mason, S. A., Schluep, J., & Kelly, J. J. (2014). Microplastic is an abundant and distinct microbial habitat in an urban river. *Environmental Science & Technology, 48*(20), 11863–11871.

Megha, K. B., Anvitha, D., Parvathi, S., Neeraj, A., Sonia, J., & Mohanan, P. V. (2025). Environmental impact of microplastics and potential health hazards. *Critical Reviews in Biotechnology, 45*(1), 97–127.

Meneses, R. A. M., Cabrera-Papamija, G., Machuca-Martínez, F., Rodríguez, L. A., Diosa, J. E., & Mosquera-Vargas, E. (2022). Plastic recycling and their use as raw material for the synthesis of carbonaceous materials. *Heliyon, 8*(3), e09028.

Mishra, A. K., Singh, J., & Mishra, P. P. (2023). Microplastics in freshwater ecosystem: A serious threat for freshwater environment. *International journal of Environmental Science and Technology, 20*(8), 9189–9204.

Nabi, N., Ahmad, I., Amin, A., Rather, M. A., Ahmed, I., Hajam, Y. A., Khursheed, S., Malik, M. M., & Abubakr, A. (2024). Understanding the sources, fate and effects of microplastics in aquatic environments with a focus on risk profiling in aquaculture systems. *Reviews in Aquaculture, 16*(4), 1947–1980.

Nguyen, M. K., Rakib, M. R. J., Lin, C., Hung, N. T. Q., Le, V. G., Nguyen, H. L., et al. (2023). A comprehensive review on ecological effects of microplastic pollution: An interaction with pollutants in the ecosystems and future perspectives. *TrAC Trends in Analytical Chemistry, 168*, 117294.

Parker, B., Andreou, D., Green, I. D., & Britton, J. R. (2021). Microplastics in freshwater fishes: Occurrence, impacts and future perspectives. *Fish and Fisheries, 22*(3), 467–488.

Pataki, A. M. (2024). The roles of cities, industries, and public as stakeholders in combating marine plastic pollution within the framework of the emerging plastic regime (plastic treaty) why are these stakeholders relevant and how should they be included in the plastic regime (Master's thesis, UiT Norges arktiske universitet).

Prasad, G. S., Rout, S. K., Malik, M. M., Karmakar, S., Amin, A., & Ahmad, I. (2023). Occurrence of xenoestrogen alkylphenols (Octylphenols and Nonylphenol) and its impact on the aquatic ecosystem. In M. A. Rather, A. Amin, Y. A. Hajam, A. Jamwal, & I. Ahmad (Eds.), *Xenobiotics in Aquatic Animals* (Reproductive and developmental impacts) (pp. 275–284). Springer.

Prasad, G. S., Rout, S. K., Amin, A., Malik, M. M., Ahmad, I., Rather, M. A., Karmakar, S., Khalil, H. S., & Abubakar, A. (2024). Impact of COVID-19 on the aquatic environment and implications for fisheries with special emphasis to Indian perspective. *Proceedings of the Indian National Science Academy, 90*(4), 821–839.

Revel, M., Yakovenko, N., Caley, T., Guillet, C., Châtel, A., & Mouneyrac, C. (2020). Accumulation and immunotoxicity of microplastics in the estuarine worm Hediste diversicolor in environmentally relevant conditions of exposure. *Environmental Science and Pollution Research, 27*, 3574–3583.

Santillo, D., Miller, K., & Johnston, P. (2017). Microplastics as contaminants in commercially important seafood species. *Integrated Environmental Assessment and Management, 13*(3), 516–521.

Scircle, A., Cizdziel, J. V., Tisinger, L., Anumol, T., & Robey, D. (2020). Occurrence of microplastic pollution at oyster reefs and other coastal sites in the Mississippi sound, USA: Impacts of freshwater inflows from flooding. *Toxics, 8*(2), 35.

Sruthisree, C., Gowda, G., Nayak, H., Harish, B. D., & Amin, A. (2015). Effect of heavy metal (lead) on oxygen consumption and ammonia excretion of iridescent shark (*Pangusius hypohthalmus*). *Journal of Experimental Zoology, India, 18*(2).

Tabassum, I., Hassan, N., Amin, A., Malik, M. M., Abubakr, A., Prasad, G. S., Latief, F., & Rather, S. A. (2025). COVID-19 and global environmental change: A threat to food Security and nutrition. In *Food Security, Nutrition and sustainability through aquaculture technologies* (pp. 437–445). Springer.

Taylor, M. L., Gwinnett, C., Robinson, L. F., & Woodall, L. C. (2016). Plastic microfibre ingestion by deep-sea organisms. *Scientific Reports, 6*(1), 33997. https://doi.org/10.1038/srep33997.

Vaid, A., Jiang, J., Sawant, A., Lerakis, S., Argulian, E., Ahuja, Y., et al. (2023). A foundational vision transformer improves diagnostic performance for electrocardiograms. *npj Digital Medicine, 6*(1), 108.

Wang, P. Y., Zhao, Z. Y., Xiong, X. B., Wang, N., Zhou, R., Zhang, Z. M., et al. (2023). Microplastics affect soil bacterial community assembly more by their shapes rather than the concentrations. *Water Research, 245*, 120581.

Wu, X., Zhao, X., Chen, R., Liu, P., Liang, W., Wang, J., et al. (2022). Wastewater treatment plants act as essential sources of microplastic formation in aquatic environments: A critical review. *Water Research, 221*, 118825.

Zhang, S., Wang, J., Liu, X., Qu, F., Wang, X., Wang, X., et al. (2019). Microplastics in the environment: A review of analytical methods, distribution, and biological effects. *TrAC Trends in Analytical Chemistry, 111*, 62–72.

Microplastics in Aquatic Systems: Sources, Accumulation, Environmental Impact and Mitigation Strategies

Kurapati Nagendrasai, Adnan Amin, and Saurav Kumar

1 Introduction

The proliferation of plastics, a hallmark of the modern industrial era, has inadvertently introduced a pervasive environmental challenge—the accumulation of microplastics in aquatic systems (Tursi et al., 2022). Microplastics (MPs), defined as plastic particles smaller than 5 mm, are a pervasive environmental pollutant. Their ubiquity and persistence have positioned them as one of the most significant threats to aquatic ecosystems and human health (Vivekanand et al., 2021; Nabi et al., 2024; Debnath et al., 2024). MPs are synthetic polymer particles (<5 mm in diameter), these minute particles originate from the degradation of larger plastic debris/items, known as secondary microplastics, or are intentionally generated for leverage in products like cosmetics and industrial abrasives, referred to as primary microplastics. The extensive production, improper disposal, and persistence of plastics in the environment have resulted in the widespread dissemination of MPs across soil, air, freshwater, and marine environments (Bamigboye et al., 2024). MPs pose severe ecological and health risks by disrupting ecosystems and entering food chains. MPs have been detected in diverse environmental matrices. In aquatic systems, they are introduced through direct dumping, runoff, and wastewater discharges. While marine environments have traditionally been the focus of MP research, the presence and impact of these contaminants in freshwater systems are garnering increasing attention. Rivers, lakes, and groundwater supplies integral to sustaining life on Earth are particularly vulnerable. MPs not only disrupt aquatic biodiversity but also

K. Nagendrasai · S. Kumar
Aquatic Environmental Management, AEHM Division, ICAR-Central Institute of Fisheries Education, Versova, Mumbai, Maharashtra, India

A. Amin (✉)
Division of Aquatic Environmental Management, Faculty of Fisheries Rangil, Ganderbal, SKUAST-Kashmir, Jammu and Kashmir, India

A. Abubakr et al. (eds.), *Microplastics: Ecological and Food Security Implications*,
https://doi.org/10.1007/978-3-031-94480-2_7

infiltrate human food chains, presenting complex challenges that require immediate action (Ory et al., 2020; Bexeitova et al., 2024). Globally, the mismanagement of plastic waste significantly contributes to MP pollution in freshwater systems. Nations with high population densities and inadequate waste management infrastructures, including India, Indonesia, and Nigeria, exhibit intense levels of MP pollution (Bexeitova et al., 2024). Industrial discharge, urban runoff, agricultural activities, and wastewater effluents are the primary contributors. Their global outreach and endurance in the environment have positioned microplastics as a complex challenge for environmental management, requiring urgent attention to mitigate their far-reaching impacts.

2 Classification of Microplastics (MPs)

These are classified into two main categories: primary and secondary MPs. Primary MPs are intentionally/purposefully manufactured small particles, commonly used in personal care products, industrial abrasives, and medical applications. For example, microbeads found in exfoliating skincare products fall into this category. Secondary MPs, yet result from the breakdown of larger plastic articles due to physical, chemical, or biological degradation processes. Common sources include discarded plastic bags, bottles, fishing nets, and tire wear particles generated during vehicular transport (Bexeitova et al., 2024).

3 Sources of Microplastics (MPs)

MPs emanate from diverse provenances that reflect the complexity of their pathways into the environment. Primary MPs are purposefully synthesized, that include microbeads used in cosmetics, industrial raw materials, and other applications. These particles are often introduced into natural ecosystems through urban runoff and wastewater discharge. On the flip side, secondary MPs arise from the degradation of larger plastic items (Fig. 1). This breakdown occurs through physical, chemical, and biological mechanisms, contributing particles from provenances such as agricultural plastic films, improperly discarded fishing gear, and everyday plastic waste (Stolte et al., 2015). These breakdown processes are accelerated by environmental factors such as ultraviolet radiation, ocean turbulence, and mechanical abrasion (Cole et al., 2011; Auta et al., 2017). Moreover, textiles and synthetic fibers released during laundry cycles, as well as abrasion from road surfaces, further exacerbate the issue (Bexeitova et al., 2024).

MPs enter hydrological ecosystems through two primary pathways:

1. *Primary Sources*: These are manufactured at microscopic sizes, such as microbeads in cosmetics, synthetic fibers from textiles, and tire wear particles.

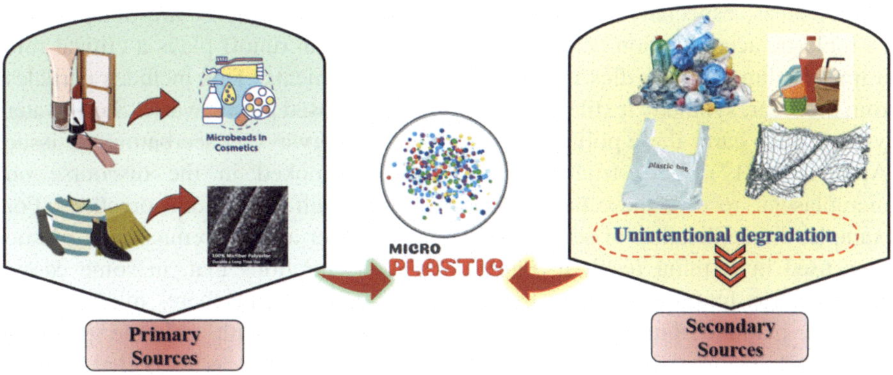

Fig. 1 Sources of MPs in the environment

They predominantly enter waterways via sewage systems and urban runoff (Tursi et al., 2022). Synthetic fibers, for instance, are released during washing processes, contributing significantly to wastewater pollution (Browne et al., 2011).

2. *Secondary Sources*: Resulting from breakdown through physical, chemical, and biological processes, factors like UV radiation, wave dynamics, and microbial activity contribute to their fragmentation, increasing their widespread presence (Andrady, 2017; Jambeck et al., 2015).

Wastewater treatment plants (WWTPs) designed to filter contaminants serve as both barriers and conduits, with treatment processes capturing some MPs but allowing smaller particles to bypass filtration systems, allowing their release into rivers and oceans (Kazour et al., 2019). In terrestrial environments, MPs accumulate from agricultural activities, including the use of plastic mulch and fertilizers, and from the atmospheric deposition of airborne particles. Household activities, such as washing synthetic textiles, release significant quantities of microfibers into wastewater systems, which subsequently flow into aquatic environments (Vivekanand et al., 2021). A single wash cycle of a synthetic garment can release approximately 1900 fibers, highlighting the critical role of domestic practices in microplastic pollution (Hartline et al., 2016). Additional contributors include road runoff containing tire wear particles, plastic pellets from industrial spillage, and degraded fishing gear.

4 Fate and Transport in Aquatic Ecosystems

Once introduced into the environment, microplastics exhibit a widespread distribution across aquatic, terrestrial, and atmospheric systems. Aquatic environments, particularly oceans, serve as major sinks for microplastics (Table 1). Rivers act as conduits, delivering significant quantities of these pollutants to marine ecosystems.

Coastal areas, especially those near densely populated regions, are hotspots for microplastic accumulation (Zhang et al., 2019a). Urban runoff plays a critical role in transporting microplastics into freshwater environments. This includes particles from tire wear, synthetic textiles, and improperly disposed plastic waste. Stormwater systems often carry these pollutants directly into waterways, exacerbating the issue (Andrady, 2017). Terrestrial systems, often overlooked in the discourse on microplastics, receive these particles primarily through agricultural practices. For example, the application of sewage sludge as fertilizer and the remnants of plastic films used in farming result in microplastic concentrations that, in some cases, surpass those found in aquatic environments. In the atmosphere, microplastics originate mainly from synthetic fibers and tire wear, with airborne particles transported over long distances and even detected in remote areas (Dris et al., 2015).

Microplastics are transported across aquatic environments through natural and anthropogenic pathways. Ocean currents, wind, and river systems distribute microplastics globally, from surface waters to deep-sea sediments. Their physical and chemical properties of MPs, including density and hydrophobicity, influence their distribution in various environmental matrices. For example, lightweight microplastics often remain suspended in surface waters, while denser particles settle into sediments. Schmidt et al. (2017) estimated that rivers contribute approximately 80–94% of the plastic load entering oceans. Additionally, biofouling processes alter the buoyancy of microplastics, causing them to sink and settle in benthic environments, where they pose risks to marine organisms and habitats (Tursi et al., 2022). In oceanic gyres such as the Great Pacific Garbage Patch, plastics aggregate due to circular current patterns, creating concentrated zones of pollution (Eriksen et al., 2014). These hotspots serve as hubs for the accumulation of microplastics, with significant ecological ramifications.

The polar regions exemplify the extensive reach of microplastics, where sea ice acts as a temporary sink. However, with global warming and ice melt, these particles are re-released into marine ecosystems, exacerbating contamination levels. Urban runoff and stormwater overflow further advance the transport/dispersion of MPs, emphasizing the interconnectedness of terrestrial and aquatic systems.

Table 1 Occurrence and distribution of various MPs in some aquatic systems

Aquatic system	Type of waters	Occurrence (Mean)	Types of MPs	References
North Sea	Surface	1.5 items/m^3	Fragments	Maes et al. (2017)
Arctic	Surface	31.3 items/L	Fibers	Barrows et al. (2018)
North Atlantic	Deep-sea	70.8 items/m^3	Fibers	Courtene-Jones et al. (2017)
Bay of Brest, Atlantic	Surface	0.24 items/m^3	Fibers, fragments	Frere et al. (2017)
Indian Ocean, India	Beaches (sand)	81 mg/kg	Fragments	Reddy et al. (2006)

5 Sampling, and Analytical Techniques for Detection of MPs

The analysis of MPs from environmental samples involves several stages, including sampling, preprocessing, and identification. Sampling is the first critical step and varies depending on the target environment and research objectives. Methods such as selective sampling are used for visible plastic fragments, while volume-reduced sampling targets smaller particles. Tools like manta nets are commonly employed for collecting surface water samples, while box samplers are utilized for sediment collection (Hidalgo-Ruz et al., 2012). Preprocessing of samples is necessary to isolate microplastics from environmental impurities. Techniques like density separation, filtration, and digestion are employed to enhance the clarity and accuracy of analyses. For instance, high-density solutions such as $ZnCl_2$ are used to separate denser plastics from sediments, increasing extraction efficiency (Nuelle et al., 2014). Once preprocessed, the identification of microplastics relies on various analytical methods. Visual inspection, though straightforward and cost-effective, is limited to larger particles and is prone to subjective error (Hidalgo-Ruz et al., 2012).

Detecting MPs in water systems requires advanced analytical techniques capable of identifying their composition, size, and quantity. Spectroscopic techniques such as Fourier Transform Infrared (FTIR) and Raman spectroscopy are widely used for detailed compositional analysis (Gong & Xie, 2020). These techniques enable researchers to determine the polymer types and assess the presence of additives or contaminants. FTIR, for example, is effective for identifying plastic particles larger than 20 μm, while Raman spectroscopy excels in detecting smaller particles but is hindered by fluorescence interference (Tagg et al., 2017). Scanning electron microscopy (SEM) and pyrolysis-gas chromatography-mass spectrometry (Py-GC-MS) offer additional insights into the morphological and chemical characteristics of MPs (Bexeitova et al., 2024). Thermal analysis, including Pyr-GC-MS, provides insights into the chemical composition of microplastics by analyzing their decomposition products. However, this method is time-intensive and limited in processing large sample volumes.

Although detection methods have improved, significant challenges persist. The absence of standardized procedures/methods for sampling, pretreatment, and analysis hinders the comparability of data across studies. Additionally, detecting nanoplastics smaller than 1 μm presents considerable technical difficulties, requiring the advancement of more precise instruments and techniques (Bexeitova et al., 2024).

6 Accumulation and Eco-environmental Impacts

The charisma of MPs in the environment offers multifaceted challenges to ecosystems and human health (Fig. 2). Once introduced into aquatic ecosystems, MPs are ingested by a broad spectrum of life, includes plankton, fish, amphibians, and others. These particles can bioaccumulate in organisms and biomagnify across food webs, disrupting ecological balance (Cole et al., 2013). MPs also act as delivery agents for hazardous pollutants, such as persistent organic pollutants (POPs) and heavy metals, which adheres to their surfaces. This process increases the toxicity of MPs, causing oxidative stress, endocrine disruption, and reproductive impairments in aquatic species (Sruthisree et al., 2015; Amin et al., 2016, 2017; Gurjar et al., 2023; Prasad et al., 2023; Bexeitova et al., 2024). Sediments act as both sinks and secondary sources of MPs, as particles can settle and later resuspend due to hydrological disturbances. In soils, MPs alter microbial activity, reduce nutrient availability, and impair plant growth by affecting root systems and biomass. Seasonal variations further influence MP concentrations, with higher levels observed during periods of heavy rainfall and runoff. The spatial variability of MP pollution highlights the need for site-specific mitigation strategies (Bexeitova et al., 2024).

MPs offer multifaceted threats to blue ecosystems, disrupting ecological balance and affecting biodiversity at all levels. Their impacts are primarily categorized as physical, chemical, and biological.

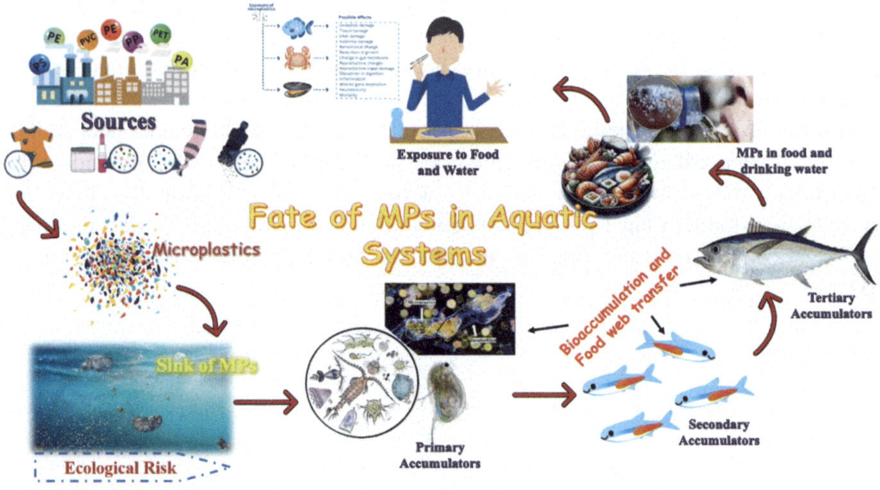

Fig. 2 Fate and accumulation of MPs in aquatic ecosystems

6.1 Physical Impacts

Physically, MPs act as foreign objects when ingested by aquatic life. Species encompassing zooplankton to large mammals mistake MPs for food, inducing ingestion. This results in gastrointestinal blockages, reduced feeding efficiency, and altered foraging behavior (Rochman et al., 2014). Fish and seabirds often accumulate microplastics in their stomachs, which can reduce their overall energy intake and cause starvation (Vivekanand et al., 2021). They also contribute to habitat degradation as floating plastics form dense aggregations, altering the natural flow of water and light penetration. This disrupts photosynthesis in aquatic plants and affects oxygen levels in the water, thereby threatening entire ecosystems.

6.2 Chemical Impacts

MPs function as haulers for a wide range of harmful chemicals, including POPs, heavy metals, and plastic additives like bisphenol-A (BPA) and phthalates. These chemicals leach into surrounding water or are absorbed by organisms ingesting microplastics. Once inside organisms, these chemicals can cause endocrine disruption, reproductive failure, and immune suppression (Halden, 2010; Prata et al., 2020). In addition, MPs able to retain hydrophobic pollutants such as pesticides and polycyclic aromatic hydrocarbons (PAHs) from surrounding waters (Andrady, 2017). These adsorbed pollutants are bioavailable to organisms ingesting microplastics, magnifying the toxic effects (Xu et al., 2021).

6.3 Biological Impacts

Biologically, MPs disrupt food webs and ecological interactions. In planktonic organisms, microplastic ingestion can impair growth, reproduction, and mobility, thereby affecting the base of the aquatic food web (Cole et al., 2013). Predatory fish and marine mammals consuming microplastic-laden prey experience bioaccumulation of toxins, leading to higher concentrations of pollutants at higher trophic levels (Browne et al., 2007). MPs also serve as delivery agents of invasive species and pathogens. They serve as substrates for microbial biofilms, including harmful bacteria and viruses. These biofilms can travel across oceans, introducing pathogens to new regions and disrupting local ecosystems. Coral reefs, vital to marine biodiversity, are particularly vulnerable as microplastics impair their health and resilience (Hall et al., 2015).

7 Toxicity of MPs to Aquatic Biota

MPs exhibit both physical and chemical toxicity to aquatic organisms, severely affecting their physiology, behavior, and survival (Table 2). MPs also act as conduits for harmful pollutants, including PAHs, heavy metals, and pesticides, which leach into the tissues of aquatic species, disrupting metabolic activities and leading to oxidative stress (Godoy et al., 2019; Qiao et al., 2019). Additionally, MPs can alter gene expression and endocrine system functions in fish such as Japanese medaka (*Oryzias latipes*), potentially impairing reproduction and growth (Rochman et al., 2014). Toxicity is often size-dependent; smaller MPs and nanoplastics pose higher risks as they penetrate tissues more readily, causing cellular damage and bioaccumulation in critical organs like the gills and liver (Lu et al., 2016; Zhang et al., 2019b). Such impacts are further exacerbated by the interaction of MPs with other stressors, including temperature and salinity, creating compounding threats to aquatic ecosystems.

8 Human Health Implications

The implications of MP pollution extend beyond environmental concerns, directly impacting human well-being. Humans are exposed to MPs primarily through ingestion (food and water), inhalation (airborne particles), and dermal contact (personal care products). Notably, seafood consumption significantly contributes to human ingestion of MPs. As MPs buildup in the tissues of aquatic species consumed as seafood, humans are exposed to these particles through dietary intake. For example,

Table 2 Toxic effects of MPs and associated pollutants on various aquatic biota

Plastic types[a]	Pollutants	Aquatic biota	Toxicological effects	References
PE	Polychlorinated biphenyls (PCBs)	*Arenicola marina* (Blow lugworm)	Bioaccumulation, altered feeding behavior	Besseling et al. (2013)
PVC	Chlorinated antidepressants	*Misgurnus anguillicaudatus* (Pond loach)	Bioaccumulation and effects the liver	Qu et al. (2018)
PS	Roxithromycin	*Daphnia magna*	Increased oxidative stress and biochemical activity	Zhang et al. (2019b)
PE	PCBs and PAHs	*Oryzias latipes*	Endocrine disruption, alteration in gene expression	Rochman et al. (2014)
PE, PS	Hydrophobic organic contaminants	*Skeletonema costatum*	Growth suppression, oxidative stress	Zhu et al. (2019)

[a]*PE* polyethylene, *PVC* polyvinyl chloride, *PS* polystyrene

bivalves including mussels and oysters often contain microplastics owing to filter-feeding (Simionov et al., 2023). The small size of MPs allows them to penetrate biological tissues, raising concerns about their perpetual effects on human well-being (Galloway, 2015; Cox et al., 2019). MPs can enter the gastrointestinal (GI) tract, bloodstream, and even the placenta, posing risks such as inflammation, oxidative stress, and immune system disruption (Xu et al., 2020; Ragusa et al., 2022). Infants and pregnant women are particularly vulnerable due to their heightened sensitivity to environmental toxins. MPs have been found in human blood, lungs, placenta, and breast milk, raising concerns about their systemic distribution and potential toxicity (Bexeitova et al., 2024). Additionally, MPs can act as carriers for pathogens and chemical additives, such as bisphenols and phthalates, further compounding their health risks (Malik et al., 2017; Prata, 2018; Bexeitova et al., 2024). The aeonic health consequences of MP exposure remain under-researched, necessitating further investigation to understand their potential impacts fully (Vivekanand et al., 2021).

9 Mitigation Strategies and Policy Interventions

Addressing microplastic pollution demands a multifaceted strategy encompassing preventive measures, enhanced waste management, technological innovation, and public engagement. Effective mitigation strategies must tackle both the origins and threats of MPs to achieve sustainable environmental outcomes.

Prevention through policy interventions These focus on reducing the generation and promulgation of MPs into the environment. Preventive measures include banning microbeads in cosmetics, promoting biodegradable alternatives, and implementing extended producer responsibility (EPR) schemes. Policy interventions play a critical role in curbing MP pollution. Regulatory interventions, such as bans on microbeads in cosmetics and targeting single-use plastics, have proven effective in curbing primary microplastic pollution (Auta et al., 2017). Eco-labeling initiatives encourage consumers to choose environmentally friendly products, fostering sustainable consumption patterns. International agreements and collaborations are necessary to address the transboundary nature of MP pollution. Educational campaigns targeting consumers and industries can further promote sustainable practices and reduce plastic consumption (Bexeitova et al., 2024).

Recycling and effective waste management Enhancing recycling rates for plastic products can significantly reduce the generation of secondary microplastics. Advanced sorting technologies and color-coded waste bins are crucial for improving recycling efficiency. The 4R concept, which includes Refuse, Reduce, Reuse, and Recycle, focuses on sustainable strategies to manage plastic use and mitigate pollution (Mallik et al., 2021). It encourages reducing reliance on single-use plastics, minimizing consumption, reusing products to extend their lifespan, and recycling materials to limit waste. Biodegradable alternatives (Bioplastics) to conventional plastics offer a promising long-term solution. These materials, derived from renewable resources, degrade more readily in natural environments, minimizing their ecological footprint (Song et al., 2009). Developing biodegradable and sustainable substitutes to traditional plastics can mitigate the generation of microplastics (Tursi et al., 2022; Prasad et al., 2024; Tabassum et al., 2025). Combining the 4R approach with bioplastics adoption provides an effective solution to tackle plastic pollution and promote sustainability.

Awareness and behavioral change in public Educational campaigns that inform communities about the environmental consequences of plastic waste can encourage responsible consumption and disposal practices. Additionally, public awareness campaigns and community-driven initiatives, like beach cleanups, are essential for fostering behavioral change and reducing plastic waste at its source (Vivekanand et al., 2021). Public education campaigns promoting reduced plastic consumption and responsible waste management are critical to long-term solutions (Deng et al., 2020).

Technological innovations Advanced oxidation processes (AOPs), membrane filtration, and microbial degradation are promising technologies for removing MPs from wastewater and drinking water. AOPs involve generating reactive oxygen species to degrade MPs into less harmful substances, while membrane filtration systems use semipermeable barriers to physically separate MPs from water (Bamigboye et al., 2024). Bioremediation, which involves using microbes ability of degrading plastics, extends a promising approach to reducing microplastic contamination. Microbial degradation, which employs bacteria and fungi capable of breaking down plastic polymers, is gaining traction as a sustainable approach. Studies have shown that strains such as *Bacillus cereus* and *Bacillus gottheilii* can degrade MPs by enzymatic action, reducing their size and toxicity (Auta et al., 2017). Species of bacteria like *Pseudomonas* and *Rhodococcus* have shown potential for breaking down plastic polymers, providing an environmentally friendly solution to the problem (Yang et al., 2015; Auta et al., 2017). Innovations in wastewater treatment technologies, including advanced filtration systems for WWTPs can apprehend MPs before they enter aquatic ecosystems, further reducing their environmental impact. Innovations in natural polymer-based materials, such as those derived from algae, represent a potential breakthrough (El Semary et al., 2022).

10 Future Directions and Recommendations

Although substantial progress has been achieved in studying microplastic pollution, numerous gaps persist. Standardized methods for sampling and analysis are urgently needed to enable comparability across studies and to build a cohesive understanding of global microplastic distribution (Hidalgo-Ruz et al., 2012). Research on MPs in non-marine environments, such as soil and air, lags behind marine studies and warrants greater attention (Zhang et al., 2019a). The aeonic ecological and health effects of MPs remain ill-defined. Investigating the pathways and impacts of MPs as they move through food chains is critical for assessing their implications for human health. Additionally, enhancing technologies for microplastic removal and exploring innovative degradation methods will be vital for mitigating their environmental impact. Investments in infrastructure for waste collection and recycling, especially in developing countries, are crucial. Additionally, the establishment of global monitoring networks can facilitate the tracking of MP pollution and the evaluation of mitigation endeavors. Policymakers should integrate scientific findings into actionable frameworks, ensuring that interventions are evidence-based and context-specific (Bexeitova et al., 2024).

11 Conclusion

Microplastics represent a pressing environmental issue with far-reaching ecological and human health impacts. Their persistence and ubiquity necessitate coordinated global efforts to mitigate their effects. Preventive measures, innovative recycling technologies, and public education are critical in addressing this challenge. While biotechnological advancements offer promising solutions, sustained investment in research and policy enforcement is essential to curb the future amassing of microplastics in aquatic environments. Effective mitigation will require collaboration among governments, industries, scientists, and communities to ensure a sustainable future for aquatic ecosystems and human health.

References

Amin, A., Indulkar, S. T., & Pai, R. (2016). Sublethal effect of Buprofezin pesticide on carcass composition of *Cyprinus carpio* communis fingerlings. *Journal of Experimental Zoology India, 20*(1), 269–272.

Amin, A., Ramachandra Naik, A. T., Priyadarshini, N., Nayak, H., & Sree, C. S. (2017). Toxic effect of heavy metal lead on oxygen consumption of rohu (*Labeo rohita*) fingerlings. *Biochemical & Cellular Archives, 17*(1), 225–228.

Andrady, A. L. (2017). The plastic in microplastics: A review. *Marine Pollution Bulletin, 119*(1), 12–22. https://doi.org/10.1016/j.marpolbul.2017.01.082

Auta, H. S., Emenike, C. U., & Fauziah, S. H. (2017). Distribution and importance of microplastics in the marine environment: A review of the sources, fate, effects, and potential solutions. *Environment International, 102*, 165–176. https://doi.org/10.1016/j.envint.2017.02.013

Bamigboye, O., Alfred, M. O., Bayode, A. A., Unuabonah, E. I., & Omorogie, M. O. (2024). The growing threats and mitigation of environmental microplastics. *Environmental Chemistry and Ecotoxicology, 6*, 259–268. https://doi.org/10.1016/j.enceco.2024.07.001

Barrows, A. P. W., Cathey, S. E., & Petersen, C. W. (2018). Marine environment microfiber contamination: Global patterns and the diversity of microparticle origins. *Environmental Pollution, 237*, 275–284. https://doi.org/10.1016/j.envpol.2018.02.062

Besseling, E., Wegner, A., Foekema, E. M., Van Den Heuvel-Greve, M. J., & Koelmans, A. A. (2013). Effects of microplastic on fitness and PCB bioaccumulation by the lugworm Arenicola marina (L.). *Environmental Science & Technology, 47*(1), 593–600. https://doi.org/10.1021/es302763x

Bexeitova, K., Baimenov, A., Varol, E. A., Kudaibergenov, K., Zhantikeyev, U., Sailaukhanuly, Y., Toshtay, K., Tauanov, Z., Azat, S., & Berndtsson, R. (2024). Microplastics in freshwater systems: A review of classification, sources, and environmental impacts. *Chemical Engineering Journal Advances, 20*, 100649. https://doi.org/10.1016/j.ceja.2024.100649

Browne, M. A., Crump, P., Niven, S. J., Teuten, E., Tonkin, A., Galloway, T., & Thompson, R. (2011). Accumulation of microplastic on shorelines woldwide: Sources and sinks. *Environmental Science & Technology, 45*(21), 9175–9179. https://doi.org/10.1021/es201811s

Browne, M. A., Galloway, T., & Thompson, R. (2007). Microplastic--An emerging contaminant of potential concern? *Integrated Environmental Assessment and Management, 3*(4), 559–566. https://doi.org/10.1002/ieam.5630030412

Cole, M., Lindeque, P., Fileman, E., Halsband, C., Goodhead, R., Moger, J., & Galloway, T. S. (2013). Microplastic ingestion by zooplankton. *Environmental Science & Technology, 47*(12), 6646–6655. https://doi.org/10.1021/es400663f

Cole, M., Lindeque, P., Halsband, C., & Galloway, T. S. (2011). Microplastics as contaminants in the marine environment: A review. *Marine Pollution Bulletin, 62*(12), 2588–2597. https://doi.org/10.1016/j.marpolbul.2011.09.025

Courtene-Jones, W., Quinn, B., Gary, S. F., Mogg, A. O., & Narayanaswamy, B. E. (2017). Microplastic pollution identified in deep-sea water and ingested by benthic invertebrates in the Rockall Trough, North Atlantic Ocean. *Environmental Pollution, 231*, 271–280. https://doi.org/10.1016/j.envpol.2017.08.026

Cox, K. D., Covernton, G. A., Davies, H. L., Dower, J. F., Juanes, F., & Dudas, S. E. (2019). Human consumption of microplastics. *Environmental Science & Technology, 53*(12), 7068–7074. https://doi.org/10.1021/acs.est.0c04032

Debnath, R., Prasad, G. S., Amin, A., Malik, M. M., Ahmad, I., Abubakr, A., Borah, S., Rather, M. A., Impellitteri, F., Tabassum, I., & Piccione, G. (2024). Understanding and addressing microplastic pollution: Impacts, mitigation, and future perspectives. *Journal of Contaminant Hydrology, 266*, 104399.

Deng, L., Cai, L., Sun, F., Li, G., & Che, Y. (2020). Public attitudes towards microplastics: Perceptions, behaviors and policy implications. *Resources, Conservation and Recycling, 163*, 105096. https://doi.org/10.1016/j.resconrec.2020.105096

Dris, R., Gasperi, J., Rocher, V., Saad, M., Renault, N., & Tassin, B. (2015). Microplastic contamination in an urban area: A case study in Greater Paris. *Environmental Chemistry, 12*(5), 592–599. https://doi.org/10.1071/EN14167

El Semary, N., Alsuhail, M., Al Amer, K., & AlNaim, A. (2022). Applications of algae for environmental sustainability: Novel bioplastic formulation method from marine green alga. *Frontiers in Marine Science, 9*, 1047284. https://doi.org/10.3389/fmars.2022.1047284

Eriksen, M., Lebreton, L. C., Carson, H. S., Thiel, M., Moore, C. J., Borerro, J. C., Galgani, F., Ryan, P. G., & Reisser, J. (2014). Plastic pollution in the world's oceans: More than 5 trillion plastic pieces weighing over 250,000 tons afloat at sea. *PLoS One, 9*(12), e111913. https://doi.org/10.1371/journal.pone.0111913

Frere, L., Paul-Pont, I., Rinnert, E., Petton, S., Jaffré, J., Bihannic, I., Soudant, P., Lambert, C., & Huvet, A. (2017). Influence of environmental and anthropogenic factors on the composition, concentration and spatial distribution of microplastics: A case study of the Bay of Brest (Brittany, France). *Environmental Pollution, 225*, 211–222. https://doi.org/10.1016/j.envpol.2017.03.023

Galloway, T. S. (2015). Micro-and nano-plastics and human health. In *Marine anthropogenic litter* (pp. 343–366). Springer. https://doi.org/10.1007/978-3-319-16510-3

Godoy, V., Blázquez, G., Calero, M., Quesada, L., & Martín-Lara, M. A. (2019). The potential of microplastics as carriers of metals. *Environmental Pollution, 255*, 113363. https://doi.org/10.1016/j.envpol.2019.113363

Gong, J., & Xie, P. (2020). Research progress in sources, analytical methods, eco-environmental effects, and control measures of microplastics. *Chemosphere, 254*, 126790. https://doi.org/10.1016/j.chemosphere.2020.126790

Gurjar, U. R., Takar, S., Amin, A., & Martin Xavier, K. A. (2023). Microplastic contamination in aquatic organisms: An ecotoxicological perspective. In *Xenobiotics in aquatic animals: Reproductive and developmental impacts* (pp. 353–367). Springer.

Halden, R. U. (2010). Plastics and health risks. *Annual Review of Public Health, 31*(1), 179–194. https://doi.org/10.1146/annurev.publhealth.012809.103714

Hall, N. M., Berry, K. L. E., Rintoul, L., & Hoogenboom, M. O. (2015). Microplastic ingestion by scleractinian corals. *Marine Biology, 162*, 725–732. https://doi.org/10.1007/s00227-015-2619-7

Hartline, N. L., Bruce, N. J., Karba, S. N., Ruff, E. O., Sonar, S. U., & Holden, P. A. (2016). Microfiber masses recovered from conventional machine washing of new or aged garments. *Environmental Science & Technology, 50*(21), 11532–11538. https://doi.org/10.1021/acs.est.6b03045

Hidalgo-Ruz, V., Gutow, L., Thompson, R. C., & Thiel, M. (2012). Microplastics in the marine environment: A review of the methods used for identification and quantification. *Environmental Science & Technology, 46*(6), 3060–3075. https://doi.org/10.1021/es2031505

Jambeck, J. R., Geyer, R., Wilcox, C., Siegler, T. R., Perryman, M., Andrady, A., Narayan, R., & Law, K. L. (2015). Plastic waste inputs from land into the ocean. *Science, 347*(6223), 768–771. https://doi.org/10.1126/science.1260352

Kazour, M., Terki, S., Rabhi, K., Jemaa, S., Khalaf, G., & Amara, R. (2019). Sources of microplastics pollution in the marine environment: Importance of wastewater treatment plant and coastal landfill. *Marine Pollution Bulletin, 146*, 608–618. https://doi.org/10.1016/j.marpolbul.2019.06.066

Lu, Y., Zhang, Y., Deng, Y., Jiang, W., Zhao, Y., Geng, J., Ding, L., & Ren, H. (2016). Uptake and accumulation of polystyrene microplastics in zebrafish (Danio rerio) and toxic effects in liver. *Environmental Science & Technology, 50*(7), 4054–4060. https://doi.org/10.1021/acs.est.6b00183

Maes, T., Van der Meulen, M. D., Devriese, L. I., Leslie, H. A., Huvet, A., Frère, L., Robbens, J., & Vethaak, A. D. (2017). Microplastics baseline surveys at the water surface and in sediments of the North-East Atlantic. *Frontiers in Marine Science, 4*, 135. https://doi.org/10.3389/fmars.2017.00135

Malik, M., Balkhi, M. H., & Abubakr, A. (2017). Comparative study on the contamination level of Dal Lake, Kashmir based on Total Viable Bacterial Counts over a period of time. *Ecology Environment and Conservation, 23*(4), 2145–2147.

Mallik, A., Xavier, K. M., Naidu, B. C., & Nayak, B. B. (2021). Ecotoxicological and physiological risks of microplastics on fish and their possible mitigation measures. *Science of the Total Environment, 779*, 146433. https://doi.org/10.1016/j.scitotenv.2021.146433

Nabi, N., Ahmad, I., Amin, A., Rather, M. A., Ahmed, I., Hajam, Y. A., Khursheed, S., Malik, M. M., & Abubakr, A. (2024). Understanding the sources, fate and effects of microplastics in aquatic environments with a focus on risk profiling in aquaculture systems. *Reviews in Aquaculture, 16*(4), 1947–1980.

Nuelle, M. T., Dekiff, J. H., Remy, D., & Fries, E. (2014). A new analytical approach for monitoring microplastics in marine sediments. *Environmental Pollution, 184*, 161–169. https://doi.org/10.1016/j.envpol.2013.07.027

Ory, N. C., Lehmann, A., Javidpour, J., Stöhr, R., Walls, G. L., & Clemmesen, C. (2020). Factors influencing the spatial and temporal distribution of microplastics at the sea surface – A year-long monitoring case study from the urban Kiel Fjord, southwest Baltic Sea. *Science of the Total Environment, 736*, 139493. https://doi.org/10.1016/j.scitotenv.2020.139493

Prasad, G. S., Rout, S. K., Amin, A., Malik, M. M., Ahmad, I., Rather, M. A., Karmakar, S., Khalil, H. S., & Abubakar, A. (2024). Impact of COVID-19 on the aquatic environment and implications for fisheries with special emphasis to Indian perspective. *Proceedings of the Indian National Science Academy, 90*(4), 821–839.

Prasad, G. S., Rout, S. K., Malik, M. M., Karmakar, S., Amin, A., & Ahmad, I. (2023). Occurrence of xenoestrogen alkylphenols (Octylphenols and Nonylphenol) and its impact on the aquatic ecosystem. In *Xenobiotics in aquatic animals: Reproductive and developmental impacts* (pp. 275–284). Springer.

Prata, J. C. (2018). Microplastics in wastewater: State of the knowledge on sources, fate and solutions. *Marine Pollution Bulletin, 129*(1), 262–265. https://doi.org/10.1016/j.marpolbul.2018.02.046

Prata, J. C., da Costa, J. P., Lopes, I., Duarte, A. C., & Rocha-Santos, T. (2020). Environmental exposure to microplastics: An overview on possible human health effects. *Science of the Total Environment, 702*, 134455. https://doi.org/10.1016/j.scitotenv.2019.134455

Qiao, R., Sheng, C., Lu, Y., Zhang, Y., Ren, H., & Lemos, B. (2019). Microplastics induce intestinal inflammation, oxidative stress, and disorders of metabolome and microbiome in zebrafish. *Science of the Total Environment, 662*, 246–253. https://doi.org/10.1016/j.scitotenv.2019.01.245

Qu, H., Ma, R., Wang, B., Zhang, Y., Yin, L., Yu, G., Deng, S., Huang, J., & Wang, Y. (2018). Effects of microplastics on the uptake, distribution and biotransformation of chiral antidepressant venlafaxine in aquatic ecosystem. *Journal of Hazardous Materials, 359*, 104–112. https://doi.org/10.1016/j.jhazmat.2018.07.016

Ragusa, A., Matta, M., Cristiano, L., Matassa, R., Battaglione, E., Svelato, A., De Luca, C., D'Avino, S., Gulotta, A., Rongioletti, M. C. A., Catalano, P., Santacroce, C., Notarstefano, V., Carnevali, O., Giorgini, E., Vizza, E., Familiari, G., & Nottola, S. A. (2022). Deeply in plasticenta: Presence of microplastics in the intracellular compartment of human placentas. *International Journal of Environmental Research and Public Health, 19*(18), 11593. https://doi.org/10.3390/ijerph191811593

Reddy, M. S., Basha, S., Adimurthy, S., & Ramachandraiah, G. (2006). Description of the small plastics fragments in marine sediments along the Alang-Sosiya ship-breaking yard, India. *Estuarine, Coastal and Shelf Science, 68*(3–4), 656–660. https://doi.org/10.1016/j.ecss.2006.03.018

Rochman, C. M., Kurobe, T., Flores, I., & Teh, S. J. (2014). Early warning signs of endocrine disruption in adult fish from the ingestion of polyethylene with and without sorbed chemical pollutants from the marine environment. *Science of the Total Environment, 493*, 656–661. https://doi.org/10.1016/j.scitotenv.2014.06.051

Schmidt, C., Krauth, T., & Wagner, S. (2017). Export of plastic debris by rivers into the sea. *Environmental Science & Technology, 51*(21), 12246–12253. https://doi.org/10.1021/acs.est.7b06377

Simionov, I. A., Călmuc, M., Iticescu, C., Călmuc, V., Georgescu, P. L., Faggio, C., & Petrea, Ş. M. (2023). Human health risk assessment of potentially toxic elements and microplastics accumulation in products from the Danube River Basin fish market. *Environmental Toxicology and Pharmacology, 104*, 104307. https://doi.org/10.1016/j.etap.2023.104307

Song, J. H., Murphy, R. J., Narayan, R., & Davies, G. B. H. (2009). Biodegradable and compostable alternatives to conventional plastics. *Philosophical Transactions of the Royal Society B: Biological Sciences, 364*(1526), 2127–2139. https://doi.org/10.1098/rstb.2008.0289

Sruthisree, C., Gowda, G., Nayak, H., Harish, B. D., & Amin, A. (2015). Effect of heavy metal (lead) on oxygen consumption and ammonia excretion of iridescent shark (*Pangusius hypohthalmus*). *Journal of Experimental Zoology India, 18*(2), 1–4.

Stolte, A., Forster, S., Gerdts, G., & Schubert, H. (2015). Microplastic concentrations in beach sediments along the German Baltic coast. *Marine Pollution Bulletin, 99*(1–2), 216–229. https://doi.org/10.1016/j.marpolbul.2015.07.022

Tabassum, I., Hassan, N., Amin, A., Malik, M. M., Abubakr, A., Prasad, G. S., Latief, F., & Rather, S. A. (2025). COVID-19 and global environmental change: A threat to food security and nutrition. In *Food security, nutrition and sustainability through aquaculture technologies* (pp. 437–445). Springer.

Tagg, A. S., Harrison, J. P., Ju-Nam, Y., Sapp, M., Bradley, E. L., Sinclair, C. J., & Ojeda, J. J. (2017). Fenton's reagent for the rapid and efficient isolation of microplastics from wastewater. *Chemical Communications, 53*(2), 372–375. https://doi.org/10.1039/C6CC08798A

Tursi, A., Baratta, M., Easton, T., Chatzisymeon, E., Chidichimo, F., De Biase, M., & De Filpo, G. (2022). Microplastics in aquatic systems, a comprehensive review: Origination, accumulation, impact, and removal technologies. *RSC Advances, 12*(44), 28318–28340. https://doi.org/10.1039/D2RA04713F

Vivekanand, A. C., Mohapatra, S., & Tyagi, V. K. (2021). Microplastics in aquatic environment: Challenges and perspectives. *Chemosphere, 282*, 131151. https://doi.org/10.1016/j.chemosphere.2021.131151

Xu, S., Ma, J., Ji, R., Pan, K., & Miao, A. J. (2020). Microplastics in aquatic environments: Occurrence, accumulation, and biological effects. *Science of the Total Environment, 703*, 134699. https://doi.org/10.1016/j.scitotenv.2019.134699

Xu, X., Zhang, L., Xue, Y., Gao, Y., Wang, L., Peng, M., Jiang, S., & Zhang, Q. (2021). Microplastic pollution characteristic in surface water and freshwater fish of Gehu Lake, China. *Environmental Science and Pollution Research, 28*, 67203–67213. https://doi.org/10.1007/s11356-021-15338-8

Yang, Y., Yang, J., Wu, W. M., Zhao, J., Song, Y., Gao, L., Yang, R., & Jiang, L. (2015). Biodegradation and mineralization of polystyrene by plastic-eating mealworms: Part 2. Role of gut microorganisms. *Environmental Science & Technology, 49*(20), 12087–12093. https://doi.org/10.1021/acs.est.5b02663

Zhang, P., Yan, Z., Lu, G., & Ji, Y. (2019a). Single and combined effects of microplastics and roxithromycin on Daphnia magna. *Environmental Science and Pollution Research, 26*, 17010–17020. https://doi.org/10.1007/s11356-019-05031-2

Zhang, S., Wang, J., Liu, X., Qu, F., Wang, X., Wang, X., Li, Y., & Sun, Y. (2019b). Microplastics in the environment: A review of analytical methods, distribution, and biological effects. *TrAC Trends in Analytical Chemistry, 111*, 62–72. https://doi.org/10.1016/j.trac.2018.12.002

Zhu, Z. L., Wang, S. C., Zhao, F. F., Wang, S. G., Liu, F. F., & Liu, G. Z. (2019). Joint toxicity of microplastics with triclosan to marine microalgae *Skeletonema costatum*. *Environmental Pollution, 246*, 509–517. https://doi.org/10.1016/j.envpol.2018.12.044

Ecotoxicological Consequences of Microplastics in Aquatic Life

Manoj Tukaram Kamble, Seema Vijay Medhe, Balasaheb Ramdas Chavan, Viraj Yeshwant Daunde, Gargi Rajesh Palekar, Sangharsh Himmat Tayade, Ndakalimwe Naftal Gabriel, and Nopadon Pirarat

1 Introduction

Microplastics (MPs), defined as plastic particles smaller than 5 mm, have become a significant environmental concern due to their ubiquity and persistence in aquatic ecosystems (Lusher, 2015). MPs are classified into two categories: primary microplastics, which are manufactured at this size and used in products such as personal care items, cleaning products, and industrial abrasives (Browne et al., 2007), and secondary microplastics, which result from the degradation of larger plastic items, such as bottles, bags, and fishing nets, through physical, chemical, and biological processes (Barnes et al., 2009).

Microplastics enter aquatic ecosystems from both land-based and water-based sources (Lebreton & Andrady, 2019; Nabi et al., 2024; Debnath et al., 2024). Land-based sources include urban runoff, improper plastic waste disposal, agricultural

M. T. Kamble · N. Pirarat (✉)
Center of Excellence in Wildlife, Exotic, and Aquatic Animal Pathology, Faculty of Veterinary Science, Chulalongkorn University, Bangkok, Thailand

S. V. Medhe (✉)
ASEAN Institute for Health Development, Mahidol University, Nakhon Pathom, Thailand

B. R. Chavan
College of Fisheries, Dr. Balasaheb Sawant Konkan Agricultural University, Ratnagiri, India

V. Y. Daunde · G. R. Palekar · S. H. Tayade
Center of Excellence in Wildlife, Exotic, and Aquatic Animal Pathology, Faculty of Veterinary Science, Chulalongkorn University, Bangkok, Thailand

International Graduate Course of Veterinary Science and Technology (VST), Faculty of Veterinary Science, Chulalongkorn University, Bangkok, Thailand

N. N. Gabriel
Department of Fisheries and Ocean Sciences, Faculty of Agriculture, Engineering and Natural Science, Sam Nujoma Campus, University of Namibia, Henties Bay, Namibia

A. Abubakr et al. (eds.), *Microplastics: Ecological and Food Security Implications*,
https://doi.org/10.1007/978-3-031-94480-2_8

activities, and microbead release from personal care products (Wang et al., 2019c). Secondary microplastics form when larger plastic debris, such as bottles and bags, break down due to environmental forces like wind, UV radiation, and mechanical abrasion (Thompson et al., 2004; Rillig, 2012). In marine environments, MPs enter oceans through rivers, ship waste disposal, and fishing industries, including discarded fishing nets and lines (Jambeck et al., 2015; Cole et al., 2011). Wastewater treatment plants also contribute significantly to microplastic pollution, with studies showing that around 25% of microplastics in the ocean come from treated water discharges (Wright et al., 2013).

The widespread presence of microplastics in aquatic ecosystems is concerning. They have been detected across various environments, including marine, freshwater, and estuarine ecosystems, as well as in deep-sea trenches (Van Cauwenberghe & Janssen, 2014; Nerland et al., 2014). In freshwater systems, concentrations are often higher than in marine environments, likely due to urban runoff and industrial discharges (Dris et al., 2015). MPs have even been found in remote areas like the Arctic and Antarctic regions, highlighting the global extent of this issue (Obbard et al., 2014). The contamination of aquatic ecosystems by MPs poses significant risks to organisms at multiple trophic levels, with MPs detected in plankton, benthic organisms, fish, and apex predators like seals and whales, either directly or through prey (Burkhardt-Holm & N'Guyen, 2019; Wright et al., 2013). As microplastic pollution grows, so does its potential to impact ecosystem health, leading to ecological and human health risks (Khan et al., 2022).

Microplastics are found in all aquatic environments, from surface waters to deep-sea sediments (Cózar et al., 2014). In marine ecosystems, they are primarily present as small fragments and microfibers, which are easily ingested by various organisms (Lusher et al., 2017). Coastal areas, especially those with high population densities and industrial activity, have the highest concentrations of MPs (Jambeck et al., 2015). MPs also accumulate in estuarine and freshwater sediments due to their higher density compared to water (Rochman et al., 2019).

The increasing prevalence of microplastics in aquatic ecosystems raises concerns about their risks to aquatic life. MPs not only pose physical harm but also chemical threats, as they can adsorb hazardous pollutants, including POPs, heavy metals, and pharmaceuticals (Koelmans et al., 2015; Prasad et al., 2023). This dual threat—acting both as pollutants and carriers of toxic chemicals-exacerbates the ecological consequences for organisms that ingest them (Rillig, 2012).

To analyze the research trends and assess the abundance of MPs in aquatic ecosystems, VOS viewer, a software tool for constructing and visualizing bibliometric networks, was utilized in this study (Fig. 1). The visualization highlights the key research areas on MPs and their interactions with aquatic environments. Three primary clusters emerged from the analysis: (i) the red cluster, associating MPs with toxicity and bioaccumulation in aquatic organisms; (ii) the blue cluster, linking MPs to environmental exposure and water pollution; and (iii) the green cluster, examining ecotoxicological effects, oxidative stress, and the role of particle size in MP toxicity. At the core of the visualization, the term "microplastics" is strongly connected to "aquatic environments" and "ecotoxicology," underscoring

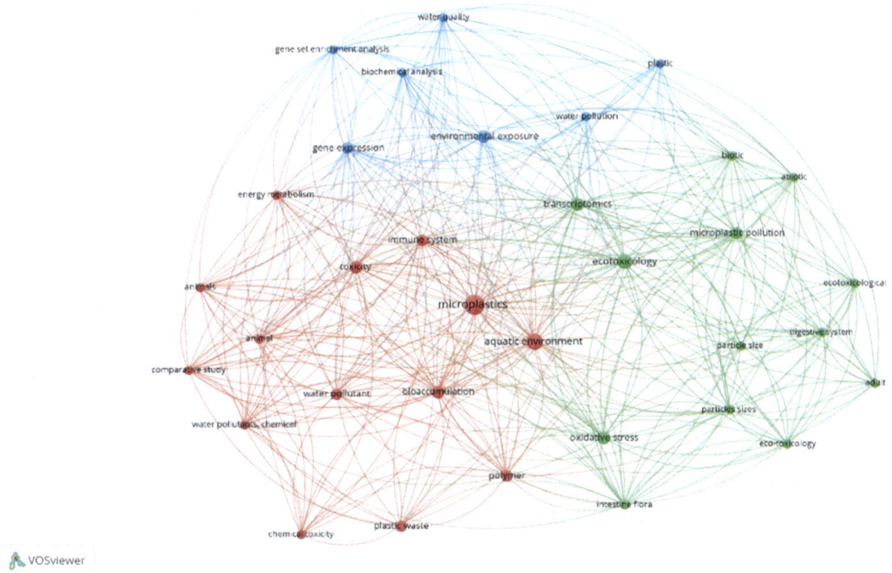

Fig. 1 Mapping the key research areas on microplastics and aquatic environments

the ecological risks and the increasing research focus on these pollutants. This bibliometric analysis emphasizes the necessity of continued research to fully understand the long-term ecological consequences of MPs in aquatic systems.

Research on the ecotoxicological effects of microplastics is vital for understanding their impact on biodiversity, ecosystem functioning, and human health (Ataguba et al., 2015). Ingested microplastics can cause mechanical damage to the digestive tract and respiratory systems, internal inflammation, and stress responses (Issac & Kandasubramanian, 2021; Wright et al., 2013). Moreover, chemical contaminants adsorbed onto MPs can bioaccumulate in organisms, leading to toxic effects that impair growth, reproduction, and immune function (Tamayo-Belda et al., 2023). Understanding these effects is crucial for assessing the ecological and human health risks posed by microplastics. Future research should focus on their long-term impacts, interactions with other pollutants, and their potential to disrupt food webs and ecosystem services. Given the complexity of the issue, an integrated approach to studying microplastic fate, transport, and toxicity is needed to inform regulatory policies and mitigation strategies (Naz et al., 2024). This chapter provides a detailed analysis of the ecotoxicological consequences of MPs on aquatic life, examining their sources, distribution, uptake mechanisms, and toxicological impacts across species. It also explores the long-term ecological risks, including food web disruption and potential human health implications, while proposing strategies for mitigating microplastic pollution in aquatic ecosystems.

2 Mechanisms of Microplastic Uptake and Toxicity in Aquatic Organisms

2.1 Uptake Pathways and Bioaccumulation of Microplastics in Aquatic Food Web

Microplastics primarily enter aquatic organisms through ingestion (Desforges et al., 2015). Many species, including zooplankton, mollusks, and fish, mistake MPs for food due to their size and resemblance to natural prey (Ziani et al., 2023). Filter-feeding organisms, such as bivalves and small invertebrates, are particularly vulnerable as they filter large volumes of water to extract food (Khan et al., 2022). This ingestion can cause physical harm, including blockages and internal injuries, and disrupt digestive processes (Zolotova et al., 2022). MPs can also be absorbed through gills and skin, especially in species constantly exposed to contaminated water (Cole et al., 2011). For fish, which are highly exposed to suspended MPs, this route is of particular concern (Pan et al., 2021). Additionally, MPs can adhere to external tissues, prolonging exposure and increasing the likelihood of ingestion through the gastrointestinal tract (Naz et al., 2024).

Once ingested or absorbed, MPs persist in the organism's body as they are not easily degraded (Borriello et al., 2023). Bioaccumulation occurs when MPs or the toxic substances they adsorb accumulate in tissues over time (Ziani et al., 2023). MPs have been found in organs like the liver and gills of fish, contributing to internal exposure and long-term health effects (Borriello et al., 2023). Due to their small size and large surface area, MPs can adsorb pollutants such as heavy metals, POPs, and chemicals like bisphenol A (BPA) and polychlorinated biphenyls (PCBs) (Ziani et al., 2023). Once ingested, these toxins can be released into the organism's system (Belmaker et al., 2024).

Biomagnification occurs as MPs and the toxic substances they carry are transferred up the food chain. As small organisms ingest MPs and are consumed by predators, such as fish and birds, the concentration of MPs in predator tissues increases (Anbumani & Kakkar, 2018). This process can result in higher concentrations of MPs and associated contaminants in top predators, potentially affecting human health through seafood consumption (Khan et al., 2022).

Trophic transfer is a key pathway for MPs and their contaminants in aquatic food webs (Walkinshaw et al., 2020). Studies show that MPs can pass from primary producers, like phytoplankton, to herbivores and then to carnivores. This transfer not only involves MPs themselves but also the pollutants they carry, amplifying the toxicological risks across the food chain (Koelmans et al., 2019). MPs ingestion and their effects have been documented in species ranging from plankton to large fish, underscoring the widespread nature of this contamination (Esposito et al., 2022).

2.2 *Toxicological Mechanisms of Microplastics in Aquatic Organisms*

2.2.1 Physical and Chemical Impacts of Microplastics on Aquatic Organisms

MPs poses significant risks to aquatic organisms due to their small size, widespread distribution, and chemical properties (Li et al., 2016). As non-biodegradable materials, MPs persist in ecosystems, accumulating in both freshwater and marine species (Akdogan & Guven, 2019). Ingested MPs can cause mechanical damage to internal organs, disrupt the digestive system, and reduce nutrient absorption (Kumar et al., 2020). Their small size allows them to be ingested by a wide range of organisms, from plankton to large marine mammals. Additionally, MPs' hydrophobic nature increases their surface area, enabling them to adsorb toxic chemicals from the environment, including POPs, heavy metals, and pharmaceuticals (Yu et al., 2019). These chemicals are transferred into organisms that consume MPs, causing toxicity that affects various physiological systems, such as the immune, reproductive, and endocrine systems (Huang et al., 2021; Yostawonkul et al., 2023, 2024).

2.2.2 Disruption of Digestive Systems, Clogging of Feeding Apparatus, and Internal Injuries

Ingestion of MPs, especially those smaller than 5 mm, can cause internal injuries, such as physical abrasion of the digestive tract (Ahrendt et al., 2020). For filter-feeding organisms like mussels and oysters, MPs can clog feeding apparatus, impair feeding efficiency, and cause starvation or stunted growth (Hirt & Body-Malapel, 2020). MPs can also obstruct gills and digestive systems, leading to chronic health issues and reduced feeding ability (Wang et al., 2025). Additionally, they may cause ulcers, lesions, and gastrointestinal blockages, resulting in significant mortality in extreme cases (Ahrendt et al., 2020; Hirt & Body-Malapel, 2020). When MPs accumulate in the gastrointestinal tract, they reduce available space for food digestion, affecting energy metabolism and leading to diminished growth and reproduction (Sharma et al., 2021). In some species, MPs can translocate to vital organs like the liver, causing systemic toxicity (Taylor et al., 2016).

2.2.3 Leaching of Toxic Chemicals from Plastics

MPs not only act as pollutants but also as carriers for toxic chemicals, including plastic additives and environmental contaminants (Gurjar et al., 2023; Rafa et al., 2024). Many plastics contain toxic additives such as plasticizers, flame retardants, and stabilizers, which leach out as the plastic degrades. These chemicals can disrupt endocrine and immune systems and cause developmental and reproductive impairments (Ullah et al., 2023). MPs also absorb hydrophobic pollutants like POPs and heavy metals from surrounding water (Rafa et al., 2024). These chemicals accumulate on MPs' surfaces and enter the food chain when ingested by organisms. Studies have shown that MPs can adsorb harmful chemicals like DDT and PCBs, which biomagnifies as they pass through the food web, leading to higher concentrations in organisms at higher trophic levels, including fish and marine mammals (Rochman et al., 2019; Prasad et al., 2024; Tabassum et al., 2025).

2.2.4 Interaction with Microorganisms and Potential for Microbial Biofilm Formation on MPs

MPs provide a surface for microbial colonization, forming biofilms that alter microbial community composition in aquatic ecosystems (Yang et al., 2020). These biofilms can enhance the persistence of harmful microorganisms on MPs, creating pathways for pathogens to enter organisms that ingest them (Zhong et al., 2023). Microbial communities can also modify MPs' surface properties, increasing their capacity to transport harmful chemicals into aquatic organisms (Bocci et al., 2024). MPs may serve as vectors for the spread of antibiotic-resistant bacteria and other harmful microorganisms (Ferheen et al., 2024). Biofilm interactions with aquatic species can cause infections, inflammation, and other health issues (Kamble et al., 2022), complicating risk assessments since the effects of these biofilms on aquatic organisms are not fully understood (Wu et al., 2024).

2.2.5 Inflammatory Responses and Oxidative Stress as Key Mechanisms

MPs exposure has been shown to induce inflammation and oxidative stress in aquatic organisms (Subaramaniyam et al., 2023). MPs can trigger the release of reactive oxygen species (ROS), leading to oxidative damage to cells, tissues, and organs. This stress impairs immune function, disrupts metabolic processes, and increases susceptibility to disease (Masud & Cable, 2023; Kamble et al., 2024e). In some species, MPs exposure alters immune responses, increasing inflammation and reducing pathogen resistance (Masud & Cable, 2023). Accumulated ROS can cause cellular damage and gene expression changes, affecting overall organism health (Lee et al., 2023b). Chronic exposure to MPs exacerbates these effects, potentially leading to long-term ecological consequences (Kadac-Czapska et al., 2024).

3 Ecotoxicological Effects on Aquatic Life

3.1 Microorganisms and Primary Producers

Planktonic organisms, including phytoplankton and zooplankton, are particularly vulnerable to microplastics due to their small size (Table 1), which makes them prone to mistaking MPs for food particles (Ma et al., 2019). Ingesting MPs can cause mechanical damage to digestive organs and block feeding structures, disrupting energy intake, and reducing growth and reproduction rates (Fig. 2). This results in diminished populations (Kvale et al., 2021). For primary producers like phytoplankton, MPs can impair photosynthesis and nutrient cycling, essential for aquatic ecosystem health (Montoya et al., 2024). Additionally, microplastic ingestion alters microbial communities in the water column, promoting the growth of bacteria that form biofilms on MPs (Naz et al., 2024). These shifts can disrupt nutrient cycling and food availability, ultimately compromising the efficiency of the food web and triggering long-term ecological changes (Wijesooriya et al., 2023; Chavan et al., 2018). Microplastics also play a role in nutrient cycling by acting as vectors for pollutants. Due to their hydrophobic surfaces, MPs adsorb harmful chemicals such as heavy metals, POPs, and pesticides (Seeley et al., 2020). When ingested, these toxins increase the toxicity of MPs and enhance the bioavailability of these contaminants to organisms higher in the food chain (Langenfeld et al., 2024). Moreover, MPs can disrupt microbial communities responsible for processing essential nutrients like nitrogen and phosphorus, altering their availability and affecting primary production and overall ecosystem productivity (Gao et al., 2021).

3.2 Invertebrates

The presence of MPs in the environment poses significant risks to aquatic organisms, particularly invertebrates (Table 1). Filter feeders like mussels and oysters, and benthic organisms such as polychaetes, are especially vulnerable due to their feeding behavior and ecological roles.

Filter feeders ingest MPs while filtering water for food, as these particles resemble planktonic organisms or organic material (Bergmann et al., 2015). Mussels, for instance, efficiently filter large quantities of water, which leads to the accidental consumption of MPs (Xu et al., 2024). Studies have shown that MPs accumulate in the digestive systems of these organisms, impairing digestion and potentially causing bioaccumulation (Marcharla et al., 2024). While some MPs are excreted through feces, smaller particles can translocate into the gills or other organs, resulting in internal damage (Liu et al., 2024). Similarly, benthic organisms like polychaetes ingest MPs from sediment or organic material on the sea floor, facing risks of ingestion, bioaccumulation, and harm to physiological functions (Sfriso et al.,

Table 1 Toxicological effects of different microplastics on aquatic life

Test organism	Microplastics			Effects	References
	Type	Size	Exposure duration		
Primary producer					
Tetraselmis suecica	Polystyrene	0.1 μm	15 d	Increased mortality	Raju et al. (2022)
Chaetoceros neogracile	Polystyrene	0.1 μm	96 h	Decrease in cell density, chlorophyll a (Chl-a) content, and protein content	Mojiri et al. (2024)
Chlorella pyrenoidosa	Polystyrene	1 μm	30 d	Inhibit growth Pyrenoid damage Decreased photosynthetic efficiency	Mao et al. (2018)
Chaetoceros neogracile	Polystyrene	2 μm	72 h	Decreased growth	Seoane et al. (2019)
Skeletonema costatum	Polystyrene	74 μm	96 h	Inhibit growth	Zhu et al. (2019b)
Chlamydomonas reinhardtii	Polystyrene	5 μm	7 d	Growth inhibition Reduced Chl-a content	Li et al. (2022)
Phaeodactylum tricornutum	Unplasticized polyvinyl chloride	25 μm	4 d	Growth rate decreased	Guo et al. (2020)
Spirulina sp.	Polyethylene and polypropylene	0.5–1 mm^2	30 d	Growth rate reduced	Hadiyanto et al. (2021)
Chaetoceros meülleri	Polystyrene	1 μm	96 h	Growth inhibition	Wang et al. (2021b)
Invertebrates					
Mytilus spp.	Polystyrene	20 μm	7 d	Induce immune response Increased oxidative stress	Cole et al. (2020)
Mytilus galloprovincialis	Polystyrene	50 nm	48 h	Delayed development Disruption in shell formation process	Balbi et al. (2017)
Emerita analoga	Polypropylene	1 mm	71 d		

				Increased mortality Reduced embryonic development Decreased retention of egg clutches	Horn et al. (2020)
Artemia salina	Polystyrene	1–10 μm	30 d	Mortality increased Induce oxidative stress and acute cell stress	Eom et al. (2020)
Mytilus galloprovincialis	PVC	<1 μm	58 d	Reduced respiration rate Low byssus production	Germishuizen (2021)
Hediste diversicolor	Polyethylene	300–355 μm	28 d	Generate neurotoxicity Decline in enzymatic activities Suppression of defense mechanism	Urban-Malinga et al. (2022)
Pinctada margaritifera	Polystyrene	10 μm	60 d	Decreased growth Digestion decreased	Gardon et al. (2018)
Perna viridis	PVC	50 μm	91 d	Decreased filtration ability Impaired respiration rates and byssus production Increased mortality	Rist et al. (2016)
Crassostrea virginica	Polystyrene	3 μm	48 h	Bioreactivity and sublethal impacts	Gaspar et al. (2018)
Mytilus edulis	HDPE	102.6 μm	52 d	Decreased filtration ability Impaired byssal threads and attachment strength	Green et al. (2019)
Fish					
Ctenopharyngdon idella	Polystyrene	470 nm	20 d	DNA damage, erythrocytes mutagenic and cytotoxic effect	Guimarães et al. (2021)
Danio rerio	Polystyrene NPs	1000 nm	12 h	Nanoplastics induce immune response in test organism	Brandts et al. (2020)
Oreochromis mossambicus	Polypropylene	–	14 d	Increased ROS levels Fluctuations in homeostasis Higher apoptosis DNA damage	Jeyavani et al. (2023)

(continued)

Table 1 (continued)

Test organism	Microplastics Type	Size	Exposure duration	Effects	References
Danio rerio	Propylene copolymer	<200 µm	21 d	Anxiety ROS generation Mitochondrial dysfunction	Félix et al. (2023)
Ctenopharyngodon idella	Polystyrene	15 µm	14 d	Induced oxidative stress Liver congestion Severe intestinal damage	Hao et al. (2023)
Pseudobagrus fulvidraco	Polyethylene	32–50 µm	96 h	Decreased RBC, hemoglobin, hematocrit, calcium, total protein and magnesium Induce antioxidant response	Lee et al. (2023a)
Oreochromis niloticus	Polyacrylamide	0.1–0.4 mm	28 d	Shorten antioxidant enzyme activity Induce oxidative stress and lipid peroxidation	Raza et al. (2023)
Oreochromis niloticus	Microplastics	100 nm	15 d	Decreased antioxidant capacity Increased oxidative stress and DNA damage.	Hamed et al. (2020)
Oryzias melastigma	Polystyrene	10 µm	60 d	Decreased antioxidant capacity Caused histological changes Increased oxidative stress	Wang et al. (2019b)
Cyprinus carpio	PVC	100–200 µm	60 d	Induced oxidative stress Decreased antioxidant activity	Xia et al. (2020)
Carassius auratus	Polyvinyl chloride	0.1–1000 µm	96 h	Increased oxidative stress Induced inflammation, hemorrhaging and necrosis	Romano et al. (2020)
Oryzias javanicus	Polystyrene	5 µm	21 d	Induced oxidative stress lipid peroxidation neurotoxicity	Usman et al. (2021)
Aquatic birds and mammals					
Aquatic birds	Polyethylene terephthalate	<0.7 µm	–	Decreased feeding capacity Lowered reproductive capacity	Carlin et al. (2020)

White-beaked dolphin	Microplastic	<5 mm	–	Increased mortality	Nelms et al. (2019)
Indo-Pacific humpback dolphins	Microplastic	<5 mm		Mortality occurred (microplastic found in intestine)	Zhu et al. (2019a)
Caretta caretta, *Lepidochelys kempi* and *Dermochelys coriacea*	Microplastic	<5 mm		Mortality occurred	Duncan et al. (2019)
Delphinapterus leucas	Microplastic	<5 mm		Mortality occurred	Moore et al. (2020)

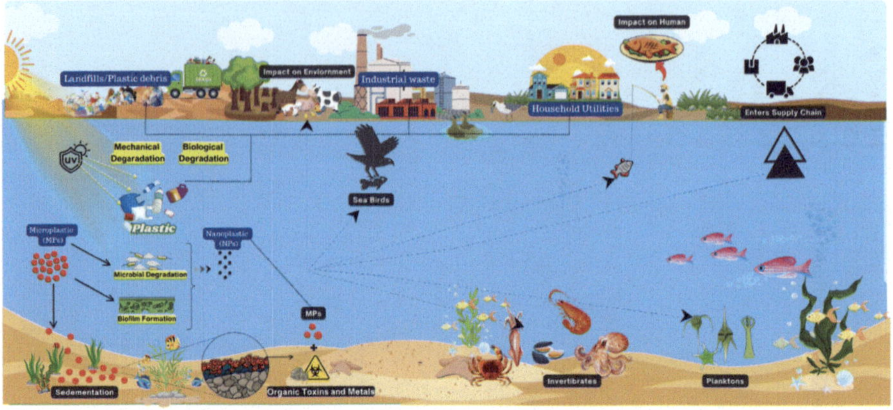

Fig. 2 Factors affecting microplastic distribution in aquatic systems

2024). Exposure to MPs can also disrupt feeding behavior, leading to reduced energy intake and impaired food consumption (Zolotova et al., 2022).

Ingestion of MPs can cause several physical effects, such as gut blockage, which impairs digestion and leads to nutritional deficiencies. The physical presence of MPs in the digestive tract can prevent nutrient absorption, reducing feeding efficiency (Borriello et al., 2023). Large amounts of MPs may cause internal injuries, inflammation, and stunted growth (Hirt & Body-Malapel, 2020). In filter-feeding species like mussels, MPs can reduce filtration capacity, further hindering feeding effectiveness (Betsill et al., 2019).

Additionally, MPs negatively impact the reproductive and growth rates of invertebrates (Sussarellu et al., 2016). In benthic organisms, exposure to MPs can disrupt reproductive processes, decreasing egg viability, hatching success, and larval development (Hodkovicova et al., 2022). Studies on polychaetes have shown reduced reproductive success and altered larval development, affecting population dynamics (Hodgson, 2018). In filter-feeding species like oysters and mussels, MPs have been linked to reduced growth rates, with MPs acting as stressors that impair physiological health, including immune function and shell formation (Sendra et al., 2021; Yuan et al., 2024; Thomas et al., 2020).

3.3 Fish

Fish play a crucial role in the aquatic food web, and MPs significantly affect their health and behavior, impacting both juveniles and adults (Alberghini et al., 2022). Exposure to MPs leads to alterations in growth, behavior, and immune responses in both marine and freshwater environments (Oza et al., 2024).

MPs affect fish in several ways, including reduced growth rates, immune suppression, and physical damage (Zarantoniello et al., 2024). Ingesting MPs can cause internal injuries in the digestive tract, disrupting nutrient absorption and stunting growth (Anbumani & Kakkar, 2018). Additionally, exposure to MPs has been linked to altered immune function, making fish more vulnerable to infections and less able to combat pathogens (Alberghini et al., 2022). MPs may induce oxidative stress, damaging cellular structures and impairing immune responses (Li et al., 2023).

The bioaccumulation of MPs in fish tissues is a significant concern, as it can result in long-term health consequences (Fig. 2). Studies show that MPs, particularly those smaller than 5 μm, are ingested by fish and accumulate in organs such as the gut, liver, and gills (Li et al., 2018). This accumulation can lead to toxicity, as MPs may carry harmful chemicals, including POPs and heavy metals, which leach into the fish's system (Barboza et al., 2018), disrupting endocrine function, metabolism, and reproductive health (Wang et al., 2019b; Lusher et al., 2017; Jabeen et al., 2017).

Behavioral changes in fish exposed to MPs are another major concern. Studies indicate that MPs reduce predatory success, alter migratory patterns, and disrupt schooling behaviors (Bhuyan, 2022). MPs interfere with fish sensory systems, such as vision and olfaction, which are essential for foraging and predator avoidance (Subaramaniyam et al., 2023). This disruption can impair a fish's ability to find food or avoid predators, decreasing survival rates (Zhang et al., 2023). Additionally, MPs can alter feeding rates and activity levels, leading to changes in natural behavior and increased energy expenditure (Chen et al., 2022).

The long-term impacts of MPs on fish health remain incompletely understood, but evidence suggests that chronic exposure may have severe consequences (Alberghini et al., 2022). In addition to the immediate effects of physical and chemical toxicity, MPs may interfere with reproduction and development, especially in juveniles (Bhuyan, 2022). For example, exposure to MPs has been shown to impair gonadal development and reduce fertilization success in species like tilapia (Ismail et al., 2021). Furthermore, the bioaccumulation of MPs through the food web may affect higher trophic levels, including human consumers, due to trophic transfer (McHale & Sheehan, 2024).

3.4 Aquatic Birds and Mammals

MPs enter the food chain primarily through ingestion by lower trophic organisms such as plankton and small fish (Bilal et al., 2023). Research has demonstrated that MPs accumulate across various species, from primary producers to apex predators, including fish, seabirds, and marine mammals (Lusher et al., 2022). As MPs are consumed, they can cause mechanical damage, such as gut blockages, and chemical toxicity due to the adsorption of pollutants like pesticides, heavy metals, and POPs on their surfaces (Navarro et al., 2023). Aquatic birds and mammals are particularly vulnerable, often feeding on contaminated prey that has ingested MPs (Wang et al., 2021a). For example, seabirds consuming fish, squid, or zooplankton containing MPs may experience bioaccumulation of both the particles and associated toxins (Lusher et al., 2022). This secondary exposure can lead to tissue damage, inflammation, and stress responses (Jeong et al., 2024). Additionally, MPs can cause the bioaccumulation of chemical additives like phthalates and BPA, known to disrupt endocrine systems (Wang et al., 2021a).

MP exposure can also impact reproductive health and development in aquatic birds and mammals. In birds, MPs have been linked to altered reproductive success, including reduced egg production and hatching success (Bilal et al., 2023). For marine mammals, MPs may disrupt hormonal regulation and immune functions, affecting reproductive health (Meaza et al., 2021). The ingestion of MPs by mothers can transfer contaminants to offspring, potentially impacting their development and survival (Rapini et al., 2024). Moreover, physical ingestion of MPs by mammals, such as seals and whales, can lead to chronic issues like gut obstructions and impaired digestion (Meaza et al., 2021). These effects are especially harmful to species reliant on fat reserves and efficient digestion for energy, leading to malnutrition and weakened immune responses. Studies also indicate that the combined effects of MPs and chemical pollutants can result in synergistic toxicity, intensifying oxidative stress, causing cellular damage, and further compromising organism health (Qaiser et al., 2023; Tanaka et al., 2013; Franzellitti et al., 2019).

4 Impact on Aquatic Ecosystem Functioning

4.1 Food Web Disruption

MPs enter the aquatic food web primarily through ingestion. A variety of aquatic organisms, from zooplankton to fish and marine mammals, mistake microplastics for food (Gallo et al., 2018). Filter-feeding organisms like mussels and oysters frequently ingest microplastics during their feeding process, leading to their accumulation in digestive systems (Barboza & Gimenez, 2015). Due to their small size and large surface area, these particles can adsorb harmful chemical contaminants, amplifying their toxic effects (Browne et al., 2013; Khan et al., 2022). Once ingested, microplastics can move up the food web, transferring from primary producers like phytoplankton to higher-level predators (Chae & An, 2017; Li et al., 2023). This movement can cause negative effects, such as gastrointestinal blockages, immune suppression, and altered behavior (Hu & Palić, 2020). The chemicals associated with microplastics, such as plasticizers and heavy metals, can exacerbate these impacts, leading to cumulative toxicity in organisms higher in the food chain (Rochman et al., 2019).

The presence of MPs in aquatic food webs significantly disrupts ecosystem functioning, particularly in marine environments (Anthony et al., 2024). As MPs accumulate across trophic levels, they impair feeding behavior, reproduction, and overall health, affecting population dynamics (Ma et al., 2020). For example, zooplankton that ingest MPs experience reduced growth and reproduction, impacting species like small fish that depend on them for food (Botterell et al., 2019). Filter feeders that consume MPs may suffer from decreased feeding efficiency and energy depletion, leading to poorer survival and growth (Gardon et al., 2018). At higher trophic levels, predators consuming small fish and invertebrates accumulate MPs and toxic contaminants, contributing to biomagnification and bioaccumulation (Paul et al., 2024). This can cause significant health problems and impact the reproductive systems of top predators, including commercially important fish species (Hasan et al., 2024; Naz et al., 2024). Over time, these disruptions can alter biodiversity, weakening ecosystem stability and resilience (de Souza Machado et al., 2018).

The effects of MPs pollution on food webs have crucial implications for fisheries and commercial aquaculture. MPs ingestion by fish and other aquatic organisms can affect growth, health, and marketability (Foley et al., 2018). For instance, exposure to MPs can induce physiological stress and immune suppression, increasing farmed fish vulnerability to disease (Jovanović, 2017; Montero et al., 2022; Kamble et al., 2024a, d). The biomagnification of toxic substances associated with microplastics also poses risks to human health through contaminated seafood (Smith et al., 2018; Barboza et al., 2018). Additionally, reductions in fish populations due to MPs ingestion can lead to lower yields for fisheries and increased costs for aquaculture operations due to disease management (Ataguba et al., 2017; Kamble et al., 2024b, c) and reduced productivity (Lusher et al., 2017; De Sá et al., 2018).

4.2 Ecosystem Services

Aquatic organisms, particularly filter feeders like bivalves, mollusks, and zooplankton, play a vital role in maintaining water quality through filtration (Gonçalves et al., 2019). When these organisms ingest microplastics, it disrupts their filtration capacity (Fraissinet et al., 2024). The accumulation of MPs in their digestive systems causes physiological stress, reducing their ability to filter water effectively (Issac & Kandasubramanian, 2021). This negatively impacts the entire aquatic system, as the removal of particulate matter is essential for maintaining water clarity and quality (Avio et al., 2017; Cole et al., 2011).

MPs indirectly affect nutrient cycling in aquatic ecosystems (Besseling et al., 2014). Organisms that ingest MPs experience disruptions in feeding and energy metabolism, impairing their growth and reproduction (Wang et al., 2019a). This alters nutrient flow, especially at lower trophic levels (Galloway et al., 2017). For example, MPs exposure stresses primary producers like phytoplankton, reducing their contribution to primary production (Sjollema et al., 2016; Yin et al., 2023). Additionally, MPs can cause pollutants to bioaccumulate, further disrupting nutrient cycling (Rochman et al., 2013; Wright et al., 2013).

MPs can alter aquatic habitats by accumulating on the seafloor or in sediments, affecting their physical properties (Harris, 2020). Organisms that depend on these habitats for shelter or breeding are vulnerable to changes caused by microplastics (Botterell et al., 2019). Benthic organisms, in particular, are at risk, as MPs can alter the sediment structure, disrupting the spaces these organisms need for survival (Haegerbaeumer et al., 2019).

MPs directly harm aquatic organisms, leading to shifts in biodiversity (Marcharla et al., 2024). Smaller organisms like plankton suffer physical damage and altered physiological functions from ingesting MPs, which cascades up the food chain, affecting larger organisms that depend on them for food (Malinowski et al., 2023). Studies show that exposure to MPs increases mortality, decreases reproductive success, and slows growth rates, all contributing to biodiversity loss (Wright et al., 2013).

Ecosystem resilience refers to the ability of an ecosystem to recover from disturbances (Keith et al., 2023). MPs weaken this resilience by interfering with natural processes like predator-prey interactions and reproduction (Ali et al., 2024). As MPs accumulate, ecosystems struggle to recover from other environmental stresses, such as climate change or overfishing (Sunil et al., 2024). Furthermore, when combined with other pollutants, MPs exacerbate toxic effects, leading to further declines in ecosystem stability (Koelmans et al., 2016).

5 Long-Term Implications and Ecotoxicological Risks

5.1 Chronic Exposure and Accumulation

Chronic exposure to MPs in aquatic environments poses long-term ecotoxicological risks to organisms, ecosystems, and potentially to human health. These persistent pollutants, characterized by their small size (less than 5 mm), have become ubiquitous in aquatic ecosystems, from surface waters to benthic sediments. As MPs degrade very slowly due to their resistant nature, their accumulation in aquatic environments continues to rise, with serious implications for the organisms that inhabit these ecosystems.

5.1.1 Long-Term Exposure Consequences for Aquatic Organisms

Chronic exposure to MPs in aquatic environments can lead to a variety of harmful effects on organisms at different trophic levels (De Sá et al., 2018). These particles are ingested by a wide range of organisms, from small plankton to larger fish and marine mammals (Lusher et al., 2017). Once ingested, MPs can cause mechanical damage, disrupt digestive processes, and lead to inflammation (Wright et al., 2013). Furthermore, MPs can interfere with energy metabolism by inducing stress responses and impairing immune function (Barboza et al., 2018). For example, Daphnia, a commonly studied freshwater organism, has shown increased oxidative stress and immune responses upon chronic exposure to MPs (Imhof et al., 2017). The persistent nature of MPs in the environment means that organisms may accumulate these pollutants over time, with potential impacts on growth, reproduction, and survival (Galloway et al., 2017). The continuous exposure to MPs can lead to bioaccumulation in tissues, especially in filter feeders like mussels and oysters, which ingest large quantities of water for feeding (Van Cauwenberghe & Janssen, 2014). Although some organisms can expel MPs through their excretory systems, others may retain them in their digestive organs, leading to long-term health consequences (Paul-Pont et al., 2016).

5.1.2 Potential for MPs to Persist and Accumulate Over Time

One of the key concerns regarding MPs is their ability to persist in the environment for extended periods. MPs are highly resistant to biodegradation and can remain in aquatic ecosystems for decades, continuously accumulating in water bodies and sediments (Andrady, 2017). This long-term persistence allows MPs to act as vectors for other environmental contaminants. Due to their large surface area and hydrophobic nature, MPs can adsorb and concentrate harmful chemicals such as pesticides, heavy metals, and POPs (Koelmans et al., 2015). These adsorbed chemicals can be transferred to aquatic organisms when they ingest MPs, exacerbating the toxicity of the MPs themselves (Conesa, 2022).

As MPs are ingested by organisms, they not only cause physical harm but also enhance the bioavailability of these toxic chemicals. The chemical additives in plastics, such as plasticizers and stabilizers, may also leach into the environment, contributing to further toxic effects (Wang et al., 2016). Studies have shown that MPs, in combination with pollutants, can lead to synergistic toxic effects, making the cumulative impact of MPs on aquatic life more complex and difficult to assess (Xayachak et al., 2024).

Moreover, the trophic transfer of MPs through food webs could lead to higher concentrations in apex predators, including fish and marine mammals, which could further increase the risk of human exposure through seafood consumption. Thus, the long-term persistence and accumulation of MPs, coupled with their ability to adsorb toxic chemicals, pose significant risks not only to aquatic life but also to the broader ecosystem and human health (Barboza et al., 2018).

5.2 Ecosystem Resilience and Recovery in the Context of MPs

5.2.1 Challenges for Ecosystem Recovery from Prolonged MPs Contamination

The widespread contamination of aquatic ecosystems by MPs presents a significant challenge to ecosystem resilience and recovery. MPs, due to their small size and persistence, have infiltrated various environmental compartments, from surface waters to deep sediments, affecting both marine and freshwater ecosystems (Li et al., 2018). The ubiquity of MPs in aquatic environments has raised concerns about their long-term ecological impacts, which are primarily linked to their accumulation and the subsequent toxicity they pose to aquatic organisms at all trophic levels (Wang et al., 2016).

One of the primary challenges in ecosystem recovery from prolonged MP contamination is the slow biodegradation of these particles. MPs are not easily broken down by environmental processes, leading to their accumulation over time. While some microorganisms can degrade plastics, the degradation process is often too slow to mitigate the environmental burden, especially in high-concentration areas (Gewert et al., 2015). Additionally, MPs can act as carriers for toxic contaminants, such as POPs, heavy metals, and other harmful chemicals, which can further exacerbate the ecological damage (Rochman et al., 2013). These contaminants, once absorbed by MPs, may increase their toxicity, making it even more difficult for ecosystems to recover (Koelmans et al., 2016).

Moreover, the physical presence of MPs can cause direct harm to aquatic organisms. Ingestion of MPs by various species, including plankton, fish, and benthic invertebrates, can lead to physical injuries, digestive blockages, and other physiological disruptions (Foley et al., 2018). The disruption of key ecological processes, such as feeding, reproduction, and nutrient cycling, can slow down the recovery of ecosystems and impede the resilience of populations affected by MP contamination (Galloway et al., 2017).

5.2.2 Role of Microbial Communities in the Degradation of MPs and Ecosystem Self-Restoration

Despite these challenges, microbial communities play a crucial role in the potential recovery of ecosystems impacted by microplastics. Recent studies have highlighted the capacity of certain bacteria and fungi to break down plastics, though the process is slow and requires favorable environmental conditions (Urbanek et al., 2018). Microbial biofilms that develop on MPs can facilitate the biodegradation of the plastic material, breaking it down into smaller, less toxic components (Amaral-Zettler et al., 2020). This process, however, is not sufficient to fully restore ecosystems in the short term, as the rate of degradation is much slower than the rate at which MPs accumulate.

Interestingly, the microbial degradation of MPs can also influence the toxicity of associated contaminants. Some studies have shown that microbes can not only break down the plastic itself but also reduce the bioavailability of toxic chemicals that have adsorbed onto the plastic surfaces (Zhang et al., 2018). This microbial action may contribute to reducing the harmful effects of MPs on aquatic organisms. However, the ability of microbial communities to effectively manage MP pollution in large, polluted environments remains uncertain and is likely limited by the concentration and diversity of plastic pollution in these ecosystems (Jacquin et al., 2019).

Furthermore, the recovery of ecosystems will depend on the balance between continued plastic input and the natural resilience of the environment. For example, the natural processes of sedimentation, water circulation, and natural degradation mechanisms (e.g., microbial degradation) can help reduce the concentration of MPs over time (Hale et al., 2020). Still, without significant intervention, such as improved waste management and plastic reduction, full recovery remains unlikely (Rochman et al., 2019).

5.3 Human Health Implications

The human health implications of MPs through the aquatic food chain are becoming a growing concern as the presence of these particles in the environment increases. Due to their ubiquity and persistence, MPs can enter the human food chain primarily through seafood consumption, posing potential risks to human health.

5.3.1 The Potential for MPs to Enter the Human Food Chain

MPs have been detected in a wide range of aquatic organisms, from plankton to large marine mammals, with fish and shellfish often serving as intermediaries between the aquatic environment and human consumers (Li et al., 2018; Barboza et al., 2018). MPs can enter the human food chain when seafood, such as fish, shellfish, and crustaceans, which often ingest these particles, are consumed by humans (Rochman et al., 2015; Santillo et al., 2017). A significant body of research has shown that MPs are ingested by aquatic organisms, including commercially important species, thereby incorporating these pollutants into the food web (Güven et al., 2017; Kühn & Van Franeker, 2020). This ingestion occurs because MPs are small and resemble natural food sources like plankton, making them easy targets for filter-feeding organisms (Cole, 2014; Setälä et al., 2014).

Once in the organisms, MPs can accumulate in tissues, including the gastrointestinal tract and liver, and even be transferred across trophic levels (Carbery et al., 2018; Provencher et al., 2019). Studies indicate that MPs are not only ingested by marine organisms but can also be accumulated in edible portions of these animals, making the potential for human exposure significant (Smith et al., 2018; EFSA Panel on Contaminants in the Food Chain (CONTAM), 2016). As humans continue to consume seafood, they may inadvertently ingest these particles, which leads to a growing concern over their bioaccumulation and the potential health risks involved (Wright & Kelly, 2017; Prata et al., 2020).

5.3.2 Possible Risks Associated with the Transfer of Toxic Chemicals from MPs to Humans

In addition to their physical presence, MPs can act as carriers of various toxic chemicals, such as heavy metals, POPs, and other industrial chemicals (Rochman et al., 2019). These substances, often added to plastics during manufacturing (e.g., plasticizers, flame retardants, stabilizers), can leach out into the environment over time, particularly under the influence of environmental stressors like UV light or heat (Zimmermann et al., 2020). When MPs are ingested by marine organisms, these chemicals can be transferred alongside the plastic particles themselves, potentially reaching humans through seafood consumption (Barboza et al., 2018).

For instance, microplastics have been shown to adsorb chemicals such as BPA, PCBs, and pesticides like DDT, all of which have been linked to endocrine disruption, carcinogenic effects, and other health issues (Xing et al., 2022). As MPs pass through the food chain, the concentration of these chemicals can increase, a process known as biomagnification (Chae & An, 2017). The cumulative exposure to these hazardous chemicals could result in a variety of health problems for humans, ranging from reproductive and developmental issues to immune system disruptions (Wright & Kelly, 2017).

Furthermore, MPs themselves may also cause direct toxicological effects. Research has shown that microplastics can cause mechanical damage to tissues, inflammation, and oxidative stress, which could contribute to more long-term health effects in humans, especially when consumed regularly in seafood (Jiang et al., 2020). These direct effects, coupled with the chemical risks from adsorbed pollutants, present a dual threat to human health through the consumption of contaminated seafood (Prata et al., 2020).

6 Strategies for Mitigating the Ecotoxicological Consequences of MPs

MPs, due to their widespread presence in aquatic ecosystems, pose significant ecological and human health risks. These pollutants, which result from both primary and secondary sources, can degrade environmental quality and disrupt the health of aquatic organisms. Addressing the ecotoxicological consequences of MPs requires an integrated approach that involves pollution prevention, technological solutions, and public awareness.

6.1 Pollution Prevention and Reduction

One of the primary strategies for mitigating the impact of MPs is reducing their production and release into aquatic environments. Prevention starts at the source, which includes reducing plastic usage, especially single-use plastics, which are a major contributor to MPs contamination in marine and freshwater systems (Borrelle et al., 2020). Encouraging the use of biodegradable plastics and designing products with longer lifespans can significantly limit the introduction of MPs into the environment (Karan et al., 2019). Additionally, promoting recycling is essential to curbing the flow of microplastics into aquatic environments. Recycling plastic waste helps reduce the need for virgin plastic production, thus limiting the generation of microplastics (Geyer et al., 2017). Proper waste management, including the effective disposal and treatment of plastics through advanced technologies, also reduces the chances of plastic fragments reaching waterways (Mitrano & Wohlleben, 2020).

6.2 *Technological and Policy Solutions*

Technological innovations are also crucial for mitigating the effects of microplastics in aquatic systems. One approach is the development of MPs removal technologies, such as filtration and centrifugation methods, that can extract MPs from wastewater before they enter aquatic ecosystems (Sun et al., 2019). For instance, advanced filtration systems in wastewater treatment plants have been shown to effectively capture MPs from effluent, significantly reducing their release into the environment (Carr et al., 2016).

The adoption of policy solutions is equally important in addressing MPs pollution. International regulatory frameworks that target plastic production and waste management, such as the European Union's plastics strategy, provide a foundation for reducing MPs pollution (Matthews et al., 2021). Policies that limit the use of microbeads in personal care products, promote the use of biodegradable materials, and encourage the use of alternative materials in packaging can significantly reduce MPs levels in aquatic environments (Lamichhane et al., 2023). Additionally, international collaboration, as seen in programs like the EPHEMARE project, helps raise awareness and foster shared technological advancements across borders (Gago et al., 2018).

6.3 *Public Awareness and Education*

Public awareness campaigns play a pivotal role in mitigating MPs pollution. Raising awareness about the sources and risks of MPs can drive consumer behavior changes, such as reducing plastic use and supporting companies that adopt sustainable practices (Hartley et al., 2018). Educating the public about proper waste disposal and recycling, as well as the impact of MPs on ecosystems, can also lead to a reduction in plastic waste generation (Anderson et al., 2016). Engaging communities through educational programs, outreach events, and social media campaigns can further amplify the importance of reducing plastic pollution and minimizing the environmental and health risks associated with MPs (Bergmann et al., 2015; Grünzner & Pahl, 2023).

7 Conclusion

MPs contamination presents severe ecotoxicological risks to aquatic organisms and ecosystems. MPs cause physical harm, act as vectors for toxic chemicals, and disrupt key biological functions in marine and freshwater species. Their persistence in the environment and ability to bioaccumulate highlight the urgent need for effective mitigation strategies, including improved waste management, policy interventions, and public awareness initiatives. Future research should focus on long-term ecological impacts, synergistic toxicity with other pollutants, and innovative solutions to reduce MP pollution in aquatic environments. Addressing this issue requires interdisciplinary collaboration and global policy efforts to safeguard ecosystem health and human well-being.

References

Ahrendt, C., Perez-Venegas, D. J., Urbina, M., Gonzalez, C., Echeveste, P., Aldana, M., Pulgar, J., & Galbán-Malagón, C. (2020). Microplastic ingestion cause intestinal lesions in the intertidal fish *Girella laevifrons*. *Marine Pollution Bulletin, 151*, 110795.

Akdogan, Z., & Guven, B. (2019). Microplastics in the environment: A critical review of current understanding and identification of future research needs. *Environmental Pollution, 254*, 113011.

Alberghini, L., Truant, A., Santonicola, S., Colavita, G., & Giaccone, V. (2022). Microplastics in fish and fishery products and risks for human health: A review. *International Journal of Environmental Research and Public Health, 20*(1), 789.

Ali, N., Khan, M. H., Ali, M., Ahmad, S., Khan, A., Nabi, G., Ali, F., Bououdina, M., & Kyzas, G. Z. (2024). Insight into microplastics in the aquatic ecosystem: Properties, sources, threats and mitigation strategies. *Science of the Total Environment, 913*, 169489.

Amaral-Zettler, L. A., Zettler, E. R., & Mincer, T. J. (2020). Ecology of the plastisphere. *Nature Reviews Microbiology, 18*(3), 139–151.

Anbumani, S., & Kakkar, P. (2018). Ecotoxicological effects of microplastics on biota: A review. *Environmental Science and Pollution Research, 25*, 14373–14396.

Anderson, A., Grose, J., Pahl, S., Thompson, R. C., & Wyles, K. J. (2016). Microplastics in personal care products: Exploring perceptions of environmentalists, beauticians and students. *Marine Pollution Bulletin, 113*(1–2), 454–460.

Andrady, A. L. (2017). The plastic in microplastics: A review. *Marine Pollution Bulletin, 119*(1), 12–22.

Anthony, J., Varalakshmi, S., Kumar Sekar, A., Thalavai Sivasankarasubbiah, K., Harikrishnan, T., Rangamaran, V. R., Gopal, D., & Ramalingam, K. (2024). Microplastics pollution in Indian marine environment: Sources, effects and solutions. *Frontiers in Marine Science, 11*, 1512802.

Ataguba, G. A., Kamble, M. T., Okayi, G. R., & Chavan, B. R. (2015). An overview and assessment of two wetlands in Eastern Thailand: Kung krabaen bay and welu wetlands in Chanthaburi province. *International Journal of Agriculture Environment and Biotechnology, 8*(1), 205–213.

Ataguba, G. A., Kamble, M. T., & Salin, K. R. (2017). Food industry by-products as protein replacement in aquaculture diets of tilapia and catfishs. In A. Anal (Ed.), *Food processing by-products and their utilization* (pp. 471–507). Wiley.

Avio, C. G., Gorbi, S., & Regoli, F. (2017). Plastics and microplastics in the oceans: From emerging pollutants to emerged threat. *Marine Environmental Research, 128*, 2–11.

Balbi, T., Camisassi, G., Montagna, M., Fabbri, R., Franzellitti, S., Carbone, C., Dawson, K., & Canesi, L. (2017). Impact of cationic polystyrene nanoparticles (PS-NH2) on early embryo development of *Mytilus galloprovincialis*: Effects on shell formation. *Chemosphere, 186*, 1–9.
Barboza, L. G. A., & Gimenez, B. C. G. (2015). Microplastics in the marine environment: Current trends and future perspectives. *Marine Pollution Bulletin, 97*(1–2), 5–12.
Barboza, L. G. A., Vethaak, A. D., Lavorante, B. R., Lundebye, A.-K., & Guilhermino, L. (2018). Marine microplastic debris: An emerging issue for food security, food safety and human health. *Marine Pollution Bulletin, 133*, 336–348.
Barnes, D. K., Galgani, F., Thompson, R. C., & Barlaz, M. (2009). Accumulation and fragmentation of plastic debris in global environments. *Philosophical Transactions of the Royal Society B: Biological Sciences, 364*(1526), 1985–1998.
Belmaker, I., Anca, E. D., Rubin, L. P., Magen-Molho, H., Miodovnik, A., & van der Hal, N. (2024). Adverse health effects of exposure to plastic, microplastics and their additives: Environmental, legal and policy implications for Israel. *Israel Journal of Health Policy Research, 13*(1), 44.
Bergmann, M., Gutow, L., & Klages, M. (2015). *Marine anthropogenic litter*. Springer.
Besseling, E., Wang, B., Lurling, M., & Koelmans, A. A. (2014). Nanoplastic affects growth of *S. obliquus* and reproduction of *D. magna*. *Environmental Science & Technology, 48*(20), 12336–12343.
Betsill, M., Gonzalez, J., & Woods, A. (2019). *The impact of microplastic ingestion on the bivalve filtration efficiency of the hooked mussel (Ischadium recurvum) from the Chesapeake Bay*. Research proposal for UH 1604. Honors Research Practices at Virginia Tech.
Bhuyan, M. S. (2022). Effects of microplastics on fish and in human health. *Frontiers in Environmental Science, 10*, 827289.
Bilal, M., Yaqub, A., Hassan, H. U., Akhtar, S., Rafiq, N., Ali Shah, M. I., Hussain, I., Salman Khan, M., Nawaz, A., & Manoharadas, S. (2023). Microplastic quantification in aquatic birds: Biomonitoring the environmental health of the panjkora river freshwater ecosystem in Pakistan. *Toxics, 11*(12), 972.
Bocci, V., Galafassi, S., Levantesi, C., Crognale, S., Amalfitano, S., Congestri, R., Matturro, B., Rossetti, S., & Di Pippo, F. (2024). Freshwater plastisphere: A review on biodiversity, risks, and biodegradation potential with implications for the aquatic ecosystem health. *Frontiers in Microbiology, 15*, 1395401.
Borrelle, S. B., Ringma, J., Law, K. L., Monnahan, C. C., Lebreton, L., McGivern, A., Murphy, E., Jambeck, J., Leonard, G. H., & Hilleary, M. A. (2020). Predicted growth in plastic waste exceeds efforts to mitigate plastic pollution. *Science, 369*(6510), 1515–1518.
Borriello, L., Scivicco, M., Cacciola, N. A., Esposito, F., Severino, L., & Cirillo, T. (2023). Microplastics, a global issue: Human exposure through environmental and dietary sources. *Food, 12*(18), 3396.
Botterell, Z. L., Beaumont, N., Dorrington, T., Steinke, M., Thompson, R. C., & Lindeque, P. K. (2019). Bioavailability and effects of microplastics on marine zooplankton: A review. *Environmental Pollution, 245*, 98–110.
Brandts, I., Garcia-Ordoñez, M., Tort, L., Teles, M., & Roher, N. (2020). Polystyrene nanoplastics accumulate in ZFL cell lysosomes and in zebrafish larvae after acute exposure, inducing a synergistic immune response in vitro without affecting larval survival in vivo. *Environmental Science: Nano, 7*(8), 2410–2422.
Browne, M. A., Galloway, T., & Thompson, R. (2007). Microplastic -- An emerging contaminant of potential concern? *Integrated Environmental Assessment and Management, 3*(4), 559.
Browne, M. A., Niven, S. J., Galloway, T. S., Rowland, S. J., & Thompson, R. C. (2013). Microplastic moves pollutants and additives to worms, reducing functions linked to health and biodiversity. *Current Biology, 23*(23), 2388–2392.
Burkhardt-Holm, P., & N'Guyen, A. (2019). Ingestion of microplastics by fish and other prey organisms of cetaceans, exemplified for two large baleen whale species. *Marine Pollution Bulletin, 144*, 224–234.

Carbery, M., O'Connor, W., & Palanisami, T. (2018). Trophic transfer of microplastics and mixed contaminants in the marine food web and implications for human health. *Environment International, 115*, 400–409.
Carlin, J., Craig, C., Little, S., Donnelly, M., Fox, D., Zhai, L., & Walters, L. (2020). Microplastic accumulation in the gastrointestinal tracts in birds of prey in central Florida, USA. *Environmental Pollution, 264*, 114633.
Carr, S. A., Liu, J., & Tesoro, A. G. (2016). Transport and fate of microplastic particles in wastewater treatment plants. *Water Research, 91*, 174–182.
Chae, Y., & An, Y.-J. (2017). Effects of micro-and nanoplastics on aquatic ecosystems: Current research trends and perspectives. *Marine Pollution Bulletin, 124*(2), 624–632.
Chavan, B., Pawar, S., Kamble, M., & Naik, S. (2018). Microorganisms in maintaining carbon/nitrogen (C/N) ratio in intensive aquaculture system: A review. *International Journal of Tropical Agriculture, 36*(4), 7.
Chen, Y., Li, W., Xiang, L., Mi, X., Duan, M., & Wu, C. (2022). Fish personality affects their exposure to microplastics. *Ecotoxicology and Environmental Safety, 233*, 113301.
Cole, M. (2014). *The impacts of microplastics on zooplankton*. University of Exeter.
Cole, M., Lindeque, P., Halsband, C., & Galloway, T. S. (2011). Microplastics as contaminants in the marine environment: A review. *Marine Pollution Bulletin, 62*(12), 2588–2597.
Cole, M., Liddle, C., Consolandi, G., Drago, C., Hird, C., Lindeque, P. K., & Galloway, T. S. (2020). Microplastics, microfibres and nanoplastics cause variable sub-lethal responses in mussels (*Mytilus* spp.). *Marine Pollution Bulletin, 160*, 111552.
Conesa, J. A. (2022). Adsorption of PAHs and PCDD/Fs in microplastics: A review. *Microplastics, 1*(3), 346–358.
Cózar, A., Echevarría, F., González-Gordillo, J. I., Irigoien, X., Úbeda, B., Hernández-León, S., Palma, Á. T., Navarro, S., García-de-Lomas, J., & Ruiz, A. (2014). Plastic debris in the open ocean. *Proceedings of the National Academy of Sciences, 111*(28), 10239–10244.
De Sá, L. C., Oliveira, M., Ribeiro, F., Rocha, T. L., & Futter, M. N. (2018). Studies of the effects of microplastics on aquatic organisms: What do we know and where should we focus our efforts in the future? *Science of the Total Environment, 645*, 1029–1039.
de Souza Machado, A. A., Kloas, W., Zarfl, C., Hempel, S., & Rillig, M. C. (2018). Microplastics as an emerging threat to terrestrial ecosystems. *Global Change Biology, 24*(4), 1405–1416.
Debnath, R., Prasad, G. S., Amin, A., Malik, M. M., Ahmad, I., Abubakr, A., Borah, S., Rather, M. A., Impellitteri, F., Tabassum, I., & Piccione, G. (2024). Understanding and addressing microplastic pollution: Impacts, mitigation, and future perspectives. *Journal of Contaminant Hydrology, 266*, 104399.
Desforges, J.-P. W., Galbraith, M., & Ross, P. S. (2015). Ingestion of microplastics by zooplankton in the Northeast Pacific Ocean. *Archives of Environmental Contamination and Toxicology, 69*, 320–330.
Dris, R., Gasperi, J., Rocher, V., Saad, M., Renault, N., & Tassin, B. (2015). Microplastic contamination in an urban area: A case study in Greater Paris. *Environmental Chemistry, 12*(5), 592–599.
Duncan, E. M., Broderick, A. C., Fuller, W. J., Galloway, T. S., Godfrey, M. H., Hamann, M., Limpus, C. J., Lindeque, P. K., Mayes, A. G., & Omeyer, L. C. (2019). Microplastic ingestion ubiquitous in marine turtles. *Global Change Biology, 25*(2), 744–752.
EFSA Panel on Contaminants in the Food Chain (CONTAM). (2016). Presence of microplastics and nanoplastics in food, with particular focus on seafood. *EFSA Journal, 14*(6), e04501.
Eom, H.-J., Nam, S.-E., & Rhee, J.-S. (2020). Polystyrene microplastics induce mortality through acute cell stress and inhibition of cholinergic activity in a brine shrimp. *Molecular & Cellular Toxicology, 16*, 233–243.
Esposito, G., Prearo, M., Renzi, M., Anselmi, S., Cesarani, A., Barcelò, D., Dondo, A., & Pastorino, P. (2022). Occurrence of microplastics in the gastrointestinal tract of benthic by–catches from an eastern Mediterranean deep–sea environment. *Marine Pollution Bulletin, 174*, 113231.

Félix, L., Carreira, P., & Peixoto, F. (2023). Effects of chronic exposure of naturally weathered microplastics on oxidative stress level, behaviour, and mitochondrial function of adult zebrafish (*Danio rerio*). *Chemosphere, 310*, 136895.

Ferheen, I., Spurio, R., & Marcheggiani, S. (2024). Emerging issues on antibiotic-resistant bacteria colonizing plastic waste in aquatic ecosystems. *Antibiotics, 13*(4), 339.

Foley, C. J., Feiner, Z. S., Malinich, T. D., & Höök, T. O. (2018). A meta-analysis of the effects of exposure to microplastics on fish and aquatic invertebrates. *Science of the Total Environment, 631*, 550–559.

Fraissinet, S., Arduini, D., Martines, A., De Benedetto, G. E., Malitesta, C., Giangrande, A., & Rossi, S. (2024). Microplastics uptake by four filter feeders. *Journal of Marine Science and Engineering, 12*(6), 1000.

Franzellitti, S., Canesi, L., Auguste, M., Wathsala, R. H., & Fabbri, E. (2019). Microplastic exposure and effects in aquatic organisms: A physiological perspective. *Environmental Toxicology and Pharmacology, 68*, 37–51.

Gago, J., Carretero, O., Filgueiras, A. V., & Viñas, L. (2018). Synthetic microfibers in the marine environment: A review on their occurrence in seawater and sediments. *Marine Pollution Bulletin, 127*, 365–376.

Gallo, F., Fossi, C., Weber, R., Santillo, D., Sousa, J., Ingram, I., Nadal, A., & Romano, D. (2018). Marine litter plastics and microplastics and their toxic chemicals components: The need for urgent preventive measures. *Environmental Sciences Europe, 30*, 13.

Galloway, T. S., Cole, M., & Lewis, C. (2017). Interactions of microplastic debris throughout the marine ecosystem. *Nature Ecology & Evolution, 1*(5), 0116.

Gao, B., Yao, H., Li, Y., & Zhu, Y. (2021). Microplastic addition alters the microbial community structure and stimulates soil carbon dioxide emissions in vegetable-growing soil. *Environmental Toxicology and Chemistry, 40*(2), 352–365.

Gardon, T., Reisser, C., Soyez, C., Quillien, V., & Le Moullac, G. (2018). Microplastics affect energy balance and gametogenesis in the pearl oyster *Pinctada margaritifera*. *Environmental Science & Technology, 52*(9), 5277–5286.

Gaspar, T. R., Chi, R. J., Parrow, M. W., & Ringwood, A. H. (2018). Cellular bioreactivity of micro-and nano-plastic particles in oysters. *Frontiers in Marine Science, 5*, 345.

Germishuizen, M. (2021). *The effects of microplastic and natural particles on the invasive mussel Mytilus galloprovincialis (Lamarck, 1819) and the native Choromytilus meridionalis*. University of Cape Town.

Gewert, B., Plassmann, M. M., & MacLeod, M. (2015). Pathways for degradation of plastic polymers floating in the marine environment. *Environmental Science: Processes & Impacts, 17*(9), 1513–1521.

Geyer, R., Jambeck, J. R., & Law, K. L. (2017). Production, use, and fate of all plastics ever made. *Science Advances, 3*(7), e1700782.

Gonçalves, C., Martins, M., Sobral, P., Costa, P. M., & Costa, M. H. (2019). An assessment of the ability to ingest and excrete microplastics by filter-feeders: A case study with the Mediterranean mussel. *Environmental Pollution, 245*, 600–606.

Green, D. S., Colgan, T. J., Thompson, R. C., & Carolan, J. C. (2019). Exposure to microplastics reduces attachment strength and alters the haemolymph proteome of blue mussels (*Mytilus edulis*). *Environmental Pollution, 246*, 423–434.

Grünzner, M., & Pahl, S. (2023). Behavior change as part of the solution for plastic pollution. In J. Kramm & C. Völker (Eds.), *Living in the plastic age* (pp. 169–196). Campus Verlag.

Guimarães, A. T. B., Estrela, F. N., Pereira, P. S., de Andrade Vieira, J. E., de Lima Rodrigues, A. S., Silva, F. G., & Malafaia, G. (2021). Toxicity of polystyrene nanoplastics in *Ctenopharyngodon idella* juveniles: A genotoxic, mutagenic and cytotoxic perspective. *Science of the Total Environment, 752*, 141937.

Guo, Y., Ma, W., Li, J., Liu, W., Qi, P., Ye, Y., Guo, B., Zhang, J., & Qu, C. (2020). Effects of microplastics on growth, phenanthrene stress, and lipid accumulation in a diatom, *Phaeodactylum tricornutum*. *Environmental Pollution, 257*, 113628.

Gurjar, U. R., Takar, S., Amin, A., & Martin Xavier, K. A. (2023). Microplastic contamination in aquatic organisms: An ecotoxicological perspective. In *Xenobiotics in aquatic animals: Reproductive and developmental impacts* (pp. 353–367). Springer.

Güven, O., Gökdağ, K., Jovanović, B., & Kıdeyş, A. E. (2017). Microplastic litter composition of the Turkish territorial waters of the Mediterranean Sea, and its occurrence in the gastrointestinal tract of fish. *Environmental Pollution, 223*, 286–294.

Hadiyanto, H., Khoironi, A., Dianratri, I., Suherman, S., Muhammad, F., & Vaidyanathan, S. (2021). Interactions between polyethylene and polypropylene microplastics and *Spirulina* sp. microalgae in aquatic systems. *Heliyon, 7*, e07676.

Haegerbaeumer, A., Mueller, M.-T., Fueser, H., & Traunspurger, W. (2019). Impacts of micro-and nano-sized plastic particles on benthic invertebrates: A literature review and gap analysis. *Frontiers in Environmental Science, 7*, 425457.

Hale, R. C., Seeley, M. E., La Guardia, M. J., Mai, L., & Zeng, E. Y. (2020). A global perspective on microplastics. *Journal of Geophysical Research: Oceans, 125*(1), e2018JC014719.

Hamed, M., Soliman, H. A., Osman, A. G., & Sayed, A. E.-D. H. (2020). Antioxidants and molecular damage in Nile Tilapia (*Oreochromis niloticus*) after exposure to microplastics. *Environmental Science and Pollution Research, 27*, 14581–14588.

Hao, Y., Sun, Y., Li, M., Fang, X., Wang, Z., Zuo, J., & Zhang, C. (2023). Adverse effects of polystyrene microplastics in the freshwater commercial fish, grass carp (*Ctenopharyngodon idella*): Emphasis on physiological response and intestinal microbiome. *Science of the Total Environment, 856*, 159270.

Harris, P. T. (2020). The fate of microplastic in marine sedimentary environments: A review and synthesis. *Marine Pollution Bulletin, 158*, 111398.

Hartley, B. L., Pahl, S., Veiga, J., Vlachogianni, T., Vasconcelos, L., Maes, T., Doyle, T., Metcalfe, R. A., Öztürk, A. A., & Di Berardo, M. (2018). Exploring public views on marine litter in Europe: Perceived causes, consequences and pathways to change. *Marine Pollution Bulletin, 133*, 945–955.

Hasan, A. M., Hamed, M., Hasan, J., Martyniuk, C. J., Niyogi, S., & Chivers, D. P. (2024). A review of the neurobehavioural, physiological, and reproductive toxicity of microplastics in fishes. *Ecotoxicology and Environmental Safety, 282*, 116712.

Hirt, N., & Body-Malapel, M. (2020). Immunotoxicity and intestinal effects of nano-and microplastics: A review of the literature. *Particle and Fibre Toxicology, 17*, 1–22.

Hodgson, D. J. (2018). *The impacts of microplastic ingestion on marine polychaete worms*. University of Exeter.

Hodkovicova, N., Hollerova, A., Svobodova, Z., Faldyna, M., & Faggio, C. (2022). Effects of plastic particles on aquatic invertebrates and fish – A review. *Environmental Toxicology and Pharmacology, 96*, 104013.

Horn, D. A., Granek, E. F., & Steele, C. L. (2020). Effects of environmentally relevant concentrations of microplastic fibers on Pacific mole crab (*Emerita analoga*) mortality and reproduction. *Limnology and Oceanography Letters, 5*(1), 74–83.

Hu, M., & Palić, D. (2020). Micro-and nano-plastics activation of oxidative and inflammatory adverse outcome pathways. *Redox Biology, 37*, 101620.

Huang, W., Song, B., Liang, J., Niu, Q., Zeng, G., Shen, M., Deng, J., Luo, Y., Wen, X., & Zhang, Y. (2021). Microplastics and associated contaminants in the aquatic environment: A review on their ecotoxicological effects, trophic transfer, and potential impacts to human health. *Journal of Hazardous Materials, 405*, 124187.

Imhof, H. K., Rusek, J., Thiel, M., Wolinska, J., & Laforsch, C. (2017). Do microplastic particles affect *Daphnia magna* at the morphological, life history and molecular level? *PLoS One, 12*(11), e0187590.

Ismail, R. F., Saleh, N. E., & Sayed, A. E.-D. H. (2021). Impacts of microplastics on reproductive performance of male tilapia (*Oreochromis niloticus*) pre-fed on *Amphora coffeaeformis*. *Environmental Science and Pollution Research, 28*, 68732–68744.

Issac, M. N., & Kandasubramanian, B. (2021). Effect of microplastics in water and aquatic systems. *Environmental Science and Pollution Research, 28*, 19544–19562.
Jabeen, K., Su, L., Li, J., Yang, D., Tong, C., Mu, J., & Shi, H. (2017). Microplastics and mesoplastics in fish from coastal and fresh waters of China. *Environmental Pollution, 221*, 141–149.
Jacquin, J., Cheng, J., Odobel, C., Pandin, C., Conan, P., Pujo-Pay, M., Barbe, V., Meistertzheim, A.-L., & Ghiglione, J.-F. (2019). Microbial ecotoxicology of marine plastic debris: A review on colonization and biodegradation by the "plastisphere". *Frontiers in Microbiology, 10*, 865.
Jambeck, J. R., Geyer, R., Wilcox, C., Siegler, T. R., Perryman, M., Andrady, A., Narayan, R., & Law, K. L. (2015). Plastic waste inputs from land into the ocean. *Science, 347*(6223), 768–771.
Jeong, E., Lee, J.-Y., & Redwan, M. (2024). Animal exposure to microplastics and health effects: A review. *Emerging Contaminants, 4*, 100369.
Jeyavani, J., Sibiya, A., Stalin, T., Vigneshkumar, G., Al-Ghanim, K. A., Riaz, M. N., Govindarajan, M., & Vaseeharan, B. (2023). Biochemical, genotoxic and histological implications of polypropylene microplastics on freshwater fish *Oreochromis mossambicus*: An aquatic eco-toxicological assessment. *Toxics, 11*(3), 282.
Jiang, B., Kauffman, A. E., Li, L., McFee, W., Cai, B., Weinstein, J., Lead, J. R., Chatterjee, S., Scott, G. I., & Xiao, S. (2020). Health impacts of environmental contamination of micro-and nanoplastics: A review. *Environmental Health and Preventive Medicine, 25*, 1–15.
Jovanović, B. (2017). Ingestion of microplastics by fish and its potential consequences from a physical perspective. *Integrated Environmental Assessment and Management, 13*(3), 510–515.
Kadac-Czapska, K., Ośko, J., Knez, E., & Grembecka, M. (2024). Microplastics and oxidative stress—Current problems and prospects. *Antioxidants, 13*(5), 579.
Kamble, M. T., Rudtanatip, T., Soowannayan, C., Nambunruang, B., Medhe, S. V., & Wongprasert, K. (2022). Depolymerized fractions of sulfated galactans extracted from *Gracilaria fisheri* and their antibacterial activity against *Vibrio parahaemolyticus* and *Vibrio harveyi*. *Marine Drugs, 20*(8), 469.
Kamble, M. T., Chaiyapechara, S., Salin, K. R., Bunphimpapha, P., Chavan, B. R., Bhujel, R. C., Medhe, S. V., Kettawan, A., Thiyajai, P., Thompson, K. D., & Pirarat, N. (2024a). Guava and Star gooseberry leaf extracts improve growth performance, innate immunity, intestinal microbial community, and disease resistance in Nile tilapia (*Oreochromis niloticus*) against *Aeromonas hydrophila*. *Aquaculture Reports, 35*, 101947.
Kamble, M. T., Gallardo, W., Salin, K. R., Pumpuang, S., Chavan, B. R., Bhujel, R. C., Medhe, S. V., Kettawan, A., Thompson, K. D., & Pirarat, N. (2024b). Effect of *Moringa oleifera* leaf extract on the growth performance, hematology, innate immunity, and disease resistance of Nile tilapia (*Oreochromis niloticus*) against *Streptococcus agalactiae* Biotype 2. *Animals, 14*(6), 953.
Kamble, M. T., Salin, K. R., Chavan, B. R., Medhe, S. V., Thompson, K. D., & Pirarat, N. (2024c). Length-weight relationship and condition factor of Nile tilapia (*Oreochromis niloticus*) fed diets supplemented with guava and star gooseberry leaf extract. *F1000Research, 13*, 540.
Kamble, M. T., Soowannayan, C., Chaicherd, S., Medhe, S. V., Rudtanatip, T., Pissuwan, D., & Wongprasert, K. (2024d). Bimetallic nanoparticles with sulfated galactan eliminate *Vibrio parahaemolyticus* in shrimp *Penaeus vannamei*. *Fish & Shellfish Immunology, 151*, 109753.
Kamble, M. T., Yostawonkul, J., Medhe, S. V., Chavan, B. R., Kumar, A., Palekar, G. R., Daunde, V. Y., Tayade, S. H., Gabriel, N. N., & Ataguba, G. A. (2024e). Innovative feed additives for sustainable aquaculture: Phytobiotics encapsulated in organic nanoparticles. In N. Gabriel, K. Abasubong, V. Erasmus, & M. Kamble (Eds.), *Sustainable feed ingredients and additives for aquaculture farming: Perspectives from Africa and Asia* (pp. 501–520). Springer.
Karan, H., Funk, C., Grabert, M., Oey, M., & Hankamer, B. (2019). Green bioplastics as part of a circular bioeconomy. *Trends in Plant Science, 24*(3), 237–249.
Keith, D. A., Benson, D. H., Baird, I. R., Watts, L., Simpson, C. C., Krogh, M., Gorissen, S., Ferrer-Paris, J. R., & Mason, T. J. (2023). Effects of interactions between anthropogenic stressors and

recurring perturbations on ecosystem resilience and collapse. *Conservation Biology, 37*(1), e13995.

Khan, F. R., Catarino, A. I., & Clark, N. J. (2022). The ecotoxicological consequences of microplastics and co-contaminants in aquatic organisms: A mini-review. *Emerging Topics in Life Sciences, 6*(4), 339–348.

Koelmans, A. A., Besseling, E., & Shim, W. J. (2015). Nanoplastics in the aquatic environment. Critical review. In M. Bergmann, L. Gutow, & M. Klages (Eds.), *Marine anthropogenic litter* (pp. 325–340). Springer.

Koelmans, A. A., Bakir, A., Burton, G. A., & Janssen, C. R. (2016). Microplastic as a vector for chemicals in the aquatic environment: Critical review and model-supported reinterpretation of empirical studies. *Environmental Science & Technology, 50*(7), 3315–3326.

Koelmans, A. A., Nor, N. H. M., Hermsen, E., Kooi, M., Mintenig, S. M., & De France, J. (2019). Microplastics in freshwaters and drinking water: Critical review and assessment of data quality. *Water Research, 155*, 410–422.

Kühn, S., & Van Franeker, J. A. (2020). Quantitative overview of marine debris ingested by marine megafauna. *Marine Pollution Bulletin, 151*, 110858.

Kumar, S., Rajesh, M., Rajesh, K., Suyani, N., Ahamed Rasheeq, A., & Pratiksha, K. (2020). Impact of microplastics on aquatic organisms and human health: A review. *International Journal of Environmental Sciences & Natural Resources, 26*(2), 59–64.

Kvale, K., Prowe, A., Chien, C.-T., Landolfi, A., & Oschlies, A. (2021). Zooplankton grazing of microplastic can accelerate global loss of ocean oxygen. *Nature Communications, 12*(1), 2358.

Lamichhane, G., Acharya, A., Marahatha, R., Modi, B., Paudel, R., Adhikari, A., Raut, B., Aryal, S., & Parajuli, N. (2023). Microplastics in environment: Global concern, challenges, and controlling measures. *International journal of Environmental Science and Technology, 20*(4), 4673–4694.

Langenfeld, D., Bucci, K., Veneruzzo, C., McNamee, R., Gao, G., Rochman, C. M., Rennie, M. D., Hoffman, M. J., Orihel, D. M., & Provencher, J. F. (2024). Microplastics at environmentally relevant concentrations had minimal impacts on pelagic zooplankton communities in a large in-lake mesocosm experiment. *Environmental Science & Technology, 58*(43), 19419–19428.

Lebreton, L., & Andrady, A. (2019). Future scenarios of global plastic waste generation and disposal. *Palgrave Communications, 5*(1), 1–11.

Lee, J.-H., Kang, J.-C., & Kim, J.-H. (2023a). Toxic effects of microplastic (polyethylene) on fish: Accumulation, hematological parameters and antioxidant responses in Korean bullhead, *Pseudobagrus fulvidraco. Science of the Total Environment, 877*, 162874.

Lee, Y., Cho, J., Sohn, J., & Kim, C. (2023b). Health effects of microplastic exposures: Current issues and perspectives in South Korea. *Yonsei Medical Journal, 64*(5), 301.

Li, W. C., Tse, H. F., & Fok, L. (2016). Plastic waste in the marine environment: A review of sources, occurrence and effects. *Science of the Total Environment, 566*, 333–349.

Li, J., Liu, H., & Chen, J. P. (2018). Microplastics in freshwater systems: A review on occurrence, environmental effects, and methods for microplastics detection. *Water Research, 137*, 362–374.

Li, Z., Dong, S., Huang, F., Lin, L., Hu, Z., & Zheng, Y. (2022). Toxicological effects of microplastics and sulfadiazine on the microalgae *Chlamydomonas reinhardtii. Frontiers in Microbiology, 13*, 865768.

Li, X., Chen, Y., Zhang, S., Dong, Y., Pang, Q., Lynch, I., Xie, C., Guo, Z., & Zhang, P. (2023). From marine to freshwater environment: A review of the ecotoxicological effects of microplastics. *Ecotoxicology and Environmental Safety, 251*, 114564.

Liu, W., Liao, H., Wei, M., Junaid, M., Chen, G., & Wang, J. (2024). Biological uptake, distribution and toxicity of micro (nano) plastics in the aquatic biota: A special emphasis on size-dependent impacts. *TrAC Trends in Analytical Chemistry, 170*, 117477.

Lusher, A. (2015). Microplastics in the marine environment: Distribution, interactions and effects. In M. Bergmann, L. Gutow, & M. Klages (Eds.), *Marine anthropogenic litter* (pp. 245–307). Springer.

Lusher, A., Hollman, P., & Mendoza-Hill, J. (2017). *Microplastics in fisheries and aquaculture: Status of knowledge on their occurrence and implications for aquatic organisms and food safety* (Fisheries and Aquaculture Technical Paper). FAO.

Lusher, A., Provencher, J. F., Baak, J. E., Hamilton, B. M., Vorkamp, K., Hallanger, I. G., Pijogge, L., Liboiron, M., Bourdages, M., & Hammer, S. (2022). Monitoring litter and microplastics in Arctic mammals and birds. *Arctic Science, 8*(4), 1217–1235.

Ma, P., Wang mu, W., Liu, H., Chen yu, F., & Xia, J. (2019). Research on ecotoxicology of microplastics on freshwater aquatic organisms. *Environmental Pollutants and Bioavailability, 31*(1), 131–137.

Ma, H., Pu, S., Liu, S., Bai, Y., Mandal, S., & Xing, B. (2020). Microplastics in aquatic environments: Toxicity to trigger ecological consequences. *Environmental Pollution, 261*, 114089.

Malinowski, C. R., Searle, C. L., Schaber, J., & Höök, T. O. (2023). Microplastics impact simple aquatic food web dynamics through reduced zooplankton feeding and potentially releasing algae from consumer control. *Science of the Total Environment, 904*, 166691.

Mao, Y., Ai, H., Chen, Y., Zhang, Z., Zeng, P., Kang, L., Li, W., Gu, W., He, Q., & Li, H. (2018). Phytoplankton response to polystyrene microplastics: Perspective from an entire growth period. *Chemosphere, 208*, 59–68.

Marcharla, E., Vinayagam, S., Gnanasekaran, L., Soto-Moscoso, M., Chen, W.-H., Thanigaivel, S., & Ganesan, S. (2024). Microplastics in marine ecosystems: A comprehensive review of biological and ecological implications and its mitigation approach using nanotechnology for the sustainable environment. *Environmental Research, 256*, 119181.

Masud, N., & Cable, J. (2023). Microplastic exposure and consumption increases susceptibility to gyrodactylosis and host mortality for a freshwater fish. *Diseases of Aquatic Organisms, 153*, 81–85.

Matthews, C., Moran, F., & Jaiswal, A. K. (2021). A review on European Union's strategy for plastics in a circular economy and its impact on food safety. *Journal of Cleaner Production, 283*, 125263.

McHale, M. E., & Sheehan, K. L. (2024). Bioaccumulation, transfer, and impacts of microplastics in aquatic food chains. *Journal of Environmental Exposure Assessment, 3*(3), 15.

Meaza, I., Toyoda, J. H., & Wise, J. P., Sr. (2021). Microplastics in sea turtles, marine mammals and humans: A one environmental health perspective. *Frontiers in Environmental Science, 8*, 575614.

Mitrano, D. M., & Wohlleben, W. (2020). Microplastic regulation should be more precise to incentivize both innovation and environmental safety. *Nature Communications, 11*(1), 5324.

Mojiri, A., Vishkaei, M. N., Zhou, J. L., Trzcinski, A. P., Lou, Z., Kasmuri, N., Rezania, S., Gholami, A., Vakili, M., & Kazeroon, R. A. (2024). Impact of polystyrene microplastics on the growth and photosynthetic efficiency of diatom *Chaetoceros neogracile*. *Marine Environmental Research, 194*, 106343.

Montero, D., Rimoldi, S., Torrecillas, S., Rapp, J., Moroni, F., Herrera, A., Gómez, M., Fernández-Montero, Á., & Terova, G. (2022). Impact of polypropylene microplastics and chemical pollutants on European sea bass (*Dicentrarchus labrax*) gut microbiota and health. *Science of the Total Environment, 805*, 150402.

Montoya, D., Rastelli, E., Casotti, R., Manna, V., Trano, A. C., Balestra, C., Santinelli, C., Saggiomo, M., Sansone, C., & Corinaldesi, C. (2024). Microplastics alter the functioning of marine microbial ecosystems. *Ecology and Evolution, 14*(11), e70041.

Moore, R., Loseto, L., Noel, M., Etemadifar, A., Brewster, J., MacPhee, S., Bendell, L., & Ross, P. (2020). Microplastics in beluga whales (*Delphinapterus leucas*) from the Eastern Beaufort Sea. *Marine Pollution Bulletin, 150*, 110723.

Nabi, N., Ahmad, I., Amin, A., Rather, M. A., Ahmed, I., Hajam, Y. A., Khursheed, S., Malik, M. M., & Abubakr, A. (2024). Understanding the sources, fate and effects of microplastics in aquatic environments with a focus on risk profiling in aquaculture systems. *Reviews in Aquaculture, 16*(4), 1947–1980.

Navarro, A., Luzardo, O. P., Gómez, M., Acosta-Dacal, A., Martínez, I., de la Rosa, J. F., Macías-Montes, A., Suárez-Pérez, A., & Herrera, A. (2023). Microplastics ingestion and chemical pollutants in seabirds of Gran Canaria (Canary Islands, Spain). *Marine Pollution Bulletin, 186*, 114434.

Naz, S., Chatha, A. M. M., Khan, N. A., Ullah, Q., Zaman, F., Qadeer, A., Khan, I. M., Danabas, D., Kiran, A., & Skalickova, S. (2024). Unraveling the ecotoxicological effects of micro and nanoplastics on aquatic organisms and human health. *Frontiers in Environmental Science, 12*, 1390510.

Nelms, S. E., Barnett, J., Brownlow, A., Davison, N., Deaville, R., Galloway, T. S., Lindeque, P. K., Santillo, D., & Godley, B. J. (2019). Microplastics in marine mammals stranded around the British coast: Ubiquitous but transitory? *Scientific Reports, 9*(1), 1075.

Nerland, I. L., Halsband, C., Allan, I., & Thomas, K. V. (2014). *Microplastics in marine environments: Occurrence, distribution and effects*. Akvaplan-niva.

Obbard, R. W., Sadri, S., Wong, Y. Q., Khitun, A. A., Baker, I., & Thompson, R. C. (2014). Global warming releases microplastic legacy frozen in Arctic Sea ice. *Earth's Future, 2*(6), 315–320.

Oza, J., Rabari, V., Yadav, V. K., Sahoo, D. K., Patel, A., & Trivedi, J. (2024). A systematic review on microplastic contamination in fishes of asia: Polymeric risk assessment and future prospectives. *Environmental Toxicology and Chemistry, 43*(4), 671–685.

Pan, Z., Zhang, C., Wang, S., Sun, D., Zhou, A., Xie, S., Xu, G., & Zou, J. (2021). Occurrence of microplastics in the gastrointestinal tract and gills of fish from Guangdong, South China. *Journal of Marine Science and Engineering, 9*(9), 981.

Paul, S., Nath, S., Bhattacharjee, S., & Mukherjee, S. (2024). Unveiling the effects of microplastics pollution on marine fauna. *Blue Biotechnology, 1*(1), 6.

Paul-Pont, I., Lacroix, C., Fernández, C. G., Hégaret, H., Lambert, C., Le Goïc, N., Frère, L., Cassone, A.-L., Sussarellu, R., & Fabioux, C. (2016). Exposure of marine mussels *Mytilus spp.* to polystyrene microplastics: Toxicity and influence on fluoranthene bioaccumulation. *Environmental Pollution, 216*, 724–737.

Prasad, G. S., Rout, S. K., Malik, M. M., Karmakar, S., Amin, A., & Ahmad, I. (2023). Occurrence of xenoestrogen alkylphenols (Octylphenols and Nonylphenol) and its impact on the aquatic ecosystem. In *Xenobiotics in aquatic animals: Reproductive and developmental impacts* (pp. 275–284). Springer.

Prasad, G. S., Rout, S. K., Amin, A., Malik, M. M., Ahmad, I., Rather, M. A., Karmakar, S., Khalil, H. S., & Abubakar, A. (2024). Impact of COVID-19 on the aquatic environment and implications for fisheries with special emphasis to Indian perspective. *Proceedings of the Indian National Science Academy, 90*(4), 821–839.

Prata, J. C., da Costa, J. P., Lopes, I., Duarte, A. C., & Rocha-Santos, T. (2020). Environmental exposure to microplastics: An overview on possible human health effects. *Science of the Total Environment, 702*, 134455.

Provencher, J., Ammendolia, J., Rochman, C., & Mallory, M. (2019). Assessing plastic debris in aquatic food webs: What we know and don't know about uptake and trophic transfer. *Environmental Reviews, 27*(3), 304–317.

Qaiser, Z., Aqeel, M., Sarfraz, W., Rizvi, Z. F., Noman, A., Naeem, S., & Khalid, N. (2023). Microplastics in wastewaters and their potential effects on aquatic and terrestrial biota. *Case Studies in Chemical and Environmental Engineering, 8*, 100536.

Rafa, N., Ahmed, B., Zohora, F., Bakya, J., Ahmed, S., Ahmed, S. F., Mofijur, M., Chowdhury, A. A., & Almomani, F. (2024). Microplastics as carriers of toxic pollutants: Source, transport, and toxicological effects. *Environmental Pollution, 343*, 123190.

Raju, P., Santhanam, P., Pandian, S. S., Divya, M., Arunkrishnan, A., Devi, K. N., Ananth, S., Roopavathy, J., & Perumal, P. (2022). Impact of polystyrene microplastics on major marine primary (phytoplankton) and secondary producers (copepod). *Archives of Microbiology, 204*(1), 84.

Rapini, C. C. S., Di Berardino, C., Peserico, A., Capacchietti, G., & Barboni, B. (2024). Can mammalian reproductive health withstand massive exposure to polystyrene micro-and nanoplastic derivatives? A systematic review. *International Journal of Molecular Sciences, 25*(22), 12166.

Raza, T., Rasool, B., Asrar, M., Manzoor, M., Javed, Z., Jabeen, F., & Younis, T. (2023). Exploration of polyacrylamide microplastics and evaluation of their toxicity on multiple parameters of *Oreochromis niloticus*. *Saudi Journal of Biological Sciences, 30*(2), 103518.

Rillig, M. C. (2012). Microplastic in terrestrial ecosystems and the soil? *Environmental Science and Technology, 46*, 6453–6454.

Rist, S. E., Assidqi, K., Zamani, N. P., Appel, D., Perschke, M., Huhn, M., & Lenz, M. (2016). Suspended micro-sized PVC particles impair the performance and decrease survival in the Asian green mussel *Perna viridis*. *Marine Pollution Bulletin, 111*(1–2), 213–220.

Rochman, C. M., Hoh, E., Kurobe, T., & Teh, S. J. (2013). Ingested plastic transfers hazardous chemicals to fish and induces hepatic stress. *Scientific Reports, 3*(1), 1–7.

Rochman, C. M., Tahir, A., Williams, S. L., Baxa, D. V., Lam, R., Miller, J. T., Teh, F.-C., Werorilangi, S., & Teh, S. J. (2015). Anthropogenic debris in seafood: Plastic debris and fibers from textiles in fish and bivalves sold for human consumption. *Scientific Reports, 5*(1), 1–10.

Rochman, C. M., Brookson, C., Bikker, J., Djuric, N., Earn, A., Bucci, K., Athey, S., Huntington, A., McIlwraith, H., & Munno, K. (2019). Rethinking microplastics as a diverse contaminant suite. *Environmental Toxicology and Chemistry, 38*(4), 703–711.

Romano, N., Renukdas, N., Fischer, H., Shrivastava, J., Baruah, K., Egnew, N., & Sinha, A. K. (2020). Differential modulation of oxidative stress, antioxidant defense, histomorphology, ion-regulation and growth marker gene expression in goldfish (*Carassius auratus*) following exposure to different dose of virgin microplastics. *Comparative Biochemistry and Physiology Part C: Toxicology & Pharmacology, 238*, 108862.

Santillo, D., Miller, K., & Johnston, P. (2017). Microplastics as contaminants in commercially important seafood species. *Integrated Environmental Assessment and Management, 13*(3), 516–521.

Seeley, M. E., Song, B., Passie, R., & Hale, R. C. (2020). Microplastics affect sedimentary microbial communities and nitrogen cycling. *Nature Communications, 11*(1), 2372.

Sendra, M., Sparaventi, E., Novoa, B., & Figueras, A. (2021). An overview of the internalization and effects of microplastics and nanoplastics as pollutants of emerging concern in bivalves. *Science of the Total Environment, 753*, 142024.

Seoane, M., González-Fernández, C., Soudant, P., Huvet, A., Esperanza, M., Cid, Á., & Paul-Pont, I. (2019). Polystyrene microbeads modulate the energy metabolism of the marine diatom *Chaetoceros neogracile*. *Environmental Pollution, 251*, 363–371.

Setälä, O., Fleming-Lehtinen, V., & Lehtiniemi, M. (2014). Ingestion and transfer of microplastics in the planktonic food web. *Environmental Pollution, 185*, 77–83.

Sfriso, A. A., Juhmani, A.-S., Tomio, Y., Sfriso, A., Rizzolio, F., Adeel, M., Wahsha, M., Munari, C., & Mistri, M. (2024). Microplastic accumulation and ecological impacts on benthic invertebrates: Insights from a microcosm experiment. *Marine Pollution Bulletin, 202*, 116231.

Sharma, S., Basu, S., Shetti, N. P., Nadagouda, M. N., & Aminabhavi, T. M. (2021). Microplastics in the environment: Occurrence, perils, and eradication. *Chemical Engineering Journal, 408*, 127317.

Sjollema, S. B., Redondo-Hasselerharm, P., Leslie, H. A., Kraak, M. H., & Vethaak, A. D. (2016). Do plastic particles affect microalgal photosynthesis and growth? *Aquatic Toxicology, 170*, 259–261.

Smith, M., Love, D. C., Rochman, C. M., & Neff, R. A. (2018). Microplastics in seafood and the implications for human health. *Current Environmental Health Reports, 5*, 375–386.

Subaramaniyam, U., Allimuthu, R. S., Vappu, S., Ramalingam, D., Balan, R., Paital, B., Panda, N., Rath, P. K., Ramalingam, N., & Sahoo, D. K. (2023). Effects of microplastics, pesticides and nano-materials on fish health, oxidative stress and antioxidant defense mechanism. *Frontiers in Physiology, 14*, 1217666.

Sun, J., Dai, X., Wang, Q., Van Loosdrecht, M. C., & Ni, B.-J. (2019). Microplastics in wastewater treatment plants: Detection, occurrence and removal. *Water Research, 152*, 21–37.

Sunil, S., Bhagwat, G., Vincent, S. G. T., & Palanisami, T. (2024). Microplastics and climate change; the global impacts of a tiny driver. *Science of the Total Environment, 946*, 174160.
Sussarellu, R., Suquet, M., Thomas, Y., Lambert, C., Fabioux, C., Pernet, M. E. J., Le Goïc, N., Quillien, V., Mingant, C., & Epelboin, Y. (2016). Oyster reproduction is affected by exposure to polystyrene microplastics. *Proceedings of the National Academy of Sciences, 113*(9), 2430–2435.
Tabassum, I., Hassan, N., Amin, A., Malik, M. M., Abubakr, A., Prasad, G. S., Latief, F., & Rather, S. A. (2025). COVID-19 and global environmental change: A threat to food security and nutrition. In *Food security, nutrition and sustainability through aquaculture technologies* (pp. 437–445). Springer.
Tamayo-Belda, M., Pérez-Olivares, A. V., Pulido-Reyes, G., Martin-Betancor, K., González-Pleiter, M., Leganés, F., Mitrano, D. M., Rosal, R., & Fernández-Piñas, F. (2023). Tracking nanoplastics in freshwater microcosms and their impacts to aquatic organisms. *Journal of Hazardous Materials, 445*, 130625.
Tanaka, K., Takada, H., Yamashita, R., Mizukawa, K., Fukuwaka, M.-a., & Watanuki, Y. (2013). Accumulation of plastic-derived chemicals in tissues of seabirds ingesting marine plastics. *Marine Pollution Bulletin, 69*(1–2), 219–222.
Taylor, M., Gwinnett, C., Robinson, L. F., & Woodall, L. C. (2016). Plastic microfibre ingestion by deep-sea organisms. *Scientific Reports, 6*(1), 33997.
Thomas, M., Jon, B., Craig, S., Edward, R., Ruth, H., John, B., Dick, V. A., Heather, L. A., & Matthew, S. (2020). The world is your oyster: Low-dose, long-term microplastic exposure of juvenile oysters. *Heliyon, 6*, e03103.
Thompson, R. C., Olsen, Y., Mitchell, R. P., Davis, A., Rowland, S. J., John, A. W., McGonigle, D., & Russell, A. E. (2004). Lost at sea: Where is all the plastic? *Science, 304*(5672), 838–838.
Ullah, S., Ahmad, S., Guo, X., Ullah, S., Ullah, S., Nabi, G., & Wanghe, K. (2023). A review of the endocrine disrupting effects of micro and nano plastic and their associated chemicals in mammals. *Frontiers in Endocrinology, 13*, 1084236.
Urbanek, A. K., Rymowicz, W., & Mirończuk, A. M. (2018). Degradation of plastics and plastic-degrading bacteria in cold marine habitats. *Applied Microbiology and Biotechnology, 102*, 7669–7678.
Urban-Malinga, B., Jakubowska-Lehrmann, M., Białowąs, M., & Hallmann, A. (2022). Microplastics cause neurotoxicity and decline of enzymatic activities in important bioturbator *Hediste diversicolor*. *Marine Environmental Research, 179*, 105660.
Usman, S., Abdull Razis, A. F., Shaari, K., Amal, M. N. A., Saad, M. Z., Mat Isa, N., & Nazarudin, M. F. (2021). Polystyrene microplastics exposure: An insight into multiple organ histological alterations, oxidative stress and neurotoxicity in Javanese medaka fish (*Oryzias javanicus* Bleeker, 1854). *International Journal of Environmental Research and Public Health, 18*(18), 9449.
Van Cauwenberghe, L., & Janssen, C. R. (2014). Microplastics in bivalves cultured for human consumption. *Environmental Pollution, 193*, 65–70.
Walkinshaw, C., Lindeque, P. K., Thompson, R., Tolhurst, T., & Cole, M. (2020). Microplastics and seafood: Lower trophic organisms at highest risk of contamination. *Ecotoxicology and Environmental Safety, 190*, 110066.
Wang, J., Tan, Z., Peng, J., Qiu, Q., & Li, M. (2016). The behaviors of microplastics in the marine environment. *Marine Environmental Research, 113*, 7–17.
Wang, J., Coffin, S., Sun, C., Schlenk, D., & Gan, J. (2019a). Negligible effects of microplastics on animal fitness and HOC bioaccumulation in earthworm *Eisenia fetida* in soil. *Environmental Pollution, 249*, 776–784.
Wang, J., Li, Y., Lu, L., Zheng, M., Zhang, X., Tian, H., Wang, W., & Ru, S. (2019b). Polystyrene microplastics cause tissue damages, sex-specific reproductive disruption and transgenerational effects in marine medaka (*Oryzias melastigma*). *Environmental Pollution, 254*, 113024.

Wang, W., Gao, H., Jin, S., Li, R., & Na, G. (2019c). The ecotoxicological effects of microplastics on aquatic food web, from primary producer to human: A review. *Ecotoxicology and Environmental Safety, 173*, 110–117.

Wang, L., Nabi, G., Yin, L., Wang, Y., Li, S., Hao, Z., & Li, D. (2021a). Birds and plastic pollution: Recent advances. *Avian Research, 12*, 1–9.

Wang, S.-c., Gao, Z.-y., Liu, F.-f., Chen, S.-q., & Liu, G.-z. (2021b). Effects of polystyrene and triphenyl phosphate on growth, photosynthesis and oxidative stress of *Chaetoceros meülleri*. *Science of the Total Environment, 797*, 149180.

Wang, Y.-F., Wang, X.-Y., Chen, B.-J., Yang, Y.-P., Li, H., & Wang, F. (2025). Impact of microplastics on the human digestive system: From basic to clinical. *World Journal of Gastroenterology, 31*(4), 100470.

Wijesooriya, M., Wijesekara, H., Sewwandi, M., Soysa, S., Rajapaksha, A. U., Vithanage, M., & Bolan, N. (2023). Microplastics and soil nutrient cycling. In M. Vithanage & M. N. V. Prasad (Eds.), *Microplastics in the ecosphere: Air, water, soil, and food* (pp. 321–338). Wiley.

Wright, S. L., & Kelly, F. J. (2017). Plastic and human health: A micro issue? *Environmental Science & Technology, 51*(12), 6634–6647.

Wright, S. L., Thompson, R. C., & Galloway, T. S. (2013). The physical impacts of microplastics on marine organisms: A review. *Environmental Pollution, 178*, 483–492.

Wu, X., He, F., Xu, X., Wu, L., Rong, J., & Lin, S. (2024). Environmental health and safety implications of the interplay between microplastics and the residing biofilm. *Environment & Health, 3*, 118–132.

Xayachak, T., Haque, N., Lau, D., & Pramanik, B. K. (2024). The missing link: A systematic review of microplastics and its neglected role in life-cycle assessment. *Science of the Total Environment, 954*, 176513.

Xia, X., Sun, M., Zhou, M., Chang, Z., & Li, L. (2020). Polyvinyl chloride microplastics induce growth inhibition and oxidative stress in *Cyprinus carpio* var. larvae. *Science of the Total Environment, 716*, 136479.

Xing, J., Zhang, S., Zhang, M., & Hou, J. (2022). A critical review of presence, removal and potential impacts of endocrine disruptors bisphenol A. *Comparative Biochemistry and Physiology Part C: Toxicology & Pharmacology, 254*, 109275.

Xu, Z., Huang, L., Xu, P., Lim, L., Cheong, K.-L., Wang, Y., & Tan, K. (2024). Microplastic pollution in commercially important edible marine bivalves: A comprehensive review. *Food Chemistry: X, 23*, 101647.

Yang, Y., Liu, W., Zhang, Z., Grossart, H. P., & Gadd, G. M. (2020). Microplastics provide new microbial niches in aquatic environments. *Applied Microbiology and Biotechnology, 104*, 6501–6511.

Yin, M., Yan, B., Wang, H., Wu, Y., Wang, X., Wang, J., Zhu, Z., Yan, X., Liu, Y., & Liu, M. (2023). Effects of microplastics on nitrogen and phosphorus cycles and microbial communities in sediments. *Environmental Pollution, 318*, 120852.

Yostawonkul, J., Kitiyodom, S., Supchukun, K., Thumrongsiri, N., Saengkrit, N., Pinpimai, K., Hajitou, A., Thompson, K. D., Rattanapinyopituk, K., Maita, M., Kamble, M. T., Yata, T., & Pirarat, N. (2023). Masculinization of red tilapia (*Oreochromis* spp.) using 17-α-methyltestosterone-loaded alkyl polyglucosides integrated into nanostructured lipid carriers. *Animals, 13*(8), 1364.

Yostawonkul, J., Kitiyodom, S., Kamble, M. T., Supchukun, K., Saengkrit, N., Sukkarun, P., Medhe, S. V., Thompson, K. D., Boonrungsiman, S., & Temisak, S. (2024). Immersion of nanostructured lipid carriers loaded with 17-alpha methyltestosterone for masculinization of red tilapia (*Oreochromis* sp.). *Aquaculture, 586*, 740780.

Yu, F., Yang, C., Zhu, Z., Bai, X., & Ma, J. (2019). Adsorption behavior of organic pollutants and metals on micro/nanoplastics in the aquatic environment. *Science of the Total Environment, 694*, 133643.

Yuan, K.-K., Yu, Y.-Y., Mo, Y.-H., Liu, Y.-J., Zhang, W.-X., Lv, J.-J., Shi, W., Liu, G.-X., Li, H.-Y., & Yang, W.-D. (2024). Exposure to microplastics renders immunity of the thick-shell

mussel more vulnerable to diarrhetic shellfish toxin-producing harmful algae. *Science of the Total Environment, 926*, 172125.
Zarantoniello, M., Cattaneo, N., Conti, F., Carrino, M., Cardinaletti, G., Şener, İ., & Olivotto, I. (2024). Mitigating dietary microplastic accumulation and oxidative stress response in European seabass (*Dicentrarchus labrax*) juveniles using a natural microencapsulated antioxidant. *Antioxidants, 13*(7), 812.
Zhang, K., Shi, H., Peng, J., Wang, Y., Xiong, X., Wu, C., & Lam, P. K. (2018). Microplastic pollution in China's inland water systems: A review of findings, methods, characteristics, effects, and management. *Science of the Total Environment, 630*, 1641–1653.
Zhang C, Wang F, Wang Q, Zou J, Zhu J (2023) Species-specific effects of microplastics on juvenile fishes. Frontiers in Physiology 14:1256005.
Zhong, H., Wu, M., Sonne, C., Lam, S. S., Kwong, R. W., Jiang, Y., Zhao, X., Sun, X., Zhang, X., & Li, C. (2023). The hidden risk of microplastic-associated pathogens in aquatic environments. *Eco-Environment & Health, 2*(3), 142–151.
Zhu, J., Yu, X., Zhang, Q., Li, Y., Tan, S., Li, D., Yang, Z., & Wang, J. (2019a). Cetaceans and microplastics: First report of microplastic ingestion by a coastal delphinid, *Sousa chinensis*. *Science of the Total Environment, 659*, 649–654.
Zhu, Z.-l., Wang, S.-c., Zhao, F.-f., Wang, S.-g., Liu, F.-f., & Liu, G.-z. (2019b). Joint toxicity of microplastics with triclosan to marine microalgae *Skeletonema costatum*. *Environmental Pollution, 246*, 509–517.
Ziani, K., Ioniță-Mîndrican, C.-B., Mititelu, M., Neacşu, S. M., Negrei, C., Moroşan, E., Drăgănescu, D., & Preda, O.-T. (2023). Microplastics: A real global threat for environment and food safety: A state of the art review. *Nutrients, 15*(3), 617.
Zimmermann, L., Dombrowski, A., Völker, C., & Wagner, M. (2020). Are bioplastics and plant-based materials safer than conventional plastics? *In vitro* toxicity and chemical composition. *Environment International, 145*, 106066.
Zolotova, N., Kosyreva, A., Dzhalilova, D., Fokichev, N., & Makarova, O. (2022). Harmful effects of the microplastic pollution on animal health: A literature review. *PeerJ, 10*, e13503.

Microplastics in Fish: An Emerging Environmental Challenge

Sonam Angmo, Keisham Geenita, Bejawada Chanikya Naidu, Zaheer Abass, Sahina Akter, Kenyum Lollen, and Muzammal Hoque

1 Introduction

Microplastic pollution has emerged as a pervasive environmental crisis, severely impacting aquatic ecosystems and human health. Global plastic production exceeds 330 million tons annually, with only 9% being recycled, while the remainder accumulates in landfills, undergoes incineration, or infiltrates natural environments (Kumar et al., 2021; OECD, 2024). Plastics constitute 60–80% of marine debris, with floating plastic waste comprising 90% of this pollution. As plastics degrade into microplastics (MPs), defined as particles <5 mm, they infiltrate marine environments, disrupting ecological balance (Bajt, 2021). Rivers transport between 1.15 and 2.41 million tons of plastic waste into the ocean annually, with Asian river systems contributing approximately 67% of this influx (Lebreton et al., 2017; Guo et al., 2020). MPs have been detected in surface waters, sediments, deep-sea trenches, and remote polar regions, highlighting their widespread distribution. Their composition varies by polymer type, with polyethylene (PE), polypropylene (PP), polystyrene (PS), polyethylene terephthalate (PET), and polyvinyl chloride (PVC) being the most prevalent (Obbard et al., 2014; Eriksen et al., 2013; Kanhai, 2019). The fate of MPs is influenced by polymer density and hydrodynamic conditions, with lighter

S. Angmo (✉) · K. Geenita · K. Lollen · M. Hoque
ICAR-Central Institute of Fisheries Education (CIFE), Mumbai, Maharashtra, India

B. C. Naidu
ICAR-Central Institute of Fisheries Technology (CIFT), Kochi, Kerala, India

Z. Abass
Division of Fisheries Resources Management, Faculty of Fisheries, Sher-e-Kashmir University of Agricultural Science and Technology of Kashmir, Shalimar, Srinagar, Jammu and Kashmir, India

S. Akter
Centre for Marine Living Resources and Ecology (CMLRE), Kochi, Kerala, India

A. Abubakr et al. (eds.), *Microplastics: Ecological and Food Security Implications*,
https://doi.org/10.1007/978-3-031-94480-2_9

MPs remaining suspended in the water column and denser MPs accumulating in sediments (Dai et al., 2018; Abidli et al., 2021). Once in the marine environment, MPs interfere with gas exchange, alter sediment structure, and disrupt nutrient cycling, ultimately affecting biodiversity. They also serve as vectors for persistent organic pollutants (POPs), heavy metals, and pathogenic microorganisms, further exacerbating their environmental and ecological risks (Wang et al., 2018; Malik et al., 2017; Yoshida et al., 2016).

The ingestion of MPs has been widely documented across marine taxa, with severe implications for aquatic biodiversity and human health. Zooplankton and filter-feeding organisms ingest MPs, leading to reduced feeding efficiency, metabolic imbalances, and trophic transfer (Borja & Elliott, 2019). Fish species exposed to MPs exhibit oxidative stress, enzymatic dysfunction, altered growth, and reproductive impairments, while sediment-dwelling organisms accumulate MPs, further driving bioaccumulation (Reisser et al., 2014; Sutton et al., 2016). MPs also promote microbial colonization, acting as reservoirs for antibiotic-resistant genes and harmful bacteria (Shim & Thomposon, 2015). Their presence in commercially valuable seafood raises significant food safety concerns, as human exposure occurs through dietary ingestion, drinking water, and airborne inhalation (Jahnke et al., 2017). MPs have been detected in human tissues, including the liver, lungs, placenta, and bloodstream, with potential effects on metabolic regulation, endocrine function, and neurotoxicity (Islam et al., 2024; Abass et al., 2025). Their ability to cross biological barriers and accumulate systemically underscores the urgent need for research on their long-term health implications. Addressing MP pollution requires a multifaceted approach, including stringent regulations on plastic waste, improved waste management strategies, and the adoption of biodegradable alternatives. Enhanced monitoring, public awareness, and policy-driven interventions are critical to mitigating the adverse effects of MPs on marine ecosystems and human well-being.

2 Microplastics

Microplastics (MPs) are synthetic solid particles or polymer matrices that constitute pervasive environmental contaminants (Reshi et al., 2023). Defined by the European Chemicals Agency and the US National Oceanic and Atmospheric Administration as plastic fragments smaller than 5 mm, MPs are further categorized by size into large microplastics (5 mm to 1 mm), small microplastics (1 mm to 1 μm), and nanoplastics (<1 μm) (Crawford & Quinn, 2017; Nabi et al., 2024; Debnath et al., 2024). These particles exist in various morphological forms, including microbeads, fibers, fragments, films, foams, pellets, and filaments (Hidalgo-Ruz et al., 2012; Wright et al., 2013). MPs originate from two primary sources: primary microplastics, which are intentionally manufactured for commercial applications such as cosmetics and industrial abrasives, and secondary microplastics, which result from the degradation of larger plastic debris due to environmental factors such as mechanical wear, photodegradation, and microbial activity (Arthur et al., 2009). The persistence and widespread distribution of MPs in marine and terrestrial environments have raised global concerns regarding their ecological and toxicological implications.

MPs are primarily composed of petroleum-derived polymers, each exhibiting distinct physicochemical properties that influence their environmental fate, persistence, and degradation pathways. Among the most prevalent polymers, polyethylene (PE) is extensively utilized in packaging materials, including plastic bottles, grocery bags, and toiletry containers. Despite its widespread use, PE's high chemical stability facilitates its prolonged environmental persistence, where exposure to abiotic and biotic factors promotes its fragmentation into MPs (Li et al., 2016; Gangadoo et al., 2020). Similarly, polypropylene (PP) is a widely used polymer known for its thermal stability, durability, and resistance to biodegradation. Found in disposable consumer goods such as face masks, drinking straws, and food packaging, PP microplastics are frequently detected in aquatic ecosystems, where they undergo degradation primarily via photo-oxidative mechanisms (Gewert et al., 2015; Bond et al., 2018; Vibhatabandhu & Srithongouthai, 2024; Zhao et al., 2023). Polyethylene terephthalate (PET), another dominant polymer, is valued for its lightweight, high mechanical strength, and resistance to moisture, making it essential in beverage packaging and textiles. However, PET's slow degradation rate exacerbates plastic pollution, necessitating improved recycling and disposal strategies to mitigate its environmental impact (Benyathiar et al., 2022; Sarda et al., 2022; Kibria et al., 2023).

Among high-impact synthetic polymers, polyvinyl chloride (PVC) is extensively employed in industrial applications due to its mechanical strength, flame retardancy, and chemical resistance. However, the presence of hazardous additives and its high chlorine content classify PVC as an environmentally persistent and toxic polymer, with its degradation releasing harmful compounds that pose significant risks to aquatic biota (Abreu et al., 2023; Li et al., 2023; Mijangos et al., 2023; Sharma & Chatterjee, 2017; Tran et al., 2023; Ahmad et al., 2023). Polystyrene (PS), widely utilized in packaging, electronics, and disposable consumer goods, undergoes degradation through interactions with environmental pollutants such as heavy metals and pesticides, leading to the formation of secondary MPs and nanoplastics (Rabot et al., 2023; Lu et al., 2023; Zhang et al., 2022b). The environmental implications of polyamide (PA) microplastics have also gained attention, particularly concerning their role in wastewater treatment systems. While most MPs inhibit anaerobic digestion, PA6 has been found to enhance methane production by releasing caprolactam, a compound that stimulates microbial methanogenesis (Chen et al., 2021). As PA undergoes weathering, its chemical properties evolve, influencing both its environmental mobility and potential toxicity (Fernandez-Gonzalez et al., 2021). Given their persistence, MPs contribute to long-term contamination, bioaccumulation, and potential adverse effects on ecosystem stability and human health, necessitating comprehensive regulatory measures and innovative mitigation strategies, behaviour and potential toxicity (Fernandez-Gonzalez et al., 2021).

3 Sources of Microplastics

MPs originate from diverse sources and are broadly classified into primary and secondary microplastics. These persistent pollutants have become a major environmental concern due to their widespread occurrence in aquatic ecosystems and their ability to undergo bioaccumulation and trophic transfer. Primary microplastics are intentionally manufactured plastic particles, including microbeads and pellets, that directly contribute to environmental contamination. Industrial wastewater is a significant pathway for these particles to enter aquatic systems, as they are widely used in air-blasting applications such as paint and engine cleaning (Park et al., 2020). Additional contributors include styrofoam used for packaging, pre-production plastic pellets in manufacturing, and synthetic fibers released from textile production (Sheavly & Register, 2007; De Falco et al., 2019). In contrast, secondary microplastics result from the degradation of larger plastic debris through thermal, photochemical, and biological processes. Prolonged environmental exposure leads to the fragmentation of plastic waste into micro-sized particles, further exacerbating the accumulation of MPs in marine and freshwater environments. Land-based sources account for approximately 80–90% of MPs in aquatic ecosystems (Duis & Coors, 2016). Urban plastic waste, including discarded bags, bottles, and packaging materials, contributes significantly to MP pollution, surpassing ocean-based sources (Naidu et al., 2022). Waste from construction, shipbuilding, and improper incineration processes also increases MP concentrations in sediments and water bodies (Yang et al., 2021). Ineffective waste disposal, inadequate landfill management, and suboptimal waste collection systems further exacerbate the release of MPs into the environment (Tursi et al., 2022). Personal care and household products are another major contributor, as many cosmetics, face washes, hand soaps, and sunscreens contain plastic microparticles as exfoliants or structural components (Guerranti et al., 2019; Prasad et al., 2024; Tabassum et al., 2025). Synthetic textile fibers shed during laundering significantly elevate MP contamination in wastewater streams, with laundering processes contributing up to 35% of total MP emissions (Carney Almroth et al., 2018). Domestic wastewater and industrial discharges also act as key pathways for MPs, as wastewater treatment plants (WWTPs) are unable to fully filter out plastic microparticles despite multi-stage treatment processes (Murphy et al., 2016; Talvitie et al., 2017). Elevated MP concentrations have been reported in both water and sediment downstream of WWTPs (Magnusson, 2016; McCormick et al., 2014). Industrial activities contribute to MPs contamination primarily through plastic flakes from packaging materials and synthetic fibers released from textiles. Studies indicate that a single wash cycle can release between 110,000 and 1,000,000 fibers, depending on textile composition, washing conditions, detergent use, and garment wear (De Falco et al., 2019). Sewage effluents also introduce significant MPs into the environment, particularly from cosmetic and personal care products containing microbeads. A single use of an exfoliating face wash can release between 4500 and 94,500 plastic particles, while toothpaste usage discharges approximately 4000 microbeads per application (Carr et al., 2016). The discharge of untreated sewage and wastewater remains a critical factor in the introduction of MPs into aquatic ecosystems.

Ocean-based sources contribute an estimated 10–20% of MPs in marine environments (Li, 2018; Karbalaei et al., 2019). Major contributors include commercial fishing, marine tourism, and offshore industrial activities. Abandoned or lost fishing equipment, including plastic monofilament lines and synthetic nets, accounts for a substantial fraction of marine plastic pollution (Naji et al., 2017). It is estimated that over 600,000 tonnes of fishing gear are discarded into the ocean annually (Good et al., 2010). Additionally, shipping-related plastic waste and pollution from offshore industries such as petrochemical operations further exacerbate MP contamination (Calero et al., 2021). Fishing activities contribute significantly to MPs pollution through the degradation of plastic ropes and nets, which experience mechanical stress during hauling operations and abrasion from seabed contact. Modern fishing gear predominantly consists of polyethylene (PE), polypropylene (PP), and polyamide (PA) (Andrady, 2011), and degradation rates vary depending on environmental factors such as ultraviolet (UV) radiation exposure and substrate composition. A study conducted in the United Kingdom reported monthly mass losses of 0.39% for PP, 1.02% for PA, and 0.45% for PE in submerged plastic ropes at a depth of 10 m (Krüger et al., 2020). Mechanical stress from hydraulic net haulers and net drums further accelerates material wear, leading to a continuous release of MPs into the marine environment.

Lost and abandoned fishing gear, commonly referred to as "ghost gear," presents an additional and long-lasting source of MPs pollution. These materials persist in marine environments for extended periods, contributing to marine litter and negatively impacting aquatic biodiversity through entanglement and ghost fishing (Macfadyen et al., 2009; Derraik, 2002). Recent estimates suggest that lost fishing gear alone is responsible for over 45,000 tonnes of plastic pollution annually (Kuczenski et al., 2022). Given the extensive contribution of both land-based and ocean-based sources to MPs contamination, urgent mitigation strategies are needed to reduce plastic emissions, enhance waste management practices, and regulate industrial discharges to minimize the long-term environmental and ecological risks associated with MPs pollution (Fig. 1).

4 Microplastics in Aquatic Environments

Microplastic contamination has emerged as a critical environmental issue in both freshwater and marine ecosystems, with its pervasive distribution posing significant ecological risks. These microscopic plastic particles have been identified across diverse aquatic compartments, including surface waters, the water column, benthic sediments, and even polar regions (Abidli et al., 2021; Wang et al., 2017). Variations in microplastic coloration provide insights into their sources, with transparent fibers predominantly linked to the degradation of fishing lines and nets, while colored fragments originate from the breakdown of plastic-based consumer products such as

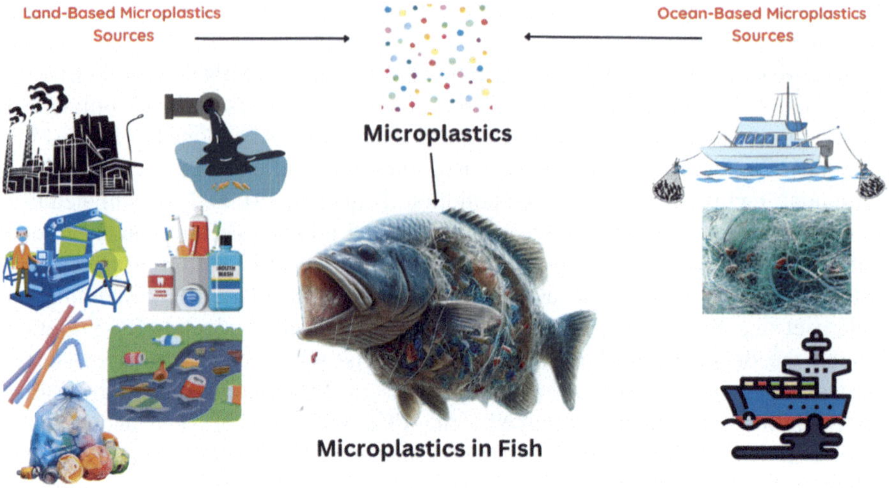

Fig. 1 Sources of microplastics in fish

textiles and packaging materials. The prevalence of fibers and fragments in global aquatic environments is largely attributed to the degradation of larger plastic debris, emphasizing the dominance of secondary microplastics as a major pollutant (Dai et al., 2018; Eriksen et al., 2013; Zhang et al., 2019). Estimates suggest that over five trillion plastic particles are suspended in the ocean, with more than 90% classified as secondary microplastics formed through fragmentation processes (Eriksen et al., 2014). The most commonly detected polymer types, including polyethylene (PE), polypropylene (PP), polystyrene (PS), polyethylene terephthalate (PET), and polyvinyl chloride (PVC), reflect their extensive industrial and consumer applications (Horton et al., 2017; Obbard et al., 2014). Inadequate waste management practices further exacerbate microplastic accumulation in aquatic ecosystems, and as plastic waste continues to enter the environment and degrade, microplastic concentrations are expected to rise over time (Barnes et al., 2009; Eriksen et al., 2014; Sruthisree et al., 2015; Amin et al., 2016, 2017).

Anthropogenic activities serve as primary contributors to microplastic pollution in aquatic environments. Domestic washing machines, for instance, release approximately 1900 microplastic fibers per wash cycle into wastewater (Browne et al., 2011). Personal care products such as facial scrubs have also been identified as a substantial source, containing up to 50,391 plastic microbeads per gram, with a single application potentially introducing between 10,000 and 100,000 microplastics into sewage systems (Cheung & Fok, 2017). Although modern wastewater treatment plants (WWTPs) are designed to remove a significant proportion of microplastics, substantial volumes still escape into the environment (Carr et al., 2016). For instance, a WWTP on the River Clyde in Glasgow was estimated to discharge approximately 65 million microplastic particles into the river each day, despite a removal efficiency exceeding 98% (Murphy et al., 2016). Research indicates that personal care products alone contribute approximately 209.7 trillion plastic microbeads (equivalent to 306.9 tonnes) to aquatic ecosystems in Mainland China annually, with over 80% entering via WWTP effluents (Cheung & Fok, 2017). Once introduced, microplastics from wastewater and land-based sources are transported by riverine systems into marine environments. Additionally, the fragmentation of larger plastic debris significantly contributes to microplastic pollution, with key sources including aquaculture, fishing, shipping, and tourism-related activities (Galloway & Lewis, 2016; Cole et al., 2011). Elevated microplastic concentrations have been documented in the surface waters of urban lakes in Wuhan, China, where fibers predominate due to the degradation of fishing nets and lines (Wang et al., 2017). Tourism has also been recognized as a significant contributor to microplastic pollution, particularly in coastal areas such as Huatulco Bay, Mexico (Retama et al., 2016). Consequently, the spatial distribution of microplastics varies across geographic regions, largely influenced by human activities, land use patterns, and population density.

The transport and fate of microplastics in aquatic ecosystems are dictated by their physicochemical properties, including shape, size, color, density, and composition (Zhang et al., 2017). Virgin plastic polymers exhibit densities ranging from 0.8 to 1.5 g/cm^3, compared to freshwater at 1.0 g/cm^3 and seawater at approximately 1.02–1.07 g/cm^3. Low-density polymers such as PE and PP remain predominantly buoyant, whereas high-density plastics like PVC and polyester tend to sink and accumulate in sediments (Kooi et al., 2017). However, biofilm formation and interactions with organic matter can alter their buoyancy over time, facilitating their movement throughout aquatic environments. Smaller microplastics often exhibit colloidal behavior, remaining suspended within the water column, while granular particles accumulate at the surface and fibers disperse throughout various depths (Filella, 2015; Zhang et al., 2017). Hydrodynamic forces, including surface currents and wind patterns, enable microplastics to travel vast distances, contributing to their widespread distribution. The extensive presence of MPs in both freshwater and marine ecosystems increase their bioavailability to a wide range of aquatic organisms, raising concerns about their ecological and potential human health risks. Given the increasing prevalence of microplastic contamination, a comprehensive understanding of its sources, transport mechanisms, and environmental consequences is essential for developing effective mitigation strategies aimed at reducing long-term ecological impacts (Fig. 2).

5 Microplastic Concentrations Across Global Seas

MPs contamination in marine environments exhibits considerable variability across different seas, influenced by anthropogenic activities, industrial discharges, oceanic currents, and geographical characteristics. Numerous recent studies have extensively documented the occurrence of microplastics in marine sediments and water bodies,

Fig. 2 Catch composition of dol net fishing with considerable quantity of plastics at Karanja Jetty, Maharashtra, India

providing insights into their sources, dominant debris types, and polymer compositions.

The Andaman Sea experiences a substantial influx of fibrous microplastics, predominantly composed of acrylic, polyethylene (PE), nylon, and polyethylene-polypropylene (PEP) co-polymers. The concentration of MPs in this region is approximately 15.36 particles/kg dry weight (DW), significantly influenced by heightened anthropogenic activities along the coastline. Factors such as uncontrolled plastic disposal, semi-treated and untreated urban effluent discharge, and intensified shipping and fishing activities contribute to the accumulation of fibrous MPs. Additionally, fragments and pellets composed of nylon, acrylic, and ionomer have been reported at concentrations reaching 45.17 particles/kg DW (Goswami et al., 2021). Similarly, the Arabian Sea has been identified as a major microplastic accumulation zone, with fibrous MPs reaching concentrations of up to 128.02 particles/kg DW. The movement of low-salinity waters from the Bay of Bengal into the Arabian Sea facilitates the transport of microplastics from coastal regions. Extensive laundry waste containing synthetic fibers, coupled with large-scale fishing and shipping operations, further exacerbates MP contamination. Near the Arabian Sea coastline, fragmented MPs composed of PE and polypropylene (PP) have been detected at an alarming concentration of 664 particles/kg DW, largely attributed to industrial discharges, urban wastewater influx, and the oxidative and mechanical degradation of plastic debris (Goswami et al., 2021; Yaranal et al., 2021).

In polar regions such as the Arctic Sea, MP contamination varies widely, with reported concentrations ranging from 42 to 6595 particles/kg DW. The dominant polymer types include PE, PP, and polyamide (PA), with thermohaline circulation playing a crucial role in transporting microplastic debris from Northern Europe into Arctic waters. Additionally, MPs trapped within sea ice are released during seasonal melting, contributing to their accumulation. The widespread use of PP and PE in packaging materials and fishing gear further elevates MP concentrations in this region (Bergmann et al., 2017). The Baltic Sea is characterized by a high prevalence of fibrous MPs composed of PE, PP, polyethylene terephthalate (PET), polydimethylsiloxane (PDMS), and polyvinyl chloride (PVC). Concentrations range between 34 and 863 particles/kg DW, influenced by wastewater discharges, sediment fractionation, wave currents, and vertical stratification within the water column. Bottom sediment deposits are significantly affected by localized erosion and accumulation processes (Zobkov & Esiukova, 2017; Chubarenko et al., 2022).

Microplastic pollution in the North Sea has been reported at concentrations of 65.95 particles/kg DW. The unique geomorphology of this region, characterized by a low flushing rate and a narrow entrance, results in the formation of tidal eddies that trap floating MPs, leading to their accumulation in seafloor sediments (Claessens et al., 2011). Similarly, the Black Sea exhibits high levels of MP contamination, predominantly in the form of fibrous particles composed of PE and PP, with concentrations ranging between 98 and 106.7 particles/kg DW. Polyamide fibers are strongly linked to maritime activities, particularly commercial fishing, while storm-induced surface mixing redistributes MPs throughout the water column and seabed. Fluvial transport from untreated wastewater and sewage sludge is a significant contributor to MP pollution in this semi-enclosed marine system (Cincinelli et al., 2021; Pojar et al., 2021).

The Bohai Sea, a semi-enclosed water body surrounded by dense industrial and urban infrastructure, exhibits MP concentrations ranging between 102.9 and 163.3 particles/kg DW. The region's weak hydrodynamic forces facilitate the deposition of fine particulate matter and small-sized MPs onto the seafloor. Reported polymer types include rayon, PET, cellulose propionate (CP), PA, PE, PP, and polycarbonate (PC) (Yu et al., 2016; Zhang et al., 2022a). In the Caribbean Sea, microplastic contamination is largely driven by tourism and dense coastal populations. Fibrous MPs have been recorded at concentrations of 261 particles/kg DW, while fragmented MPs reach 1109 particles/kg DW (Bosker et al., 2018; Rangel-Buitrago et al., 2021). The high influx of single-use plastics is a primary contributor to this extensive pollution.

Microplastic pollution in the Caspian Sea is predominantly attributed to riverine input from the Gorganrud, Nokandeh, and Qarasu rivers, which transport synthetic debris into the marine environment. Fibers composed of polystyrene (PS) and PET are dominant, with an average concentration of 196.67 particles/kg DW. Oceanic currents further facilitate MP dispersion throughout the region (Manbohi et al., 2021). Finally, the South China Sea exhibits one of the highest recorded MP contamination levels, with granular MPs composed of polyester, rayon, and nylon reaching concentrations of 7705 particles/kg DW. The primary sources of MPs in this region are riverine discharges and coastal anthropogenic activities (Cui et al., 2022). Given the rising microplastic contamination across global marine environments, urgent mitigation strategies, improved waste management policies, and stricter regulatory frameworks are essential to mitigate the long-term ecological and environmental risks associated with MPs.

6 MPs Contamination in Fish

Microplastic contamination in fish has been extensively documented worldwide, with ingestion reported in over 728 fish species across diverse aquatic ecosystems (Hossain & Olden, 2022; Kasamesiri & Thaimuangphol, 2020; Talley et al., 2020; Valente et al., 2019; Yuan et al., 2019; Gurjar et al., 2023). Microplastic particles have been detected in fish from wild populations, aquaculture systems, and even commercial market samples, where they accumulate in the gastrointestinal tract, gills, and tissues (Rochman et al., 2015; Wu et al., 2020), as outlined in Table 1. The extent of microplastic ingestion is influenced by various factors, including pollutant bioavailability, exposure duration, feeding behavior, and the physiological characteristics of the fish species (Akhbarizadeh et al., 2018).

Fish ingest microplastics through two primary pathways:

Primary Ingestion Fish directly ingest microplastics either by mistaking them for food or through incidental consumption while foraging. The resemblance of MPs to natural prey, often due to their varied coloration and size, increases the likelihood of ingestion. This similarity leads to misidentification by zooplankton and fish, facilitating the entry of microplastics into the marine food web. Over time, these particles accumulate in commercially important fish species consumed by humans, raising concerns about potential health risks (Shaw & Day, 1994) (Fig. 3).

Secondary Ingestion Microplastics also enter fish through the consumption of prey that has already accumulated these particles, resulting in bioaccumulation and biomagnification (Farrell & Nelson, 2013; Provencher et al., 2019; Zhang et al., 2019). As microplastics move through trophic levels, their concentrations can increase in higher-order predators, exacerbating their potential ecological and physiological impacts. These findings underscore the growing concern regarding microplastic pollution and its implications for marine biodiversity, trophic interactions, and ecosystem health (Fig. 3).

7 Impacts of MPs Contamination on Fishes

Microplastic (MP) pollution has been shown to affect a wide range of fish species, influencing their growth, metabolism, survival, and physiological functions. The severity of these effects varies across species, depending on exposure levels, feeding habits, and biological characteristics. At the cellular level, MPs induce oxidative stress, DNA damage, disruptions in protein activity, and cell death, leading to

Table 1 MPs in fishes reported globally

Species name	Area	Habitat	Type	References
Mugil cephalus	South Africa urban harbour	Benthopelagic	Fiber, fragment, foam, film, & line	Naidoo et al. (2016)
Limanda limanda, Platichthys flesu, Gadus morhua, Clupea harengus, Scomber scombrus	North Sea and Baltic Sea	Benthopelagic	PE (40%), PA, PP, PS, PET, PU, & rubber	Rummel et al. (2017)
Engraulis japonicas	Tokyo Bay, Japan	Benthopelagic	PE (52%), PP (43.3%), PS, etc.	Tanaka and Takada (2016)
Platichthys flesu, Osmerus eperlanus	River Thames, UK	Demersal Pelagic	PE, PET, PA, acrylic, etc	McGoran et al. (2017)
Platycephalus indicus, Saurida tumbil, Sillago sihama, Cynoglossus abbreviatus	Musa Estuary & Persian Gulf	Demersal Demersal Pelagic Mesopelagic	Fiber & fragment	Abbasi et al. (2018)
Rastrilliger kanagurta, Epinephalus merra	Tuticorin coast, India	Pelagic Reef-associated	PE & PP	Kumar et al. (2018)
Forty-six fish species	Amazon River Estuary	Demersal	PA (97.4%) & PE	Pegado et al. (2018)
Solea solea	Adriatic Sea, Italy	Benthic	PVC, PP, PE, PET, and PAM	Pellini et al. (2018)
M. surmuletus	NW Iberian shelf, Spain	Benthic		Filgueiras et al. (2020)
24 fish species	Beibu Gulf, South China Sea	16 Demersal 8 Pelagic	PES, PP, PE, Acrylics, and Rayon	Koongolla et al. (2020)
Mullus barbatus	Adriatic Sea, Tyrrhenian Sea, and Sardinia, Italy	Benthic	PET/ polyester	Valente et al. (2022)
Cirrhinus reba, Glossogobius giuris	Brahmaputra river, Bangladesh	Benthic	PE, PAM	Ferdous et al. (2023)
Leiognathus brevirostris, Siganus canaliculatus	Tamil Nadu, India	Benthic	–	Keerthika et al. (2023)
Pterotolithus macolatus, Sillaginopsis panijus, Panna microdon, Polynemus paradise		Demersal		Matluba et al. (2023)
		Benthic	PE, PET	

(continued)

Table 1 (continued)

Species name	Area	Habitat	Type	References
Ammodytes personatus and *Gobiidae*	Jiaozhou Bay, China			Zhang et al. (2023)
Platichthys flesus	River Thames and River Stour, UK	Riverine	PP	Horton et al. (2024)
Microgadus tomcod, Ictalurus punctatu	SLRE, Canada	Benthic	PET, PE	Kabir et al. (2025)

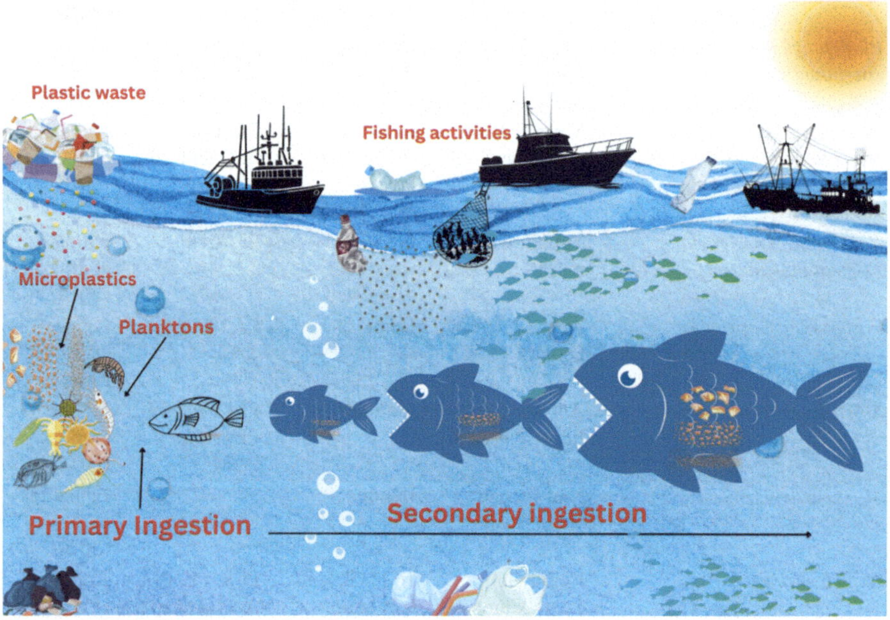

Fig. 3 Ingestion of microplastic in fish

compromised physiological functions (Xu et al., 2021; Roda et al., 2020; Mazurais et al., 2015; Duan et al., 2020; Prasad et al., 2023). As MPs accumulate in organs and tissues, they induce histopathological changes, gut dysbiosis, metabolic imbalances, and neuromuscular dysfunction, impairing digestion and overall health (Hamed et al., 2021; Cao, 2024; Oliveira et al., 2013; Luís et al., 2015). These physiological disruptions translate into negative impacts at the individual level, where several fish species experience growth and metabolic impairments. Studies have reported reduced survival rates and poor body condition in *Ambassis dussumieri* (Naidoo & Glassom, 2019), slowed growth in *Menidia beryllina* (Athey et al., 2020), and impaired feeding and swimming behaviors in *Lates calcarifer* (Guven et al., 2018). MPs can physically damage the digestive system, causing internal injuries and biochemical disturbances. In *Carassius auratus*, gastrointestinal (GIT) damage, oxidative stress, and weight loss have been observed following MP exposure (Jabeen et al., 2017). Additionally, the ingestion of MPs by ichthyoplankton can lead to false satiation or starvation, further affecting early developmental stages (Mallik et al., 2024). MPs also significantly alter protein levels and blood chemistry in multiple fish species, leading to behavioral and neurological impairments that disrupt feeding efficiency, swimming patterns, and predatory instincts. Microplastics pose a significant threat to fish gill function, as their inhalation through water can lead to physical abrasion, inflammation, and oxidative stress. These pollutants accumulate in gill filaments, causing structural damage, excessive mucus secretion, and impaired gas exchange. Prolonged exposure may result in respiratory distress, reduced oxygen uptake, and morphological alterations in the gills, ultimately compromising fish health. In addition to their physiological effects, MPs also have profound consequences on fish reproduction and development. Disruptions in protein signalling and germline development have been observed in both larval and adult stages, often leading to developmental delays and reduced reproductive success (Rochman et al., 2014; Duan et al., 2020). In female fish, MPs have been shown to interfere with vitellogenesis, a critical process in oocyte maturation (Bourguiba et al., 2003), while in males, exposure to MPs has been linked to hormonal imbalances, reduced sperm quality, and altered reproductive function (Sharifinia et al., 2020).

At the population level, prolonged MP exposure contributes to declining fish populations, disrupted population dynamics, and forced relocations due to habitat degradation. These effects extend to the community level, where MPs alter species interactions, disrupt trophic relationships, and favor pollution-tolerant species, ultimately impacting biodiversity. At the ecosystem level, MP pollution weakens resilience, reduces biodiversity, and makes aquatic environments more susceptible to additional stressors. While certain species exhibit resilience to MPs, indicating species-specific tolerance mechanisms (Malinich et al., 2018), the widespread presence of MPs in fish raises concerns regarding food web stability, bioaccumulation, and biomagnification, with potential repercussions for human health (Wright et al., 2013). Addressing MP pollution is crucial for sustaining fish populations, preserving aquatic biodiversity, and ensuring the long-term health of marine and freshwater ecosystems.

Predatory fish species are particularly vulnerable to microplastic contamination due to their higher trophic positions, as MPs are transferred through the food web (Mahamud et al., 2022). The ingestion and retention of MPs are influenced by their physical and chemical properties, including size, shape, and buoyancy (Walkinshaw et al., 2020). Once consumed, MPs can accumulate in the gastrointestinal tract, leading to intestinal blockages, oxidative stress, and translocation to other tissues such as the liver, gills, and muscle (Fossi et al., 2018; Wright & Kelly, 2017; Pagano et al., 2020). Research by Wright and Kelly (2017) demonstrated that ultrafine plastic particles can penetrate biological membranes and be transported via vascular and lymphatic pathways, raising concerns about their long-term effects on fish health, reproductive success (Sharifinia et al., 2020), and overall ecosystem stability.

Studies worldwide have provided insights into the frequency of occurrence (FO) of MPs, their mean abundance, and the types of polymers ingested by fish species across different environments. Research on the geographical distribution of MP contamination spans multiple continents, with significant contributions from Brazil, the USA, China, South Africa, and the UK. In Brazil, ingestion rates have been reported to range from 7.9% to over 83% across different fish species. Similarly, studies conducted in the USA have documented MP ingestion in various species, with some reports indicating a 100% FO (McNeish et al., 2018; Ryan et al., 2019; Hurt et al., 2020). In China, ingestion rates have been recorded as high as 95.7% FO (Jabeen et al., 2017). The frequency and mean ingestion rates of MPs vary widely, with some studies reporting relatively low FO values, such as 7.5% in Switzerland (Faure et al., 2015) and 18.8% in Germany (Roch et al., 2019), while others have found high prevalence rates exceeding 50%, such as 73.5% in Canada (Campbell et al., 2017) and 76.7% in Egypt (Khan et al., 2020). Notably, studies in South Africa, Argentina, and China have recorded 100% FO values (Naidoo et al., 2017; Pazos et al., 2017; Blettler et al., 2019). Various polymer types, including PE, PP, PET, and PA, have been detected in ingested MPs. Among the most frequently reported are PE and PP, which have been found in fish species examined in China (Zhang et al., 2019), Portugal (Bessa et al., 2018), and Egypt (Khan et al., 2020). Additionally, synthetic fibers such as rayon (RAY) and polyester (PES) have been identified in fish samples from the UK (McGoran et al., 2017) and Australia (Halstead et al., 2018). These findings underscore the extensive spread of MP contamination in global fisheries and its potential implications for biodiversity, ecosystem health, and human consumption. A comprehensive understanding of MP ingestion pathways in fish is essential for assessing ecological risks and developing effective mitigation strategies to reduce the impact of microplastic pollution in aquatic environments.

8 Conclusion

MPs contamination in marine and freshwater ecosystems poses a critical threat to fish populations, biodiversity, and ecosystem stability. At an ecological scale, MPs disrupt trophic dynamics by altering predator-prey interactions, favouring pollution-tolerant species, and destabilizing food webs. The potential for bioaccumulation and biomagnification further exacerbates these concerns, as MPs and their adsorbed contaminants become increasingly concentrated in higher trophic organisms. Predatory fish, which play a pivotal role in ecosystem balance, are particularly susceptible to trophic transfer, raising concerns about biodiversity loss and human exposure through seafood consumption. The ingestion of MPs by commercially and ecologically significant fish species underscores the urgent need for comprehensive risk assessments to evaluate their long-term implications. Scientific evidence demonstrates that MPs and their associated chemical pollutants induce oxidative stress, metabolic dysregulation, reproductive impairments, and physiological dysfunctions in fish, with cascading effects on aquatic food chains and ecosystem resilience. Addressing this challenge necessitates an integrated approach combining advanced scientific research, evidence-based policy interventions, and sustainable fisheries management. Regulatory measures should focus on reducing plastic waste at its source through stricter controls on plastic production, enhanced waste management strategies, and the development of biodegradable fishing gear. Additionally, international collaboration is essential for establishing standardized monitoring frameworks, facilitating data-sharing, and formulating global mitigation policies. Future research should prioritize the advancement of detection methodologies, such as high-resolution imaging and chemical fingerprinting, to trace MPs within biological systems and elucidate their toxicological interactions with persistent organic pollutants (POPs). Ensuring the sustainability of global fisheries and protecting aquatic ecosystems from MP contamination is imperative for preserving biodiversity, maintaining ecological resilience, and securing vital food resources for future generations. Addressing this escalating environmental threat requires a coordinated effort from researchers, policymakers, industry stakeholders, and the public. Only through collaborative and sustained action can we mitigate the ecological and health risks posed by MPs, safeguard aquatic environments, and promote a more sustainable and responsible approach to fisheries management.

References

Abass, Z., Chuskit, D., Shah, T. H., Wanjari, R. N., Ahmad, I., Bala, I. M., et al. (2025). Microplastic an emerging threat to nutritional and food security. In *Food security, nutrition and sustainability through aquaculture technologies* (pp. 397–415). Springer.

Abbasi, S., et al. (2018). Microplastics in different tissues of fish and prawn from the Musa Estuary, Persian Gulf. *Chemosphere, 205*, 80–87.

Abidli, S., Akkari, N., Lahbib, Y., & Trigui El Menif, N. (2021). First evaluation of microplastics in two commercial fish species from the lagoons of Bizerte and Ghar El Melh (Northern Tunisia). *Regional Studies in Marine Science, 41*, 101581. https://doi.org/10.1016/j.rsma.2020.101581

Abreu, C. M. R., Rezende, T. C., Serra, A. C., Fonseca, A. C., Braslau, R., & Coelho, J. F. J. (2023). Convenient and industrially viable internal plasticization of poly (vinyl chloride): Copolymerization of vinyl chloride and commercial monomers. *Polymer, 267*, 125688. https://doi.org/10.1016/j.polymer.2023.125688

Ahmad, T., Liu, X., & Guria, C. (2023). Preparation of polyvinyl chloride (PVC) membrane blended with acrylamide grafted bentonite for oily water treatment. *Chemosphere, 310*, 136840. https://doi.org/10.1016/j.chemosphere.2022.136840

Akhbarizadeh, R., Moore, F., & Keshavarzi, B. (2018). Investigating a probable relationship between microplastics and potentially toxic elements in fish muscles from northeast of Persian Gulf. *Environmental Pollution, 232*, 154–163. https://doi.org/10.1016/j.envpol.2017.09.028

Amin, A., Indulkar, S. T., & Pai, R. (2016). Sublethal effect of Buprofezin pesticide on carcass composition of *Cyprinus carpio* communis fingerlings. *Journal of Experimental Zoology India, 20*(1), 269–272.

Amin, A., Ramachandra Naik, A. T., Priyadarshini, N., Nayak, H., & Sree, C. S. (2017). Toxic effect of heavy metal lead on oxygen consumption of rohu (*Labeo rohita*) fingerlings. *Biochemical and Cellular Archives, 17*(1), 225–228.

Andrady, A. L. (2011). Microplastics in the marine environment. *Marine Pollution Bulletin, 62*(8), 1596–1605.

Arthur, C., Baker, J. E., & Bamford, H. A. (2009). Proceedings of the International Research Workshop on the Occurrence, Effects, and Fate of Microplastic Marine Debris, September 9–11, 2008, University of Washington Tacoma, Tacoma, WA, USA.

Athey, S. N., Albotra, S. D., Gordon, C. A., Monteleone, B., Seaton, P., Andrady, A. L., ... & Brander, S. M. (2020). Trophic transfer of microplastics in an estuarine food chain and the effects of a sorbed legacy pollutant. *Limnology and Oceanography Letters, 5*(1), 154–162.

Bajt, O. (2021). From plastics to microplastics and organisms. *FEBS Open Bio, 11*(4), 954–966. https://doi.org/10.1002/2211-5463.13120

Barnes, D. K. A., et al. (2009). Accumulation and fragmentation of plastic debris in global environments. *Philosophical Transactions of the Royal Society B: Biological Sciences, 364*, 1985–1998.

Benyathiar, P., Kumar, P., Carpenter, G., Brace, J., & Mishra, D. K. (2022). Polyethylene terephthalate (PET) bottle-to-bottle recycling for the beverage industry: A review. *Polymers, 14*, 2366. https://doi.org/10.3390/polym14122366

Bergmann, M., Wirzberger, V., Krumpen, T., et al. (2017). High quantities of microplastic in arctic deep-sea sediments from the HAUSGARTEN observatory. *Environmental Science & Technology, 51*, 11000–11010. https://doi.org/10.1021/acs.est.7b03331

Bessa, F., Barría, P., Neto, J. M., Frias, J. P. G. L., Otero, V., Sobral, P., & Marques, J. C. (2018). Occurrence of microplastics in commercial fish from a natural estuarine environment. *Marine Pollution Bulletin, 128*, 575–584. https://doi.org/10.1016/j.marpolbul.2018.01.044

Blettler, M. C. M., Garello, N., Ginon, L., Abrial, E., Espinola, L. A., & Wantzen, K. M. (2019). Massive plastic pollution in a mega-river of a developing country: Sediment deposition and ingestion by fish (Prochilodus lineatus). *Environmental Pollution, 255*, 113348. https://doi.org/10.1016/j.envpol.2019.113348

Bond, T., Ferrandiz-Mas, V., Felipe-Sotelo, M., & van Sebille, E. (2018). The occurrence and degradation of aquatic plastic litter based on polymer physicochemical properties: A review. *Critical Reviews in Environmental Science and Technology, 48*, 685–722. https://doi.org/10.1080/10643389.2018.1483155

Borja, A., & Elliott, M. (2019). So when will we have enough papers on microplastics and ocean litter? *Marine Pollution Bulletin, 146*, 312–316. https://doi.org/10.1016/j.marpolbul.2019.05.069

Bosker, T., Guaita, L., & Behrens, P. (2018). Microplastic pollution on Caribbean beaches in the Lesser Antilles. *Marine Pollution Bulletin, 133*, 442–447. https://doi.org/10.1016/j.marpolbul.2018.05.060

Bourguiba, S., Genissel, C., Lambard, S., Bouraıma, H., & Carreau, S. (2003). Regulation of aromatase gene expression in Leydig cells and germ cells. *The Journal of steroid biochemistry and molecular biology, 86*(3–5), 335–343.

Browne, M. A., et al. (2011). Accumulation of microplastic on shorelines woldwide: Sources and sinks. *Environmental Science & Technology, 45*, 9175–9179.

Calero, M., et al. (2021). Green strategies for microplastics reduction. *Current Opinion in Green and Sustainable Chemistry, 28*, 100442. https://doi.org/10.1016/j.cogsc.2020.100442

Campbell, S. H., Williamson, P. R., & Hall, B. D. (2017). Microplastics in the gastrointestinal tracts of fish and the water from an urban prairie creek. *FACETS, 2*, 395–409. https://doi.org/10.1139/facets-2017-0008

Cao, Y., Zhao, Q., Jiang, F., Geng, Y., Song, H., Zhang, L., ... & Tian, S. (2024). Interactions between inhalable aged microplastics and lung surfactant: Potential pulmonary health risks. *Environmental Research, 245*, 117803.

Carney Almroth, B. M., et al. (2018). Quantifying shedding of synthetic fbers from textiles; a source of microplastics released into the environment. *Environmental Science and Pollution Research, 25*, 1191–1199. https://doi.org/10.1007/s11356-017-0528-7

Carr, S. A., et al. (2016). Transport and fate of microplastic particles in wastewater treatment plants. *Water Research, 91*, 174–182.

Chen, H., Tang, M., Yang, X., Tsang, Y. F., Wu, Y., Wang, D., & Zhou, Y. (2021). Polyamide 6 microplastics facilitate methane production during anaerobic digestion of waste activated sludge. *Chemical Engineering Journal, 408*, 127251. https://doi.org/10.1016/j.cej.2020.127251

Cheung, P. K., & Fok, L. (2017). Characterisation of plastic microbeads in facial scrubs and their estimated emissions in Mainland China. *Water Research, 122*, 53–61.

Chubarenko, I., Esiukova, E., Zobkov, M., & Isachenko, I. (2022). Microplastics distribution in bottom sediments of the Baltic Sea Proper. *Marine Pollution Bulletin, 179*, 113743. https://doi.org/10.1016/j.marpolbul.2022.113743

Cincinelli, A., Scopetani, C., Chelazzi, D., et al. (2021). Microplastics in the Black Sea sediments. *Science of the Total Environment, 760*, 143898. https://doi.org/10.1016/j.scitotenv.2020.143898

Claessens, M., Meester, S. D., Landuyt, L. V., et al. (2011). Occurrence and distribution of microplastics in marine sediments along the Belgian coast. *Marine Pollution Bulletin, 62*, 2199–2204. https://doi.org/10.1016/j.marpolbul.2011.06.030

Cole, M., et al. (2011). Microplastics as contaminants in the marine environment: A review. *Marine Pollution Bulletin, 62*, 2588–2597.

Crawford, C.B., Quinn, B., 2017. Plastic production, waste and legislation. *Microplastic Pollution, 30*, 39–56.

Cui, Y., Liu, M., Selvam, S., et al. (2022). Microplastics in the surface waters of the South China sea and the western Pacifc Ocean: Diferent size classes refecting various sources and transport. *Chemosphere, 299*, 134456. https://doi.org/10.1016/j.chemosphere.2022.134456

Dai, Z., Zhang, H., Zhou, Q., Tian, Y., Chen, T., Tu, C., Fu, C., & Luo, Y. (2018). Occurrence of microplastics in the water column and sediment in an inland sea affected by intensive anthropogenic activities. *Environmental Pollution, 242*, 1557–1565.

De Falco, F., Di Pace, E., Cocca, M., & Avella, M. (2019). The contribution of washing processes of synthetic clothes to microplastic pollution. *Scientific Reports, 9*, 6633. https://doi.org/10.1038/s41598-019-43023-x

Debnath, R., Prasad, G. S., Amin, A., Malik, M. M., Ahmad, I., Abubakr, A., Borah, S., Rather, M. A., Impellitteri, F., Tabassum, I., & Piccione, G. (2024). Understanding and addressing microplastic pollution: Impacts, mitigation, and future perspectives. *Journal of Contaminant Hydrology, 266*, 104399.

Derraik, J. G. (2002). The pollution of the marine environment by plastic debris: A review. *Marine Pollution Bulletin, 44*, 842–852.

Duan, Z., Duan, X., Zhao, S., Wang, X., Wang, J., Liu, Y., Peng, Y., Gong, Z., & Wang, L. (2020). Barrier function of zebrafish embryonic chorions against microplastics and nanoplastics and its impact on embryo de velopment. *Journal of Hazardous Materials, 395*, 122621. https://doi.org/10.1016/j.jhazmat.2020.122621

Duis, K., & Coors, A. (2016). Microplastics in the aquatic and terrestrial environment: Sources (with a specifc focus on personal care products), fate and efects. *Environmental Sciences Europe, 28*, 1–25. https://doi.org/10.1186/s12302-015-0069-y

Eriksen, M., Mason, S., Wilson, S., Box, C., Zellers, A., Edwards, W., Farley, H., & Amato, S. (2013). Microplastic pollution in the surface waters of the Laurentian Great Lakes. *Marine Pollution Bulletin, 77*(1–2), 177–182.

Eriksen, M., et al. (2014). Plastic pollution in the world's oceans: More than 5 trillion plastic pieces weighing over 250,000 tons afloat at sea. *PLoS One, 9*, e111913.

Farrell, P., & Nelson, K. (2013). Trophic level transfer of microplastic: Mytilus edulis (L.) to Carcinus maenas (L.). *Environmental Pollution, 177*, 1–3. https://doi.org/10.1016/j.envpol.2013.01.046

Faure, F., Saini, C., Potter, G., Galgani, F., De Alencastro, L. F., & Hagmann, P. (2015). An evaluation of surface micro-and mesoplastic pollution in pelagic ecosystems of the Western Mediterranean Sea. *Environmental Science and Pollution Research, 22*, 12190–12197.

Ferdous, S. R., Amin, A., Hasan, J., et al. (2023). Prevalence of microplastics in commonly consumed fish species of the river old Brahmaputra, Bangladesh. *Environmental Science and Pollution Research, 30*, 85639–85654. https://doi.org/10.1007/s11356-023-28514-9

Fernandez-Gonzalez, V., Andrade, J. M., Ferreiro, B., Lopez-Mahía, P., & Muniategui Lorenzo, S. (2021). Monitorization of polyamide microplastics weathering using attenuated total reflectance and microreflectance infrared spectrometry. *Spectrochimica Acta. Part A, Molecular and Biomolecular Spectroscopy, 263*, 120162. https://doi.org/10.1016/j.saa.2021.120162

Filella, M. (2015). Questions of size and numbers in environmental research on microplastics: Methodological and conceptual aspects. *Environment and Chemistry, 12*, 527.

Filgueiras, A. V., Preciado, I., Carton, A., & Gago, J. (2020). Microplastic ingestion by pelagic and benthic fish and diet composition: A case study in the NW Iberian shelf. *Marine Pollution Bulletin, 160*, 111623. https://doi.org/10.1016/j.marpolbul.2020.111623

Fossi, M. C., Pedà, C., Compa, M., et al. (2018). Bioindicators for monitoring marine litter ingestion and its impacts on Mediterranean biodiversity. *Environmental Pollution, 237*, 1023–1040. https://doi.org/10.1016/j.envpol.2017.11.019

Galloway, T. S., & Lewis, C. N. (2016). Marine microplastics spell big problems for future generations. *Proceedings. National Academy of Sciences. United States of America, 113*, 2331–2333.

Gangadoo, S., Owen, S., Rajapaksha, P., Plaisted, K., Cheeseman, S., Haddara, H., Truong, V. K., Ngo, S. T., Vu, V. V., Cozzolino, D., Elbourne, A., Crawford, R., Latham, K., & Chapman, J. (2020). Nano-plastics and their analytical characterisation and fate in the marine environment: From source to sea. *Science of the Total Environment, 732*, 138792. https://doi.org/10.1016/j.scitotenv.2020.138792

Gewert, B., Plassmann, M. M., & MacLeod, M. (2015). Pathways for degradation of plastic polymers floating in the marine environment. *Environmental Science. Processes & Impacts, 17*, 1513–1521. https://doi.org/10.1039/C5EM00207A

Good, T. P., et al. (2010). Derelict fishing nets in Puget Sound and the Northwest Straits: Patterns and threats to marine fauna. *Marine Pollution Bulletin, 60*, 39–50. https://doi.org/10.1016/j.marpolbul.2009.09.005

Goswami, P., Vinithkumar, N. V., & Dharani, G. (2021). Microplastics particles in seafoor sediments along the Arabian Sea and the Andaman Sea continental shelves: First insight on the occurrence, identifcation, and characterization. *Marine Pollution Bulletin, 167*, 112311. https://doi.org/10.1016/j.marpolbul.2021.112311

Guerranti, C., et al. (2019). Microplastics in cosmetics: Environmental issues and needs for global bans. *Environmental Toxicology and Pharmacology, 68*, 75–79. https://doi.org/10.1016/j.etap.2019.03.007

Guo, X., Hu, G., Fan, X., & Jia, H. (2020). Sorption properties of cadmium on microplastics: The common practice experiment and a two-dimensional correlation spectroscopic study. *Ecotoxicology and Environmental Safety, 190*, 110118. https://doi.org/10.1016/j.ecoenv.2019.110118

Gurjar, U. R., Takar, S., Amin, A., & Martin Xavier, K. A. (2023). Microplastic contamination in aquatic organisms: An ecotoxicological perspective. In *Xenobiotics in aquatic animals: Reproductive and developmental impacts* (pp. 353–367). Springer.

Guven, O., Bach, L., Munk, P., Dinh, K. V., Mariani, P., & Nielsen, T. G. (2018). Microplastic does not magnify the acute effect of PAH pyrene on predatory performance of a tropical fish (Lates calcarifer). *Aquatic Toxicology, 198*, 287–293.

Halstead, J. E., Smith, J. A., Carter, E. A., Lay, P. A., & Johnston, E. L. (2018). Assessment tools for microplastics and natural fibres ingested by fish in an urbanised estuary. *Environmental Pollution, 234*, 552–561. https://doi.org/10.1016/j.envpol.2017.11.085

Hamed, M., Soliman, H. A., Badrey, A. E., & Osman, A. G. (2021). Microplastics induced histopathological lesions in some tissues of tilapia (Oreochromis niloticus) early juveniles. *Tissue and Cell, 71*, 101512.

Hidalgo-Ruz, V., Gutow, L., Thompson, R. C., & Thiel, M. (2012). Microplastics in the marine environment: a review of the methods used for identification and quantification. *Environmental science & technology, 46*(6), 3060–3075.

Horton, A. A., Walton, A., Spurgeon, D. J., Lahive, E., & Svendsen, C. (2017). Microplastics in freshwater and terrestrial environments: Evaluating the current understanding to identify the knowledge gaps and future research priorities. *The Science of the Total Environment, 586*, 127–141. https://doi.org/10.1016/j.scitotenv.2017.01.190

Horton, A. A., Weerasinghe, K. D. I., Mayor, D. J., & Lampitt, R. (2024). Microplastics in commercial marine fish species in the UK – A case study in the River Thames and the River Stour (East Anglia) estuaries. *Science of the Total Environment, 915*, 170170. https://doi.org/10.1016/j.scitotenv.2024.170170

Hossain, M. A., & Olden, J. D. (2022). Global meta-analysis reveals diverse effects of microplastics on freshwater and marine fishes. *Fish and Fisheries, 23*(6), 1439–1454. https://doi.org/10.1111/faf/12701

Hurt, R., O'Reilly, C. M., & Perry, W. L. (2020). Microplastic prevalence in two fish species in two U.S. reservoirs. *Limnology and Oceanography Letters, 5*, 147–153. https://doi.org/10.1002/lol2.10140

Islam, M. S., Lee, Z., Shaleh, A., & Soo, H. S. (2024). The United Nations Environment Assembly resolution to end plastic pollution: Challenges to effective policy interventions. *Environment, Development and Sustainability, 26*, 10927–10944.

Jabeen, K., Su, L., Li, J., Yang, D., Tong, C., Mu, J., & Shi, H. (2017). MPs and mesoplastics in fish from coastal and fresh waters of China. *Environmental Pollution, 221*, 141–149.

Jahnke, A., Arp, H. P. H., Escher, B. I., Gewert, B., Gorokhova, E., Kühnel, D., Ogonowski, M., Potthoff, A., Rummel, C., Schmitt-Jansen, M., & Toorman, E. (2017). Reducing uncertainty and confronting ignorance about the possible impacts of weathering plastic in the marine environment. *Environmental Science & Technology Letters, 4*(3), 85–90.

Kabir, A. E., Michon, E., Mingelbier, M., Robert, D., Soubaneh, Y. D., Xie, H., & Lu, Z. (2025). Microplastics in the benthic fish from the Canadian St. Lawrence River and Estuary: Occurrence, spatial distribution and ecological risk assessment. *Marine Pollution Bulletin, 212*, 117509.

Kanhai, L. D. (2019). *Microplastic abundance, distribution and composition in the Atlantic and Arctic oceans*. University of Plymouth.

Karbalaei, S., et al. (2019). Abundance and characteristics of microplastics in commercial marine fish from Malaysia. *Marine Pollution Bulletin, 148*, 5–15. https://doi.org/10.1016/j.marpolbul.2019.07.072

Kasamesiri, P., & Thaimuangphol, W. (2020). Microplastics ingestion by freshwater fish in the Chi River Thailand. *International Journal of Geomate, 18*, 114–119. https://doi.org/10.21660/2020.67.9110

Keerthika, K., Padmavathy, P., Rani, V., et al. (2023). Microplastics accumulation in pelagic and benthic species along the Thoothukudi Coast, South Tamil Nadu, India. *Marine Pollution Bulletin, 189*, 114735. https://doi.org/10.1016/j.marpolbul.2023.114735

Khan, F. R., Shashoua, Y., Crawford, A., Drury, A., Sheppard, K., Stewart, K., & Sculthorp, T. (2020). "The Plastic Nile": First evidence of microplastic contamination in fish from the Nile River (Cairo, Egypt). *Toxics, 8*, 22. https://doi.org/10.3390/toxics8020022

Kibria, M. G., Masuk, N. I., Safayet, R., Nguyen, H. Q., & Mourshed, M. (2023). Plastic waste: Challenges and opportunities to mitigate pollution and effective management. *International Journal of Environmental Research, 17*, 20. https://doi.org/10.1007/s41742-023-00507-z

Kooi, M., et al. (2017). Ups and downs in the ocean: Effects of biofouling on vertical transport of microplastics. *Environmental Science & Technology, 51*, 7963–7971.

Koongolla, J. B., Lin, L., Pan, Y. F., Yang, C. P., Sun, D. R., Liu, S., Xu, X. R., Maharana, D., Huang, J. S., & Li, H. X. (2020). Occurrence of microplastics in gastrointestinal tracts and gills of fish from Beibu Gulf, South China Sea. *Environmental Pollution, 258*, 113734.

Krüger, L., Casado-Coy, N., Valle, C., Ramos, M., Sánchez-Jerez, P., Gago, J., et al. (2020). Plastic debris accumulation in the seabed derived from coastal fish farming. *Environmental Pollution, 257*, 113336.

Kuczenski, B., Vargas Poulsen, C., Gilman, E. L., Musyl, M., Geyer, R., & Wilson, J. (2022). Plastic gear loss estimates from remote observation of industrial fishing activity. *Fish and Fisheries, 23*(1), 22–33.

Kumar, V. E., et al. (2018). Occurrence of microplastics in fishes from two landing sites in Tuticorin, South east coast of India. *Marine Pollution Bulletin, 135*, 889–894.

Kumar, M., Chen, H., Sarsaiya, S., Qin, S., Liu, H., Awasthi, M. K., Kumar, S., Singh, L., Zhang, Z., Bolan, N. S., Pandey, A., Varjani, S., & Taherzadeh, M. J. (2021). Current research trends on micro- and nano-plastics as an emerging threat to global environment: A review. *Journal of Hazardous Materials, 409*, 124967. https://doi.org/10.1016/j.jhazmat.2020.124967

Lebreton, L. C. M., Van Der Zwet, J., Damsteeg, J. W., Slat, B., Andrady, A., & Reisser, J. (2017). River plastic emissions to the world's oceans. *Nature Communications, 8*, 15611. https://doi.org/10.1038/ncomms15611

Li, W. C. (2018). The occurrence fate and efects of microplastics in the marine environment. In *Microplastic contamination in aquatic environments* (pp. 133–173). Elsevier.

Li, W. C., Tse, H. F., & Fok, L. (2016). Plastic waste in the marine environment: A review of sources, occurrence and effects. *Science of the Total Environment, 566–567*, 333–349. https://doi.org/10.1016/j.scitotenv.2016.05.084

Li, W., Bai, Z., Zhang, T., Jia, Y., Hou, Y., Chen, J., Guo, Z., Kong, L., Bai, J., & Li, W. (2023). Comparative study on pyrolysis behaviors and chlorine release of pure PVC polymer and commercial PVC plastics. *Fuel, 340*, 127555. https://doi.org/10.1016/j.fuel.2023.127555

Lu, J., Lu, Q., Di, L., Zhou, Y., & Zhou, Y. (2023). Iron-based biochar as efficient persulfate activation catalyst for emerging pollutants removal: A review. *Chinese Chemical Letters, 34*, 108357. https://doi.org/10.1016/j.cclet.2023.108357

Luís, L. G., Ferreira, P., Fonte, E., Oliveira, M., & Guilhermino, L. (2015). Does the presence of microplastics influence the acute toxicity of chromium (VI) to early juveniles of the common goby (Pomatoschistus microps)? A study with juveniles from two wild estuarine populations. *Aquatic Toxicology, 164*, 163–174.

Macfadyen, G., Huntington, T., & Cappell, R. (2009). *Abandoned, lost or otherwise discarded fishing gear* (p. 115). United Nations Environment Programme: Food and Agriculture Organization of the United Nations.

Magnusson, K. (2016). *Microlitter in sewage treatment systems: A Nordic perspective on waste water treatment plants as pathways for microscopic anthropogenic particles to marine systems*. Nordic Council of Ministers.

Mahamud, A. S. U., Anu, M. S., Baroi, A., Datta, A., Khan, M. S. U., Rahman, M., Tabassum, T., Tanwi, J. T., & Rahman, T. (2022). Microplastics in fishmeal: A threatening issue for sustainable aquaculture and human health. *Aquaculture Reports, 25*, 101205.

Malik, M., Balkhi, M. H., & Abubakr, A. (2017). Comparative study on the contamination level of Dal Lake, Kashmir based on Total Viable Bacterial Counts over a period of time. *Ecology, Environment and Conservation, 23*(4), 2145–2147.

Mallik, A., Chakraborty, P., Ramteke, K. K., Bhavan, S. G., Jaiswar, A. K., Nayak, B. B., ... & Bhushan, S. (2024). Diet characteristics of tidal creek-associated fishes of the northeastern Arabian Sea with special reference to microplastic ingestion. *Chemosphere, 363*, 142886.

Malinich, T. D., Chou, N., Sepúlveda, M. S., & Höök, T. O. (2018). No evidence of microplastic impacts on consumption or growth of larval Pimephales promelas. *Environmental Toxicology and Chemistry, 37*, 2912–2918. https://doi.org/10.1002/etc.4257

Manbohi, A., Mehdinia, A., Rahnama, R., et al. (2021). Spatial distribution of microplastics in sandy beach and inshore-offshore sediments of the southern Caspian Sea. *Marine Pollution Bulletin, 169*, 112578. https://doi.org/10.1016/j.marpolbul.2021.112578

Matluba, M., Ahmed, M. K., Chowdhury, K. A., Khan, N., Ashiq, M. A. R., & Islam, M. S. (2023). The pervasiveness of microplastic contamination in the gastrointestinal tract of fish from the western coast of Bangladesh. *Marine Pollution Bulletin, 193*, 115145.

Mazurais, D., Ernande, B., Quazuguel, P., Severe, A., Huelvan, C., Madec, L., Mouchel, O., Soudant, P., Robbens, J., Huvet, A., & Zambonino Infante, J. (2015). Evaluation of the impact of polyethylene mi crobeads ingestion in European sea bass (Dicentrarchus labrax) larvae. *Marine Environmental Research, 112*, 78–85. https://doi.org/10.1016/j.marenvres.2015.09.009

McCormick, A., Hoellein, T. J., Mason, S. A., Schluep, J., & Kelly, J. J. (2014). Microplastic is an abundant and distinct microbial habitat in an urban river. *Environmental Science & Technology, 48*, 11863–11871.

McGoran, A. R., Clark, P. F., & Morritt, D. (2017). Presence of microplastic in the digestive tracts of European flounder, Platichthys flesus, and European smelt, Osmerus eperlanus, from the River Thames. *Environmental Pollution, 220*, 744–751. https://doi.org/10.1016/j.envpol.2016.09.078

McNeish, R. E., Kim, L. H., Barrett, H. A., Mason, S. A., Kelly, J. J., & Hoellein, T. J. (2018). Microplastic in riverine fish is connected to species traits. *Scientific Reports, 2018*(8), 11639. https://doi.org/10.1038/s41598-018-29980-9

Mijangos, C., Calafel, I., & Santamaría, A. (2023). Poly(vinyl chloride), a historical polymer still evolving. *Polymer, 266*, 125610. https://doi.org/10.1016/j.polymer.2022.125610

Murphy, F., Ewins, C., Carbonnier, F., & Quinn, B. (2016). Wastewater treatment works (WwTW) as a source of microplastics in the aquatic environment. *Environmental science & technology, 50*(11), 5800–5808.

Nabi, N., Ahmad, I., Amin, A., Rather, M. A., Ahmed, I., Hajam, Y. A., Khursheed, S., Malik, M. M., & Abubakr, A. (2024). Understanding the sources, fate and effects of microplastics in aquatic environments with a focus on risk profiling in aquaculture systems. *Reviews in Aquaculture, 16*(4), 1947–1980.

Naidoo, T., et al. (2016). Plastic ingestion by estuarine mullet Mugil cephalus (Mugilidae) in an urban harbour, KwaZulu-Natal, South Africa. *African Journal of Marine Science, 38*, 145–149.

Naidoo, T., Goordiyal, K., & Glassom, D. (2017). Are nitric acid (HNO3) digestions efficient in isolating microplastics from juvenile fish? *Water, Air, and Soil Pollution, 228*, 1–11. https://doi.org/10.1007/s11270-017-3654-4

Naidoo, T., & Glassom, D. (2019). Decreased growth and survival in small juvenile fish, after chronic exposure to environmentally relevant concentrations of microplastic. *Marine Pollution Bulletin, 145*, 254–259.

Naidu, B. C., Xavier, K. M., Shukla, S. P., Jaiswar, A. K., & Nayak, B. B. (2022). Comparative study on the microplastics abundance, characteristics, and possible sources in yellow clams of different demographic regions of the northwest coast of India. *Journal of Hazardous Materials Letters, 3*, 100051.

Naji, A., et al. (2017). Plastic debris and microplastics along the beaches of the Strait of Hormuz, Persian Gulf. *Marine Pollution Bulletin, 114*, 1057–1062. https://doi.org/10.1016/j.marpolbul.2016.11.032

Obbard, R. W., et al. (2014). Global warming releases microplastic legacy frozen in Arctic Sea ice. *Earth's Future, 2*, 315–320.

Oliveira, M., Ribeiro, A., Hylland, K., & Guilhermino, L. (2013). Single and combined effects of microplastics and pyrene on juveniles (0+ group) of the common goby Pomatoschistus microps (Teleostei, Gobiidae). *Ecological indicators, 34*, 641–647.

Pagano, M., Stara, A., Aliko, V., & Faggio, C. (2020). Impact of neonicotinoids to aquatic invertebrates—In vitro studies on Mytilus galloprovincialis: A review. *Journal of Marine Science and Engineering, 8*(801), 801–814. https://doi.org/10.3390/jmse8100801

Park, E.-J., et al. (2020). Repeated-oral dose toxicity of polyethylene microplastics and the possible implications on reproduction and development of the next generation. *Toxicology Letters, 324*, 75–85. https://doi.org/10.1016/j.toxlet.2020.01.008

Pazos, R. S., Maiztegui, T., Colautti, D. C., Paracampo, A. H., & Gómez, N. (2017). Microplastics in gut contents of coastal freshwater fish from Río de la Plata estuary. *Marine Pollution Bulletin, 122*, 85–90. https://doi.org/10.1016/j.marpolbul.2017.06.007

Pegado, T., et al. (2018). First evidence of microplastic ingestion by fishes from the Amazon River estuary. *Marine Pollution Bulletin, 133*, 814–821.

Pellini, G., Gomiero, A., Fortibuoni, T., Ferrà, C., Grati, F., Tassetti, A. N., et al. (2018). Characterization of microplastic litter in the gastrointestinal tract of Solea solea from the Adriatic Sea. *Environmental Pollution, 234*, 943–952. https://doi.org/10.1016/j.envpol.2017.12.038

Pojar, I., Stănică, A., Stock, F., et al. (2021). Sedimentary microplastic concentrations from the Romanian Danube River to the Black Sea. *Scientific Reports, 11*, 2000. https://doi.org/10.1038/s41598-021-81724-4

Prasad, G. S., Rout, S. K., Malik, M. M., Karmakar, S., Amin, A., & Ahmad, I. (2023). Occurrence of xenoestrogen alkylphenols (Octylphenols and Nonylphenol) and its impact on the aquatic ecosystem. In *Xenobiotics in aquatic animals: Reproductive and developmental impacts* (pp. 275–284). Springer.

Prasad, G. S., Rout, S. K., Amin, A., Malik, M. M., Ahmad, I., Rather, M. A., Karmakar, S., Khalil, H. S., & Abubakar, A. (2024). Impact of COVID-19 on the aquatic environment and implications for fisheries with special emphasis to Indian perspective. *Proceedings of the Indian National Science Academy, 90*(4), 821–839.

Provencher, J. F., Ammendolia, J., Rochman, C. M., & Mallory, M. L. (2019). Assessing plastic debris in aquatic food webs: What we know and don't know about uptake and trophic transfer. *Environmental Reviews, 27*, 304–317. https://doi.org/10.1139/er-2018-0079

Rabot, C., Chen, Y., Lin, S.-Y., Miller, B., Chiang, Y.-M., Oakley, C. E., Oakley, B. R., Wang, C. C. C., & Williams, T. J. (2023). Polystyrene upcycling into fungal natural products and a biocontrol agent. *Journal of the American Chemical Society, 145*, 5222–5230. https://doi.org/10.1021/jacs.2c12285

Rangel-Buitrago, N., Arroyo-Olarte, H., Trilleras, J., et al. (2021). Microplastics pollution on Colombian Central Caribbean beaches. *Marine Pollution Bulletin, 170*, 112685. https://doi.org/10.1016/j.marpolbul.2021.112685

Reisser, J., Shaw, J., Hallegraeff, G., Proietti, M., Barnes, D. K., Thums, M., Wilcox, C., Hardesty, B. D., & Pattiaratchi, C. (2014). Millimeter-sized marine plastics: A new pelagic habitat for microorganisms and invertebrates. *PLoS One, 9*, e100289.

Retama, I., et al. (2016). Microplastics in tourist beaches of Huatulco Bay, Pacific coast of southern Mexico. *Marine Pollution Bulletin, 113*, 530–535.

Roch, S., Walter, T., Ittner, L. D., Friedrich, C., & Brinker, A. (2019). A systematic study of the microplastic burden in freshwater fishes of south-western Germany-Are we searching at the right scale?. *Science of the total environment, 689*, 1001–1011.

Rochman, C. M., Kurobe, T., Flores, I., & Teh, S. J. (2014). Early warn ing signs of endocrine disruption in adult fish from the ingestion of polyethylene with and without sorbed chemical pollutants from the marine environment. *Science of the Total Environment, 493*, 656–661. https://doi.org/10.1016/j.scitotenv.2014.06.051

Rochman, C. M., Tahir, A., Williams, S. L., Baxa, D. V., Lam, R., Miller, J. T., Teh, F. C., Werorilangi, S., & Teh, S. J. (2015). Anthropogenic debris in seafood: Plastic debris and fibers from textiles in fish and bivalves sold for human consumption. *Scientific Reports, 5*, 14340. https://doi.org/10.1038/srep14340

Roda, J. F. B., Lauer, M. M., Risso, W. E., dos Reis, B., & Martinez, C. (2020). Microplastics and copper effects on the neotropical te leost Prochilodus lineatus: Is there any interaction? *Comparative Biochemistry and Physiology Part A: Molecular & Integrative Physiology, 242*, 110659. https://doi.org/10.1016/j.cbpa.2020.110659

Rummel, C. D., Jahnke, A., Gorokhova, E., Kühnel, D., & Schmitt-Jansen, M. (2017). Impacts of biofilm formation on the fate and potential effects of microplastic in the aquatic environment. *Environmental science & technology letters, 4*(7), 258–267.

Ryan, M. G., Watkins, L., & Walter, M. T. (2019). Hudson River juvenile blueback herring avoid ingesting microplastics. *Marine Pollution Bulletin, 146*, 935–939. https://doi.org/10.1016/j.marpolbul.2019.07.004

Sarda, P., Hanan, J. C., Lawrence, J. G., & Allahkarami, M. (2022). Sustainability performance of polyethylene terephthalate, clarifying challenges and opportunities. *Journal of Polymer Science, 60*, 7–31. https://doi.org/10.1002/pol.20210495

Sharifinia, M., Bahmanbeigloo, Z. A., Keshavarzifard, M., Khanjani, M. H., & Lyons, B. P. (2020). Microplastic pollution as a grand challenge in marine research: A closer look at their adverse impacts on the immune and reproductive systems. *Ecotoxicology and Environmental Safety, 204*, 111109.

Sharma, S., & Chatterjee, S. (2017). Microplastic pollution, a threat to marine ecosystem and human health: A short review. *Environmental Science and Pollution Research, 24*, 21530–21547.

Shaw, D. G., & Day, R. H. (1994). Colour- and form-dependent loss of plastic micro-debris from the North Pacific Ocean. *Marine Pollution Bulletin, 28*, 39–43.

Sheavly, S. B., & Register, K. M. (2007). Marine debris & plastics: Environmental concerns, sources, impacts and solutions. *Journal of Polymers and the Environment, 15*, 301–305.

Shim, W. J., & Thompson, R. C. (2015). Microplastics in the ocean. *Archives of Environmental Contamination and Toxicology, 69*, 265–268. https://doi.org/10.1007/s00244-015-0216-x

Sruthisree, C., Gowda, G., Nayak, H., Harish, B. D., & Amin, A. (2015). Effect of heavy metal (lead) on oxygen consumption and ammonia excretion of iridescent shark (*Pangusius hypohthalmus*). *Journal of Experimental Zoology India, 18*(2), 1–4.

Sutton, R., Mason, S. A., Stanek, S. K., Willis-Norton, E., Wren, I. F., & Box, C. (2016). Microplastic contamination in the San Francisco Bay, California, USA. *Marine Pollution Bulletin, 109*, 230–235. https://doi.org/10.1016/j.marpolbul.2016.05.077

Tabassum, I., Hassan, N., Amin, A., Malik, M. M., Abubakr, A., Prasad, G. S., Latief, F., & Rather, S. A. (2025). COVID-19 and global environmental change: A threat to food security and nutrition. In *Food security, nutrition and sustainability through aquaculture technologies* (pp. 437–445). Springer.

Talley, T. S., Venuti, N., & Whelan, R. (2020). Natural history matters: Plastics in estuarine fish and sediments at the mouth of an urban watershed. *PLoS One, 15*, e0229777. https://doi.org/10.1371/journal.pone.0229777

Talvitie, J., et al. (2017). Solutions to microplastic pollution – Removal of microplastics from wastewater efuent with advanced wastewater treatment technologies. *Water Research, 123*, 401–407. https://doi.org/10.1016/j.watres.2017.07.005

Tanaka, K., & Takada, H. (2016). Microplastic fragments and microbeads in digestive tracts of planktivorous fish from urban coastal waters. *Scientific Reports, 6*, 34351.

The Organization for Economic Cooperation and Development (OECD). (2024). *Global plastics outlook: Plastic waste by end-of-life fate and region – Projections*. OECD Environment Statistics (database). https://doi.org/10.1787/e4e8c086-en

Tran, T. V., Jalil, A. A., Nguyen, T. M., Nguyen, T. T. T., Nabgan, W., & Nguyen, D. T. C. (2023). A review on the occurrence, analytical methods, and impact of microplastics in the environment. *Environmental Toxicology and Pharmacology, 102*, 104248. https://doi.org/10.1016/j.etap.2023.104248

Tursi, A., Baratta, M., Easton, T., Chatzisymeon, E., Chidichimo, F., Biase, M. D., & Filpo, G. D. (2022). Microplastics in aquatic systems, a comprehensive review: Origination, accumulation, impact, and removal technologies. *RSC Advances, 12*, 28318–28340. https://doi.org/10.1039/D2RA04713F

Valente, T., Sbrana, A., Scacco, U., Jacomini, C., Bianchi, J., Palazzo, L., de Lucia, G. A., Silvestri, C., & Matiddi, M. (2019). Exploring microplastic ingestion by three deep-water elasmobranch species: A case study from the Tyrrhenian Sea. *Environmental Pollution, 253*, 342–350. https://doi.org/10.1016/j.envpol.2019.07.001

Valente, T., Pelamatti, T., Giacomo, C., Camedda, A., Letizia, M., Andrea, G., Lucia, D., Jacomini, C., Piermarini, R., Regoli, F., Sbrana, A., Ventura, D., Silvestri, C., & Matiddi, M. (2022). One is not enough: Monitoring microplastic ingestion by fish needs a multispecies approach. *Marine Pollution Bulletin, 184*, 114133. https://doi.org/10.1016/j.marpolbul.2022.114133

Vibhatabandhu, P., & Srithongouthai, S. (2024). Abundance and characteristics of microplastics contaminating the surface water of the Inner Gulf of Thailand. *Research Square*. https://doi.org/10.21203/rs.3.rs-400647/v1

Walkinshaw, C., Lindeque, P. K., Thompson, R., Tolhurst, T., & Cole, M. (2020). Microplastics and seafood: Lower trophic organisms at highest risk of contamination. *Ecotoxicology and Environmental Safety, 190*, 110066. https://doi.org/10.1016/j.ecoenv.2019.110066

Wang, W., et al. (2017). Microplastics pollution in inland freshwaters of China: A case study in urban surface waters of Wuhan, China. *Science of the Total Environment, 575*, 1369–1374.

Wang, W., Yuan, W., Chen, Y., & Wang, J. (2018). Microplastics in surface waters of dongting lake and hong lake, China. *Science of the Total Environment, 633*, 539–545.

Wright, S. L., & Kelly, F. J. (2017). Plastic and human health: A micro issue? *Environmental Science & Technology, 51*, 6634–6647. https://doi.org/10.1021/acs.est.7b00423

Wright, S. L., Thompson, R. C., & Galloway, T. S. (2013). The physical impacts of microplastics on marine organisms: A review. *Environmental Pollution, 178*, 483–492.

Wu, F., Wang, Y., Leung, J. Y. S., Huang, W., Zeng, J., Tang, Y., Chen, J., Shi, A., Yu, X., Xu, X., Zhang, H., & Cao, L. (2020). Accumulation of microplastics in typical commercial aquatic species: A case study at a productive aquaculture site in China. *Science of the Total Environment, 708*, 135432. https://doi.org/10.1016/j.scitotenv.2019.135432

Xu, K., Zhang, Y., Huang, Y., & Wang, J. (2021). Toxicological effects of microplastics and phenanthrene to zebrafish (Danio rerio). *Science of the Total Environment, 757*, 143730.

Yang, Z., et al. (2021). Is incineration the terminator of plastics and microplastics? *Journal of Hazardous Materials, 401*, 123429. https://doi.org/10.1016/j.jhazmat.2020.123429

Yaranal, N. A., Subbiah, S., & Mohanty, K. (2021). Distribution and characterization of microplastics in beach sediments from Karnataka (India) coastal environments. *Marine Pollution Bulletin, 169*, 112550. https://doi.org/10.1016/j.marpolbul.2021.112550

Yoshida, S., Hiraga, K., Takehana, T., Taniguchi, I., Yamaji, H., Maeda, Y., Toyohara, K., Miyamoto, K., Kimura, Y., & Oda, K. (2016). A bacterium that degrades and assimilates poly ethylene terephthalate. *Science, 351*, 1196–1199.

Yu, X., Peng, J., Wang, J., et al. (2016). Occurrence of microplastics in the beach sand of the Chinese inner sea: The Bohai Sea. *Environmental Pollution, 214*, 722–730. https://doi.org/10.1016/j.envpol.2016.04.080

Yuan, W., Liu, X., Wang, W., Di, M., & Wang, J. (2019). Microplastic abundance, distribution and composition in water, sediments, and wild fish from Poyang Lake, China. *Ecotoxicology and Environmental Safety, 170*, 180–187. https://doi.org/10.1016/j.ecoenv.2018.11.126

Zhang, K., et al. (2017). Occurrence and characteristics of microplastic pollution in Xiangxi Bay of three Gorges reservoir, China. *Environmental Science & Technology, 51*, 3794–3801.

Zhang, F., Wang, X., Xu, J., Zhu, L., Peng, G., Xu, P., & Li, D. (2019). Foodweb transfer of microplastics between wild caught fish and crustaceans in East China Sea. *Marine Pollution Bulletin, 146*, 173–182. https://doi.org/10.1016/j.marpolbul.2019.05.061

Zhang, M., Lin, Y., Booth, A. M., et al. (2022a). Fate, source and mass budget of sedimentary microplastics in the Bohai Sea and the Yellow Sea. *Environmental Pollution, 294*, 118640. https://doi.org/10.1016/j.envpol.2021.118640

Zhang, Q., He, Y., Cheng, R., Li, Q., Qian, Z., & Lin, X. (2022b). Recent advances in toxicological research and potential health impact of microplastics and nanoplastics in vivo. *Environmental Science and Pollution Research, 29*, 40415–40448. https://doi.org/10.1007/s11356-022-19745-3

Zhang, K., Zheng, S., Liang, J., et al. (2023). Microplastic load of benthic fauna in Jiaozhou Bay, China. *Environmental Pollution, 320*, 121073. https://doi.org/10.1016/j.envpol.2023.121073

Zhao, H., Li, P., He, X., & Ning, J. (2023). Microplastics pollution and risk assessment in selected surface waters of the Wei River Plain, China. *Exposure and Health, 15*, 745–755. https://doi.org/10.1007/s12403-022-00522-z

Zobkov, M., & Esiukova, E. (2017). Microplastics in Baltic bottom sediments: Quantifcation procedures and frst results. *Marine Pollution Bulletin, 114*, 724–732. https://doi.org/10.1016/j.marpolbul.2016.10.060

Impact of Microplastics and Nanomaterials on Fish Health: Oxidative Stress and Antioxidant Defense Mechanisms

Muhammad Mudassar Shahzad, Aasia Karim, Fatima Yasin, Fariha Latif, Muhammad Umer, Waseem Abbas, Muhammad Sabtain Khan, and Muhammad Usman

1 Introduction

Microplastics and nanoscale plastic particles are not only contaminating our water bodies particularly rivers and seas but also invading the bodies of fish and causing damage (Ahmed et al., 2023). The particles often found in the fish originate from various consumer and technical products because the negative effects of the plastic particles on fish are multifaceted and can range from cell and tissue damage to severe health effects. Fish respond to this damaging effect with protective mechanisms, but when there is too much plastic particles, the protective mechanisms are overloaded. This leads to an exacerbation of tissue damage and disease development in the fish therefore it is essential to address this issue on top priority basis. We need to find ways to prevent these tiny pollutants from entering our waters, so we need to clean them up and protect the fish (Akarsu et al., 2020).

Microplastics and nanomaterials have emerged as pervasive pollutants in aquatic ecosystems, posing significant threats to fish health and the broader ecological balance. These synthetic particles, ranging in size from nanometres to millimetres, originate from a variety of sources and have infiltrated water bodies, sediments, and aquatic organisms. Their presence in the environment is a direct consequence of human activity, and their impacts on fish physiology and behaviour are becoming

M. M. Shahzad (✉) · F. Yasin · M. Umer · W. Abbas · M. S. Khan · M. Usman
Department of Zoology, Division of Science and Technology, University of Education, Township, Lahore, Pakistan

A. Karim
Department of Zoology, Sardar Bahadur Khan Women's University, Quetta, Pakistan

F. Latif
Institute of Zoology, Bahauddin Zakariya University, Multan, Pakistan

A. Abubakr et al. (eds.), *Microplastics: Ecological and Food Security Implications*,
https://doi.org/10.1007/978-3-031-94480-2_10

increasingly evident through scientific research (Sruthisree et al., 2015; Amin et al., 2016; Amin et al., 2017; Malik et al., 2017; Alomar et al., 2021).

Microplastics, defined as plastic particles smaller than 5 millimetres, and nanomaterials, which are less than 100 nanometres in size, enter aquatic ecosystems through both primary and secondary sources. Primary sources include microbeads intentionally manufactured for use in cosmetics and personal care products, such as exfoliating scrubs and toothpaste, which are washed into wastewater systems and eventually discharged into natural water bodies. Industrial processes also contribute significantly, as nanomaterials are widely used in paints, coatings, and textiles, with particles being released during production, application, and wear. In agriculture, fertilizers and pesticides often contain microplastics as carriers, while irrigation systems can shed plastic particles into the environment. Secondary sources arise from the degradation of larger plastic items, such as bags, bottles, and fishing nets, which break down into smaller fragments due to UV radiation, mechanical abrasion, and microbial activity. Additionally, synthetic fibres from textiles, such as polyester and nylon, are released during washing and enter aquatic systems through wastewater (De Falco et al., 2019; Nabi et al., 2024; Debnath et al., 2024).

Once introduced into aquatic environments, microplastics and nanomaterials interact with fish through ingestion or absorption. Fish may ingest these particles directly, mistaking them for food, or indirectly through contaminated prey. Nanomaterials, due to their small size, can also be absorbed through the gills or skin. The fate of these particles within fish depends on several factors, including their size, shape, and chemical properties. Smaller particles, particularly nanomaterials, are more easily absorbed and can penetrate deeper into tissues, while larger microplastics tend to accumulate in the gastrointestinal tract. The chemical composition of these particles also plays a critical role; hydrophobic particles, for example, adhere to biological membranes, while surface charges influence their interactions with cells. The size and species of the fish further determine the extent of exposure, with larger fish often ingesting more microplastics due to their higher food intake, while smaller fish are more susceptible to the effects of nanomaterials due to their physiological characteristics (Bhuyan, 2022).

The presence of microplastics and nanomaterials in fish tissues triggers a cascade of physiological disruptions, primarily through the induction of oxidative stress. Nanomaterials, with their high surface area and reactivity, are particularly potent in generating reactive oxygen species (ROS), which are chemically reactive molecules that can damage cellular components. Microplastics, on the other hand, often act as carriers for toxic substances such as heavy metals and persistent organic pollutants, which amplify their harmful effects. Once inside the fish, these particles generate ROS that attack DNA, causing mutations and strand breaks, as well as proteins, leading to misfolding and loss of function. Lipids in cell membranes are also targeted, resulting in compromised membrane integrity and cellular dysfunction. The accumulation of ROS overwhelms the fish's natural antioxidant defences, leading to oxidative stress, which in turn triggers an inflammatory response. Damaged cells release pro-inflammatory cytokines, activating the immune system. While this response is initially protective, chronic inflammation caused by prolonged exposure to these pollutants can result in tissue damage and impaired organ function (Alberghini et al., 2022).

To counteract oxidative stress, fish rely on a complex system of antioxidant defences. These include enzymatic antioxidants such as superoxide dismutase (SOD), catalase (CAT), and glutathione peroxidase (GPx), which work together to neutralize ROS. SOD converts superoxide radicals into hydrogen peroxide, which is then broken down into water and oxygen by catalase. Glutathione peroxidase reduces hydrogen peroxide and lipid peroxides, preventing further damage to cellular components. Non-enzymatic antioxidants, such as glutathione, vitamin C, and vitamin E, also play a critical role in scavenging free radicals and protecting cells from oxidative damage (Prasad et al., 2023; Wang et al., 2024). At the molecular level, oxidative stress-responsive genes such as Nrf2 (nuclear factor erythroid 2-related factor 2) and Keap1 (Kelch-like ECH-associated protein 1) regulate the expression of antioxidant proteins, enhancing the fish's ability to respond to oxidative stress. However, the continuous exposure to microplastics and nanomaterials can overwhelm these defence mechanisms, leaving fish vulnerable to cellular damage and physiological dysfunction (Sayed et al., 2023).

The impacts of microplastics and nanomaterials on fish health extend beyond oxidative stress and inflammation. Physical damage to tissues, particularly in the gastrointestinal tract, is a common consequence of microplastic ingestion. Sharp or irregularly shaped particles can cause abrasions and blockages, impairing nutrient absorption and digestion. In the liver and muscles, the accumulation of ROS and toxic substances leads to cellular damage and reduced organ function. These physiological effects have broader implications for fish growth and reproduction. Microplastics can disrupt endocrine signalling, altering growth rates and reducing reproductive success. Nanomaterials, due to their ability to penetrate deeper into tissues, can interfere with gamete development and fertilization, further compromising reproductive outcomes. Behavioural changes have also been observed, with fish exposed to these pollutants exhibiting altered feeding patterns, reduced appetite, and impaired predator avoidance. These changes not only affect individual fish but also have cascading effects on population dynamics and ecosystem stability (Alomar et al., 2020).

The scale and complexity of the problem necessitate a multifaceted approach to mitigation. Policy interventions, such as banning microbeads in cosmetics and implementing stricter regulations on plastic waste, are critical first steps. Upgrading wastewater treatment plants to capture microplastics and nanomaterials before they enter natural water systems can significantly reduce pollution. Bioremediation, which involves the use of microorganisms or enzymes to degrade microplastics, offers a promising solution for addressing existing contamination. For fish already affected by these pollutants, antioxidant-based therapies could provide relief. By supplementing fish diets with antioxidants such as vitamin C, vitamin E, or glutathione precursors, it may be possible to enhance their natural defences against oxidative stress and mitigate the physiological damage caused by microplastics and nanomaterials (Mattsson et al., 2017).

To better understand the effects of these pollutants on fish, researchers employ a range of experimental techniques. Controlled exposure studies allow scientists to observe how microplastics and nanomaterials are absorbed, distributed, and excreted in fish. Oxidative stress assays, using fluorescent probes to measure ROS levels or biochemical tests to assess the activity of antioxidant enzymes, provide insights into the cellular response to these pollutants. Histological analysis of tissue samples reveals physical damage and inflammation, while gene expression studies, using techniques such as qPCR or RNA sequencing, shed light on the activation of oxidative stress-responsive genes. These methods are essential for unravelling the complex interactions between microplastics, nanomaterials, and fish physiology, and for developing effective strategies to mitigate their impacts (Sayed et al., 2023).

The pervasive presence of microplastics and nanomaterials in aquatic ecosystems represents a significant challenge for environmental and ecological health. These pollutants, born of human activity, have far-reaching consequences for fish and other aquatic organisms, disrupting physiological processes, impairing growth and reproduction, and altering behaviour (Bhagat et al., 2020). Addressing this issue requires a comprehensive understanding of the sources, mechanisms, and impacts of these pollutants, as well as the development of innovative solutions to reduce their presence in the environment. By combining scientific research, policy interventions, and technological advancements, it may be possible to mitigate the effects of microplastics and nanomaterials and protect the health of aquatic ecosystems for future generations (Wang et al., 2024).

2 Oxidative Stress and Antioxidant Defense Mechanism

Oxidative stress is a condition resulting from an imbalance between the generation and accumulation of reactive oxygen species (ROS) in cells and tissues and the capacity of a biological system to neutralize these reactive products (Pizzino et al., 2017; Gurjar et al., 2023). Superoxide radicals (O2•−), hydrogen peroxide (H_2O_2), hydroxyl radicals (•OH), and singlet oxygen (1O2) are typically classified as reactive oxygen species (ROS), produced as metabolic by-products by biological systems (Navarro-Yepes et al., 2014). Processes such as protein phosphorylation, activation of various transcription factors, apoptosis, immunology, and differentiation rely on appropriate ROS formation and maintenance at low levels within cells (Rajendran et al., 2014). Increased creation of reactive oxygen species (ROS) begins to adversely affect critical cellular components such as proteins, lipids, and nucleic acids (Wu et al., 2013). Reactive oxygen species (ROS) are predominantly generated by mitochondria under both healthy and pathological settings. Specifically, superoxide anion (O2•−) can be produced by cellular respiration, as well as by lipoxygenases (LOX) and cyclooxygenases (COX) during arachidonic acid metabolism, and by endothelial and inflammatory cells (Al-Gubory et al., 2012).

Numerous studies have been undertaken on fish and shellfish (Birnie-Gauvin et al., 2017; Prego-Faraldo et al., 2017; Ji et al., 2018) to assess their antioxidant

Table 1 Enzymatic and non-enzymatic antioxidant defences

Enzymatic antioxidants	Non-enzymatic antioxidants
Superoxide dismutase	Vitamin E
Catalase	Vitamin C
Glutathione peroxidises	Vitamin A
Glutathione reductase	Glutathione
Glutathione-S-transferases	

systems, demonstrating that they have evolved mechanisms to either prevent or mitigate the consequences of oxidative stress (Table 1).

To prevent the detrimental effects of naturally occurring reactive oxygen species (ROS), living organisms have evolved an antioxidant defense system comprising two distinct categories: (1) the enzymatic antioxidant system, which includes various enzymes such as superoxide dismutase (SOD), glutathione peroxidase (GPx), glutathione reductase (GR), and catalase (CAT); and (2) the non-enzymatic antioxidants, which encompass glutathione, thioredoxin, vitamin C, and vitamin E (Mishra et al., 2015). The components of the antioxidant defense system can maintain equilibrium between the generation and elimination of reactive oxygen species under standard physiological settings (Martínez-Álvarez et al., 2005).

Because superoxide is the predominant reactive oxygen species generated by cells, the dismutation of reactive oxygen species by superoxide dismutase is crucial in both fish and shellfish. Superoxide dismutase is prevalent in both the mitochondria (Mn2þ-SOD) and the cytoplasm (Cu2þ-SOD and Zn2þ-SOD) (Cao et al., 2019). The generated H_2O_2 can be neutralized by CAT or GPx, or it may be subsequently released into the cytoplasm (Schieber & Chandel, 2014; Li et al., 2016).

The enzymatic defence comprises a collection of enzymes that neutralise distinct activities of reactive oxygen species (ROS) (Sies, 2015). Catalase, functioning in both mitochondria and peroxisomes, facilitates the decomposition of H_2O_2 into H_2O and O_2, so preserving the equilibrium of reactive oxygen species generation and removal, which is crucial for the operation of innate immunity (Devi et al., 2019). Modulation of catalase is crucial for various activities, including proliferation, differentiation, cell migration, and apoptosis (Wang et al., 2013a, b). Hydrogen peroxide can be neutralized by GPx in the presence of glutathione (GSH), either within the mitochondria or the cytoplasm. Glutathione peroxidase is crucial in preventing lipid peroxidation and protecting cell membranes from harm (Monteiro et al., 2006). Glutathione reductase activity is essential for the regeneration of reduced glutathione, which was previously oxidized by the action of GPx (Parolini et al., 2019). Glutathione S-transferase (GST) facilitates the conjugation of reduced glutathione to various substrates to enhance the solubility of these molecules in water. Glutathione S-transferase is the most soluble antioxidant and is extensively distributed throughout all cellular compartments (Birnie-Gauvin et al., 2017). The antioxidant system reliant on GSH functions in conjunction with selenium, a crucial mineral that has been extensively researched in fish. The glutaredoxin system utilizes the reduction status of glutathione to maintain and regulate cellular redox homeostasis and redox-dependent signaling pathways. The Keap1-Nrf2-ARE (Kelch-like

ECH-Associating Protein 1 - Nuclear Factor Erythroid 2 Related Factor 2 - Antioxidant Response Element) modulates the signaling pathways governing the transcription of antioxidant enzymes (Chen et al., 2021a; Li et al., 2022).

In contrast to enzymatic redox regulation and antioxidant molecules, numerous non-enzymatic antioxidants can directly and non-specifically neutralize the chemical properties of diverse reactive oxygen species (ROS). In higher animals, these antioxidants are often derived from dietary sources, however in lesser animals and certain fish species, tiny antioxidants such ascorbic acid are produced (Jones, 2008). Higher animals have lost the ability to synthesize ascorbic acid due to the absence of the enzyme L-Gulonolactone oxidase or L-Gulonolactone synthase in their evolutionary development. In addition to ascorbic acid (vitamin C), additional minor non-enzymatic antioxidants include glutathione (GSH), vitamin A (retinol, β-carotene), vitamin E (tocopherol), and polyphenolic chemicals, among others. It is important to note that all the aforementioned antioxidants function independently or in succession to safeguard the cell from the detrimental effects of reactive oxygen species (ROS); otherwise, the cell undergoes oxidative stress, observable in animals, including fish fishes (Bal et al., 2022).

Vitamin C (Vit C), or L-ascorbic acid, is a vital micronutrient among non-enzymatic antioxidants, needed for mitigating oxidative stress (Narra et al., 2015). Most fish are unable of synthesizing vitamin C autonomously; however, they can obtain it from external sources. Vitamin C is a potent antioxidant as it may be oxidized by various free radicals in aqueous solutions, transforming into less reactive compounds. Vitamin C functions by attaching to reactive oxygen species (ROS) in the body and neutralizing free radicals by hydrogen ion (H+) donation. In cellular membranes, it collaborates with vitamin E (Vit E), renewing Vit E following its contact with ROS (Cheng et al., 2018). Vitamin C functions as a reducing agent, primarily diminishing transition metals like Fe and Cu ions that interact with H_2O_2 to generate the hydroxyl radical (Hossain et al., 2018). Eventually, Vitamin E, in vivo, serves as a powerful inhibitor of lipid peroxidation by functioning as a proton donor for the peroxyl radical, so disrupting the chain reaction of free radical generation. The donation of H+ results in the functional impairment of vitamin E until vitamin C intervenes to restore it (Monteiro et al., 2006).

Both external and internal stressors, such as exposure to pollutants, toxicants, pesticides, microplastics, and food scarcity, contribute to the increased production of reactive oxygen species (ROS), leading to oxidative stress in fish (Thakur et al., 2022). Elevated oxidative stress can impede development, productivity, and reproduction in animals, including fish. Consequently, the aforementioned parameters may be utilized to assess the health condition of aquatic animals broadly and fish specifically under various stressors, including microplastics, pesticides, and harmful nanoparticles (Panda et al., 2022; Pati et al., 2023).

3 Microplastics Induced Oxidative Stress in Fishes

Microplastics (MPs) are a major threat to aquatic life and are primarily found in aquatic habitats (Wang et al., 2024). Through physical, chemical, and biological processes, plastic waste decomposes into MPs with a diameter of less than 5 mm (Prata et al., 2020).Textile fibers and microbeads that were initially smaller than 5 mm are likewise considered MPs (Nava et al., 2023). Oceans, beaches, deep sea and coastal sediments, and surface waters from the Arctic to Antarctica are all host to MPs, which are prevalent in aquatic environments (Karbalaei et al., 2018). MPs have grown to be a significant environmental problem as plastic waste increases. According to earlier research, MPs have a number of harmful impacts on fish, benthic animals, zooplankton, phytoplankton, and other aquatic species (Cao et al., 2022; He et al., 2022; Wang et al., 2023).

MPs, tiny, persistent plastic particles under 5 mm, pose a significant environmental concern due to their potential harm to aquatic and human health (Rochman et al., 2019). Primary and secondary microplastics are produced by various goods, with secondary microplastics ending up in the ocean due to littering, improper disposal, product wear and tear, and unintentional spills (Bakir et al., 2023). MPs, when introduced into aquatic ecosystems, pose a significant threat to various marine life species (Cai et al., 2021; Prasad et al., 2024; Tabassum et al., 2025). Oxidative stress occurs when the body's antioxidants and reactive oxygen species (ROS) are not in balance (Bhatti et al., 2018). ROS, produced by the body, can be elevated by pollutants or poisons, causing inflammation and negative health effects. Decreases in fish populations impact diets, global customs, and economies (Liang et al., 2023). Oxidative stress, triggered by excessive ROS generation or compromised antioxidant defenses, can lead to negative health outcomes in fish, including tissue damage, organ malfunction, and weakened immunological function (Padmini & Udayakumar, 2021).

Environmental pollutants like pesticides and microplastics cause various stresses, with oxidative stress being the most extensively researched (Cole et al., 2011; Bhatti et al., 2018; Cole et al., 2021; Liang et al., 2023). Although it has been discovered that different nanoparticles cause ROS in fish (Paital et al., 2019), nanotechnology is also useful for managing fish health. Additionally, some of the nanoparticles appear to be helpful in treating fish illnesses (Xiao et al., 2023). Microplastics' oxidative stress in fish depends on the type, size, and form of the plastic, as well as the species and life stage of the fish (Subaramaniyam et al., 2023). Liu et al. (2019) and Liu et al. (2019) found that microplastics cause oxidative stress in common carp by releasing excess reactive oxygen species and reducing antioxidant enzyme activity. Yu et al. (2020a, b) found that nylon microplastics cause oxidative stress in crucian carp gills by disrupting mitochondrial function and increasing ROS production. Polyethylene microplastics damage cell membranes, increase lipid peroxidation, and cause oxidative stress in tilapia liver Chen et al. (2021a, b).

A study found 49% of fish had microplastics in their gills, gastrointestinal tract, and dorsal muscle, causing higher brain AChE activity and lipid peroxidation (LPO).

Consuming these fish could expose humans to 842 microplastics annually (Subaramaniyam et al., 2023). Fang et al. (2021) found that crucian carp's liver was damaged by acrylic microplastics, resulting in increased levels of ROS and LPO. Li et al. (2020) found that polypropylene microplastics increased oxidative stress in grass carp intestines, leading to inflammation and immunological activation. Microplastics can harm fish health by damaging kidneys, liver, and intestines, as they accumulate in fish organs and tissues post-consumption (Dawson et al., 2018). According to Zhang et al. (2021), ingesting polycarbonate microplastics causes apoptosis, fibrosis, inflammation, increased lipid buildup, liver damage, and structural damage in grass carp. Wang et al. (2019) found that polyurethane microplastics changed the composition of the gut microbiota in zebrafish, causing inflammation and oxidative stress.

According to Zhang et al. (2020), tilapia exposed to polystyrene microplastics experience inflammation, increased oxidative stress, changes in organ function, and damage to their liver and kidneys. Microplastic exposure in fish can lead to behavioral changes, oxidative stress, and organ damage, as shown in juvenile perch behavior and increased risk of dangerous behaviors like open water swimming (Lönnstedt & Eklöv, 2016). Wright et al. (2020) found that exposure to polyvinyl chloride microplastics reduces brown trout swimming activity and circadian rhythms, impacting their survival, reproductive success, and ecosystem role. This can lead to decreased growth, reproduction, predator-prey interactions, and increased mortality rates (Fig. 1).

3.1 Plastic Additives in Fish Health

Plastic additives, tiny plastic particles or stabilizers, are added to various consumer products like cleaning supplies, cosmetics, and personal care items to enhance their properties or capabilities (Koelmans et al., 2019). Additives, such as microbeads, microfibers, and microgranules, are made from various plastics like polyethylene, polypropylene, and polystyrene (Ramachandraiah et al., 2022). Plastic additives can enter the environment through stormwater runoff, wastewater discharge, and unintentional spills (Yusuf et al., 2022). Plastic additives can harm human health by consuming marine life and accumulating in the food chain, as per research by Rochman et al. (2019). Up to 11,000 microplastic particles may be consumed by people who frequently eat seafood each year (Cox et al., 2019). According to Su et al. (2020), physical harm to fish tissues is another mechanism that causes inflammation and an increase in ROS generation. Plastic additives can cause fish oxidative stress, potentially disrupting the balance between ROS and antioxidant levels, as compounds from microplastics release (Chae & An, 2020).

Plastic additives can harm fish's liver, kidney, and intestines, as well as stress liver cells in crucian carp and zebrafish embryos (Yang et al., 2020). Furthermore, fish have been shown to store plastics in their guts, which can cause inflammation and physical harm (Brennecke et al., 2016). Plastic additives have been found to alter

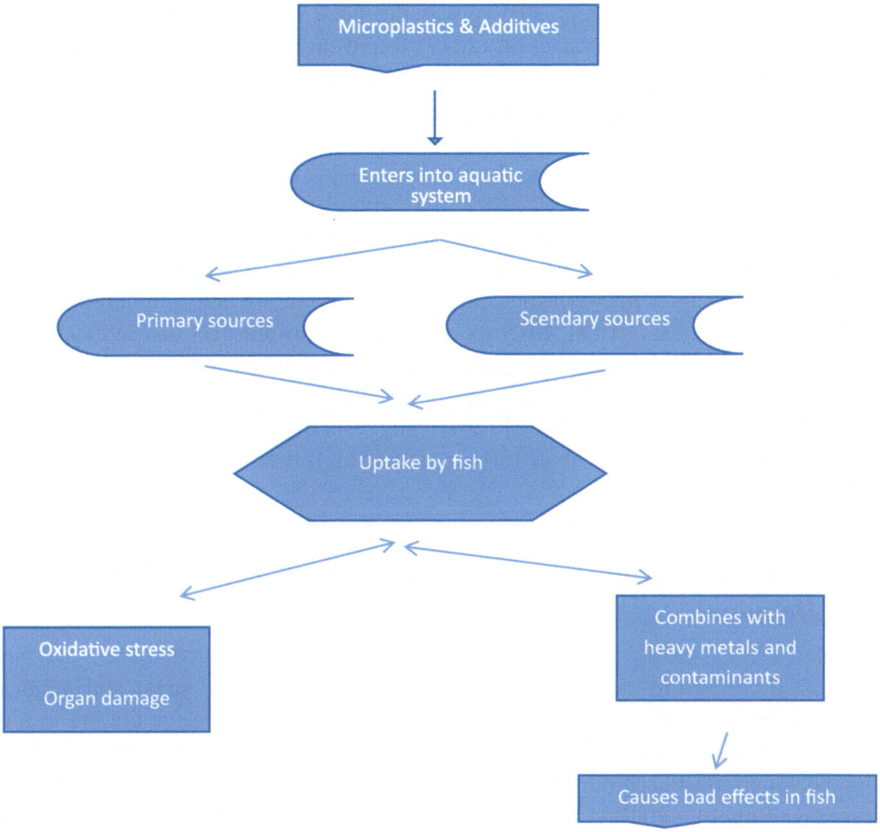

Fig. 1 The effects of microplastics on fish health

fish behavior, such as feeding and swimming patterns, potentially impacting their survival and reproduction (Chen et al., 2020; Qiang & Cheng, 2021). Adeleye et al. (2014) found that plastic additives, particularly microplastics, can contribute to fish oxidative stress, as evidenced by various research studies. Boháčková et al. (2023) study found that rainbow trout exposed to plastic additives showed a significant rise in oxidative stress biomarkers. Wang et al.'s, 2019 study revealed that zebrafish liver tissues were damaged by exposure to plastics, which are plastic additives that can be released into the environment. Iheanacho et al. (2020) studied the neurotoxicity and oxidative stress in *Clarias gariepinus* fish exposed to plastic additives compared to control groups. Polyvinyl chloride consumption in *European sea bass* led to liver damage, increased AST and ALT levels, and hypoproteinemia, resulting in impaired immune activities (Espinosa et al., 2018).

Barboza et al.'s (2018a) study found that *Dipterarchus labrax* juveniles' liver and gills show elevated mercury levels due to exposure to microplastics, causing oxidative stress. Zitouni et al. (2021) found 68 plastic additives in Dicentrarchus labrax tissues, including plasticizers, fiber-reinforced plastics, curing agents, flame retardants, and heat stabilizers. PEVA was the most common tissue microplastic type, followed by PA. Alomar et al. (2017) found no oxidative damage in *Mullus surmuletus*' liver after microplastic ingestion, suggesting low levels may trigger detoxification, despite slight GST activity increase.

Fish, a nutrient-rich food, can pose health risks when contaminated with microplastics, particularly in areas where fish is a staple or heavily polluted (Barboza et al., 2018b). Schwabl et al.'s (2019) study found microplastics in human stool samples for the first time, indicating human consumption and excretion of these particles.Factors like shape, polymer type, size, surface charge, chemistry, and additional chemicals in microplastics influence their retention and elimination within the body (Smith et al., 2018). Microplastics, like nano-plastics, can interact with proteins, phospholipids, and carbohydrates, leading to adverse effects upon cellular uptake, similar to nano-plastics (Lehner et al., 2019). The EFSA Panel on Contaminants in the Food Chain CONTAM (2016) suggests that the human body can only absorb microplastics smaller than 150 μm, requiring further investigation.

3.2 Pathways Involved in Microplastics-Mediated Oxidative Damage

Microplastic exposure triggers toxicity pathways like oxidative stress and inflammation, affecting specific areas, blood, organs, tissues, and bioaccumulating in the body (Ferrante et al., 2022). Through immune system reactions or mitochondrial damage, microplastics produce reactive oxygen species (ROS), which neutralize dangerous foreign particles (Yang et al., 2020). Microplastic contamination leads to oxidative stress in proteins, lipids, and DNA by altering antioxidant defense mechanisms and gene expression in marine vertebrates and invertebrates, causing neurotoxicity, immunological responses, endocrine system disruption, reproductive abnormities, genomic instability, embryotoxicity, and trans-generational toxicity (Alimba & Faggio, 2019).

Microplastic exposure in seabream heads altered the JNK (c-Jun N-terminal kinase) and ERK (extracellular signal-regulated kinase) signaling pathways involved in fish detoxification, according to a study by Tang et al. (2018). Cocci et al. (2022) found that microplastics in red mullet and European hake guts increase antioxidant enzyme (catalase and superoxide dismutase and cytokine (interleukin-1β, 10, and interferon) levels, leading to ROS generation and immune cell infiltration. Microplastics cause oxidative stress, leading to liver apoptosis in golden pompano. High polystyrene exposure causes gene (Grp78, Xbp-1, Eif-2α, and chop) upregulation, endoplasmic reticulum stress, and cell death (Yao et al., 2023).

Carp (*Cyprinus carpio*) immune systems were impacted by polystyrene microplastics, which also activated the TLR2 signaling pathway. According to the study, carp hepatopancreas and hepatocytes had higher levels of TLR2, COX2, TNF-α, tumor necrosis factor receptor-associated factor 6 (TRAF6), IL-1β,

NF-κB, iNOS, and p65 in their mRNA and protein expressions (Cui et al., 2023). Wang et al. (2022) discovered that prolonged exposure to polystyrene microplastics reduced the expression of genes involved in the Keap1-Nrf2 signaling pathway in juvenile loach (*Paramisgurnus dabryanus*). Exposure to polystyrene microplastic increased the expression of p53, gadd45ba, and caspase3b, which in turn caused apoptosis.Additionally, TNF-α and PTGS2A, two crucial gene markers in the zebrafish inflammatory mechanism, were elevated (Umamaheswari et al., 2021b). In zebrafish, subchronic exposure to polystyrene led to decreased spermatocyte counts and increased ROS generation in the testis and ovaries (Qiang & Cheng, 2021). Furthermore, the thickness of the basement membrane decreased and apoptosis (BAX, p-53, caspase 7, 8, and 9) increased in male gonads. In the same species, prolonged (35 days) exposure to polystyrene microplastics changed the gene regulation patterns and metabolic pathways triggered by oxidative stress and ROS generation (Umamaheswari et al., 2021a).

Polyethylene microplastics increased carp gill gene expression, triggering overactive NLRP3 (nucleotide-binding oligomerization domain-like receptor protein 3) inflammasomes. The NF-κB pathway increased inflammatory factors like TNF-α, IFN-γ, IL-2, IL-6, IL-8, and IL-1β, leading to immune disorders, while anti-inflammatory factors decreased.The NF-κB pathway initiated oxidative stress, apoptosis, and NLRP3 inflammasome, thereby promoting inflammatory and immunological responses (Cao et al., 2023). Santos et al. (2022) found that combined exposure of microplastics with copper heavy metal in zebrafish gills induces oxidative stress, lipid peroxidation, upregulation of genes, and downregulation of apoptosis and serotonin synthesis (Fig. 2).

Exposure to MPs considerably reduces fish growth and promotes oxidative damage; in particular, MPs exposure raises malondialdehyde levels. The diversity of fish gut microbiota, however, is not significantly impacted by MPs (Wang et al., 2024). Growth impairment, oxidative stress, neurotoxicity, reproductive toxicity, and modifications to metabolism and energy budget are only a few of the detrimental effects of MPs on fish that have been noted in earlier reviews (Liang et al., 2023; Mallik et al., 2021).Variations in biological organisms' levels of malondialdehyde (MDA) are frequently thought to as markers of cellular damage and oxidative stress (Henning & Weber, 2021).

4 Pesticides Induced Oxidative Stress in Fish Health

Fish being a nutrient rich food is used widely all over the world and is bound to live in water for its survival. Due to the human activities, most of the water bodies become highly polluted with pesticides residues. Most of these pesticides may produce reactive oxygen species (ROS) and thus induce oxidative stress in fish. Normally, ROS maintain homeostasis and cell signaling but in case of excess production, body cannot metabolize ROS and these species cause oxidative stress. Oxidative stress may cause cell death or can harm DNA, sperm, and other parts of

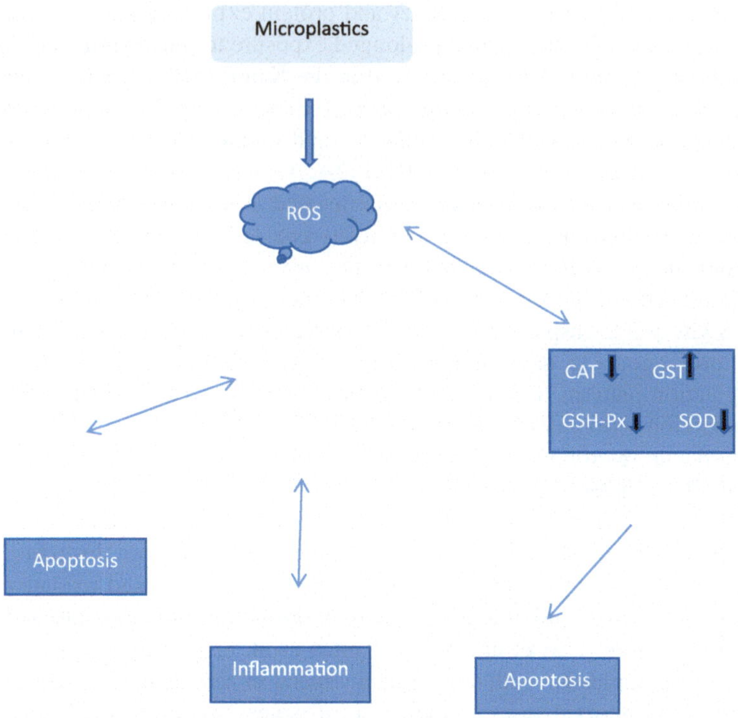

Fig. 2 Fish pathways impacted by exposure to microplastics

the body. Most of the pesticides induce production of reactive oxygen species by different biochemical mechanisms such as facilitation of Fenton reaction, the impairment of membrane-bound electron transport and subsequent accumulation of reduced intermediates redox cycling, photosensitization, inactivation of antioxidant enzymes and depletion of free radical scavengers. Pesticides induce toxicity of reproductive and neurological functions mainly depend on Oxidative stress. However, the action of a specific pesticide may vary for different organisms and may affect different species of fish differently.

In fish, pesticides can cause embryotoxicity and teratogenicity also in addition with oxidative stress. Most of the pesticides affected liver, gill and muscle in fish. Major pesticides that are harmful enough to induce oxidative stress include organochlorine pesticides, organofluorine pesticides, organophosphates, carbamates, pyrethroids, bipyridyl herbicides, triazine herbicides and chloroacetanilide herbicides.

(a) *Organochlorine pesticides*

These pesticides are banned in most of the countries for being persistent organic pollutants (POPs). Their residues persist in the environment up to this day. DDT and methylmercury are hepatotoxic pesticides and cause hepatocyte necrotic cell death in fish even in very low concentrations.

(b) *Organofluorine pesticides*

These pesticides are used as acaricides. Inhibition of Acetylcholine esterase (AChE) and a significant increase in MDA level is directly related with etoxazole concentrations.

(c) *Organophosphates*

In *context* to the toxicity of these pesticides, they can cause inhibition of acetylcholine esterase (AChE) and its activity, which initiates lipid peroxidation by the accumulation of free radicals. Organophosphates cause changes in ATPase activity and initiate cell injury.

(d) *Carbamates*

These are used as insecticides and can suppress growth by decreasing release of GSH content in liver and kidney of fish. Carbaryl increases GST level by inducing hepatic cytochrome P450 (CYP1A.

(e) *Pyrethroids*

These are most popular insecticides which are being used widely all over the world. Deltamethrin is one of the synthetic pyrethroids. Exposure to these insecticides increases level of LPO significantly in liver, gills and kidneys. A considerable increase in the amount of total thiol, protein thiol and non-protein thiol is also observed in liver, while a major reduction in these parameters is experiential in the gills of fish after the exposure of these insecticides.

(f) *Bipyridyl*

These herbicides may damage cell membranes. Acute exposure to triazine may increase plasma haemoglobin and induce oxidative stress in fish. It is necessary to measure tissue GSH and MDA level for the assessment of triazine exposure and toxicity level.

With reference to mode of action against target species, pesticides are categorized into many groups i.e. Insecticides, Herbicides, Fungicides, Nematicides, Rodenticide, Molluscicides, Acaricides, Homicides, Ovicides and so. The three major pesticides are insecticides (insect control), herbicides (weed control), and fungicides (Mycoticontrol). The amount of a specific pesticide required to kill an organism is called lethal dose and the amount of pesticide that kill 50% of a fish population within a given period of time is called lethal concentration 50 (LC50).

4.1 *Insecticides and Fish Health*

Various intensive agriculture activities and their run off drainage are the main reason of water contamination with insecticides. Fish are the most vulnerable species to water contamination. These insecticides can impaired many biochemical and physiological processes in fish and thus can damage health of fish.

4.1.1 Alterations in Blood Biochemical Parameters

Blood parameters indicate abnormality level of specific enzymes released by injured tissues and cells into the plasma of fish after the exposure to the insecticides. Pesticides may reduce synthesis of biochemical processes in cells of fish liver.

4.1.2 Tissue and Organ Damage

Many insecticides results in the reduction of growth and thus survival of fish by increasing susceptibility level to pathological alterations especially of reproductive organs. Incidence of tissue damage mainly depends on the amount and duration of insecticides exposure. Many insecticides cause specific or non-specific histopathological damage.

4.1.3 Reproductive Dysfunction

A slight change in environmental conditions may alter reproductive success in fish. Insecticides can reduce levels of fertilization, fecundity, hatching rate and rate of survivability of larva and may alter ovarian and testicular histology and process of vitellogenesis. Insecticides can delay maturation time of gonads and cause impairment of steroidogenesis. In most of the fishes, these chemicals alter behavior and pattern of reproductive migrations and also cause impairment of olfactory response as well as disruption in coordinating courtship behavior of male and female. Some of the insecticides are identified as chemicals that disrupt normal function of endocrine system in fish.

4.1.4 Development Disorders

In viviparous fish, insecticides directly affect larvae and embryos and reduce the chances of fish survival. Deformities in Spinal morphology are most common symptom of developmental disorders by insecticides. Reduced feeding rate and metabolism suppress fish growth and cause energy to be waste by insecticides exposure.

4.1.5 Neurotoxicity

Insecticides such as organophosphate and carbamate cause inhibition of ACHE or and butyryl cholinesterase (BACHE). These insecticides also suppress the action of γ-aminobutrate (GABA) Neurotransmitters. These insecticides firstly change transport of ions across the nerve cell membranes and consequently prevent nerve stimulations. The synthetic pyrethroids block GABA receptors by disrupting ion channels in the nerve cell membrane and alter normal function of neuron.

4.1.6 Behavioral Alterations

Insecticides may alter fish behavior by stoping the development of response in accordance with the identified situation. Many insecticides interfere with nervous systems and sensory receptors and cause a drastic change in behavior such as feeding behavior, swimming patterns, competition and predation. The effect of certain insecticides on the activity of ACHE may lead to decreased mobility of the fish.

4.1.7 Genotoxicity

Some insecticides may interact with genetic materials and can damage DNA by DNA inactivation. These chemical substances cause mutation by causing alteration in genetic material in the cell genotype. These alterations are heritable and may transfer in next generation. Genotoxic proprieties of insecticides may form tumors in various fish. Insecticides exposure cause chromosomal aberration in eggs and larvae of fish.

4.1.8 Immuno-Suppression

In fish, immune system is of great importance to fight against infectious diseases. Insecticides destroy defense mechanisms in fish by altering innate and acquired immunity. This change results in uncontrolled proliferation of cells and immune-suppression in fish. Due to the exposure of insecticides fish become susceptible to infectious diseases.

4.1.9 Effect on Growth of Fish

Disruption in the synthesis of nucleic acid and its metabolism is the main action of insecticide toxicity and may lead to decreased RNA production. Organo-phosphate alter biosynthesis of nucleic acid.

4.1.10 Histopathological Alteration

Histopathological study is a basic tool to identify health status and functionality of any organ. Nature of any toxicant can be recognized easily by tissue alterations in fish, exposed to different insecticides. Tissue damages in organs as a result of insecticide exposure cause reduced growth and survival. Tissue alterations make fish susceptible to infectious diseases. The pathological changes in different tissues such as gills, liver, kidney and spleen of fish treated with different insecticides can disturb homeostasis and lead to physiological disorders in fish.

4.2 Herbicides and Fish Health

Herbicides are the chemicals, which are widely used to agricultural lands, lawns and gardens to control nuisance growth of algae. Regarding the effects of herbicides on fish, their application may cause death of fish, although fish is not target of herbicides. Herbicides are toxic and their residues may remain in water for a long period of time. Some of the commonly used herbicides are Copper sulfate, Fluridone, Sonar, 2–4-D, Glyphosphate, Rodeo, Diquat, Weedtrine, Endothall, Aquathol, and Hydrothol. Other negative effects of herbicides on fish include:

4.2.1 Chronic Toxicity

One of the main effect of herbicides on fish and other aquatic animal is chronic toxicity due to buildup in tissues. Fibrosis and degeneration of liver tissue is the most common symptom of herbicide toxicity.

4.2.2 Genotoxicity

Herbicides may alter fish DNA and thus cause health issues resulting in imbalance in fish populations by disrupting genetic diversity. Chromosomal abnormalities and mutations are obvious in fish after the application of herbicides.

4.2.3 Behavioral Changes

Fish behavior can be worsened by herbicides. Fish have trouble feeding, swimming, and reproducing.

4.2.4 Reproductive Issues

Herbicides elevate level of lipid in ovary and atretic follicles, and thus cause abnormalities in gonads and the reproductive organs of fish.

4.2.5 Food Chain Disruption

Herbicides disrupt food chain by killing important plants and small plant eating insects and thus create stress in fish for forage.

4.2.6 Fish Kills

In most of the cases, application of herbicides creates shortage of oxygen in water due to the utilization of available oxygen by decomposed plants. Fish death occurs from suffocation.

They can also be carried into streams by rainwater runoff or leached into underground water supplies.

4.3 Fungicides and Fish Health

Fungal organisms damage crops and produce plant diseases. Fungicides are used to prevent the growth of these fungal organisms. Use of fungicides is very dangerous to non-target organisms such as fish. Generally, fungicides are not as toxic to aquatic animals as insecticides, though; mercury containing fungicides are banned for agriculture use in most of the countries of the world due to their harmful effects to the environment. They accumulate in the environment and cause poisoning for beneficial invertebrates. Fungicides affect fish health through impairment of metabolism, oxidative damage, inhibition of ACHE activity, biochemical changes, histopathological changes as well as hematological and developmental changes, mutagenesis and carcinogenicity.

4.3.1 Biochemical Effect

Fungicides cause disruption in glutaminases including phos-phate activated glutaminase enzymes which are associated with brain function.

4.3.2 Hematological Effect

Certain fungicides effect erythropoietic organs such as spleen and kidneys and might cause a reduced number of erythrocytes and haemoglobin content. As a result fish die due to anemia.

4.3.3 Histological Effect

Fish organs such as the gills, heart, kidney, liver, muscle, and spleen are all at risk from commonly used fungicides. In the gills, fungicides resulted in epithelial hyperplasia with lamellar fusion, oedema, general necrosis, degeneration of the main and secondary gill lamellae, and epithelial hypertrophy. There were notable impacts on the cardiac muscle, which is part of the body's circulatory system. Toxic liver cells are distinguished from normal liver cells by enlarged sinusoids, nuclear hypertrophy, atrophic areas, many vacuoles, and larger sinusoids and atrophic areas.

4.3.4 Genotoxic Effect

Tissue-specific damage in nucleic acid content may be due to differential effects of Fenvalerate or its metabolite(s) on synthesis and/degradation of DNA or RNA in gill and kidney cells of the fish.

4.4 Pathways Involved in Pesticides Mediated Oxidative Damage

4.4.1 TNFR1/TNF-α Pathway

Tumor necrosis factor receptor (TNFR) facilitates communication signals of tumor necrosis factor-alpha (TNF-α). As a result caspase-8 become activated and cleaves effector caspase-3 through mitochondrial route. During the process of acute inflammation, TNF-α is produced by macrophages or monocytes which is a dominant pro-inflammatory cytokine and enables signaling events within cells to initiate apoptosis or necrosis.

4.4.2 Nurr1 and the NF-κB Pathway

Dopaminergic neurone metabolism is significantly impacted by the transcription factor orphan nuclear receptor-related 1 (Nurr1), which is a member of the nuclear receptor subfamily 4 group A member 2 (NR4A2) protein family. Nurr1 impairment is one of the main causes of Parkinson's disease. Nurr1 has anti-inflammatory properties and inhibit transcription factor NF-κB in brain tissue.

4.4.3 Protein Kinase C Signaling Pathway

The proteolytic cleavage of protein kinase C (PKC)δ by caspase-3 aided in the dopaminergic PC12 cells' death called apoptosis. Through the phosphorylation of hydroxyl groups on residues of threonine and serine, these protein kinase enzymes can alter the target protein's function, cellular location, or connection with other proteins.

4.4.4 NF-κB Signaling Pathway

Endosulfan is an insecticide which is used for a variety of insects on crops like grains, fruits, and vegetables. In neuroblastoma cells, endosulfan exposure can raised intracellular H_2O_2 remarkably and induce oxidative stress in time and dose dependent manner. Oxidative stress can be detected by the expression and activity of this ubiquitous transcription factor NF-κB.

4.4.5 Endoplasmic Reticulum (ER) Stress

The ER plays various important roles including the the transport of proteins and posttranslational modification. Protein folding capacity of the ER cause ER stress overwhelmed due to the misfolded proteins that are accumulated within the lumen of ER.

4.4.6 Nonmitochondrial Apoptosis Pathway

Apoptosis is a process of programmed cell death that is used to eradicate discarded cells. Pesticides initiate apoptosis by various intrinsic and extrinsic pathways that cause DNA and mitochondria damage and trigger signaling pathways.

4.4.7 Mitochondrial Apoptosis Pathway

Another potential route for pesticide-induced oxidative stress is the mitochondrial apoptotic pathway. BCL2-associated X (Bax) and B cell lymphoma 2 (Bcl-2) are the primary regulators of mitochondrial integrity in this pathway. They also affect caspase activation and cytochrome c release. Two well-known proteins with opposing roles that are linked to cell death are Bcl-2 and Bax. ER stress and apoptosis are frequently linked to mitochondrial malfunction in fish and humans.

4.4.8 Autophagy

Effect oxidative stress can be reduced by the process of autophagy. Autophagy permanently removes damaged parts of eukaryotic cell.

4.4.9 Mitogen-Activated Protein Kinases (MAPKs)

The mitogen-activated protein kinases (MAPKs) activate or deactivate their target by phosphorylating dual-serine and threonine residues on their substrates or on their own (autophosphorylation). Different mechanisms of cells such as gene expression, apoptosis, cell survival, immune response and stress response are regulated by these MAPKs.

4.4.10 Keap1/Nrf2/ARE Pathway and Ca^{2+} Signaling

The Keap1/Nrf2/ARE pathway has a major impact on the regulation of cytoprotective responses to endogenous and exogenous stresses induced by ROS. The Nrf2/Keap1 pathway is composed of four components: (a) the actin-binding Kelch-like ECH-associated protein 1 (Keap1), (b) the nuclear factor erythroid2-related factor 2 (Nrf2), (c) a group of small musculoaponeurotic fibrosarcoma (Maf) proteins, and (d) the antioxidant response element (ARE), which is essential for the antioxidant response in this pathway. When the transcription factor Nrf2 binds to the antioxidant response element (ARE) in DNA, it triggers the expression of a variety of detoxifying enzymes and antioxidant proteins/enzymes.

5 Nano-forms of Chemicals for Fish Health and Disease Management

Aquaculture is undoubtedly a flourishing industry that ensures food security for the increasing population. However, this industry is under severe threat from bacterial, viral, parasitic, and fungal infections besides environmental contamination. Hence, in the recent days, new technologies are surfacing to confront these challenges efficiently. Nanotechnology is one of the novel and pioneering skills with a wide continuum of applications ranging from aquatic pond sterilization, water treatment, nutrient delivery, diagnosis, drug delivery, aquatic disease control, etc. (Dasari et al., 2024). Hence, there is an urgent need to solve such kind of outbreaks in the aquaculture industry and take precautionary measures to avoid further damage (Khursheed et al., 2023).

Nanotechnology opens enormous opportunities in different fields including the aquaculture sector. Nanoparticles (NPs) have unique physical and chemical properties such as high surface zone, high reactivity, tunable pore size, and particle morphology which are useful in many industrial applications (Varaprasad et al., 2022). Specifically, nanotechnology has evolved as a new avenue to solve the issues raised in the sustainability of aquaculture sectors (Abinaya et al., 2023). The fascination with nanotechnology is a result of the intriguing and extraordinary qualities demonstrated by nanoparticles, including their physical, chemical and biological activities, and their expanded applicability when compared to their bulk components (Joudeh & Linke, 2022). Several research works proved the role of different nanoparticles in fish-pathogen interactions (Shaalan et al., 2016). Rapid development of metal and metal-oxide nanoparticles and their usage in several fields including electronics, energy, biomedicine etc., is a frequent practice in recent times besides understanding their nature (Khan et al., 2024). In general, metal-oxide nanoparticles are more popular due to their high reactivity and effectivity than metal NPs. The mode of action of various nanoparticles makes them desirable as an antimicrobial tool in fish disease management.

Gold nanoparticles (Au NPs) are typically prepared in a wide variety of shapes such as spherical, tube, rods, etc., and possess size and shape-tunable properties (Pierre et al., 2024). Gold nanoparticles have typical chemical, thermal, mechanical, electrical and optical properties. Gold NPs can be detected by magnetic resonance imaging and their surface electrons show plasmon resonance properties. Hence, these are used as contrast agents on multiple imaging platforms and used in biomedical research (Luo et al., 2021). Their unique chemical and physical properties make them promising candidates in drug target identification (Chen & Feng, 2022). Gold NPs are inert and non-toxic, and thus have gained attention in cancer research including drug delivery etc., and the combination of Au2+ and Au3+ proved to enhance the activity of antitumor drugs (Huang et al., 2023).

Copper is an essential trace element found in food nutrients or supplements (Smriti et al., 2023). Similarly, copper nanoparticles (Cu NPs) are useful for food, biomedical and various other industrial applications (Woźniak-Budych et al., 2023). At low concentration, copper NPs are non-toxic to several organisms including aquatic species (Delavari et al., 2022). Moreover, copper nanoparticles fused with other elements showed effective antimicrobial activities (Zhang et al., 2021). Copper oxide nanoparticles (CuO NPs) are also largely used in many industries such as food, paints, ceramics, electronics, pharmaceutical, biomedical, etc. (Letchumanan et al., 2021). Copper oxide nanoparticles are used as biocides and also as antimicrobial agents (Malhotra et al., 2020). Specifically, at low dosage, CuO nanoparticles are useful in antimicrobial activities (Balasubramanian et al., 2024). Zinc oxide nanoparticles (ZnO NPs) serve as a source for plastic, ceramic, glass, paint, and food industries (Shaalan et al., 2017). Zinc oxide nanoparticles also have the potential to diagnose and control various diseases (Abinaya et al., 2023). In addition, zinc oxide nanoparticles stand as an example for biocompatible and biodegradable nano-platforms that are promising in reducing the side effects (Awad et al., 2019). Although ZnO is helpful as a food supplement at low concentrations, it becomes toxic when over accumulated (Pati et al., 2023). Zinc oxide NPs are widely used in drug delivery systems due to their low toxicity and beneficial qualities (Rasheed et al., 2023). The role of ZnO NPs in antimicrobial, anticancer, anti-inflammatory and anti-diabetic activities including immune-modulatory effects was studied by various researchers (Mandal et al., 2024). Selenium is an essential trace element for microbial organisms and animals including fish and is available in both inorganic and organic forms. Selenium nanoparticles (Se NPs) are used in various fields due to their biocompatibility, bioavailability and low toxicity (Dawood et al., 2021). These nanoparticles fight against fungal, bacterial, and parasitic infections including diabetes and cancer (Dash et al., 2022). Selenium functions as an excellent antioxidant and promotes a healthy immune system. Also, Se NPs act as chemopreventive and anti-inflammatory agents (Yazhiniprabha et al., 2022). Different sources of selenium NPs boost final weight, relative gain rate, and antioxidant status (Zhu et al., 2024). Inclusion of selenium NPs avoid cell damage and play active role in reactive oxygen species (ROS) protection, growth, fertility and immunity of fish (Ni et al., 2023).

Fish diseases and management is a well-known fact that fish is the major contributor to ensuring global food and nutrtion security (Ahmed et al., 2020). Diseases are considered as key universal problems faced by the fishes which cause mortality in turn reduce the yield. Pathogenic and non-pathogenic fish diseases are the two main categories that affect fish growth and development (Deenathayalan et al., 2024). Pathogenic fish diseases are considered to be highly perilous compared to non-pathogenic diseases. Nonpathogenic diseases are those that are not communicable to other fish and arise as a result of nutritional deficiencies, environmental pollutants, genetic anomalies, and widespread aeration, whereas pathogenic diseases are highly communicable to other fishes, resulting in mass mortality rates (El-Sayed Ali et al., 2018). Pathogenic diseases occur due to four types of infectious agents, which include bacteria (Aeromonas, Streptococcus, Pseudomonas, Vibrio, etc., species), viruses (Rhabdovirus, Birnavirus, Iridovirus, Herpesvirus, Reovirus, Retrovirus, Orthomyxovirus, etc.), fungi *(Saprolegnia parasitica, Aphanomyces invadans, Fusarium oxysporum, Aspergillus, Batrachochytrium, Cladosporium*, etc., species) and parasites (Gyrodactylus, Dactylogyrus, Ergasilus, Argulus, Caligus, Lerneae, etc., species) (Idowu et al., 2017). Several fishes have specialized defense mechanisms against pathogens i.e., specific and non-specific defense (Chaplin, 2010). Specific defenses include adaptive immune responses against particular pathogens. Based on these defenses, vaccines are developed against certain pathogens. However, nonspecific defenses include mucus cells of the epidermis skin, and scales which trap microbial organisms and arrest their growth (Sharma et al., 2012). The pathogens that succeed in breaking these barriers/defenses can be further killed by the inflammatory reactions produced by the fish through the transportation of increased blood flow and white blood cells to infected areas. The fish immune system plays an important role in defending the fish against any disease-causing agent. Fishes have two types of immunity i.e. innate and adaptive. Innate immunity is the first line of defense against pathogen invasion, whereas adaptive immunity is much more pathogen-specific (Sharma et al., 2012). Non-specific cellular (eosinophils, cytotoxic cells, macrophages, and neutrophils) and humoral elements (C-reactive proteins, complement, interferons, lysozyme, transferrins, and lectins) make up innate immunity, whereas adaptive immunity also includes humoral (IgM, IgD, and IgT antibodies) and cellular components (T-lymphocytes) that destroy pathogens (Firdaus-Nawi & Zamri-Saad, 2016). To combat the invaded pathogens, both immune systems must function effectively together. Fish health management has several requirements and needs to observe all aquaculture practices. Before that, fish farmers should be trained properly in aquaculture practices. Initially, it is a must to provide healthy seeds and young fish, proper nutrition, optimal water quality, hygienic conditions, and appropriate waste management practices (Anita & Khati, 2019). Fish diseases in the fisheries are managed by the usage of both synthetic and natural chemicals. Synthetic chemicals include antibiotics (sulfonamides and tetracyclines), vitamins, hormones, disinfectants (green malachite and hydrogen peroxide), and antihelminthics (pyrethroids and avermectin) (Citarasu, 2010). Natural chemicals include essential oils, herbal medicines, and probiotics. Generally, bacterial toxicity causes organ failure in localized or systemic infections (Kothari et al., 2019). There are various bacterial diseases in fish that can be controlled only through antibiotics. These antibiotics act as antimicrobial agents or chemo-medical agents (Romero et al., 2012). However, over usage of antibiotics may increase biological carrying in water and may also develop antibiotic resistance in bacteria. Probiotics are a natural method of treatment using beneficial microorganisms which survive in the digestive tract of fish to enhance growth, improve nutrient digestion, and inhibit pathogens. However, they raise histamine levels in fish and cause inflammation (Kothari et al., 2019).

It was found that the nano-formulations of various chemicals induced oxidative stress in fish. For example, ZnO nanoparticles are found to increase the toxic load in the liver of teleost fish via mitochondrial-dependent pathways. More particularly, the above nanoparticle induces lipid accumulation, low lipolysis and activates mitophagy (Chen et al., 2022). Copper ferrite nanoparticles increase lipid peroxidation (LPO), glutathione-S-transferase (GST), and glutathione peroxidase (GPx) and alleviate catalase (CAT) and glutathione (GSH) level, and are finally considered cytotoxic in the ovaries of catfish (Srikanth & Nutalapati, 2022).

Maghemite (gamma-Fe2O3) nanoparticles at 40 and 60 ppm hampers developments and induces oxidative stress in Zebrafish Embryos/Larvae (Thirumurthi et al., 2022). Another metallic nanoparticle, namely magnetic nanoparticles, is noted to be neurotoxic in nature and induces oxidative stress by inhibiting the activity of SOD, CAT, GPx, and GSH values in the brain of rainbow trout (Ucar et al., 2022). Polystyrene nanoforms are not only responsible for generating oxidative stress by stimulating LPO generation but also induce cardiotoxicity in carps (Wu et al., 2022). Nano-forms of various plastics damage the reproductive cells and increase oxidative stress in Zebrafish, as documented by Sarasamma et al. (2020). Similarly, some other non-metal nanoparticles such as graphene oxide in zebrafish, in *Anabas testudineus* (Paital et al., 2019), fullerene in freshwater fish *Carassius auratus* (Zhu et al., 2008), pyrethroid nano-forms in carp, tilapia, and trout (Yang et al., 2020) are proved to induce oxidative stress, primarily by deregulating the antioxidant enzymes. Since fish is regarded as an important source of protein, essential fatty acids, vitamins, and minerals such as selenium and iodine that are usually absent in meat or other crops, stress and disease management are very important. As with most aquatic organisms, cultured fish species around the world experience a variety of infectious and non-infectious diseases due to contaminated water from industrial and agricultural sources and intensive aquaculture. Additionally, because fish in captivity have a reduced immune system, stress on them increases the spread of diseases like bacteria, fungi, and viruses. Fish infections have so received much attention in recent years. Due to illnesses brought on by numerous infections, the aquaculture sector experiences significant losses each year. Therefore, it is crucial to effectively detect and manage diseases in order to maximize productivity and guarantee the high quality of the finished product (Ninawe et al., 2017).

5.1 Role of Nano-particles in Disease Diagnosis

Fish disease identification is an indispensable part of modern aquaculture, and rapid and real-time diagnosis is an essential part of the early and precise treatment of diseases. However, the farmed fish will be affected by viruses, bacteria, parasites, metal pollution, and fishing damage (Low et al., 2017). Fish diseases are caused by a combination of different pathogens. Conventional methods involve detection via fish tissue dissection, which is destructive, time-consuming, and costly (Singh et al., 2021). The direct diagnosis of diseased fish underwater requires a high level of technology, while the large-scale and rapid spread of fish diseases have to limit the required time for diagnosis. The diversity and heterogeneity of fish diseases increase the difficulty of diagnosis, and the accuracy of diagnosis using these various physiological indicators is low (Fig. 3).

The potential uses of nanotechnology in the aquaculture and fisheries sectors are numerous. Typically, a nanoparticle (NP) is defined as a structure with a size between 0.1 and 100 nm (1/ 1,000,000 mm). Nanoparticles have attracted much attention in diagnosing various fish diseases due to their sensitiveness and preciseness in diagnosing various bacterial, viral and fungal diseases (Munawar et al., 2021). Various kinds of nanoparticles, like ZnO, TiO2, Cu-based nanoparticles, silver nanoparticles, etc., are used in the diagnosis of fish diseases. Ravikumar et al. (2012) reported that nanoparticles of CeO2 could be used for control of several fish diseases associated with *Vibrio parahaemolyticus*, *Serratia spp*., *Aeromonas hydrophila*, *Vibrio harveyi*, *Bacillus subtilis*, and *Serratia sp*. Saad et al. (2022) investigated the effects of biological selenium nanoparticles on bacterial load and heavy metal accumulation in Nile tilapia fish and reported that those nanoparticles could able to reduce the load of *Aeromonas hydrophila* by 28%–45% and cadmium and mercury levels by 50%–87% and 57%–73%, respectively in fish organs. Reported that nano-biosensors can detect fish contaminants and very low concentrations of pathogens such as bacteria, viruses, and parasites. They may be based on various nanomaterials, such as carbon nanotubes.

In fish, nanotechnology is used for the synthesis of nanoparticles, which have numerous applications for diagnosis. These include the use of gold nanorods, and the application of carbon nanotubes, for quick and affordable disease detection, making illness identification feasible at an early stage. Nanoparticles are also crucial for sensitive and specific diagnosis of viral, bacterial, and fungal diseases in aquaculture and fish, and they can even cross the blood-brain barrier in fish. Instruments of sub-micrometer sizes are used for diagnosis, prevention, and better cure of illness as well as to improve the overall life quality of patients. Nanotechnology is used to accelerate the development of regenerative drugs, allowing for the generation of

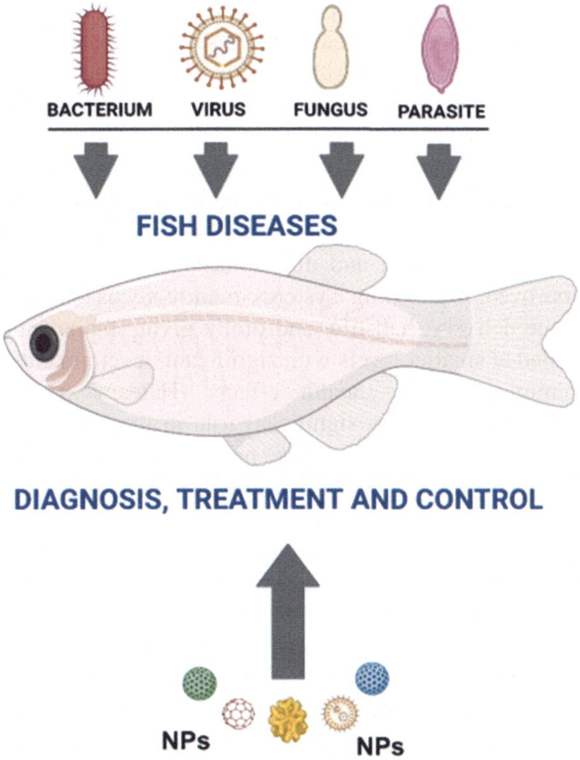

Fig. 3 Schematic representation of the present study

bone, cartilage, prosthetic skin, and other tissues to treat serious injuries or organ insufficiency in patients. The use of this technology changes cellular functions more effectively to emulate the effects of organs and natural tissues (Haleem et al., 2023). In recent years, medical imaging has become increasingly important for diagnosing a wide range of diseases. The advancement and progress in computer tomography and MRI technology have been remarkable, providing highly sensitive and accurate diagnostic tools for both in vivo and in vitro applications that surpass the capabilities of modern equipment. The progress in magnetic resonance and computer tomography is impressive. However, nanotechnology offers instruments for in vitro and in vivo diagnostics that are sensitive and very accurate, well beyond modern equipment's capabilities. The ultimate goal of these advancements is to facilitate early disease diagnosis, with nanotechnology potentially revolutionizing cellular and sub-cellular diagnostics. Different types of nanoparticles can be used in fish cells to deliver medication effectively. Once they reach the target area, nanoparticles start working at the molecular and cellular levels. Gills in fish are responsible for exchanging gases between their internal and external water environments. Various substances, including organic compounds and metal nanoparticles, can interact with fish gills and eventually enter their bloodstream (Zhang et al., 2021). Nanoparticles have been found to reduce the death rate in fish by influencing cell behavior.

Nanomedicines are expected to bring significant benefits to drug delivery and treatment. Physicians can use these particles to target the source of illness more effectively, reducing side effects and increasing treatment efficacy. They also provide new methods for controlling drug release. Due to the various types of nano-drugs available, as well as their applications in drug delivery and diagnostics, nano-medicines offer substantial benefits in the fields of instrumentation and pharmaceutical manufacturing. The production of nanoparticles and treatments based on nanotechnology is being developed as a research field. In the recent decade, by using imaging, therapies, and diagnostics, manufacture of nano-drugs has significantly improved. Nano-drug systems mainly focus on upgrading the bioavailability of tissue delivery, half-life, and orally giving therapeutic items. Nano-drugs are controlled at smaller levels with significant developments in their medicinal effects and minimize adverse health effects (Haleem et al., 2023). In pharmacology, nanoparticles play a significant role in the nutrient supply of fish, such as in the synthesis of vaccines. These nanoparticles have broad applicability due to their numerous advantages, such as their ability to protect antigens from destruction by enzymes. When biological materials combine with nanotechnology, there is a significant increase in innovative nanoparticles that are synthesized effectively at a suitable cost. Nanoparticle synthesis is also a niche for researchers, but it is a fast-growing and impacting key economic field, providing new nano-labeled products with novel and unique functions. Nanoparticles improve newly developed nano-labeled products, which are important factors for the success of the nanotechnology industry (Munawar et al., 2021). By utilizing the intrinsic qualities of different types of nanoparticles, the increasing science of nano-medicine shows numerous opportunities to improve fish health. Nano-silver and zinc oxide have been employed by prior research as antibacterial and therapeutic nanomaterials to decrease infections in aquaculture systems. In short, the development of nanomedicine has vast, and non-specific use in the field of animal health (Kakakhel et al., 2021). The application of metal oxide nanoparticles in aquaculture disease management has been extensively studied, and it has been found that nanoparticles such as ZnO-NPs, CuO-NPs, Au-NPs, Ag-NPs, and TiO2-NPs possess the ability to fight against various disease-causing pathogens. ZnO-NPs are reported to inhibit the growth of various pathogenic bacteria viz. *Staphylococcus aureus, Flavobacteriam branchiophilum, Aeromonas hydrophila, Vibrio species, Pseudomonas aeruginosa, Edward seillatarda, Bacillus cereus* and *Citrobacter spp* (Swain et al., 2014), whereas CuO-NPs are effective against *Saprolegnia sp.* from the host white fish and also used as strong antifungal agent. Au-NPs have been tested against bacterial pathogens of fish and observed to possess antibacterial properties. Ag-NPs are used as an effective antibacterial agent as it release Ag + ions which bind with the proteins cell membrane of the bacterial cells and damage bacterial cell membranes. Also, Ag-NPs are found to be effective against multidrug resistant bacteria among them, methicillin-resistant *Staphylococcus aureus* is one.

5.2 Role of Nano-particles in Water Treatment

In addition to this, the microorganisms and heavy metals present in the watercourse retards the growth rate of fishes leading to a great economic loss in the fishery sector (Shah & Parveen, 2022). Nanotechnology has been used very effectively to remove toxins from water. Poor performance and harm to the organisms result from water quality loss (Boyd & Tucker, 2012). Globally, the use of nanotechnology in water treatment is on the rise, with a wide range of potential customer-specific uses. The capacity to include diverse qualities in multifunctional materials is one of the most beneficial features of these systems, including nanoadsorbents, nanometals, nanomembranes, and nanophotocatalysts, among others. For instance, nanomaterials can simultaneously remove particles and pollutants and increase process efficiency (Gehrke et al., 2015). Water treatment is one of the most crucial foundational elements needed for sustained aquaculture. The excessive use of antibiotics and other synthetic chemical substances in fisheries and the discharge of waste materials from cities, businesses, and agriculture, have made water poisoning leading to an indirect impact on aquatic individuals as well as causing foodborne illnesses when consumed by human beings. These waters contain heavy metals that cause fish to grow more slowly and eventually die. Aquaculture uses nanotechnology mostly for treating water to provide a suitable environment for fish growth and reproduction. In this light, the scientific community supports adsorption and photocatalysis as the most effective and reasonably priced methods of water purification. Activated forms of carbon or alumina, along with either zeolite or iron-containing materials can be used effectively to hold both aerobic and anaerobic biofilm to remove various pollutants like nitrates, nitrites and ammonia. When silver nanoparticles are used to cure fungi in water directly, young trout have been found to be affected, although rainbow trout fish kept in fish farms may not become infected by fungi if their water is filtered through a device coated with silver nanoparticles. The ultrafine form of iron particles can be used as a cleaning agent for less harmful compounds present in water courses like dioxins, carbon tetrachloride, trichloroethane, or polychlorinated biphenyls (Rather et al., 2011). TiO2 has been used effectively to treat wastewater due to its non-toxicity property, biological and chemical stability and photocatalytic activities. Many previous studies have documented the photocatalytic activities of TiO2 against various disease-causing bacterial, algal, and viral agents (Ouyang et al., 2016). Kuang et al. (2017) reported that graphene oxide and graphene nanosheets can be used as efficient water source cleaning agent. The graphene oxide -TiO2 combination has been used for removing various heavy metal ions and organic dyes from polluted water (Atchudan et al., 2017). Also, silver nanoparticles are gaining much more importance in cleaning the water source that is used for fish rearing due to their advanced cleaning system (Munawar et al., 2021).

Silver nanoparticle are employed for the cleaning fresh water inhabiting fishes due to advanced cleaning system and special plates that cleans each drop of water in each direction of movement of water. Silver nanoparticles to improve water quality in aquaculture. The intensive intervention which aquaculture practices produced, impact directly over the environment, because it needs higher water quantities, and this source every day is scarcer around the world. Also, not consumed food for fishes, excretion products, feces, chemical products, and antibiotics, generated higher quantities of waste during organism's production and was released to environment around these production farming. Hence, Nanotechnology has become the most significant for human life in treating and diagnosis of diseases (Vanni & Layne, 1997). Silver nanoparticles also used to remove the toxic heavy metals in water that are main threat to life of fishes. These nanoparticles bond to the heavy metals and remove them in specific concentration. Through the use of silver based nanoparticles, heavy metals and environmental pollutants can be easily removed at high concertation through integrative approach. Water quality is, of course, a critical factor in fish health. The standard concerns include ensuring water quality for the immediate needs of the species e.g. dissolved oxygen levels, temperature, salinity), removal of nitrogen wastes as well as the interactions of these parameters (Salze & Davis, 2015).

Chitosan [(1, 4)-2-amino-2-deoxy-D-glucan] is a linear polysaccharide which is an abundant natural bio-polymer like cellulose. When chitin undergoes deacetylation, it turns into chitosan, and the extent of deacetylation affects its properties and applications (Abdel-Razek, 2019). Chitosan nanoparticles (CS NPs) find various uses across agriculture, food protection, cosmetics, medicine, pharmaceuticals, drug delivery, drug formulations, and wastewater treatment (Bhoopathy et al., 2021). Chitosan NPs popularity in these fields is mainly due to its biocompatibility and biodegradability and low toxicity (Zhang et al., 2023). Chitosan has antimicrobial properties which includes antibacterial, antifungal, antiviral and antiparasitic activities (Khan et al., 2024).

5.3 Effect of Nano-particles of Fish Health and Growth

Dietary supplementation of micronutrients in the form of nanoparticles as feed supplements has been reported to enhance the productivity and meat quality of fish, e.g., nFe, nSe, nAg, and nZnO (Khosravi-Katuli et al., 2017). Nano supplementation of micronutrients offers several advantages, such as greater bioavailability, easy absorption, enhanced utilization, and promoting cellular functions. Selenium is a trace element and has an important role in the development and immune status of organisms (Dawit Moges et al., 2022). Several researchers have conducted selenium supplementation for enhancing the productivity and growth performance in *Carassius auratus gibelio*, *Ctenopharyngodon idellus*, improved productivity and growth in stress-resilient fish improved immune status in *Cyprinus carpio* and juvenile *Lates Calcarifer* Block. However, studies on the assessment of Se-Supplementation on the growth performance in tilapia are lacking. The study assessed the growth performance, length-weight relationship, condition factor (FCF), fatty acid analysis and nutrition quality index, and dietary levels of Se-Nps on the fingerlings of *O. niloticus* to improve the fish production and yield level and quality.

The use of nanoparticles of trace elements is one of the promising approaches for feed supplementation in a variety of fish, which serves as a handy way of delivery and easy availability to the fish. The present study showed that the administration of a specific dose of Se-Nps enhanced the growth performance and nutrient quality index in Nile tilapia that serves as one of the staple fish food across the continents and can meet the nutrition requirement of the growing population. Studied, that the experimental and the control groups showed negative allometric growth ($b < 3$); however, tilapia fish fed 1 mg/kg of Se-Nps inclusion showed nearly isometric growth with b value of 2.81. Some studies have reported significantly higher growth in terms of weight gain in the diet containing 1 mg/kg of Se-Nps in fish, which was in line with our findings (Rathore et al., 2021).

Nile tilapia demonstrated enhanced growth performance upon nanoselenium supplementation. Growth parameters RGR and SGR values were significantly high and low in the 1 mg/kg and 2 mg/kg Se-Nps inclusion groups, respectively in comparison to others. The low growth performance in 2 mg/kg Se-Nps inclusion group could be attributed to the toxic effect of the over-dose supplementation of Se-Nps. In addition, the fish fed 1 mg/kg Se-Nps revealed better k values (condition factor) than the other treatment groups. Several studies also reported low FCR, SGR, and RGR obtained for 2 mg/kg Se-Nps in different fish varieties. Fish fed 1 mg/kg Se-Nps had by far lower FCR (1.07) than the control group (2.53) ($p < 0.05$). According to, the higher the FCR, the lower is the growth efficiency and inverse is the link between SGR and FCR (Iqbal et al., 2020).

However, some studies reported enhanced growth performance at higher Se-Nps supplementation dose, such as at 3.67 mg/kg Se inclusion in African catfish. On the contrary, optimum Se-Nps dose concentration was low for rainbow trout (0.38 mg/kg), channel catfish (0.25 mg/kg), and for groupers (0.77 mg/kg). Such findings demonstrate that the optimum dose of Se-Nps supplementation for enhancing growth and productivity varies with fish type, which could be attributed to the bioaccumulation or bioavailability of Se inside the fish tissues.

Unquestionably, nutraceuticals are recognized to be highly effective in enhancing fish development and immune characteristics. However, the costs associated with their incorporation are higher than the minimum requirements. Many studies supported the use of nanotechnology in fisheries for the efficient delivery of nutraceuticals and dietary supplements. By making nutrients more soluble and protecting those from the harsh gut environment, these systems primarily aim to increase the bioavailability, bio-accessibility, and, subsequently, effectiveness of the nutrients. Previous studies also reported a faster growth rate and higher final body weight in *Carassius auratus gibelio* young carp when fed with diet supplemented with iron and selenium nanoparticles (ETC, 2003; Frederick et al., 2010). Iron nanoparticles have been proven to have a growth-promoting effect on young carp and sturgeon. Additionally, it was discovered that adding nano-selenium to a fish's diet could increase its weight, relative growth performance, antioxidant activity, glutathione peroxidase activity, and concentrations of selenium in muscle of crucian carp (*C. auratus gibelio*) (Bhupinder, 2014). Khosravi-Katuli et al. (2017) reported that using nanoparticles in the fisheries sector enhanced immunomodulation, digestion and reproduction compared to other forms of Se. Wang et al. (2013a, b) also reported that nano supplementation could help in increasing fish and animal productivity by reducing oxidative stress. Studies reported that diets supplemented with 0.2 mg/kg selenium nanoparticles had rapid weight gain rate and a reduced feed coefficient than feed with 0, 0.1, 0.4, 0.8, and 1.6 mg/kg selenium nanoparticles (Qin et al., 2016). Feed supplemented with selenium nanoparticles increases immunity and disease resistance while providing resistance to the effects of hypoxic stress (Kumar et al., 2017). When delivering biological materials, nanoparticles are less expensive than other materials. Additionally, nanoparticles possess adjuvant qualities that can raise antigen effectiveness. The cellular uptake of nanoparticles is facilitated by their diminutive size, as they are only capable of entering living cells via cellular endocytosis (Zhao et al., 2014). Adjuvants lessen administration frequency while enhancing immune response (Evensen, 2009). Luis et al. (2019) reported that the stability, solubility, targeting, biocompatibility, and permeability of vaccines can be enhanced by the application of nanotechnology. In conclusion, it can be asserted that the implementation of nanotechnology in aquaculture systems, the administration of pharmaceuticals through porous nanostructures in fish feed, and the utilization of nano sensors for pathogen detection in aquaculture systems hold promising prospects for impacting fish health.

Dawit Moges et al. (2022) studied the nano selenium supplementation enhanced the growth performance and nutrition quality of Nile tilapia at a specific doses. Selenium exhibits diverse roles in the productivity and immune status of fish; hence, nano selenium supplementation increased the bioavailability and overall growth parameters like length, weight, omega 3 fatty acid content, and overall growth of tilapia. The findings highlight the utility of nanotechnology for the enhanced production of Nile tilapia that could be utilized for meeting the increasing demand of malnutrition across the global population. Industries related to fish farming could explore the development of nano-feed supplementations with nano selenium and other micronutrients at optimum levels for enhancing the growth performance and productivity.

6 Conclusion

Contaminants like microplastics and pesticides have harmed the nutritional value and general health of many marine creatures, particularly fish. Humans can profit from fish in numerous ways, including food and financial gain. Fish health is impacted by microplastics and pesticides either by themselves or in conjunction with other environmental pollutants. It has also been discovered that toxic nanomaterials cause fish to experience oxidative stress. Fish exposed to these pollutants experience reductions in growth, oxidative stress, immunotoxicity, neurotoxicity, inflammation, organ damage, physical harm, and behavioral changes.

Furthermore, pesticide and microplastic residues in aquatic organisms have the potential to spread across food webs and have a negative impact on human health. This study offers a thorough understanding of how pesticides and microplastics impact fish signaling pathways and induce oxidative stress. There is still much to learn about the long-term impacts of microplastic exposure on aquatic life and human health, despite a recent surge in study in this area. According to studies, microplastics may build up in human tissues and organs, which could result in oxidative stress, inflammation, and harm to the liver, kidneys, brain, and stomach, among other health issues. Nevertheless, more research is required to fully comprehend the risks mentioned and create plans to reduce any exposure. It has been suggested that using nanotechnology in the fishing sector could help with a number of issues, especially those related to fish illnesses. In order to assess the prevalence of microplastics and pesticides in fish that enter food webs, as well as their possible detrimental effects on personal health and general quality of life over the course of their lives, future study should incorporate comparative and protracted monitoring studies.

References

Abdel-Razek, N. (2019). Antimicrobial activities of chitosan nanoparticles against pathogenic microorganisms in Nile tilapia, Oreochromis niloticus. *Aquaculture International, 27*(5), 1315–1330.

Abinaya, M., Shanthi, S., Palmy, J., Al-Ghanim, K. A., Govindarajan, M., & Vaseeharan, B. (2023). Exopolysaccharides-mediated ZnO nanoparticles for the treatment of aquatic diseases in freshwater fish Oreochromis mossambicus. *Toxics, 11*(4), 313.

Adeleye, A. S., Conway, J. R., Perez, T., Rutten, P., & Keller, A. A. (2014). Influence of extracellular polymeric substances on the long-term fate, dissolution, and speciation of copper-based nanoparticles. *Environmental Science & Technology, 48*(21), 12561–12568.

Ahmed, F., Soliman, F. M., Adly, M. A., Soliman, H. A., El-Matbouli, M., & Saleh, M. (2020). In vitro assessment of the antimicrobial efficacy of chitosan nanoparticles against major fish pathogens and their cytotoxicity to fish cell lines. *Journal of Fish Diseases, 43*(9), 1049–1063.

Ahmed, A. S., Billah, M. M., Ali, M. M., Bhuiyan, M. K. A., Guo, L., Mohinuzzaman, M., & Cai, W. (2023). Microplastics in aquatic environments: A comprehensive review of toxicity, removal, and remediation strategies. *Science of the Total Environment, 876*, 162414.

Akarsu, C., Kumbur, H., Gökdağ, K., Kıdeyş, A. E., & Sanchez-Vidal, A. (2020). Microplastics composition and load from three wastewater treatment plants discharging into Mersin Bay, north eastern Mediterranean Sea. *Marine Pollution Bulletin, 150*, 110776.

Alberghini, L., Truant, A., Santonicola, S., Colavita, G., & Giaccone, V. (2022). Microplastics in fish and fishery products and risks for human health: A review. *International Journal of Environmental Research and Public Health, 20*(1), 789.

Al-Gubory, K. H., Garrel, C., Faure, P., & Sugino, N. (2012). Roles of antioxidant enzymes in corpus luteum rescue from reactive oxygen species-induced oxidative stress. *Reproductive Biomedicine Online, 25*(6), 551–560.

Alimba, C. G., & Faggio, C. (2019). Microplastics in the marine environment: Current trends in environmental pollution and mechanisms of toxicological profile. *Environmental Toxicology and Pharmacology, 68*, 61–74.

Alomar, C., Sureda, A., Capó, X., Guijarro, B., Tejada, S., & Deudero, S. (2017). Microplastic ingestion by Mullus surmuletus Linnaeus, 1758 fish and its potential for causing oxidative stress. *Environmental Research, 159*, 135–142.

Alomar, C., Deudero, S., Compa, M., & Guijarro, B. (2020). Exploring the relation between plastic ingestion in species and its presence in seafloor bottoms. *Marine Pollution Bulletin, 160*, 111641.

Alomar, C., Sanz-Martín, M., Compa, M., Rios-Fuster, B., Álvarez, E., Ripolles, V., et al. (2021). Microplastic ingestion in reared aquaculture fish: Biological responses to low-density polyethylene controlled diets in Sparus aurata. *Environmental Pollution, 280*, 116960.

Amin, A., Indulkar, S. T., & Pai, R. (2016). Sublethal effect of Buprofezin pesticide on carcass composition of *Cyprinus carpio* communis fingerlings. *Journal of Experimental Zoology India, 20*(1), 269–272.

Amin, A., Ramachandra Naik, A. T., Priyadarshini, N., Nayak, H., & Sree, C. S. (2017). Toxic effect of heavy metal lead on oxygen consumption of rohu (*Labeo rohita*) fingerlings. *Biochemical & Cellular Archives, 17*(1), 225–228.

Anita, & Khati, A. (2019). Biomedicines and their role in fish health management. *International Journal of Fauna and Biological Studies, 6*, 10–14.

Atchudan, R., Edison, T. N. J. I., Perumal, S., Karthikeyan, D., & Lee, Y. R. (2017). Effective photocatalytic degradation of anthropogenic dyes using graphene oxide grafting titanium dioxide nanoparticles under UV-light irradiation. *Journal of Photochemistry and Photobiology A: Chemistry, 333*, 92–104.

Awad, A., Zaglool, A. W., Ahmed, S. A., & Khalil, S. R. (2019). Transcriptomic profile change, immunological response and disease resistance of Oreochromis niloticus fed with conventional and Nano-Zinc oxide dietary supplements. *Fish & Shellfish Immunology, 93*, 336–343.

Bakir, A., Doran, D., Silburn, B., Russell, J., Archer-Rand, S., Barry, J., et al. (2023). A spatial and temporal assessment of microplastics in seafloor sediments: A case study for the UK. *Frontiers in Marine Science, 9*, 1093815.

Bal, A., Pati, S. G., Panda, F., & Paital, B. (2022). Dehydration induced hypoxia and its role on mitochondrial respiratory enzymes and oxidative stress responses in liver of Asian stinging catfish Heteropneustes fossilis. *Comparative Biochemistry and Physiology Part C: Toxicology & Pharmacology, 256*, 109300.

Balasubramanian, S., Haneen, M. A., Sharma, G., & Perumal, E. (2024). Acute copper oxide nanoparticles exposure alters zebrafish larval microbiome. *Life Sciences, 336*, 122313.

Barboza, L. G. A., Vethaak, A. D., Lavorante, B. R., Lundebye, A. K., & Guilhermino, L. (2018a). Marine microplastic debris: An emerging issue for food security, food safety and human health. *Marine Pollution Bulletin, 133*, 336–348.

Barboza, L. G. A., Vieira, L. R., Branco, V., Carvalho, C., & Guilhermino, L. (2018b). Microplastics increase mercury bioconcentration in gills and bioaccumulation in the liver, and cause oxidative stress and damage in Dicentrarchus labrax juveniles. *Scientific Reports, 8*(1), 15655.

Bhagat, J., et al. (2020). Microplastics induce oxidative stress and cytotoxicity in fish cells. *Aquatic Toxicology, 229*, 105688. https://doi.org/10.1016/j.aquatox.2020.105688

Bhatti, F. U. R., Kim, S. J., Yi, A. K., Hasty, K. A., & Cho, H. (2018). Cytoprotective role of vitamin E in porcine adipose-tissue-derived mesenchymal stem cells against hydrogen-peroxide-induced oxidative stress. *Cell and Tissue Research, 374*, 111–120.

Bhoopathy, S., Inbakandan, D., Thirugnanasambandam, R., Kumar, C., Sampath, P., Bethunaickan, R., et al. (2021). A comparative study on chitosan nanoparticle synthesis methodologies for application in aquaculture through toxicity studies. *IET Nanobiotechnology, 15*(4), 418–426.

Bhupinder, S. S. (2014). Nanotechnology in agri-food production: An overview. *Nanotechnology, Science and Applications, 7*, 31–53. https://doi.org/10.2147/NSA.S39406

Bhuyan, M. S. (2022). Effects of microplastics on fish and in human health. *Frontiers in Environmental Science, 10*, 827289.

Birnie-Gauvin, K., Costantini, D., Cooke, S. J., & Willmore, W. G. (2017). A comparative and evolutionary approach to oxidative stress in fish: A review. *Fish and Fisheries, 18*(5), 928–942.

Boháčková, J., Havlíčková, L., Semerád, J., Titov, I., Trhlíková, O., Beneš, H., & Cajthaml, T. (2023). In vitro toxicity assessment of polyethylene terephthalate and polyvinyl chloride microplastics using three cell lines from rainbow trout (Oncorhynchus mykiss). *Chemosphere, 312*, 136996.

Boyd, C. E., & Tucker, C. S. (2012). *Pond aquaculture water quality management*. Springer Science & Business Media.

Brennecke, D., Duarte, B., Paiva, F., Caçador, I., & Canning-Clode, J. (2016). Microplastics as vector for heavy metal contamination from the marine environment. *Estuarine, Coastal and Shelf Science, 178*, 189–195.

Cai, L., Zhao, Y., Shi, H., Zhang, W., Zhang, S., Xu, X., et al. (2021). Ecotoxicological effects of microplastics on marine organisms: A review. *Ecotoxicology and Environmental Safety, 214*, 112104. https://doi.org/10.1016/j.ecoenv.2021.112104

Cao, H., Yu, R., Zhang, Y., Hu, B., Jian, S., Wen, C., et al. (2019). Effects of dietary supplementation with β-glucan and Bacillus subtilis on growth, fillet quality, immune capacity, and antioxidant status of Pengze crucian carp (Carassius auratus var. Pengze). *Aquaculture, 508*, 106–112.

Cao, Q., Sun, W., Yang, T., Zhu, Z., Jiang, Y., Hu, W., et al. (2022). The toxic effects of polystyrene microplastics on freshwater algae Chlorella pyrenoidosa depends on the different size of polystyrene microplastics. *Chemosphere, 308*, 136135.

Cao, J., Xu, R., Wang, F., Geng, Y., Xu, T., Zhu, M., et al. (2023). Polyethylene microplastics trigger cell apoptosis and inflammation via inducing oxidative stress and activation of the NLRP3 inflammasome in carp gills. *Fish & Shellfish Immunology, 132*, 108470.

Chae, Y., & An, Y. J. (2020). Effects of food presence on microplastic ingestion and egestion in Mytilus galloprovincialis. *Chemosphere, 240*, 124855.

Chaplin, D. D. (2010). Overview of the immune response. *Journal of Allergy and Clinical Immunology, 125*(2), S3–S23.

Chen, Y., & Feng, X. (2022). Gold nanoparticles for skin drug delivery. *International Journal of Pharmaceutics, 625*, 122122.

Chen, Q., Lackmann, C., Wang, W., Seiler, T. B., Hollert, H., & Shi, H. (2020). Microplastics lead to hyperactive swimming behaviour in adult zebrafish. *Aquatic Toxicology, 224*, 105521.

Chen, Q., Li, M., Li, L., Li, J., Li, Y., & Luo, Y. (2021a). Polyethylene microplastics cause oxidative stress and DNA damage in the liver of tilapia (Oreochromis niloticus). *Environmental Pollution, 272*, 115964.

Chen, Z., Huang, Z., Liu, J., Wu, E., Zheng, Q., & Cui, L. (2021b). Phase transition of mg/Al-flocs to mg/Al-layered double hydroxides during flocculation and polystyrene nanoplastics removal. *Journal of Hazardous Materials, 406*, 124697.

Chen, G. H., Song, C. C., Zhao, T., Hogstrand, C., Wei, X. L., Lv, W. H., et al. (2022). Mitochondria-dependent oxidative stress mediates ZnO nanoparticle (ZnO NP)-induced mitophagy and lipotoxicity in freshwater teleost fish. *Environmental Science & Technology, 56*(4), 2407–2420.

Cheng, C. H., Liang, H. Y., Luo, S. W., Wang, A. L., & Ye, C. X. (2018). The protective effects of vitamin C on apoptosis, DNA damage and proteome of pufferfish (Takifugu obscurus) under low temperature stress. *Journal of Thermal Biology, 71*, 128–135.

Citarasu, T. (2010). Herbal biomedicines: A new opportunity for aquaculture industry. *Aquaculture International, 18*(3), 403–414.

Cocci, P., Gabrielli, S., Pastore, G., Minicucci, M., Mosconi, G., & Palermo, F. A. (2022). Microplastics accumulation in gastrointestinal tracts of Mullus barbatus and Merluccius merluccius is associated with increased cytokine production and signaling. *Chemosphere, 307*, 135813.

Cole, M., Lindeque, P., Halsband, C., & Galloway, T. S. (2011). Microplastics as contaminants in the marine environment: A review. *Marine Pollution Bulletin, 62*(12), 2588–2597.

Cole, M., Lindeque, P., Fileman, E., Halsband, C., Goodhead, R., & Moger, J. (2021). Microplastics cause widespread oxidative stress in fish. *Aquatic Toxicology, 232*, 105717.

Cox, K. D., Covernton, G. A., Davies, H. L., Dower, J. F., Juanes, F., & Dudas, S. E. (2019). Human consumption of microplastics. *Environmental Science & Technology, 53*(12), 7068–7074.

Cui, J., Zhang, Y., Liu, L., Zhang, Q., Xu, S., & Guo, M. Y. (2023). Polystyrene microplastics induced inflammation with activating the TLR2 signal by excessive accumulation of ROS in hepatopancreas of carp (Cyprinus carpio). *Ecotoxicology and Environmental Safety, 251*, 114539.

Dasari, R., Vankara, A. P., Khateef, R., Shegu, V. R. T., Degati, V., & Thummala, C. (2024). Role of nanoparticles in fish disease management: A review. *Biocatalysis and Agricultural Biotechnology, 58*, 103218.

Dash, J. P., Mani, L., & Nayak, S. K. (2022). Antibacterial activity of Blumea axillaris synthesized selenium nanoparticles against multidrug resistant pathogens of aquatic origin. *Egyptian Journal of Basic and Applied Sciences, 9*(1), 65–76.

Dawit Moges, F., Hamdi, H., Al-Barty, A., Zaid, A. A., Sundaray, M., Parashar, S. K. S., et al. (2022). Effects of selenium nanoparticle on the growth performance and nutritional quality in Nile Tilapia, Oreochromis niloticus. *PloS One, 17*(6), e0268348.

Dawood, M. A., Basuini, M. F. E., Yilmaz, S., Abdel-Latif, H. M., Kari, Z. A., Abdul Razab, M. K. A., et al. (2021). Selenium nanoparticles as a natural antioxidant and metabolic regulator in aquaculture: A review. *Antioxidants, 10*(9), 1364.

Dawson, A. L., Kawaguchi, S., King, C. K., Townsend, K. A., King, R., Huston, W. M., & Bengtson Nash, S. M. (2018). Turning microplastics into nanoplastics through digestive fragmentation by Antarctic krill. *Nature Communications, 9*(1), 1001.

De Falco, F., Di Pace, E., Cocca, M., & Avella, M. (2019). The contribution of washing processes of synthetic clothes to microplastic pollution. *Scientific Reports, 9*(1), 6633.

Debnath, R., Prasad, G. S., Amin, A., Malik, M. M., Ahmad, I., Abubakr, A., Borah, S., Rather, M. A., Impellitteri, F., Tabassum, I., & Piccione, G. (2024). Understanding and addressing microplastic pollution: Impacts, mitigation, and future perspectives. *Journal of Contaminant Hydrology, 266*, 104399.

Deenathayalan, U., Nandita, R., Kavithaa, K., Kavitha, V. S., Govindasamy, C., Al-Numair, K. S., et al. (2024). Evaluation of developmental toxicity and oxidative stress caused by zinc oxide nanoparticles in zebra fish embryos/larvae. *Applied Biochemistry and Biotechnology, 196*(8), 4954–4973.

Delavari, N. M., Gharaei, A., Mirdar, H. J., Davari, A., & Rastiannasab, A. (2022). Modulatory effect of dietary copper nanoparticles and vitamin C supplementations on growth performance, hematological and immune parameters, oxidative status, histology, and disease resistance against Yersinia ruckeri in rainbow trout (Oncorhynchus mykiss). *Fish Physiology and Biochemistry, 48*, 1–19.

Devi, G., Harikrishnan, R., Paray, B. A., Al-Sadoon, M. K., Hoseinifar, S. H., & Balasundaram, C. (2019). Effect of symbiotic supplemented diet on innate-adaptive immune response, cytokine gene regulation and antioxidant property in Labeo rohita against Aeromonas hydrophila. *Fish & Shellfish Immunology, 89*, 687–700.

EFSA Panel on Contaminants in the Food Chain (CONTAM). (2016). Presence of microplastics and nanoplastics in food, with particular focus on seafood. *EFSA Journal, 14*(6), e04501.

El-Sayed Ali, T., El-Sayed, A. M., Eissa, M. A. R., & Hanafi, H. M. (2018). Effects of dietary Biogen and sodium butyrate on hematological parameters, immune response, and histological characteristics of Nile tilapia (Oreochromis niloticus) fingerlings. *Aquaculture International, 26*, 139–150.

Espinosa, C., Beltrán, J. M. G., Esteban, M. A., & Cuesta, A. (2018). In vitro effects of virgin microplastics on fish head-kidney leucocyte activities. *Environmental Pollution, 235*, 30–38.

ETC (2003). *Action group on erosion, technology and concentration down on the farm: The impact of nanoscale technologies on food and agriculture*. Avaliable at: http://www.etcgroup.org/en/materials/publications.html?pub_id=80

Evensen, O. (2009). Development in fish vaccinology with focus on delivery methodologies, adjuvants and formulations. *Options Mediterraneennes, 86*, 177–186.

Fang, C., Zheng, R., Hong, F., Jiang, Y., Chen, J., Lin, H., et al. (2021). Microplastics in three typical benthic species from the Arctic: Occurrence, characteristics, sources, and environmental implications. *Environmental Research, 192*, 110326.

Ferrante, M. C., Monnolo, A., Del Piano, F., Mattace Raso, G., & Meli, R. (2022). The pressing issue of micro-and nanoplastic contamination: Profiling the reproductive alterations mediated by oxidative stress. *Antioxidants, 11*(2), 193.

Firdaus-Nawi, M., & Zamri-Saad, M. (2016). Major components of fish immunity: A review. *Pertanika Journal of Tropical Agricultural Science, 39*(4).

Frederick, K., Kuhn, M. E., & Tarver, T. (2010). Newinsights into food and health. *Food Technology-Champaign Then Chic, 64*(5), 44–49. https://doi.org/10.13057/nusbiosci/n140213

Gehrke, I., Geiser, A., & Somborn-Schulz, A. (2015). Innovations in nanotechnology for water treatment. *Nanotechnology, Science and Applications, 8*, 1–17.

Gurjar, U. R., Takar, S., Amin, A., & Martin Xavier, K. A. (2023). Microplastic contamination in aquatic organisms: An ecotoxicological perspective. In *Xenobiotics in aquatic animals: Reproductive and developmental impacts* (pp. 353–367). Springer Nature Singapore.

Haleem, A., Javaid, M., Singh, R. P., Rab, S., & Suman, R. (2023). Applications of nanotechnology in medical field: A brief review. *Global Health Journal, 7*(2), 70–77.

He, M., Yan, M., Chen, X., Wang, X., Gong, H., Wang, W., & Wang, J. (2022). Bioavailability and toxicity of microplastics to zooplankton. *Gondwana Research, 108*, 120–126.

Henning, T., & Weber, D. (2021). Redox biomarkers in dietary interventions and nutritional observation studies-from new insights to old problems. *Redox Biology, 41*, 101922.

Hossain, M. S., Koshio, S., Ishikawa, M., Yokoyama, S., Sony, N. M., Dossou, S., & Wang, W. (2018). Influence of dietary inosine and vitamin C supplementation on growth, blood chemistry, oxidative stress, innate and adaptive immune responses of red sea bream, Pagrus major juvenile. *Fish & Shellfish Immunology, 82*, 92–100.

Huang, H., Liu, R., Yang, J., Dai, J., Fan, S., Pi, J., et al. (2023). Gold nanoparticles: Construction for drug delivery and application in cancer immunotherapy. *Pharmaceutics, 15*(7), 1868.

Idowu, T. A., Adedeji, H. A., & Sogbesan, O. A. (2017). Fish disease and health management in aquaculture production. *International Journal Environmental & Agricultural Science, 1*(2).

Iheanacho, S. C., Igberi, C., Amadi-Eke, A., Chinonyerem, D., Iheanacho, A., & Avwemoya, F. (2020). Biomarkers of neurotoxicity, oxidative stress, hepatotoxicity and lipid peroxidation in Clarias gariepinus exposed to melamine and polyvinyl chloride. *Biomarkers, 25*(7), 603–610.

Iqbal, S., Atique, U., Mahboob, S., Haider, M. S., Iqbal, H. S., Al-Ghanim, K. A., et al. (2020). Effect of supplemental selenium in fish feed boosts growth and gut enzyme activity in juvenile tilapia (Oreochromis niloticus). *Journal of King Saud University-Science, 32*(5), 2610–2616.

Ji, Y., Qiu, J., Xie, T., McCarron, P., & Li, A. (2018). Accumulation and transformation of azaspiracids in scallops (Chlamys farreri) and mussels (Mytilus galloprovincialis) fed with Azadinium poporum, and response of antioxidant enzymes. *Toxicon, 143*, 20–28.

Jones, D. P. (2008). Radical-free biology of oxidative stress. *American Journal of Physiology-Cell Physiology, 295*(4), C849–C868.

Joudeh, N., & Linke, D. (2022). Nanoparticle classification, physicochemical properties, characterization, and applications: A comprehensive review for biologists. *Journal of Nanobiotechnology, 20*(1), 262.

Kakakhel, M. A., Wu, F., Sajid, W., Zhang, Q., Khan, I., Ullah, K., & Wang, W. (2021). Long-term exposure to highconcentration silver nanoparticles induced toxicity, fatality, bioaccumulation, and histological alteration in fish (Cyprinus carpio). *Environmental Sciences Europe, 33*(1), 1–11.

Karbalaei, S., Hanachi, P., Walker, T. R., & Cole, M. (2018). Occurrence, sources, human health impacts and mitigation of microplastic pollution. *Environmental Science and Pollution Research, 25*, 36046–36063.

Khan, Z., Maqsood, Q., Baradoke, A., Ferreira, L. F. R., Franco, M., Schmidt, J. E., & Hussain, N. (2024). Environmental and toxicological implications of chitosan nanostructures. In *Advances in chemical pollution, environmental management and protection* (Vol. 10, pp. 137–172). Elsevier.

Khosravi-Katuli, K., Prato, E., Lofrano, G., Guida, M., Vale, G., & Libralato, G. (2017). Effects of nanoparticles in species of aquaculture interest. *Environmental Science and Pollution Research, 24*, 17326–17346.

Khursheed, S., Dutta, J., Ahmad, I., Rather, M. A., Badroo, I. A., Bhat, T. A., et al. (2023). Biogenic silver nanoparticles: Synthesis, applications and challenges in food sector with special emphasis on aquaculture. *Food Chemistry, X*, 101051.

Koelmans, A. A., Nor, N. H. M., Hermsen, E., Kooi, M., Mintenig, S. M., & De France, J. (2019). Microplastics in freshwaters and drinking water: Critical review and assessment of data quality. *Water Research, 155*, 410–422.

Kothari, D., Patel, S., & Kim, S. K. (2019). Probiotic supplements might not be universally-effective and safe: A review. *Biomedicine & Pharmacotherapy, 111*, 537–547.

Kuang, L., Liu, Y., Fu, D., & Zhao, Y. (2017). FeOOH-graphene oxide nanocomposites for fluoride removal from water: Acetate mediated nano FeOOH growth and adsorption mechanism. *Journal of Colloid and Interface Science, 490*, 259–269.

Kumar, N., Krishnani, K. K., Gupta, S. K., & Singh, N. P. (2017). Selenium nanoparticles enhanced thermal tolerance and maintain cellular stress protection of Pangasius hypophthalmus reared under lead and high temperature. *Respiratory Physiology & Neurobiology, 246*, 107–116.

Lehner, R., Weder, C., Petri-Fink, A., & Rothen-Rutishauser, B. (2019). Emergence of nanoplastic in the environment and possible impact on human health. *Environmental Science & Technology, 53*(4), 1748–1765.

Letchumanan, D., Sok, S. P., Ibrahim, S., Nagoor, N. H., & Arshad, N. M. (2021). Plant-based biosynthesis of copper/copper oxide nanoparticles: An update on their applications in biomedicine, mechanisms, and toxicity. *Biomolecules, 11*(4), 564.

Li, H., Jiang, W., Liu, Y., Jiang, J., Zhang, Y., Wu, P., et al. (2016). The metabolites of glutamine prevent hydroxyl radical-induced apoptosis through inhibiting mitochondria and calcium ion involved pathways in fish erythrocytes. *Free Radical Biology and Medicine, 92*, 126–140.

Li, S., Jiang, Y., Sun, Q., Coffin, S., Chen, L., Qiao, K., et al. (2020). Tebuconazole induced oxidative stress related hepatotoxicity in adult and larval zebrafish (Danio rerio). *Chemosphere, 241*, 125129.

Li, B., Wang, Y., Zhao, H., Yin, K., Liu, Y., Wang, D., et al. (2022). Oxidative stress is involved in the activation of NF-κB signal pathway and immune inflammatory response in grass carp gill induced by cypermethrin and/or sulfamethoxazole. *Environmental Science and Pollution Research, 29*, 1–14.

Liang, W., Li, B., Jong, M. C., Ma, C., Zuo, C., Chen, Q., & Shi, H. (2023). Process-oriented impacts of microplastic fibers on behavior and histology of fish. *Journal of Hazardous Materials, 448*, 130856.

Liu, J., Wu, S., Zhao, H., Zhang, J., & Jiang, L. (2019). Microplastics as carriers of bacterial pathogens: An in vitro study on the adsorption of Bacillus cereus by polystyrene microplastics. *Environmental Pollution, 248*, 982–988.

Lönnstedt, O. M., & Eklöv, P. (2016). RETRACTED: Environmentally relevant concentrations of microplastic particles influence larval fish ecology. *Science, 352*(6290), 1213–1216.

Low, C. F., Rozaini, M. Z. H., Musa, N., & Syarul Nataqain, B. (2017). Current knowledge of metabolomic approach in infectious fish disease studies. *Journal of Fish Diseases, 40*(10), 1267–1277.

Luis, A. I. S., Campos, E. V. R., de Oliveira, J. L., & Fraceto, L. F. (2019). Trends in aquaculture sciences: From now to use of nanotechnology for disease control. *Reviews in Aquaculture, 11*(1), 119–132.
Luo, D., Wang, X., Burda, C., & Basilion, J. P. (2021). Recent development of gold nanoparticles as contrast agents for cancer diagnosis. *Cancers, 13*(8), 1825.
Malhotra, N., Ger, T. R., Uapipatanakul, B., Huang, J. C., Chen, K. H. C., & Hsiao, C. D. (2020). Review of copper and copper nanoparticle toxicity in fish. *Nanomaterials, 10*(6), 1126.
Malik, M., Balkhi, M. H., & Abubakr, A. (2017). Comparative study on the contamination level of Dal Lake, Kashmir based on Total viable bacterial counts over a period of time. *Ecology Environment and Conservation, 23*(4), 2145–2147.
Mallik, A., Xavier, K. M., Naidu, B. C., & Nayak, B. B. (2021). Ecotoxicological and physiological risks of microplastics on fish and their possible mitigation measures. *Science of the Total Environment, 779*, 146433.
Mandal, A. H., Ghosh, S., Adhurjya, D., Chatterjee, P., Samajdar, I., Mukherjee, D., et al. (2024). Exploring the impact of zinc oxide nanoparticles on fish and fish-food organisms: A review. *Aquaculture Reports, 36*, 102038.
Martínez-Álvarez, R. M., Morales, A. E., & Sanz, A. (2005). Antioxidant defenses in fish: Biotic and abiotic factors. *Reviews in Fish Biology and Fisheries, 15*, 75–88.
Mattsson, K., Johnson, E. V., Malmendal, A., Linse, S., Hansson, L. A., & Cedervall, T. (2017). Brain damage and behavioural disorders in fish induced by plastic nanoparticles delivered through the food chain. *Scientific Reports, 7*(1), 11452.
Mishra, V., Shah, C., Mokashe, N., Chavan, R., Yadav, H., & Prajapati, J. (2015). Probiotics as potential antioxidants: A systematic review. *Journal of Agricultural and Food Chemistry, 63*(14), 3615–3626.
Monteiro, D. A., De Almeida, J. A., Rantin, F. T., & Kalinin, A. L. (2006). Oxidative stress biomarkers in the freshwater characid fish, Brycon cephalus, exposed to organophosphorus insecticide Folisuper 600 (methyl parathion). *Comparative Biochemistry and Physiology Part C: Toxicology & Pharmacology, 143*(2), 141–149.
Munawar, N., Aleem, M., Abru, S., Laraib, T., Zainab, S. N., Manzoor, S., & Ali, A. (2021). Interaction and applications of nanoparticles in fishes and aquaculture. *Scholars Bulletin, 7*(6), 150–155.
Nabi, N., Ahmad, I., Amin, A., Rather, M. A., Ahmed, I., Hajam, Y. A., Khursheed, S., Malik, M. M., & Abubakr, A. (2024). Understanding the sources, fate and effects of microplastics in aquatic environments with a focus on risk profiling in aquaculture systems. *Reviews in Aquaculture, 16*(4), 1947–1980.
Narra, M. R., Rajender, K., Reddy, R. R., Rao, J. V., & Begum, G. (2015). The role of vitamin C as antioxidant in protection of biochemical and haematological stress induced by chlorpyrifos in freshwater fish Clarias batrachus. *Chemosphere, 132*, 172–178.
Nava, V., Chandra, S., Aherne, J., Alfonso, M. B., Antão-Geraldes, A. M., Attermeyer, K., et al. (2023). Plastic debris in lakes and reservoirs. *Nature, 619*(7969), 317–322.
Navarro-Yepes, J., Zavala-Flores, L., Anandhan, A., Wang, F., Skotak, M., Chandra, N., et al. (2014). Antioxidant gene therapy against neuronal cell death. *Pharmacology & Therapeutics, 142*(2), 206–230.
Ni, J., Ren, L., Ma, Y., Xiong, H., & Jian, W. (2023). Selenium nanoparticles coated with polysaccharide-protein complexes from abalone viscera improve growth and enhance resistance to diseases and hypoxic stress in juvenile Nile tilapia (Oreochromis niloticus). *Fish & Shellfish Immunology, 134*, 108624.
Ninawe, A. S., Hameed, A. S., & Selvin, J. (2017). Advancements in diagnosis and control measures of viral pathogens in aquaculture: An Indian perspective. *Aquaculture International, 25*, 251–264.
Ouyang, J., Sun, F., Feng, W., Sun, Y., Qiu, X., Xiong, L., et al. (2016). Quercetin is an effective inhibitor of quorum sensing, biofilm formation and virulence factors in Pseudomonas aeruginosa. *Journal of Applied Microbiology, 120*(4), 966–974.

Padmini, K., & Udayakumar, K. (2021). Oxidative stress in fish and its impact on fish health. *Advanced Fish Health Management, 2021*, 227–245.

Paital, B., Guru, D., Mohapatra, P., Panda, B., Parida, N., Rath, S., et al. (2019). Ecotoxic impact assessment of graphene oxide on lipid peroxidation at mitochondrial level and redox modulation in fresh water fish Anabas testudineus. *Chemosphere, 224*, 796–804.

Panda, F., Pati, S. G., Anwar, T. N., Samanta, L., & Paital, B. (2022). Seasonal variation of water quality modulated redox regulatory system in the apple snail Pila globosa and its use as a bioindicator species in freshwater ecosystems across India. *Water, 14*(20), 3275.

Parolini, M., Iacobuzio, R., De Felice, B., Bassano, B., Pennati, R., & Saino, N. (2019). Age-and sex-dependent variation in the activity of antioxidant enzymes in the brown trout (Salmo trutta). *Fish Physiology and Biochemistry, 45*, 145–154.

Pati, S. G., Panda, F., Samanta, L., & Paital, B. (2023). Spatio-temporal changes in oxidative stress physiology parameters in apple snail Pila globosa as a function of soil Mg, Ca, organic carbon and aquatic physico-chemical factors. *Environmental Geochemistry and Health, 45*(5), 2591–2610.

Pierre, L., Bruna Bugueño, J. E., Leyton Bongiorno, P. A., Torres Mediano, A., & Rodríguez-Mercado, F. J. (2024). Colorimetric indicator based on gold nanoparticles and sodium alginate for monitoring fish spoilage. *Polymers, 16*(6), 829.

Pizzino, G., Irrera, N., Cucinotta, M., Pallio, G., Mannino, F., Arcoraci, V., et al. (2017). Oxidative stress: Harms and benefits for human health. *Oxidative Medicine and Cellular Longevity, 2017*(1), 8416763.

Prasad, G. S., Rout, S. K., Malik, M. M., Karmakar, S., Amin, A., & Ahmad, I. (2023). Occurrence of xenoestrogen alkylphenols (Octylphenols and Nonylphenol) and its impact on the aquatic ecosystem. In *Xenobiotics in aquatic animals: Reproductive and developmental impacts* (pp. 275–284). Springer.

Prasad, G. S., Rout, S. K., Amin, A., Malik, M. M., Ahmad, I., Rather, M. A., Karmakar, S., Khalil, H. S., & Abubakar, A. (2024). Impact of COVID-19 on the aquatic environment and implications for fisheries with special emphasis to Indian perspective. *Proceedings of the Indian National Science Academy, 90*(4), 821–839.

Prata, J. C., da Costa, J. P., Lopes, I., Duarte, A. C., & Rocha-Santos, T. (2020). Environmental exposure to microplastics: An overview on possible human health effects. *Science of the Total Environment, 702*, 134455.

Prego-Faraldo, M. V., Vieira, L. R., Eirin-Lopez, J. M., Méndez, J., & Guilhermino, L. (2017). Transcriptional and biochemical analysis of antioxidant enzymes in the mussel Mytilus galloprovincialis during experimental exposures to the toxic dinoflagellate Prorocentrum lima. *Marine Environmental Research, 129*, 304–315.

Qiang, L., & Cheng, J. (2021). Exposure to polystyrene microplastics impairs gonads of zebrafish (Danio rerio). *Chemosphere, 263*, 128161.

Qin, F., Shi, M., Yuan, H., Yuan, L., Lu, W., Zhang, J., et al. (2016). Dietary nano-selenium relieves hypoxia stress and, improves immunity and disease resistance in the Chinese mitten crab (Eriocheir sinensis). *Fish & Shellfish Immunology, 54*, 481–488.

Rajendran, P., Nandakumar, N., Rengarajan, T., Palaniswami, R., Gnanadhas, E. N., Lakshminarasaiah, U., et al. (2014). Antioxidants and human diseases. *Clinica Chimica Acta, 436*, 332–347.

Ramachandraiah, K., Ameer, K., Jiang, G., & Hong, G. P. (2022). Micro-and nanoplastic contamination in livestock production: Entry pathways, potential effects and analytical challenges. *Science of the Total Environment, 844*, 157234.

Rasheed, A., Iqbal, K. J., Safdar, A., Nasir, A., Jabeen, R., Tara, N., et al. (2023). Toxicological effects of zinc oxide nanoparticles on hemato-biochemical profile of common carp (Cyprinus carpio). *Journal of King Saud University-Science, 35*(7), 102835.

Rather, M. A., Sharma, R., Aklakur, M., Ahmad, S., Kumar, N., Khan, M., et al. (2011). Nanotechnology: A novel tool for aquaculture and fisheries development. A prospective mini-review. *Fish. Aquac. J. FAJ, 16*, 1–5. https://doi.org/10.1016/j.envpol.2018.08.045

Rathore, S. S., Murthy, H. S., Girisha, S. K., Nithin, M. S., Nasren, S., Mamun, M. A. A., et al. (2021). Supplementation of nano-selenium in fish diet: Impact on selenium assimilation and immune-regulated selenoproteome expression in monosex Nile tilapia (Oreochromis niloticus). *Comparative Biochemistry and Physiology Part C: Toxicology & Pharmacology, 240*, 108907.

Ravikumar, S., Gokulakrishnan, R., & Raj, J. A. (2012). Nanoparticles as a source for the treatment of fish diseases. *Asian Pacific Journal of Tropical Disease, 2*, S703–S706.

Rochman, C. M., Brookson, C., Bikker, J., Djuric, N., Earn, A., Bucci, K., et al. (2019). Rethinking microplastics as a diverse contaminant suite. *Environmental Toxicology and Chemistry, 38*(4), 703–711.

Romero, J., Feijoó, C. G., & Navarrete, P. (2012). Antibiotics in aquaculture–use, abuse and alternatives. *Health and Environment in Aquaculture, 159*(1), 159–198.

Saad, A. M., Sitohy, M. Z., Sultan-Alolama, M. I., El-Tarabily, K. A., & El-Saadony, M. T. (2022). Green nanotechnology for controlling bacterial load and heavy metal accumulation in Nile tilapia fish using biological selenium nanoparticles biosynthesized by Bacillus subtilis AS12. *Frontiers in Microbiology, 13*, 1015613.

Salze, G. P., & Davis, D. A. (2015). Taurine: A critical nutrient for future fish feeds. *Aquaculture, 437*, 215–229.

Santos, D., Luzio, A., Félix, L., Bellas, J., & Monteiro, S. M. (2022). Oxidative stress, apoptosis and serotonergic system changes in zebrafish (Danio rerio) gills after long-term exposure to microplastics and copper. *Comparative Biochemistry and Physiology Part C: Toxicology & Pharmacology, 258*, 109363.

Sarasamma, S., Audira, G., Siregar, P., Malhotra, N., Lai, Y. H., Liang, S. T., et al. (2020). Nanoplastics cause neurobehavioral impairments, reproductive and oxidative damages, and biomarker responses in zebrafish: Throwing up alarms of wide spread health risk of exposure. *International Journal of Molecular Sciences, 21*(4), 1410.

Sayed, A. E. H., et al. (2023). Antioxidant therapies for fish exposed to microplastics. *Environmental Science and Pollution Research.* https://doi.org/10.1007/s11356-022-23789

Schieber, M., & Chandel, N. S. (2014). ROS function in redox signaling and oxidative stress. *Current Biology, 24*(10), R453–R462.

Schwabl, P., Köppel, S., Königshofer, P., Bucsics, T., Trauner, M., Reiberger, T., & Liebmann, B. (2019). Detection of various microplastics in human stool: A prospective case series. *Annals of Internal Medicine, 171*(7), 453–457.

Shaalan, M., Saleh, M., El-Mahdy, M., & El-Matbouli, M. (2016). Recent progress in applications of nanoparticles in fish medicine: A review. *Nanomedicine: Nanotechnology, Biology and Medicine, 12*(3), 701–710.

Shaalan, M. I., El-Mahdy, M. M., Theiner, S., El-Matbouli, M., & Saleh, M. (2017). In vitro assessment of the antimicrobial activity of silver and zinc oxide nanoparticles against fish pathogens. *Acta Veterinaria Scandinavica, 59*, 1–11.

Shah, Z. U., & Parveen, S. (2022). Oxidative, biochemical and histopathological alterations in fishes from pesticide contaminated river Ganga, India. *Scientific Reports, 12*(1), 3628.

Sharma, M., Shrivastav, A. B., Sahni, Y. P., & Pandey, G. (2012). Overviews of the treatment and control of common fish diseases. *International Research Journal of Pharmacy, 3*, 123–127.

Sies, H. (2015). Oxidative stress: A concept in redox biology and medicine. *Redox Biology, 4*, 180–183.

Singh, A., Gupta, H., Srivastava, A., Srivastava, A., Joshi, R. C., & Dutta, M. K. (2021). A novel pilot study on imaging-based identification of fish exposed to heavy metal (Hg) contamination. *Journal of Food Processing and Preservation, 45*(6), e15571.

Smith, M., Love, D. C., Rochman, C. M., & Neff, R. A. (2018). Microplastics in seafood and the implications for human health. *Current Environmental Health Reports, 5*, 375–386.

Smriti, A. A., Lodhi, S., & Shukla, S. (2023). Copper toxicity in aquatic ecosystem: A review. *International Journal of Fisheries and Aquatic Studies, 11*(4), 134–138.

Srikanth, K., & Nutalapati, V. (2022). Copper ferrite nanoparticles induced cytotoxicity and oxidative stress in channel catfish ovary cells. *Chemosphere, 287*, 132166.

Sruthisree, C., Gowda, G., Nayak, H., Harish, B. D., & Amin, A. (2015). Effect of heavy metal (lead) on oxygen consumption and ammonia excretion of Iridescent Shark (*Pangusius hypohthalmus*). *Journal of Experimental Zoology India, 18*(2).

Su, L., Nan, B., Craig, N. J., & Pettigrove, V. (2020). Temporal and spatial variations of microplastics in roadside dust from rural and urban Victoria, Australia: Implications for diffuse pollution. *Chemosphere, 252*, 126567.

Subaramaniyam, U., Allimuthu, R. S., Vappu, S., Ramalingam, D., Balan, R., Paital, B., et al. (2023). Effects of microplastics, pesticides and nano-materials on fish health, oxidative stress and antioxidant defense mechanism. *Frontiers in Physiology, 14*, 1217666.

Swain, P., Nayak, S. K., Sasmal, A., Behera, T., Barik, S. K., Swain, S. K., et al. (2014). Antimicrobial activity of metal based nanoparticles against microbes associated with diseases in aquaculture. *World Journal of Microbiology and Biotechnology, 30*, 2491–2502.

Tabassum, I., Hassan, N., Amin, A., Malik, M. M., Abubakr, A., Prasad, G. S., Latief, F., & Rather, S. A. (2025). COVID-19 and global environmental change: A threat to food security and nutrition. In *Food security, nutrition and sustainability through aquaculture technologies* (pp. 437–445). Springer Nature Switzerland.

Tang, J., Ni, X., Zhou, Z., Wang, L., & Lin, S. (2018). Acute microplastic exposure raises stress response and suppresses detoxification and immune capacities in the scleractinian coral Pocillopora damicornis. *Environmental Pollution, 243*, 66–74.

Thakur, S., Mathur, S., Patel, S., & Paital, B. (2022). Microplastic accumulation and degradation in environment via biotechnological approaches. *Water, 14*(24), 4053.

Thirumurthi, N. A., Raghunath, A., Balasubramanian, S., & Perumal, E. (2022). Evaluation of maghemite nanoparticles–induced developmental toxicity and oxidative stress in zebrafish embryos/larvae. *Biological Trace Element Research, 200*, 1–16.

Ucar, A., Parlak, V., Ozgeris, F. B., Yeltekin, A. C., Arslan, M. E., Alak, G., et al. (2022). Magnetic nanoparticles-induced neurotoxicity and oxidative stress in brain of rainbow trout: Mitigation by ulexite through modulation of antioxidant, anti-inflammatory, and antiapoptotic activities. *Science of the Total Environment, 838*, 155718.

Umamaheswari, S., Priyadarshinee, S., Bhattacharjee, M., Kadirvelu, K., & Ramesh, M. (2021a). Exposure to polystyrene microplastics induced gene modulated biological responses in zebrafish (Danio rerio). *Chemosphere, 281*, 128592.

Umamaheswari, S., Priyadarshinee, S., Kadirvelu, K., & Ramesh, M. (2021b). Polystyrene microplastics induce apoptosis via ROS-mediated p53 signaling pathway in zebrafish. *Chemico-Biological Interactions, 345*, 109550.

Vanni, M. J., & Layne, C. D. (1997). Nutrient recycling and herbivory as mechanisms in the "top–down" effect of fish on algae in lakes. *Ecology, 78*(1), 21–40.

Varaprasad, D., Raghavendra, P., Raga Sudha, N., Sarma, L. S., Parveen, S. N., Chandana, P. S., Chandra, S., & Chandrasekhar, T. (2022). Augmentation of ethanol production in green alga Chlorococcum minutum through graphene oxide-supported platinum-ruthenium (Pt-Ru/RGO) nanoparticles. *Bioenergy Research, 15*, 280–288.

Wang, C., Yue, X., Lu, X., & Liu, B. (2013a). The role of catalase in the immune response to oxidative stress and pathogen challenge in the clam Meretrix meretrix. *Fish & Shellfish Immunology, 34*(1), 91–99.

Wang, Y., Yan, X., & Fu, L. (2013b). Effect of selenium nanoparticles with different sizes in primary cultured intestinal epithelial cells of crucian carp, Carassius auratus gibelio. *International Journal of Nanomedicine, 8*, 4007–4013.

Wang, J., Li, Y., Lu, L., Zheng, M., Zhang, X., Tian, H., et al. (2019). Polystyrene microplastics cause tissue damages, sex-specific reproductive disruption and transgenerational effects in marine medaka (Oryzias melastigma). *Environmental Pollution, 254*, 113024.

Wang, J., Gao, X., Liu, F., Dong, J., & Zhao, P. (2022). Difenoconazole causes cardiotoxicity in common carp (Cyprinus carpio): Involvement of oxidative stress, inflammation, apoptosis and autophagy. *Chemosphere, 306*, 135562.

Wang, S., Zheng, L., Shen, M., Zhang, L., Wu, Y., Li, G., et al. (2023). Habitual feeding patterns impact polystyrene microplastic abundance and potential toxicity in edible benthic mollusks. *Science of the Total Environment, 866*, 161341.

Wang, J., Wu, F., Dong, S., Wang, X., Ai, S., Liu, Z., & Wang, X. (2024). Meta-analysis of the effects of microplastic on fish: Insights into growth, survival, reproduction, oxidative stress, and gut microbiota diversity. *Water Research, 267*, 122493.

Woźniak-Budych, M. J., Staszak, K., & Staszak, M. (2023). Copper and copper-based nanoparticles in medicine—Perspectives and challenges. *Molecules, 28*(18), 6687.

Wright, S. L., Ulke, J., Font, A., Chan, K. L. A., & Kelly, F. J. (2020). Atmospheric microplastic deposition in an urban environment and an evaluation of transport. *Environment International, 136*, 105411.

Wu, J. Q., Kosten, T. R., & Zhang, X. Y. (2013). Free radicals, antioxidant defense systems, and schizophrenia. *Progress in Neuro-Psychopharmacology and Biological Psychiatry, 46*, 200–206.

Wu, H., Guo, J., Yao, Y., & Xu, S. (2022). Polystyrene nanoplastics induced cardiomyocyte apoptosis and myocardial inflammation in carp by promoting ROS production. *Fish & Shellfish Immunology, 125*, 1–8.

Xiao, K., Song, L., Li, Y., Li, C., & Zhang, S. (2023). Dietary intake of microplastics impairs digestive performance, induces hepatic dysfunction, and shortens lifespan in the annual fish Nothobranchius guentheri. *Biogerontology, 24*(2), 207–223.

Yang, C., Lim, W., & Song, G. (2020). Mediation of oxidative stress toxicity induced by pyrethroid pesticides in fish. *Comparative Biochemistry and Physiology Part C: Toxicology & Pharmacology, 234*, 108758.

Yao, F. C., Gu, Y., Jiang, T., Wang, P. F., Song, F. B., Zhou, Z., et al. (2023). The involvement of oxidative stress mediated endoplasmic reticulum pathway in apoptosis of Golden Pompano (Trachinotus blochii) liver under PS-MPs stress. *Ecotoxicology and Environmental Safety, 249*, 114440.

Yazhiniprabha, M., Gopi, N., Mahboob, S., Al-Ghanim, K. A., Al-Misned, F., Ahmed, Z., et al. (2022). The dietary supplementation of zinc oxide and selenium nanoparticles enhance the immune response in freshwater fish Oreochromis mossambicus against aquatic pathogen Aeromonas hydrophila. *Journal of Trace Elements in Medicine and Biology, 69*, 126878.

Yu, N., Zhang, M., Xu, X., Xu, L., & Liu, S. (2020a). Exposure to nylon microplastics causes oxidative stress and reduces immune function in the gills of crucian carp (Carassius auratus). *Aquatic Toxicology, 226*, 105557.

Yu, X., Zhao, L., Li, J., Zhang, J., & Li, D. (2020b). Effects of nylon microplastics on oxidative stress and mitochondrial dysfunction in the gills of crucian carp (Carassius auratus). *Science of the Total Environment, 717*, 137156.

Yusuf, A., Sodiq, A., Giwa, A., Eke, J., Pikuda, O., Eniola, J. O., et al. (2022). Updated review on microplastics in water, their occurrence, detection, measurement, environmental pollution, and the need for regulatory standards. *Environmental Pollution, 292*, 118421.

Zhang, Q., Li, D., Zhao, Y., Xu, Y., Li, X., et al. (2020). Polystyrene microplastics induce gut microbiota dysbiosis, inflammation, and metabolic disorder in larval zebrafish. *Environmental Pollution, 261*, 114130.

Zhang, J., Chen, J., Huang, Q., MacKinnon, B., Nekouei, O., Liu, H., et al. (2021). Copper/carbon core/shell nanoparticles: A potential material to control the fish pathogen Saprolegnia parasitica. *Frontiers in Veterinary Science, 8*, 689085.

Zhang, W., Zhao, Z., Zhou, J., Wang, W., Su, J., & Yuan, G. (2023). Carboxymethyl chitosan nanoparticles loaded with Ctenopharyngodon idella interferon-γ2 (CiIFN-γ2) enhance protective efficacy against bacterial infection in grass carp. *Aquaculture, 572*, 739554.

Zhao, L., Seth, A., Wibowo, N., Zhao, C. X., Mitter, N., Yu, C., & Middelberg, A. P. (2014). Nanoparticle vaccines. *Vaccine, 32*(3), 327–337.

Zhu, X., Zhu, L., Lang, Y., & Chen, Y. (2008). Oxidative stress and growth inhibition in the freshwater fish Carassius auratus induced by chronic exposure to sublethal fullerene aggregates. *Environmental Toxicology and Chemistry: An International Journal, 27*(9), 1979–1985.

Zhu, C., Wu, Z., Liu, Q., Wang, X., Zheng, L., He, S., et al. (2024). Selenium nanoparticles in aquaculture: Unique advantages in the production of Se-enriched grass carp (Ctenopharyngodon idella). *Animal Nutrition, 16*, 189–201.

Zitouni, N., Bousserrhine, N., Missawi, O., Boughattas, I., Chèvre, N., Santos, R., et al. (2021). Uptake, tissue distribution and toxicological effects of environmental microplastics in early juvenile fish Dicentrarchus labrax. *Journal of Hazardous Materials, 403*, 124055.

Advanced Techniques for Microplastic Collection, Separation, and Identification in Environmental Samples

Zaheer Abass, Tasaduq Hussain Shah, Karankumar Kishorkumar Ramteke, Dechen Chuskit, Ishtiyaq Ahmad, and Rinkesh Nemichand Wanjari

1 Introduction

One of the biggest developments in recent decades has been the extensive usage of plastics, which have drastically changed contemporary life because of their cost, durability, and resistance to deterioration. The need for these adaptable materials is anticipated to increase as the world's population continues to expand (Thompson et al., 2009). Because they can be moulded into a variety of sizes and forms, plastics are crucial to the production of a large number of goods (Halden, 2020). Plastics are essential to contemporary life, appearing in everything from commonplace products like clothes, electronics, and cosmetics to cutting-edge technology like automobiles, aeroplanes, and rocket parts (Shahul Hamid et al., 2018). These materials are made up of synthetic or semi-synthetic compounds that are frequently produced by polymerising raw materials and monomers based on petroleum. Around 4% of the world's annual petroleum production is dedicated to plastic manufacturing (Sharma et al., 2023). Global production of commodity plastics has grown dramatically since the 1950s, and by 2020, it will have reached 368 million metric tonnes (MMT). However, output decreased somewhat as a result of the COVID-19 pandemic, with a

Z. Abass · T. H. Shah · D. Chuskit
Division of Fisheries Resources Management, Faculty of Fisheries, Sher-e-Kashmir University of Agricultural Science and Technology of Kashmir, Shalimar, Srinagar, India

K. K. Ramteke
ICAR-Central Institute of Fisheries Education (CIFE), Mumbai, Maharashtra, India

I. Ahmad
Division of Fish Genetics and Biotechnology, Faculty of Fisheries, Sher-e-Kashmir University of Agricultural Science and Technology of Kashmir, Shalimar, Srinagar, India

R. N. Wanjari (✉)
Department of Fisheries Resource Management, School of Fisheries, Centurion University of Technology and Management, Paralakhemundi, Odisha, India

A. Abubakr et al. (eds.), *Microplastics: Ecological and Food Security Implications*,
https://doi.org/10.1007/978-3-031-94480-2_11

0.3% decline anticipated in 2020 (Statista, 2020; Prasad et al., 2024; Tabassum et al., 2025). Nevertheless, the pandemic brought to light the crucial role that plastics play, particularly in the extensive usage of disposable hygiene products and personal protective equipment (PPE) (Parashar & Hait, 2021). By 2025, plastic manufacturing is expected to exceed 600 million tonnes, and by 2050, it might reach one billion tonnes (Lusher et al., 2017; Plastics Europe, 2021). Approximately 8.3 billion tonnes of plastic have been manufactured worldwide since its introduction (Geyer et al., 2017). With an estimated worth of $600 billion and an annual growth rate of 8.7%, the plastics sector has emerged as a major economic force. The growing quantity of plastic debris entering the seas, mostly from land-based sources, is one of the major environmental effects of this expansion (Thompson et al., 2009). Globally, freshwater and marine ecosystems are seriously threatened by plastic contamination (Cole et al., 2011). Plastics are the most prevalent kind of marine pollution, accounting up almost 80% of marine debris (IUCN, 2018). Over half of the approximately 8 million tonnes of plastic garbage that enter the seas each year are single-use products, which adds to their persistence in the ecosystem because they are not biodegradable (Napper et al., 2015). Additionally, plastic debris has started to enter deep-sea habitats in recent years, increasing significant ecological concerns (Abass et al., 2025; Bergmann & Klages, 2012). While early research focused on the effects of large-scale plastic debris on marine ecosystems (Cole et al., 2011), recent attention has shifted to microplastic pollution. The U.S. National Oceanic and Atmospheric Administration and the European Chemicals Agency have identified microplastics, plastic particles smaller than 5 mm. The definition offered by Ragusa et al. (2021) is comparable. Macroplastics (>25 mm), mesoplastics (5–25 mm), and microplastics (<5 mm) are the three size classes into which plastic trash is often divided (Fendall & Sewell, 2009; Browne et al., 2010; Crawford & Quinn, 2017; Ng et al., 2018; Gurjar et al., 2023). Large, obvious objects like bottles, bags, and containers are examples of macroplastics, whereas mesoplastics are mid-sized. Depending on where they come from, microplastics are further separated into primary and secondary categories. Pellets consisting of polyethylene and polypropylene are examples of primary microplastics that are often used in manufacturing and frequently find their way into the environment through wastewater from homes and businesses (Eerkes-Medrano et al., 2015; Yao et al., 2020). When bigger plastic debris fragments into smaller ones, between micrometres and nanometres, secondary microplastics are created (Chen et al., 2020). The environment contains a variety of microplastics, such as fibres, pieces, and films. Fibrous microplastics, among the most prevalent forms observed, are especially difficult to detect due to their diameter and polymer makeup. These comprise polymers like rayon, polyester, polyamide (nylon), polyethylene, and polypropylene (Thompson et al., 2004; Lusher et al., 2013; Ballent et al., 2016). Microfibers, which resemble natural wool, fibers and are primarily discharged from garments during washing, are particularly prevalent in river systems and have a significant role in microplastic contamination (Barrows et al., 2018; Re, 2019). The leading cause of microfiber contamination in freshwater systems is wastewater containing these fibres (Browne et al., 2011). Watershed microplastics may build up in wetlands and be carried into rivers by surface runoff, especially during periods of high precipitation (Kumar, 2021a, b). When microplastics enter aquatic ecosystems, they are frequently consumed by creatures that confuse them for food, which causes bioaccumulation in fish and other aquatic animals. Fish gill obstruction (McNeish et al., 2018), increased mercury bioaccumulation (Barboza et al., 2018), and impaired nutrient absorption (Subaramaniyam et al., 2023) are just a few of the physiological and ecological issues that can result from this bioaccumulation. Furthermore, toxic compounds such as pesticides (Felten et al., 2020), insecticides, heavy metals (Shruthisree et al., 2015; Amin et al., 2016, 2017; Kumar et al., 2022a; Prasad et al., 2023), and other pollutants are transported by microplastics, which intensifies their negative impacts on aquatic life (Riani & Cordova, 2022). In aquatic species, microplastics also cause oxidative stress, mechanical stress, disruption of reproduction, and impairment of enzyme production (Zantis et al., 2023). Ecosystems and human health are seriously threatened by their capacity to absorb heavy metals and persistent organic pollutants (Kumar et al., 2022b; Verma et al., 2023; Malik et al., 2017).

A methodical methodology that includes sampling, separation, and characterisation is used to quantify microplastics in the environment (Fu et al., 2020). Getting environmental samples from water, soil, sediment, air, or biota is the first step in sampling, which makes sure the samples accurately reflect the amount of microplastics in the environment. The goals of the investigation are taken into consideration while selecting variables including location, depth, and sampling frequency (Bakir et al., 2020). Since microplastics are frequently found in trace amounts, they must be isolated from other ambient contaminants after collection. Filters can be used to separate microplastics from water samples, and density gradient separation is used to separate materials based on buoyancy in sediment or soil samples (Han et al., 2020; Wang et al., 2018). Organic matter is eliminated from biota samples by enzymatic or chemical digestion, leaving only microplastics for additional examination (Bai et al., 2022; Ugwu et al., 2021). In order to concentrate microplastics for in-depth investigation, this step is essential.

The physical and chemical characteristics of microplastics are examined after they have been separated (Lv et al., 2021). Although technique has limits in detecting very small microplastics, visual identification using optical microscopy evaluates the particle's size, shape, and colour. For accurate polymer composition determination, more sophisticated methods such as pyrolysis gas chromatography-mass spectrometry (GC-MS), Raman spectroscopy, and Fourier-transform infrared spectroscopy (FTIR) are employed (Möller et al., 2022). Particle size analysis follows chemical identification, and size distributions are frequently measured using image analysis software (Mariano et al., 2021). For quantification, the bulk of microplastics may be weighed to determine concentrations, and particle counts can be done manually or using automated software for larger samples (Tan et al., 2022). This thorough procedure offers important new information about the level of microplastic pollution in ecosystems.

2 Types of Plastic Material

The ecological impact of plastics has been exacerbated by rising pollution levels and their pervasive effects, making them one of the biggest environmental threats to living things. Despite a significant increase in production and consumption, only 5% of plastics are recycled, despite the fact that they have significantly improved everyday life, industrial processes, and agricultural practices (De Souza Machado et al., 2018). Due to their high concentrations in soil, water, and air, microplastics (MPs) pose a major threat to ecosystem health and are attracting a lot of scientific attention (Guo et al., 2020; Roy et al., 2022). According to studies, common polymers found in aquatic environments include polyethylene terephthalate (PET), polyvinyl chloride (PVC), polyethylene (PE), polypropylene (PP), polystyrene (PS), and polyamide (PA) (Tang et al., 2021; Zhang et al., 2020; Khalid et al., 2021; Nabi et al., 2024). Furthermore, the most prevalent microplastic in soil ecosystems has been shown to be polyethylene (PE) (Li et al., 2020). Figure 1 provides a graphic representation of the plastic materials that customers frequently utilise.

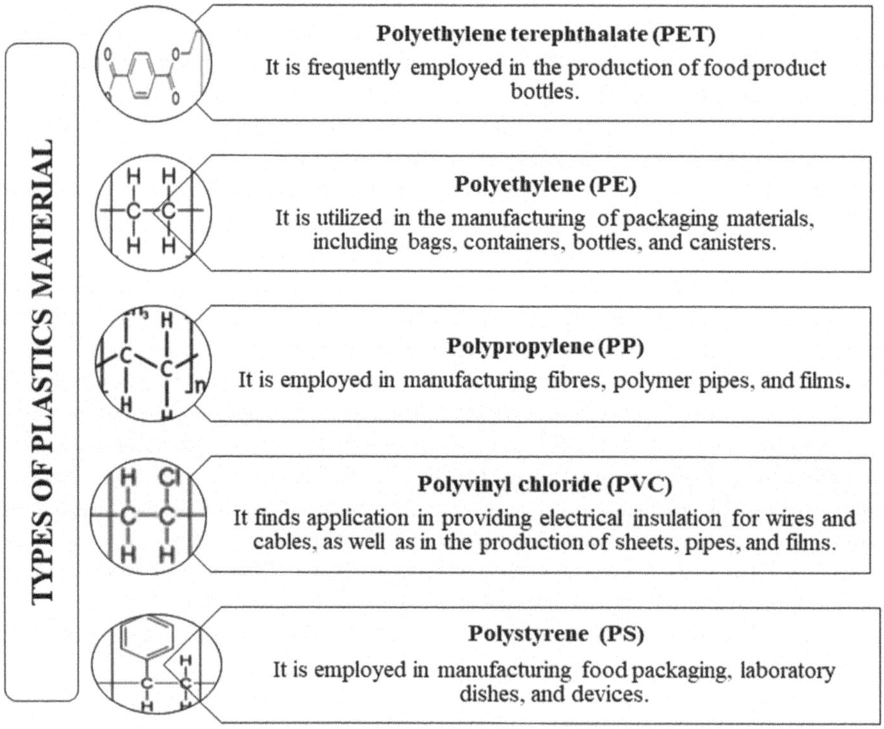

Fig. 1 Types of plastic material

3 MPs in the Aquatic Environment

Microplastics (MPs), which are a crucial field of study in environmental sciences, are a growing problem in water and wastewater systems as a result of the breakdown or release of plastic components into the environment (Adegoke et al., 2023; Debnath et al., 2024). MPs get into water systems through a variety of biological pathways, uncontrolled industrial wastewater discharge, and the breakdown of plastic debris (Menon et al., 2023). Consequently, MPs are present in a variety of settings, such as soil (Moeck et al., 2023), water (Ahmad et al., 2024), aquatic systems (Elizalde-Velázquez & Gómez-Oliván, 2021), and even the food chain (Al Mamun et al., 2023). MPs can accumulate along the food chain and possibly make their way into the human diet by acting as a food source for creatures in rivers, lakes, and seas. As a result, there may be direct and indirect health concerns associated with these tiny plastic particles (Abass et al., 2025; Zhou et al., 2021). Moreover, MPs can be inhaled by people when they are present in the air (Amato-Lourenço et al., 2020). These particles have the ability to enter the lungs by inhalation, where they may cause respiratory infections and other respiratory-related illnesses.

4 Advanced Methods of Microplastic Collection, Separation, and Characterization

Environmental scientists and researchers must analyse and detect microplastics (MPs) in aquatic habitats (Mai et al., 2018). These studies provide important information about the forms, impacts, and prevalence of MP pollution (Ashrafy et al., 2023). A variety of methods are employed to identify and describe MPs, such as:

1. **Filtration and Microscopy**: This involves visual identification and microscopic counting of MPs.
2. **Spectroscopic Techniques**: These include Fourier-transform infrared spectroscopy (FTIR), Raman spectroscopy, and scanning electron microscopy (SEM) combined with energy dispersive spectroscopy (EDS).
3. **Thermal Analysis Methods**: Techniques like pyrolysis gas chromatography-mass spectrometry (Pyro GC-MS), thermogravimetric analysis-mass spectrometry (TGA-MS), and thermodesorption gas chromatography-mass spectrometry (TED-GC-MS) fall into this category.
4. **Combined Techniques**: Advanced methods that integrate tagging, liquid chromatography, and SEM dispersive X-ray spectrometry to enhance the precision of MP detection and identification.

   The information gathered from these analyses is essential for assessing the effects on the environment, keeping an eye on the quality of the water, guaranteeing the safety of food, and safeguarding public health (Du et al., 2021). These

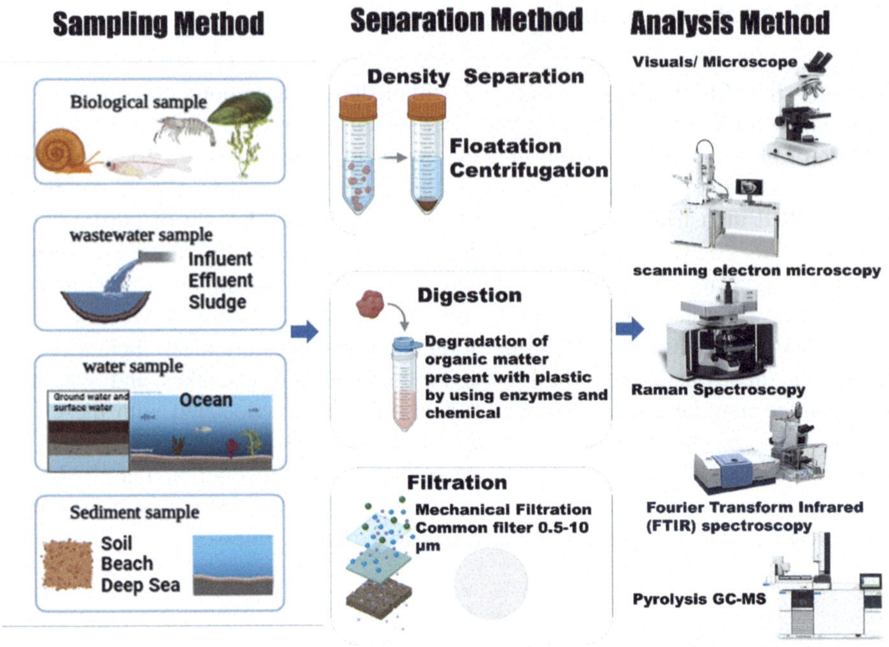

Fig. 2 Advancing microplastic research: a multidisciplinary framework for sampling, separation, and characterization

techniques also aid in educating the public and politicians about the seriousness of MP pollution and the pressing need to mitigate it (Deng et al., 2020; Mitrano & Wohlleben, 2020). These methods are also useful for determining the sources of pollution, comprehending long-term environmental repercussions, and examining the biological effects of MPs on ecosystems (Deme et al., 2022). Analysing MPs is essential for sustainability and environmental protection initiatives, as shown in Fig. 2. These techniques are essential for protecting human health (Singh et al., 2023), identifying pollution sources (Iheanacho et al., 2023), and comprehending the possible ecological implications of MPs (Rai et al., 2021) and their wider environmental effects (Mousavi & Khodadoost, 2019). In order to evaluate the spread of MPs in soils and aquatic habitats as well as their wider environmental consequences, they are essential (Yuan et al., 2022; Özgenç, 2024). Moreover, these techniques help understand how MPs might enter the human food chain (Al Mamun et al., 2023), assess their health impacts (Jenner et al., 2022; Yuan et al., 2022), and guide the development of effective policies and management strategies (Kumar et al., 2020; Zhou et al., 2020).

5 Microplastic Separation

Research on the existence and elimination of microplastics has attracted a lot of interest, especially in the creation of different treatment techniques meant to eliminate them. Numerous research have focused on the possible hazards that microplastics may pose to daily life as well as the necessity for efficient methods to identify and measure micron-sized polymeric particles in sediments and aquatic ecosystems. Plastics' distinct qualities relative to the surrounding materials are frequently used in the process of extracting them from intricate environmental samples. In wet environmental matrices, plastics are typically (1) less dense and (2) more hydrophobic than other components. Additionally, size-based filtration techniques can effectively extract plastics in matrices with minimal solid concentration.

6 Density-Based Separation

Polymers are frequently separated from denser sediments using a technique called flotation, which usually involves adding salt solutions to make the polymers more buoyant. Selecting the right salt is essential since it must strike a balance between aspects including environmental impact, processing expenses, and recovery efficiency. Plastics either float to the top on their own or with the help of mechanisms like elutriation. Although it has the potential to dissolve or harm some polymers, particularly smaller ones, ethanol can also be utilised as a floating medium. Plastics in the millimetre size range can be separated using flotation quite effectively, but smaller particles are harder to recover from the air-liquid interface, making flotation less effective. For the finest plastic fractions, flotation is also less effective because surface fouling can drastically change particle density and the buoyant force diminishes. Bubbles sticking to non-plastic particles presents another difficulty because denser materials float to the top and make separation more difficult. Although they are frequently employed in other scientific domains, density separation techniques are less frequently used in studies on plastic pollution. For example, the biological sciences commonly use density gradient centrifugation with Percoll, a medium made of suspended colloidal silica nanoparticles, which may be modified for the separation of plastics (Nguyen et al., 2019).

7 Size-Based/Filtration Separation

Filtration techniques for separating plastics require careful consideration of the trade-offs between capturing small particles and avoiding filter clogging or reduced sample throughput. Sequential filtration, which uses filters with progressively smaller pore sizes, can help mitigate clogging. For example, filter papers with varying pore sizes can be used in a series to isolate microplastics, with larger particles being captured by the first filter and smaller ones passing through to subsequent filters. However, this method primarily confirms the presence of nanometer-scale plastics rather than providing an accurate count, as a significant portion of these tiny particles is often lost due to attachment to filter surfaces. Unlike traditional "dead-end" filtration, crossflow filtration offers a more efficient alternative by passing samples tangentially across porous sleeves, reducing clogging and allowing for the concentration of small particles. This process can be further refined using asymmetric flow field-flow fractionation, which separates particles by size, providing a more precise fractionation of tiny plastic particles. This combination and appropriate filter papers enhance the ability to study nanometer-scale plastics with greater accuracy and reliability (Nguyen et al.,2019).

8 Separation from a Biological Sample

The advancement of environmental toxicological research, especially in risk assessment and management, depends on the precise and effective measurement of plastics in biological samples obtained in the field. The majority of studies (90%) in an examination of peer-reviewed publications from 2012 to 2018 examined aquatic organisms, with 35% looking at fish, 19% looking at mussels, 13% looking at plankton, and less than 6% looking at other species. Acid or base digestion is frequently employed to separate microplastics from biological tissues. Hydrogen peroxide (H_2O_2), potassium hydroxide (KOH), and nitric acid (HNO_3) are commonly used to dissolve tissues and eliminate tough substances. Strong acids like hydrochloric acid and bases like sodium hydroxide can also act as digestants. However, these treatments can degrade or damage certain plastics, particularly when temperatures exceed 80 °C. For example, polyethylene (PE) and polyvinyl chloride (PVC) are resistant to these digestants, resulting in inconsistent recoveries. Enzymes such as proteinase K, chitinase, and cellulase have also been successfully used for more selective tissue digestion, either alone or in combination with chemical digestants (Nguyen et al., 2019).

9 Microplastic Characterization

Understanding the distribution and presence of microplastic particles in the environment requires microplastic characterisation, which entails locating, measuring, and examining these particles in environmental samples (Tirkey & Upadhyay, 2021). To identify the many kinds, sizes, forms, and possible origins of microplastics, a multi-step procedure is used (Shim et al., 2018; Jung et al., 2021). Filtration and density separation methods are first used to separate microplastics from environmental samples, and then microscopy is used for examination. Particle sizes, shapes, and colours are usually examined first using optical microscopy (Asamoah et al., 2021). Advanced methods including pyrolysis gas chromatography-mass spectrometry (GC–MS), Fourier-transform infrared (FTIR) spectroscopy, Raman spectroscopy, and scanning electron microscopy (SEM) are employed for more thorough investigation. While FTIR and Raman spectroscopy are used to identify polymer types by analysing the vibrational spectra of microplastic particles, SEM offers high-resolution imaging to evaluate the surface morphology of particles (Corradini et al., 2019; Chen et al., 2020; Corami et al., 2020; Zhang et al., 2020). In order to determine the functional groups in microplastics and provide information about their polymer composition, FTIR measures the absorption of infrared light. In contrast, Raman spectroscopy uses scattered light to analyse molecular vibrations and provide extremely precise information about a particle's chemical structure (Xu et al., 2019). In pyrolysis GC-MS, the sample is heated to release volatile chemicals, which are subsequently separated and examined. Pyrolysis GC-MS offers rapid identification and is non-destructive, but FTIR and Raman spectroscopy offer more detailed information about polymer composition and additives. Each technique has its limitations, such as potential interference from sample matrices and the requirement for skilled operators. The selection of methods depends on the research goals, sample characteristics, and available equipment. In order to differentiate between various kinds of microplastics, such as fragments, fibres, or beads, image analysis software is also utilised to measure and categorise microplastic particles by highlighting them in pictures and figuring out their sizes and shapes (Jung et al., 2021; Mariano et al., 2021). The distribution, abundance, and diversity of microplastics in samples are evaluated statistically, offering important information on the degree of contamination and possible sources (Chen et al., 2020). To ensure consistency and comparability among research, standardised methodology and quality control processes are essential (Almuhtaram & Andrews, 2022) (Table 1).

Microscopic analysis of microplastics (MPs) is essential for determining their morphology by visual methods such as polarising microscopy or light. These techniques have been the mainstay of early research on microplastic counts and physical characterisation (Talvitie et al., 2017). Based on their outward appearance, MPs are usually divided into three primary categories by visual identification: fibres, pieces, and pellets/microbeads.

Light microscopy is frequently employed to examine microplastics, especially those with a diameter of several hundred micrometres. Since microplastics frequently lack lustre, physical traits like flexibility and hardness are necessary for identification. Environmental samples often contain common forms of microplastics, including fibres, pieces, and beads (Abadi et al., 2021). However, because it is frequently impractical to fully remove sediments and biological materials by chemical degradation, precisely identifying microplastics under a microscope is difficult. According to studies, more than 20% of particles that resemble plastic and more than 70% of transparent particles can be mistaken for other substances because of their similarities. This problem is most noticeable when distinguishing between natural

Table 1 Comparative overview of methods for microplastic identification and analysis

Analysis method	Scope	Advantages	Limitations	References
Microscopy	Examine the material using a microscope. Commonly used microscopes range from optical types to electron microscopes, includingTransmission Electron Microscopes (TEM) and Scanning Electron Microscopes (SEM).	Allows for direct and easy observation of the surface with minimal sample preparation.	The decision-making phase is the most time-consuming.	Watteau et al. (2018), Morgado et al. (2021)
Microscopy technique	**Description**	**Advantages**	**Limitations**	
Visual examination/ microscopic counting	Direct observation of MPs using optical microscopes allow for quick identification.	Precision in detection, morphological characterization, ensuring comparability of results, sample diversity	Time-consuming, difficulty in identifying types, operator dependency, difficulty in determining size distribution, automation challenges	Randhawa (2023)
Scanning Electron Microscopy (SEM)	Provides highly magnified, detailed images of microplastics and nanoplastics. Visualizes surface textures, distinguishing organic matter from microplastics.	High-resolution imaging. Effective for analyzing surface texture and morphology. Can identify nanoplastics.	High equipment cost. Labor-intensive sample preparation. Low throughput limits sample analysis.	Randhawa (2023)
Transmission Electron Microscopy (TEM)	Offers highly magnified visual detection of microplastics. It is often used with **Energy-Dispersive X-ray Spectroscopy (EDS)** to determine elemental composition.	Extremely high resolution for detailed analysis. Can provide structural information at the nanoscale.	Requires expert interpretation and standardized protocols to minimize error. Expensive and time-intensive.	Randhawa (2023)
Analysis method	**Scope**	**Advantages**	**Limitations**	**References**
Spectroscopy	Analyze the surface of a material using optical technology.	The process preserves the integrity of the sample and allows for the simultaneous analysis of multiple	It is challenging to distinguish between multiple samples, and the recognition process is time-consuming.	Kniggendorf et al. (2019), Zhou et al. (2020)

		samples, utilizing an existing data library for reference.		
Spectroscopy techniques	**Description**	**Advantages**	**Limitations**	
FTIR	Identification based on molecular structure, providing detailed information about polymer types.	Chemical identification, multiple materials analysis, high-sensitivity and resolution, fast analysis, standardized protocols	Analysis of small particles, contamination risk, analysis of MP mixtures, training requirement, surface analysis challenges	Chen et al. (2020)
Raman Spectroscopy	Molecular identification through laser-induced scattering is effective for diverse polymer types.	Chemical identification, high sensitivity and selectivity, fast analysis, reduced interference issues, non-destructive analysis	Weak signals, fluorescent background, analysis depth, equipment cost, sample fluorescence	Araujo et al. (2018)
Analysis method	**Scope**	**Advantages**	**Limitations**	References
Thermal analysis	Use thermal analysis to examine a material's surface and investigate its physical and chemical properties.	Multiple samples can be analyzed simultaneously, and the analysis can be conducted using an existing data library.	Since the samples are not preserved, it is not possible to conduct a quantitative analysis of each sample.	Zainuddin and Syuhada (2020), Liu et al. (2021)
Thermal techniques	**Description**	**Advantages**	**Limitations**	
TGA-MS	The method involves examining thermal behavior and mass changes, thereby providing information on polymer composition.	Quantitative data on thermal degradation, high sensitivity, thermal decomposition analysis, chemical composition determination, multiple component analysis, reduced the use of organic solvent	Limited in distinguishing complex polymer mixtures, sample preparation challenges, operator dependency, equipment cost, analysis time	Mansa and Zou (2021), de la Fuente et al. (2022)
Analysis method	**Scope**	**Advantages**	**Limitations**	References
Gas Chromatography-Mass Spectrometry (GC-MS)	Analyzes the thermal degradation products of microplastics after pyrolysis. It helps determine polymer types based on distinctive pyrolysis products.	High sensitivity. Provides detailed chemical identification of polymers.	Requires sample pyrolysis, which is destructive. Limited in detecting highly complex mixtures	Zhou et al. (2020), Li et al. (2021)

(continued)

Table 1 (continued)

Spectroscopy techniques	**Description**	**Advantages**	**Limitations**	
Liquid Chromatography–Mass Spectrometry (LC-MS)	Identifies polymers by analyzing the monomers or breakdown products created during hydrolysis. Effective for specific polymer types like Nylon, PET, and polycarbonate.	Can analyze heat-labile and non-volatile compounds. Effective for quantifying polymer monomers.	Requires sample depolymerization, which is destructive. Limited by the absence of unique monomers for some plastics.	Gros et al. (2006), Ng et al. (2020)
Pyrolysis-GC-MS (Py-GC-MS)	Combines pyrolysis and GC-MS to analyze the thermal degradation products of microplastics, offering detailed polymer composition.	High precision and sensitivity. Capable of detecting complex polymer mixtures.	Expensive instrumentation. Destructive sample preparation.	Ceccarini et al. (2018)
Analysis method	**Scope**	**Advantages**	**Limitations**	References
Emerging technology	Analyze using emerging technologies that are gaining popularity, offering potential advantages over traditional analytical methods.	Advancements focused on addressing the limitations of current technologies.	The amount of data collected is smaller compared to that obtained using modern technologies.	Erni-Cassola et al. (2017)
Cutting-edge technology	**Description**	**Advantages**	**Limitations**	
Atomic force microscopy-infrared and Raman	Combines Atomic Force Microscopy (AFM) with Infrared Spectroscopy (IR) to measure thermal expansion-induced cantilever vibrations. Provides spatial resolution of 50–100 nm.	High spatial resolution (50–100 nm). Simultaneous imaging and chemical characterization.	Locating single nanoscale particles is time-consuming and challenging. Requires expertise and precision.	Luo et al. (2021a, b)
Nile Red Staining for Microplastic Detection	A fluorescent dye (9-diethylamino-5H-benzophenoxazine-5-one) is used to detect hydrophobic microplastics by binding to neutral lipids in hydrophobic environments.	Short staining time (10–-30 minutes). High recovery efficiency (up to 96%). Quick bleach wash enhances process. Effective for transparent particles.	Potential co-staining of organic molecules. Extensive and time-consuming cleaning needed. Pretreatment depends on plastic density variations.	Prata et al. (2019)

and synthetic fibres, such as cotton. Destructive testing techniques, such melting polymers with heated needles, have been created to address this (Shim et al., 2018; Hendrickson et al., 2018). An alternate technique for detecting microplastics is fluorescence microscopy. Certain plastics can glow, which makes them easier to identify, particularly those that contain additives like bleaching chemicals used in the plastic and textile industries. According to Dehghani et al. (2017), several minerals and chemicals used in paper whitening have the ability to self-fluoresce, which could result in inaccurate results. Optical microscopy makes it simple to recognise coloured microplastics, especially those that contain industrial dyes, but it is still very difficult to distinguish colourless or amorphous particles smaller than 100 μm. The issue is further exacerbated by insufficient particle separation in samples, which makes it challenging to identify small or transparent microplastics (Löder & Gerdts, 2015).

Scanning Electron Microscopy (SEM) provides incredibly detailed and enlarged images of tiny particles, including nanoplastics, making it an effective tool for analysing microplastics (MPs) (Tunali et al., 2020). Since SEM provides high-resolution visualisation, the surface roughness of particles can be essential for differentiating microplastics from organic matter (Hossain et al., 2019). Because plastic particles have distinct qualities including shape and texture that are not usually present in natural materials, researchers can identify them thanks to the detailed surface features.

Transmission Electron Microscopy (TEM) is another cutting-edge method for identifying microplastics. Though it works on a somewhat different premise and concentrates on transmission through thin samples, TEM offers high-resolution imaging, just like SEM. Although the analyst's experience may have a role in the visual detection of microplastics using TEM, cross-validation with established techniques reduces errors and increases identification accuracy. Additionally, SEM and TEM are frequently used in conjunction with Energy-Dispersive X-ray Spectroscopy (EDS) to examine the elemental makeup of particles. EDS assists in determining whether a particle is plastic by identifying components that are frequently found in additives or antioxidants, such as calcium (Ca), sodium (Na), aluminium (Al), magnesium (Mg), and silicon (Si). To differentiate microplastics from other ambient components, these elemental markers are crucial (Watteau et al., 2018; Wagner et al., 2019). Based on the elemental makeup of the surface, SEM and TEM can distinguish between inorganic particles and carbon-rich polymers, which is one of their main benefits (Abadi et al., 2021; Sabri et al., 2021). For identifying different kinds of microplastics and comprehending their possible sources, this study is especially helpful.

Notwithstanding their efficacy, SEM and TEM have drawbacks, including expensive equipment, time-consuming sample preparation, and low throughput. These limitations limit how many samples can be examined with these sophisticated imaging methods, which makes them less useful for extensive research. As a result, even while SEM and TEM offer insightful information about the properties of microplastics, they might not be the best option for extensive environmental monitoring.

10 Spectroscopy-Based Analyses of Microplastics

There are two types of spectroscopy-based approaches for examining microplastics: qualitative and quantitative approaches that enable material identification and characterisation without causing harm. Fourier Transform Infrared Spectroscopy (FTIR) and Raman spectroscopy are the two most widely utilised spectroscopic techniques. Raman spectroscopy is particularly useful for detecting and analysing tiny microplastics, whereas FTIR is generally used to identify microplastics larger than 20 μm (Araujo et al., 2018; Schymanski et al., 2018; Prata et al., 2019). To give a more thorough study of microplastics, these techniques are frequently used in conjunction (Käppler et al., 2016; Prata et al., 2019; Wright et al., 2019). Spectroscopy is now commonly used in conjunction with other methods to increase the precision and effectiveness of microplastic measurement.

11 Fourier Transform Infrared Spectroscopy (FTIR) for Microplastic Identification

A popular technique for examining microplastics is FTIR spectroscopy, which finds certain chemical connections inside the particles. Because plastics have distinct spectral patterns because to their chemical makeup and bonding patterns, it works especially well for identifying carbon-based polymers (Avio et al., 2015; Garaba & Dierssen, 2018). By differentiating non-plastic particles, the use of a comprehensive polymer spectrum library enhances plastic identification and helps prevent false positives (Jung et al., 2018). In addition to providing information about the material's origins and state of degradation, FTIR may identify the type of polymer. Carbonyl groups in the infrared spectrum, for example, can reveal the degree of oxidation in microplastics (Hebner & Maurer-Jones, 2020). Micro-FTIR, a kind of FTIR that combines spectroscopy and microscopy, allows for the accurate detection of minute plastic particles on a single platform (Morgado et al., 2021). Transmittance, reflectance, and attenuated total reflectance (ATR) are among the modes of operation for FTIR. Reflectance and ATR work well with thicker, opaque plastics without further preparation, whereas transmittance necessitates sample preparation. Nevertheless, ATR has drawbacks, including trouble identifying particles smaller than 50 μm and unstable spectra on uneven microplastic surfaces. Because each particle needs to be examined separately, ATR can be time-consuming. Additionally, fragile microplastics may be harmed by the pressure exerted by the ATR probe (Silva et al., 2018). Advanced methods such as focal plane array (FPA) detectors are used to increase accuracy. FPA detectors can distinguish between fibres and microplastic fragments by concurrently capturing infrared spectra from several pixel arrays. A sample may need to be cleaned before analysis because of inadequate spectrum library data or interference from additives, surfactants, and adsorbed contaminants (such heavy metals or microorganisms) (Mintenig et al., 2017;

Primpke et al., 2019). Microplastics as small as 10 μm can now be detected because to advancements in FTIR sensitivity such as micro-FTIR and FPA-based reflectance imaging (Li et al., 2020; Löder & Gerdts, 2015). These developments make FTIR an essential technique in microplastic research since they enable accurate detection and characterisation of microplastics in complicated environmental samples when paired with microscopy.

12 Raman Spectroscopy for Microplastic Analysis

Based on the distinct spectra that each polymer produces when subjected to a laser, Raman spectroscopy is a potent analytical technique for identifying microplastics. A unique "fingerprint" for identifying plastics is created by the frequency shift in the backscattered light, which correlates to the material's atomic composition and molecular structure (Witzig et al., 2024). Like FTIR, Raman spectroscopy offers comprehensive profiles of polymer composition and can be improved with the use of algorithms and data libraries (Anger et al., 2019). Since the laser beam diameter of Raman spectroscopy enables the detection of particles as small as a few micrometres, its capacity to detect extremely small microplastics is one of its main advantages. Furthermore, the non-contact nature of Raman spectroscopy protects the integrity of the microplastic samples, allowing for additional investigation following the initial inspection. Micro-Raman spectroscopy is a non-destructive technique for identifying and measuring microplastics since it blends chemical analysis with visual observation through microscopy (Kniggendorf et al., 2019). Raman spectroscopy does have several drawbacks, though. The Raman signal can be distorted by additives, dyes, and the surface curvature of microplastics, making it more difficult to identify particular polymer types (Dowarah & Devipriya, 2019). Misidentifications can also happen, especially because the spectra of polyethylene (PE) and surfactants like sodium dodecyl sulphate (SDS) are similar. A trustworthy reference spectral database is necessary for accurate identification, and mistakes in this database could lead to false positives (Witzig et al., 2024). The high expense of Raman spectroscopy equipment and the requirement for knowledgeable operators to decipher the intricate spectra represent further constraints. Only about 5000 particles may be examined in a single run, which further limits the method's capacity (Schymanski et al., 2018; Tsering et al., 2022). Notwithstanding these difficulties, Raman spectroscopy has a lot to offer in terms of chemical analysis and microplastic measurement. Raman spectroscopy will become more and more important in microplastic research as the field develops since it allows for the simultaneous chemical analysis and quantification of microplastics in a variety of environmental samples.

13 Mass Spectrometry-Based Analyses of Microplastics

By identifying the unique products created during hydrolysis or pyrolysis, mass spectrometry-based techniques, such as gas and liquid chromatography combined with mass spectrometry, are effective tools for analysing microplastics (Fischer & Scholz-Böttcher, 2017; Wang et al., 2019; Zhou et al., 2020; Li et al., 2021). High sensitivity is provided by these methods, which are very helpful for identifying and measuring microplastics, including nanoplastics. Mass spectrometry can shed light on the characteristics and makeup of microplastics by examining the particular chemical fragments created during these activities. Mass spectrometry is therefore a vital tool for environmental research and monitoring, aiding in the understanding of microplastic pollution.

14 Liquid Chromatography–Tandem Mass Spectrometry (LC–MS/MS) for Microplastic Analysis

In disciplines like medicine, pharmacology, and environmental studies, liquid chromatography-tandem mass spectrometry (LC–MS/MS) is a popular and trustworthy analytical method, especially for identifying non-volatile and heat-sensitive compounds. Particularly in sewage sludge samples, it has been successfully used to detect and measure certain polymers such as nylon, PET, and polycarbonate-based microplastics (MPs) (Ng et al., 2020; Zhang et al., 2021; Peng et al., 2020). In order to determine the mass and quantity of monomers produced during the depolymerisation of macromolecules, LC–MS/MS is utilised. In contrast to non-destructive approaches that concentrate on counting, sizing, and characterising the colour or form of MPs and nanoplastics (NPs), this technique is destructive even if it is quite effective for some polymers. Furthermore, size fractionation can be included in LC-MS/MS, providing important information about the size distribution of microplastic particles. LC-MS/MS is not without its limits, though. It is difficult to analyse a wide variety of polymer types because many of them do not decompose into discrete products that are appropriate for mass spectrometry. LC–MS/MS is currently unable to offer an accurate assessment of the overall quantity of microplastics in environmental samples because microplastics frequently consist of a mixture of different polymers (Wang et al., 2017; Peng et al., 2020; Li et al., 2021; Zhang et al., 2021).

15 PyrolysisGasChromatography–MassSpectrometry(Py-GC–MS)

Pyrolysis-gas chromatography-mass spectrometry (Py-GC–MS) is an alternate technique for examining pyrolysed gases from different polymers (Ceccarini et al., 2018). This method determines whether plastic is present in a sample by comparing its pyrograms to established polymer standards. Before being separated and subjected to GC–MS analysis, plastic particle samples are occasionally pyrolysed at temperatures greater than those employed in thermogravimetric analysis (TGA), usually around 700 °C (Shim et al., 2018). Py-GC-MS sample sizes typically fall between 0.3 and 7 mg. Certain polymers, including polyvinyl acetate (PVA), styrene-butadiene rubber (SBR), and poly(acrylonitrile-butadiene-styrene), have been effectively identified by py-GC-MS in environmental materials including soil and suspended particles. Nevertheless, this method works best for detecting microplastics made of polystyrene (PS) and its byproducts. To create molecular markers for recognising a larger variety of plastics, more investigation is required (Watteau et al., 2018).

16 Thermal-Based Analyses of Microplastics

Understanding the functional link between temperature and microplastic characteristics requires the use of thermal-based analysis. By examining how materials react to temperature changes, these investigations offer important insights about their properties. Microplastics absorb a lot of heat when environmental samples are heated, which causes phase transformations like solid to liquid or gas. At particular temperatures where the material experiences a phase change, these transitions produce endothermic peaks. By comparing the obtained thermograms to normal polymer thermograms, it is possible to determine the type and content of microplastics as well as their additives by examining their thermal behaviours, as various polymers have varied thermal stabilities (Majewsky et al., 2016).

17 Thermogravimetric Analysis (TGA)

A popular method in thermal analysis, thermogravimetric analysis (TGA) measures weight loss while a sample is heated at a regulated pace under particular temperature and ambient conditions to produce both qualitative and quantitative results. Microplastics in wastewater have been analysed using TGA, which provides accurate results for differentiating between polypropylene (PP) and polyethylene (PE). Nevertheless, it has trouble recognising other polymers like polyester (PES), polyamide (PA), and polyvinyl chloride (PVC). Furthermore, polyethylene terephthalate

(PET) and polyurethane (PU) phase transition signals cannot be distinguished using TGA (Majewsky et al., 2016; Sun et al., 2019).

TGA is a rapid and effective addition to other analytical methods since it can analyse samples without requiring laborious sample preparation, which is one of its main benefits. Despite these advantages, TGA is a destructive technique, which means that the physical properties of microplastics, including their shape or appearance, cannot be preserved. Because of this, it is mostly utilised for chemical component identification and microplastic quantification; however, its use is restricted in situations where preserving the integrity of environmental samples is crucial (Rocha-Santos & Duarte, 2015; Majewsky et al., 2016; Shim et al., 2018; Huppertsberg & Knepper, 2018; Silva et al., 2018).

18 Challenges in Microplastic and Nanoplastic Detection

The detection of extremely minute microplastics, particularly those at the nanoscale, is becoming more difficult as analytical technology develop. Even while existing techniques can identify microplastics as small as a few micrometres, new detection techniques are still required because of growing awareness of the dangers posed by nanoplastics, including their potential toxicity and environmental persistence. Because they are much smaller, nanoplastics are more challenging to properly isolate, purify, and concentrate. To overcome these obstacles, sophisticated detection systems must be continuously developed. To improve current techniques for more precise identification of these minuscule particles, as well as to optimise protocols for the extraction and analysis of nanoplastics, innovations are required. These developments are essential for improving our knowledge of the distribution, behaviour, and possible health hazards associated with nanoplastics. Better control of plastic pollution, particularly at the nanoscale, and more thorough environmental monitoring across a range of plastic particle sizes will depend on such advancements.

19 Atomic Force Microscopy-Infrared and Raman

Atomic Force Microscopy (AFM) has a lot of promise for studying nanoplastics, particularly when paired with Raman or infrared (IR) spectroscopy. AFM produces high-resolution pictures with accuracy down to a few nanometres and can function in both contact and non-contact modes. AFM additionally enables thorough chemical investigation of the sample when combined with IR or Raman spectroscopy. For example, thermal expansion brought on by infrared light absorption causes vibrations in the AFM cantilever. The frequency and amplitude of these vibrations are then measured using Fourier transform techniques, which reveal information on the material's chemical makeup (Luo et al., 2021a, b).

With a spatial resolution of 50–100 nm, AFM-IR may simultaneously acquire IR absorption spectra and pictures, enabling in-depth nanoscale material research. Polystyrene (PS) beads and other particles as small as 100 nm have been successfully studied to produce both IR and AFM spectra. However, utilising AFM-IR to separate a single nanoscale plastic particle from an unknown material is still difficult and time-consuming. The time-consuming procedure of manually looking for plastic particles is one of the main challenges in conventional microplastic analysis. Small, transparent particles may be lost due to the effectiveness of the pretreatment process, which separates and purifies samples; this is especially true when manual identification techniques are employed.

To mitigate these challenges, automated FTIR or Raman mapping can be employed, which allows for more efficient identification of particles. However, these methods require costly equipment that is not widely accessible in many microplastic research labs, making them less practical for broader use.

20 Nile Red Staining for Microplastic Detection

A simple and efficient technique for identifying tiny, transparent microplastic particles is Nile Red staining. Nile Red (9-diethylamino-5H-benzophenoxazine-5-one) has emerged as a dependable choice for microplastic investigation in contrast to other dyes like Eosin B, Rose Bengal, and Oil Red EGN, which are restricted. Because of its high affinity for neutral lipids, this fluorescent dye only binds to hydrophobic microplastics and produces intense fluorescence in hydrophobic settings (Prata et al., 2019). The effectiveness of Nile Red staining is a major benefit; recovery rates of up to 96% can be attained in about 10–30 minutes. If necessary, a brief bleach wash might improve the procedure even more.

An efficient first step for detecting microplastics is Nile Red staining, especially prior to the use of more sophisticated spectroscopic techniques. With the use of a fluorescent filter for spectrum confirmation, it can be linked with fluorescence microscopy to create FTIR microscopy. In addition to providing a quicker option than spectroscopy alone, this combination lowers the possibility of overlooking microplastics in situ (Erni-Cassola et al., 2017; Simmerman & Coleman Wasik, 2020). But a major drawback of Nile Red staining is the possibility of organic molecules co-staining, which makes sample preparation extremely important. Organic compounds can be difficult and time-consuming to remove during pretreatment.

In order to tackle this problem, methods such as density separation with reagents like Fenton's reagent or sodium chloride (NaCl) or hydrogen peroxide ($H2O_2$) treatment have been explored. While these methods can help, they are limited by the diverse densities of plastic materials (Al-Azzawi et al., 2020; Li et al., 2020). Despite these challenges, Nile Red staining remains a valuable tool for the preliminary screening of microplastics, especially when paired with advanced confirmation methods like FTIR microscopy (Sancataldo et al., 2020; Vermeiren et al., 2020).

21 Future Perspectives and Challenges

A comprehensive strategy that incorporates developments in sampling, separation, and characterisation techniques is needed to improve microplastic quantification methodologies (Jung et al., 2021; Zhang et al., 2023). Standardising sample procedures for different habitats, such as freshwater, marine ecosystems, and soil, is an essential first step towards advancement. To get consistent, trustworthy data, representative sampling sites must be established and bias must be reduced. Automated sampling tools and continuous monitoring systems may provide real-time data, exposing dynamic transport patterns and temporal variations in microplastic concentrations (Choran & Örmeci, 2023).

Future studies on microplastics stand to gain from new sampling techniques. For example, real-time monitoring of microplastics across large geographic areas may be made possible by remote sensing technologies such sophisticated imaging and spectroscopic instruments installed on drones or satellites. This would enable thorough monitoring of microplastic pollution, even in areas that are difficult to access. Analytical instruments are also expected to undergo significant advances. Improved Raman spectroscopy devices and ultra-sensitive mass spectrometers will increase the identification and measurement of traces of microplastics, offering hitherto unheard-of insights into their ambient presence. Furthermore, improved extraction methods and cutting-edge analytical instruments will contribute to a deeper comprehension of the distribution and environmental destiny of microplastics.

In order to improve the dependability of results, it will be essential to produce standardised environmental reference materials for technique validation, quality control, and inter-laboratory comparisons. Additionally, automation will be essential for improving productivity, decreasing human error, and optimising sample preparation and analysis. In conclusion, new developments in the study of microplastics suggest that more precise identification and measurement of microplastics in environmental samples will result from the combination of sophisticated detection methods, such as Raman spectroscopy, with machine learning algorithms. These developments will greatly improve our capacity to track and manage microplastic pollution globally.

22 Conclusion

The pervasive issue of microplastic pollution necessitates urgent scientific attention, as the widespread use of plastics continues to reshape our environment. This chapter has highlighted advanced techniques for collecting, separating, and identifying microplastics in various environmental samples, underscoring their ecological impacts. As global plastic production is projected to escalate, the accumulation of microplastics in freshwater and marine ecosystems poses significant threats to biodiversity and human health. Innovative methods, including density-based separation and enzymatic digestion, enhance our ability to isolate microplastics from environmental matrices. At the same time, advanced characterization techniques such as FTIR and Raman spectroscopy allow for precise identification of their polymeric composition. These methodologies are crucial for understanding microplastics' size, shape, and distribution and assessing their potential harm. Looking to the future, the chapter emphasizes the importance of standardized sampling protocols and the integration of automated monitoring systems to improve data reliability. Innovations in remote sensing and analytical technologies will further enhance our capacity to detect and quantify microplastics, enabling more effective environmental management strategies. By addressing the complexities of microplastic pollution through these advanced techniques, researchers and policymakers can work collaboratively to mitigate its impacts and safeguard ecosystems for future generations. The ongoing development of improved methodologies will be essential in our efforts to combat this pervasive environmental challenge.

References

Abadi, Z. T. R., Abtahi, B., Grossart, H. P., & Khodabandeh, S. (2021). Microplastic content of Kutum fish, Rutilus frisii kutum in the southern Caspian Sea. *Science of the Total Environment, 752*, 141542.

Abass, Z., Chuskit, D., Shah, T. H., Wanjari, R. N., Ahmad, I., Bala, I. M., & Barman, M. (2025). Microplastic an Emerging Threat to Nutritional and Food Security. In Food Security, Nutrition and Sustainability Through Aquaculture Technologies (pp. 397–415). Cham: Springer Nature Switzerland.

Adegoke, K. A., Adu, F. A., Oyebamiji, A. K., Bamisaye, A., Adigun, R. A., Olasoji, S. O., & Ogunjinmi, O. E. (2023). Microplastics toxicity, detection, and removal from water/wastewater. *Marine Pollution Bulletin, 187*, 114546.

Ahmad, M. F., Fauzi, M. A., Ahmad, M. H., Wider, W., & Lee, C. K. (2024). Are we eating plastic? Science mapping of microplastic pollution in the aquatic food chain. In *Integrated environmental assessment and management* (Vol. 20, pp. 1800–1811). Wiley.

Al Mamun, A., Prasetya, T. A. E., Dewi, I. R., & Ahmad, M. (2023). Microplastics in human food chains: Food becoming a threat to health safety. *Science of the Total Environment, 858*, 159834.

Al-Azzawi, M. S., Kefer, S., Weißer, J., Reichel, J., Schwaller, C., Glas, K., & Drewes, J. E. (2020). Validation of sample preparation methods for microplastic analysis in wastewater matrices—Reproducibility and standardization. *Water, 12*(9), 2445.

Almuhtaram, H., & Andrews, R. C. (2022). Sampling microplastics in water matrices: A need for standardization. *ACS ESandT Water, 2*(8), 1276–1278.

Amato-Lourenço, L. F., dos Santos Galvão, L., de Weger, L. A., Hiemstra, P. S., Vijver, M. G., & Mauad, T. (2020). An emerging class of air pollutants: Potential effects of microplastics to respiratory human health? *Science of the Total Environment, 749*, 141676.

Amin, A., Indulkar, S. T., & Pai, R. (2016). Sublethal effect of Buprofezin pesticide on carcass composition of *Cyprinus carpio* communis fingerlings. *Journal of Experimental Zoology, India, 20*(1), 269–272.

Amin, A., Ramachandra Naik, A. T., Priyadarshini, N., Nayak, H., & Sree, C. S. (2017). Toxic effect of heavy metal lead on oxygen consumption of rohu (*Labeo rohita*) fingerlings. *Biochemical & Cellular Archives, 17*(1), 225.

Anger, P. M., Prechtl, L., Elsner, M., Niessner, R., & Ivleva, N. P. (2019). Implementation of an open source algorithm for particle recognition and morphological characterisation for

microplastic analysis by means of Raman microspectroscopy. *Analytical Methods, 11*(27), 3483–3489.
Araujo, C. F., Nolasco, M. M., Ribeiro, A. M., & Ribeiro-Claro, P. J. (2018). Identification of microplastics using Raman spectroscopy: Latest developments and future prospects. *Water Research, 142*, 426–440.
Asamoah, B. O., Uurasjärvi, E., Räty, J., Koistinen, A., Roussey, M., & Peiponen, K. E. (2021). Towards the development of portable and in situ optical devices for detection of micro-and nanoplastics in water: A review on the current status. *Polymers, 13*(5), 730.
Ashrafy, A., Liza, A. A., Islam, M. N., Billah, M. M., Arafat, S. T., Rahman, M. M., & Rahman, S. M. (2023). Microplastics pollution: A brief review of its source and abundance in different aquatic ecosystems. *Journal of Hazardous Materials Advances, 9*, 100215.
Avio, C. G., Gorbi, S., & Regoli, F. (2015). Experimental development of a new protocol for extraction and characterization of microplastics in fish tissues: First observations in commercial species from Adriatic Sea. *Marine Environmental Research, 111*, 18–26.
Bai, C. L., Liu, L. Y., Hu, Y. B., Zeng, E. Y., & Guo, Y. (2022). Microplastics: A review of analytical methods, occurrence and characteristics in food, and potential toxicities to biota. *Science of the Total Environment, 806*, 150263.
Bakir, A., Desender, M., Wilkinson, T., Van Hoytema, N., Amos, R., Airahui, S., et al. (2020). Occurrence and abundance of meso and microplastics in sediment, surface waters, and marine biota from the South Pacific region. *Marine Pollution Bulletin, 160*, 111572.
Ballent, A., Corcoran, P. L., Madden, O., Helm, P. A., & Longstaffe, F. J. (2016). Sources and sinks of microplastics in Canadian Lake Ontario nearshore, tributary and beach sediments. *Marine Pollution Bulletin, 110*(1), 383–395.
Barboza, L. G. A., Vethaak, A. D., Lavorante, B. R., Lundebye, A. K., & Guilhermino, L. (2018). Marine microplastic debris: An emerging issue for food security, food safety and human health. *Marine Pollution Bulletin, 133*, 336–348.
Barrows, A. P. W., Cathey, S. E., & Petersen, C. W. (2018). Marine environment microfiber contamination: Global patterns and the diversity of microparticle origins. *Environmental Pollution, 237*, 275–284.
Bergmann, M., & Klages, M. (2012). Increase of litter at the Arctic deep-sea observatory HAUSGARTEN. *Marine Pollution Bulletin, 64*(12), 2734–2741.
Browne, M. A., Galloway, T. S., & Thompson, R. C. (2010). Spatial patterns of plastic debris along estuarine shorelines. *Environmental Science and Technology, 44*(9), 3404–3409.
Browne, M. A., Crump, P., Niven, S. J., Teuten, E., Tonkin, A., Galloway, T., & Thompson, R. (2011). Accumulation of microplastic on shorelines woldwide: sources and sinks. *Environmental science & technology, 45*(21), 9175–9179.
Ceccarini, A., Corti, A., Erba, F., Modugno, F., La Nasa, J., Bianchi, S., & Castelvetro, V. (2018). The hidden microplastics: new insights and figures from the thorough separation and characterization of microplastics and of their degradation byproducts in coastal sediments. *Environmental science & technology, 52*(10), 5634–5643.
Chen, G., Feng, Q., & Wang, J. (2020). Mini-review of microplastics in the atmosphere and their risks to humans. *Science of the Total Environment, 703*, 135504.
Choran, N., & Örmeci, B. (2023). Micro-flow imaging for in-situ and real-time enumeration and identification of microplastics in water. *Frontiers in Water, 5*, 1148379.
Cole, M., Lindeque, P., Halsband, C., & Galloway, T. S. (2011). Microplastics as contaminants in the marine environ.: A review. *Marine Pollution Bulletin, 62*(12), 2588–2597.
Corami, F., Rosso, B., Bravo, B., Gambaro, A., & Barbante, C. (2020). A novel method for purification, quantitative analysis and characterization of microplastic fibers using Micro-FTIR. *Chemosphere, 238*, 124564.
Corradini, F., Meza, P., Eguiluz, R., Casado, F., Huerta-Lwanga, E., & Geissen, V. (2019). Evidence of microplastic accumulation in agricultural soils from sewage sludge disposal. *Science of the Total Environment, 671*, 411–420.

Crawford, C. B., & Quinn, B. (2017). Microplastic separation techniques. In *Microplastic Pollut* (pp. 203–218). Elsevier Science.

de la Fuente, A. M., Marhuenda-Egea, F. C., Ros, M., Pascual, J. A., Saez-Tovar, J. A., Martinez-Sabater, E., & Peñalver, R. (2022). Thermogravimetry coupled with mass spectrometry successfully used to quantify polyethylene and polystyrene microplastics in organic amendments. *Environmental Research, 213*, 113583.

de Souza Machado, A. A., Kloas, W., Zarfl, C., Hempel, S., & Rillig, M. C. (2018). Microplastics as an emerging threat to terrestrial ecosystems. *Global Change Biology, 24*(4), 1405–1416.

Debnath, R., Prasad, G. S., Amin, A., Malik, M. M., Ahmad, I., Abubakr, A., Borah, S., Rather, M. A., Impellitteri, F., Tabassum, I., & Piccione, G. (2024). Understanding and addressing microplastic pollution: Impacts, mitigation, and future perspectives. *Journal of Contaminant Hydrology, 266*, 104399.

Dehghani, S., Moore, F., & Akhbarizadeh, R. (2017). Microplastic pollution in deposited urban dust, Tehran metropolis, Iran. *Environmental Science and Pollution Research, 24*, 20360–20371.

Deme, G. G., Ewusi-Mensah, D., Olagbaju, O. A., Okeke, E. S., Okoye, C. O., Odii, E. C., et al. (2022). Macro problems from microplastics: Toward a sustainable policy framework for managing microplastic waste in Africa. *Science of the Total Environment, 804*, 150170.

Deng, L., Cai, L., Sun, F., Li, G., & Che, Y. (2020). Public attitudes towards microplastics: Perceptions, behaviors and policy implications. *Resources, Conservation and Recycling, 163*, 105096.

Dowarah, K., & Devipriya, S. P. (2019). Microplastic prevalence in the beaches of Puducherry, India and its correlation with fishing and tourism/recreational activities. *Marine Pollution Bulletin, 148*, 123–133.

Du, H., Xie, Y., & Wang, J. (2021). Microplastic degradation methods and corresponding degradation mechanism: Research status and future perspectives. *Journal of Hazardous Materials, 418*, 126377.

Eerkes-Medrano, D., Thompson, R. C., & Aldridge, D. C. (2015). Microplastics in freshwater systems: A review of the emerging threats, identification of knowledge gaps and prioritisation of research needs. *Water Research, 75*, 63–82.

Elizalde-Velázquez, G. A., & Gómez-Oliván, L. M. (2021). Microplastics in aquatic environments: A review on occurrence, distribution, toxic effects, and implications for human health. *Science of the Total Environment, 780*, 146551.

Erni-Cassola, G., Gibson, M. I., Thompson, R. C., & Christie-Oleza, J. A. (2017). Lost, but found with Nile red: A novel method for detecting and quantifying small microplastics (1 mm to 20 μm) in environmental samples. *Environmental Science and Technology, 51*(23), 13641–13648.

Felten, V., Toumi, H., Masfaraud, J. F., Billoir, E., Camara, B. I., & Férard, J. F. (2020). Microplastics enhance Daphnia magna sensitivity to the pyrethroid insecticide deltamethrin: Effects on life history traits. *Science of the Total Environment, 714*, 136567.

Fendall, L. S., & Sewell, M. A. (2009). Contributing to marine pollution by washing your face: Microplastics in facial cleansers. *Marine Pollution Bulletin, 58*(8), 1225–1228.

Fischer, M., & Scholz-Böttcher, B. M. (2017). Simultaneous trace identification and quantification of common types of microplastics in environmental samples by pyrolysis-gas chromatography–mass spectrometry. *Environmental Science and Technology, 51*(9), 5052–5060.

Fu, Z., Chen, G., Wang, W., & Wang, J. (2020). Microplastic pollution research methodologies, abundance, characteristics and risk assessments for aquatic biota in China. *Environmental Pollution, 266*, 115098.

Garaba, S. P., & Dierssen, H. M. (2018). An airborne remote sensing case study of synthetic hydrocarbon detection using short wave infrared absorption features identified from marine-harvested macro-and microplastics. *Remote Sensing of Environment, 205*, 224–235.

Geyer, R., Jambeck, J. R., & Law, K. L. (2017). Production, use, and fate of all plastics ever made. *Science Advances, 3*(7), e1700782.

Gros, M., Petrović, M., & Barceló, D. (2006). Development of a multi-residue analytical methodology based on liquid chromatography–tandem mass spectrometry (LC–MS/MS) for screening and trace level determination of pharmaceuticals in surface and wastewaters. *Talanta, 70*(4), 678–690.

Guo, J. J., Huang, X. P., Xiang, L., Wang, Y. Z., Li, Y. W., Li, H., et al. (2020). Source, migration and toxicology of microplastics in soil. *Environment International, 137*, 105263.

Gurjar, U. R., Takar, S., Amin, A., & Martin Xavier, K. A. (2023). Microplastic contamination in aquatic organisms: An ecotoxicological perspective. In *Xenobiotics in aquatic animals: Reproductive and developmental impacts* (pp. 353–367). Springer Nature Singapore.

Halden, R. (2020). *Environment*. Bloomsbury Publishing.

Han, M., Niu, X., Tang, M., Zhang, B. T., Wang, G., Yue, W., et al. (2020). Distribution of microplastics in surface water of the lower Yellow River near estuary. *Science of the Total Environment, 707*, 135601.

Hebner, T. S., & Maurer-Jones, M. A. (2020). Characterizing microplastic size and morphology of photodegraded polymers placed in simulated moving water conditions. *Environmental Science: Processes and Impacts, 22*(2), 398–407.

Hendrickson, E., Minor, E. C., & Schreiner, K. (2018). Microplastic abundance and composition in western Lake Superior as determined via microscopy, Pyr-GC/MS, and FTIR. *Environmental Science and Technology, 52*(4), 1787–1796.

Hossain, M. S., Sobhan, F., Uddin, M. N., Sharifuzzaman, S. M., Chowdhury, S. R., Sarker, S., & Chowdhury, M. S. N. (2019). Microplastics in fishes from the Northern Bay of Bengal. *Science of the Total Environment, 690*, 821–830.

Huppertsberg, S., & Knepper, T. P. (2018). Instrumental analysis of microplastics—Benefits and challenges. *Analytical and Bioanalytical Chemistry, 410*, 6343–6352.

Iheanacho, S., Ogbu, M., Bhuyan, M. S., & Ogunji, J. (2023). Microplastic pollution: An emerging contaminant in aquaculture. *Aquaculture and Fisheries, 8*(6), 603–616.

IUCN. (2018). https://www.iucn.org/resources/issues-brief/marine-plastic-pollution

Jenner, L. C., Rotchell, J. M., Bennett, R. T., Cowen, M., Tentzeris, V., & Sadofsky, L. R. (2022). Detection of microplastics in human lung tissue using μFTIR spectroscopy. *Science of the Total Environment, 831*, 154907.

Jung, M. R., Horgen, F. D., Orski, S. V., Rodriguez, V., Beers, K. L., Balazs, G. H., et al. (2018). Validation of ATR FT-IR to identify polymers of plastic marine debris, including those ingested by marine organisms. *Marine Pollution Bulletin, 127*, 704–716.

Jung, J. W., Park, J. W., Eo, S., Choi, J., Song, Y. K., Cho, Y., et al. (2021). Ecological risk assessment of microplastics in coastal, shelf, and deep sea waters with a consideration of environmentally relevant size and shape. *Environmental Pollution, 270*, 116217.

Käppler, A., Fischer, D., Oberbeckmann, S., Schernewski, G., Labrenz, M., Eichhorn, K. J., & Voit, B. (2016). Analysis of environmental microplastics by vibrational microspectroscopy: FTIR, Raman or both? *Analytical and Bioanalytical Chemistry, 408*, 8377–8391.

Khalid, N., Aqeel, M., Noman, A., Khan, S. M., & Akhter, N. (2021). Interactions and effects of microplastics with heavy metals in aquatic and terrestrial environments. *Environmental Pollution, 290*, 118104.

Kniggendorf, A. K., Wetzel, C., & Roth, B. (2019). Microplastics detection in streaming tap water with Raman spectroscopy. *Sensors, 19*(8), 1839.

Kumar, M., Xiong, X., He, M., Tsang, D. C., Gupta, J., Khan, E., et al. (2020). Microplastics as pollutants in agricultural soils. *Environmental Pollution, 265*, 114980.

Kumar, R., Sharma, P., Manna, C., & Jain, M. (2021a). Abundance, interaction, ingestion, ecological concerns, and mitigation policies of microplastic pollution in riverine ecosystem: A review. *Science of the Total Environment, 782*, 146695.

Kumar, R., Sharma, P., Verma, A., Jha, P. K., Singh, P., Gupta, P. K., et al. (2021b). Effect of physical characteristics and hydrodynamic conditions on transport and deposition of microplastics in riverine ecosystem. *Water, 13*(19), 2710.

Kumar, R., Ivy, N., Bhattacharya, S., Dey, A., & Sharma, P. (2022a). Coupled effects of microplastics and heavy metals on plants: Uptake, bioaccumulation, and environmental health perspectives. *Science of the Total Environment, 836*, 155619.

Kumar, R., Sinha, R., Rakib, M. R. J., Padha, S., Ivy, N., Bhattacharya, S., et al. (2022b). Microplastics pollution load in Sundarban delta of Bay of Bengal. *Journal of Hazardous Materials Advances, 7*, 100099.

Li, C., Busquets, R., & Campos, L. C. (2020). Assessment of microplastics in freshwater systems: A review. *Science of the Total Environment, 707*, 135578.

Li, C., Gan, Y., Zhang, C., He, H., Fang, J., Wang, L., et al. (2021). "Microplastic communities" in different environments: Differences, links, and role of diversity index in source analysis. *Water Research, 188*, 116574.

Liu, Y., Li, R., Yu, J., Ni, F., Sheng, Y., Scircle, A., & Zhou, Y. (2021). Separation and identification of microplastics in marine organisms by TGA-FTIR-GC/MS: A case study of mussels from coastal China. *Environmental Pollution, 272*, 115946.

Löder, M. G., & Gerdts, G. (2015). Methodology used for the detection and identification of microplastics—A critical appraisal. In *Marine anthropogenic litter* (pp. 201–227). Springer.

Luo, H., Zeng, Y., Zhao, Y., Xiang, Y., Li, Y., & Pan, X. (2021a). Effects of advanced oxidation processes on leachates and properties of microplastics. *Journal of Hazardous Materials, 413*, 125342.

Luo, Z., Zhou, X., Su, Y., Wang, H., Yu, R., Zhou, S., et al. (2021b). Environmental occurrence, fate, impact, and potential solution of tire microplastics: Similarities and differences with tire wear particles. *Science of the Total Environment, 795*, 148902.

Lusher, A. L., Mchugh, M., & Thompson, R. C. (2013). Microplastics occur in the gastrointestinal tract of pelagic and demersal fish from the English Channel. *Marine Pollution Bulletin, 67*(1–2), 94–99.

Lusher, A., Hollman, P., & Mendoza-Hill, J. (2017). *Microplastics in fisheries and aquaculture: Status of knowledge on their occurrence and implications for aquatic organisms and food safety*. FAO.

Lv, L., Yan, X., Feng, L., Jiang, S., Lu, Z., Xie, H., et al. (2021). Challenge for the detection of microplastics in the environment. *Water Environment Research, 93*(1), 5–15.

Mai, L., Bao, L. J., Shi, L., Wong, C. S., & Zeng, E. Y. (2018). A review of methods for measuring microplastics in aquatic environments. *Environmental Science and Pollution Research, 25*, 11319–11332.

Majewsky, M., Bitter, H., Eiche, E., & Horn, H. (2016). Determination of microplastic polyethylene (PE) and polypropylene (PP) in environmental samples using thermal analysis (TGA-DSC). *Science of the Total Environment, 568*, 507–511.

Malik, M., Balkhi, M. H., & Abubakr, A. (2017). Comparative study on the contamination level of Dal Lake, Kashmir based on Total Viable Bacterial Counts over a period of time. *Ecology, Environment and Conservation, 23*(4), 2145–2147.

Mansa, R., & Zou, S. (2021). Thermogravimetric analysis of microplastics: A mini review. *Environmental Advances, 5*, 100117.

Mariano, S., Tacconi, S., Fidaleo, M., Rossi, M., & Dini, L. (2021). Micro and nanoplastics identification: Classic methods and innovative detection techniques. *Frontiers in Toxicology, 3*, 636640.

McNeish, R. E., Kim, L. H., Barrett, H. A., Mason, S. A., Kelly, J. J., & Hoellein, T. J. (2018). Microplastic in riverine fish is connected to species traits. *Scientific Reports, 8*(1), 11639.

Menon, V., Sharma, S., Gupta, S., Ghosal, A., Nadda, A. K., Jose, R., et al. (2023). Prevalence and implications of microplastics in potable water system: An update. *Chemosphere, 317*, 137848.

Mintenig, S. M., Int-Veen, I., Löder, M. G., Primpke, S., & Gerdts, G. (2017). Identification of microplastic in effluents of waste water treatment plants using focal plane array-based micro-Fourier-transform infrared imaging. *Water Research, 108*, 365–372.

Mitrano, D. M., & Wohlleben, W. (2020). Microplastic regulation should be more precise to incentivize both innovation and environmental safety. *Nature Communications, 11*(1), 5324.

Moeck, C., Davies, G., Krause, S., & Schneidewind, U. (2023). Microplastics and nanoplastics in agriculture—A potential source of soil and groundwater contamination? *Grundwasser, 28*(1), 23–35.

Möller, J. N., Heisel, I., Satzger, A., Vizsolyi, E. C., Oster, S. J., Agarwal, S., et al. (2022). Tackling the challenge of extracting microplastics from soils: A protocol to purify soil samples for spectroscopic analysis. *Environmental Toxicology and Chemistry, 41*(4), 844–857.

Morgado, V., Palma, C., & da Silva, R. J. B. (2021). Microplastics identification by infrared spectroscopy–evaluation of identification criteria and uncertainty by the Bootstrap method. *Talanta, 224*, 121814.

Mousavi, S. A., & Khodadoost, F. (2019). Effects of detergents on natural ecosystems and wastewater treatment processes: a review. *Environmental Science and Pollution Research, 26*, 26439–26448.

Nabi, N., Ahmad, I., Amin, A., Rather, M. A., Ahmed, I., Hajam, Y. A., Khursheed, S., Malik, M. M., & Abubakr, A. (2024). Understanding the sources, fate and effects of microplastics in aquatic environments with a focus on risk profiling in aquaculture systems. *Reviews in Aquaculture, 16*(4), 1947–1980.

Napper, I. E., Bakir, A., Rowland, S. J., & Thompson, R. C. (2015). Characterisation, quantity and sorptive properties of microplastics extracted from cosmetics. *Marine Pollution Bulletin, 99*(1–2), 178–185.

Ng, E. L., Lwanga, E. H., Eldridge, S. M., Johnston, P., Hu, H. W., Geissen, V., & Chen, D. (2018). An overview of microplastic and nanoplastic pollution in agroecosystems. *Science of the Total Environment, 627*, 1377–1388.

Ng, W., Minasny, B., & McBratney, A. (2020). Convolutional neural network for soil microplastic contamination screening using infrared spectroscopy. *Science of the Total Environment, 702*, 134723.

Nguyen, B., Claveau-Mallet, D., Hernandez, L. M., Xu, E. G., Farner, J. M., & Tufenkji, N. (2019). Separation and analysis of microplastics and nanoplastics in complex environmental samples. *Accounts of Chemical Research, 52*(4), 858–866.

Özgenç, E. (2024). Advanced analytical techniques for assessing and detecting microplastic pollution in water and wastewater systems. *Environmental Quality Management., 34*, e22217.

Parashar, N., & Hait, S. (2021). Plastics in the time of COVID-19 pandemic: Protector or polluter? *Science of the Total Environment, 759*, 144274.

Peng, G., Bellerby, R., Zhang, F., Sun, X., & Li, D. (2020). The ocean's ultimate trashcan: Hadal trenches as major depositories for plastic pollution. *Water Research, 168*, 115121.

Plastics Europe. (2021). https://plasticseurope.org/wp-content/uploads/2021/12/Plastics-the-Facts-2021-web-final.pdf. Accessed on 22nd November 2021.

Prasad, G. S., Rout, S. K., Malik, M. M., Karmakar, S., Amin, A., & Ahmad, I. (2023). Occurrence of xenoestrogen alkylphenols (Octylphenols and Nonylphenol) and its impact on the aquatic ecosystem. In *Xenobiotics in aquatic animals: Reproductive and developmental impacts* (pp. 275–284). Springer Nature Singapore.

Prasad, G. S., Rout, S. K., Amin, A., Malik, M. M., Ahmad, I., Rather, M. A., Karmakar, S., Khalil, H. S., & Abubakar, A. (2024). Impact of COVID-19 on the aquatic environment and implications for fisheries with special emphasis to Indian perspective. *Proceedings of the Indian National Science Academy, 90*(4), 821–839.

Prata, J. C., da Costa, J. P., Lopes, I., Duarte, A. C., & Rocha-Santos, T. (2019). Effects of microplastics on microalgae populations: A critical review. *Science of the Total Environment, 665*, 400–405.

Primpke, S., Dias, P. A., & Gerdts, G. (2019). Automated identification and quantification of microfibres and microplastics. *Analytical Methods, 11*(16), 2138–2147.

Ragusa, A., Svelato, A., Santacroce, C., Catalano, P., Notarstefano, V., Carnevali, O., Papa, F., Rongioletti, M. C. A., Baiocco, F., Draghi, S., D'Amore, E., Rinaldo, D., Matta, M., & Giorgini, E. (2021). Plasticenta: First evidence of microplastics in human placenta. *Environment International, 146*, 1.

Rai, P. K., Lee, J., Brown, R. J., & Kim, K. H. (2021). Micro-and nano-plastic pollution: Behavior, microbial ecology, and remediation technologies. *Journal of Cleaner Production, 291*, 125240.

Randhawa, J. S. (2023). Advanced analytical techniques for microplastics in the environment: A review. *Bulletin of the National Research Centre, 47*(1), 174.

Re, V. (2019). Shedding light on the invisible: Addressing the potential for groundwater contamination by plastic microfibers. *Hydrogeology Journal, 27*(7), 2719–2727.

Riani, E., & Cordova, M. R. (2022). Microplastic ingestion by the sandfish Holothuria scabra in Lampung and Sumbawa, Indonesia. *Marine Pollution Bulletin, 175*, 113134.

Rocha-Santos, T., & Duarte, A. C. (2015). A critical overview of the analytical approaches to the occurrence, the fate and the behavior of microplastics in the environment. *TrAC Trends in Analytical Chemistry, 65*, 47–53.

Roy, P., Mohanty, A. K., & Misra, M. (2022). Microplastics in ecosystems: Their implications and mitigation pathways. *Environmental Science: Advances, 1*(1), 9–29.

Sabri, N. H., Muhammad, A., Rahim, N. A., Roslan, A., & Talip, A. A. (2021). Feasibility study on co-pyrolyzation of microplastic extraction in conventional sewage sludge for the cementitious application. *Materials Today: Proceedings, 46*, 2112–2117.

Sancataldo, G., Avellone, G., & Vetri, V. (2020). Nile Red lifetime reveals microplastic identity. *Environmental Science: Processes and Impacts, 22*(11), 2266–2275.

Schymanski, D., Goldbeck, C., Humpf, H. U., & Fürst, P. (2018). Analysis of microplastics in water by micro-Raman spectroscopy: Release of plastic particles from different packaging into mineral water. *Water Research, 129*, 154–162.

Shahul Hamid, F., Bhatti, M. S., Anuar, N., Anuar, N., Mohan, P., & Periathamby, A. (2018). Worldwide distribution and abundance of microplastic: How dire is the situation? *Waste Management and Research, 36*(10), 873–897.

Sharma, S., Sharma, V., & Chatterjee, S. (2023). Contribution of plastic and microplastic to global climate change and their conjoining impacts on the environment-a review. *Science of the Total Environment, 875*, 162627.

Shim, W. J., Hong, S. H., & Eo, S. (2018). Marine microplastics: Abundance, distribution, and composition. In *Microplastic contamination in aquatic environments* (pp. 1–26). Elsevier.

Silva, A. B., Bastos, A. S., Justino, C. I., da Costa, J. P., Duarte, A. C., & Rocha-Santos, T. A. (2018). Microplastics in the environment: Challenges in analytical chemistry-a review. *Analytica Chimica Acta, 1017*, 1–19.

Simmerman, C. B., & Coleman Wasik, J. K. (2020). The effect of urban point source contamination on microplastic levels in water and organisms in a cold water stream. *Limnology and Oceanography Letters, 5*(1), 137–146.

Singh, S., Chakma, S., Alawa, B., Kalyanasundaram, M., & Diwan, V. (2023). Identification, characterization, and implications of microplastics in soil–a case study of Bhopal, Central India. *Journal of Hazardous Materials Advances, 9*, 100225.

Sruthisree, C., Gowda, G., Nayak, H., Harish, B. D., & Amin, A. (2015). Effect of heavy metal (lead) on oxygen consumption and ammonia excretion of Iridescent Shark (*Pangusius hypohthalmus*). *Journal of Experimental Zoology, India, 18*(2), 637.

Statista. (2020). https://www.statista.com/statistics/282732/global-production-of-plastics-since1950/#:~:text=Global%20plastics%20production%20was%20estimated,19's%20impacts%20on%20the%20industry. Accessed on 22nd November 2021.

Subaramaniyam, U., Allimuthu, R. S., Vappu, S., Ramalingam, D., Balan, R., Paital, B., et al. (2023). Effects of microplastics, pesticides and nano-materials on fish health, oxidative stress and antioxidant defense mechanism. *Frontiers in Physiology, 14*, 1217666.

Sun, J., Dai, X., Wang, Q., Van Loosdrecht, M. C., & Ni, B. J. (2019). Microplastics in wastewater treatment plants: Detection, occurrence and removal. *Water Research, 152*, 21–37.

Tabassum, I., Hassan, N., Amin, A., Malik, M. M., Abubakr, A., Prasad, G. S., Latief, F., & Rather, S. A. (2025). COVID-19 and global environmental change: A threat to food security and nutrition. In *Food security, nutrition and sustainability through aquaculture technologies* (pp. 437–445). Springer Nature Switzerland.

Talvitie, J., Mikola, A., Koistinen, A., & Setälä, O. (2017). Solutions to microplastic pollution–remove microplastics from wastewater effluent with advanced wastewater treatment technologies. *Water Research, 123*, 401–407.

Tan, Y., Dai, J., Wu, X., Wu, S., & Zhang, J. (2022). Characteristics, occurrence and fate of non-point source microplastic pollution in aquatic environments. *Journal of Cleaner Production, 341*, 130766.

Tang, Y., Liu, Y., Chen, Y., Zhang, W., Zhao, J., He, S., et al. (2021). A review: Research progress on microplastic pollutants in aquatic environments. *Science of the Total Environment, 766*, 142572.

Thompson, R. C., Olsen, Y., Mitchell, R. P., Davis, A., Rowland, S. J., John, A. W., et al. (2004). Lost at sea: Where is all the plastic? *Science, 304*(5672), 838–838.

Thompson, R. C., Swan, S. H., Moore, C. J., & Vom Saal, F. S. (2009). Our plastic age. *Philosophical Transactions of the Royal Society B: Biological Sciences, 364*(1526), 1973–1976.

Tirkey, A., & Upadhyay, L. S. B. (2021). Microplastics: An overview on separation, identification and characterization of microplastics. *Marine Pollution Bulletin, 170*, 112604.

Tsering, T., Sillanpää, M., Viitala, M., & Reinikainen, S. P. (2022). Variation of microplastics in the shore sediment of high-altitude lakes of the Indian Himalaya using different pretreatment methods. *Science of the Total Environment, 849*, 157870.

Tunali, M., Uzoefuna, E. N., Tunali, M. M., & Yenigun, O. (2020). Effect of microplastics and microplastic-metal combinations on growth and chlorophyll a concentration of Chlorella vulgaris. *Science of the Total Environment, 743*, 140479.

Ugwu, K., Herrera, A., & Gómez, M. (2021). Microplastics in marine biota: A review. *Marine Pollution Bulletin, 169*, 112540.

Verma, K. K., Song, X. P., Xu, L., Huang, H. R., Liang, Q., Seth, C. S., & Li, Y. R. (2023). Nano-microplastic and agro-ecosystems: A mini-review. *Frontiers in Plant Science, 14*, 1283852.

Vermeiren, P., Muñoz, C., & Ikejima, K. (2020). Microplastic identification and quantification from organic rich sediments: A validated laboratory protocol. *Environmental Pollution, 262*, 114298.

Wagner, J., Wang, Z. M., Ghosal, S., Murphy, M., Wall, S., Cook, A. M., et al. (2019). Nondestructive extraction and identification of microplastics from freshwater sport fish stomachs. *Environmental Science and Technology, 53*(24), 14496–14506.

Wang, W., Ndungu, A. W., Li, Z., & Wang, J. (2017). Microplastics pollution in inland freshwaters of China: A case study in urban surface waters of Wuhan, China. *Science of the Total Environment, 575*, 1369–1374.

Wang, W., Yuan, W., Chen, Y., & Wang, J. (2018). Microplastics in surface waters of dongting lake and hong lake, China. *Science of the Total Environment, 633*, 539–545.

Wang, J., Liu, X., Li, Y., Powell, T., Wang, X., Wang, G., & Zhang, P. (2019). Microplastics as contaminants in the soil environment: A mini-review. *Science of the Total Environment, 691*, 848–857.

Watteau, F., Dignac, M. F., Bouchard, A., Revallier, A., & Houot, S. (2018). Microplastic detection in soil amended with municipal solid waste composts as revealed by transmission electronic microscopy and pyrolysis/GC/MS. *Frontiers in Sustainable Food Systems, 2*, 81.

Witzig, C. S., Fiener, P., König, M., & Zumbülte, N. (2024). Temporal variability of microplastic concentrations in inland waters: An automated, Semicontinuous sampling of microplastics$\geq$ 11 μm in a stream in Southern Germany. *ACS ESandT Water, 4*(4), 1443–1450.

Wright, S. L., Levermore, J. M., & Kelly, F. J. (2019). Raman spectral imaging for the detection of inhalable microplastics in ambient particulate matter samples. *Environmental Science and Technology, 53*(15), 8947–8956.

Xu, X., Jian, Y., Xue, Y., Hou, Q., & Wang, L. (2019). Microplastics in the wastewater treatment plants (WWTPs): Occurrence and removal. *Chemosphere, 235*, 1089–1096.

Yao, L., Hui, L., Yang, Z., Chen, X., & Xiao, A. (2020). Freshwater microplastics pollution: Detecting and visualizing emerging trends based on Citespace II. *Chemosphere, 245*, 125627.

Yuan, W., Christie-Oleza, J. A., Xu, E. G., Li, J., Zhang, H., Wang, W., et al. (2022). Environmental fate of microplastics in the world's third-largest river: Basin-wide investigation and microplastic community analysis. *Water Research, 210*, 118002.

Zainuddin, Z., & Syuhada. (2020, June). Study of analysis method on microplastic identification in bottled drinking water. *In Macromolecular Symposia 391*(1), 1900195.

Zantis, L. J., Borchi, C., Vijver, M. G., Peijnenburg, W., Di Lonardo, S., & Bosker, T. (2023). Nano-and microplastics commonly cause adverse impacts on plants at environmentally relevant levels: A systematic review. *Science of the Total Environment, 867*, 161211.

Zhang, Y., Pu, S., Lv, X., Gao, Y., & Ge, L. (2020). Global trends and prospects in microplastics research: A bibliometric analysis. *Journal of Hazardous Materials, 400*, 123110.

Zhang, J., Guo, N., Ding, W., Han, B., Zhao, M., Wang, X., et al. (2023). Microplastic pollution and the related ecological risks of organic composts from different raw materials. *Journal of Hazardous Materials, 458*, 131911.

Zhang, L., Liu, J., Xie, Y., Zhong, S., & Gao, P. (2021). Occurrence and removal of microplastics from wastewater treatment plants in a typical tourist city in China. *Journal of Cleaner Production, 291*, 125968.

Zhou, Y., Wang, J., Zou, M., Jia, Z., Zhou, S., & Li, Y. (2020). Microplastics in soils: A review of methods, occurrence, fate, transport, ecological and environmental risks. *Science of the Total Environment, 748*, 141368.

Zhou, A., Zhang, Y., Xie, S., Chen, Y., Li, X., Wang, J., & Zou, J. (2021). Microplastics and their potential effects on the aquaculture systems: A critical review. *Reviews in Aquaculture, 13*(1), 719–733.

The Lifecycle of MPs: Understanding the Aging Process of Microplastics in Marine and Freshwater Habitats

Ifra Tabassum Giri, Adnan Amin, Monisa M. Malik, Adnan Abubakr, Sajad Ahmad Rather, Gora Shiva Prasad, and Ishfaq Hamid

1 Introduction

Plastics made from petroleum have seen rapid growth in manufacturing, and plastic products are now widely used in a wide range of sectors. China's plastics manufacturing industry is currently expanding, as seen by the country's 80.04 million tons of plastic goods produced, which increased by 5.9% yearly till 2021 (Ma et al., 2022). Twenty percent of plastics were released into the environment untreated, whereas thirty percent were recycled (Duan et al., 2021). Through radiation, wear, biodegradation, and natural breakdown, plastic items become MP contaminants (Amobonye et al., 2021). Plastic trash may be broadly divided into three categories based on size: macroplastics (>25 mm), mesoplastics (5–25 mm), and MPs (<5 mm). At the moment, MP pollution—which is divided into primary and secondary MPs-should get the greatest attention. Particles of plastic or resin that are present in commonplace objects or industrial raw materials are known as primary MPs. Through sewage treatment facilities and rivers, these contaminants are immediately discharged into the aquatic environment. According to González-Pleiter et al. (2019), secondary MPs are macroplastic pollutants that decompose and age as a result of physical, chemical, or biological processes, decreasing their volume and becoming MP pollutants. A great deal of concern has been raised in society by the widespread detection of MP contaminants in water systems, such as rivers, lakes, groundwater, oceans, and sediments, as a result of the inappropriate handling and poor management of plastics. MPs in water are now known to originate from two

I. T. Giri · A. Amin (✉) · M. M. Malik · A. Abubakr · S. A. Rather · I. Hamid
Division of Aquatic Environmental Management, Faculty of Fisheries Rangil, Ganderbal, SKUAST-Kashmir, Jammu and Kashmir, India

G. S. Prasad
Department of Aquatic Environmental Management, Faculty of Fishery Sciences, West Bengal University of Animal and Fishery Sciences, Kolkata, India

A. Abubakr et al. (eds.), *Microplastics: Ecological and Food Security Implications*,
https://doi.org/10.1007/978-3-031-94480-2_12

sources: solid land-based sources and freshwater (Xu et al., 2020; Debnath et al., 2024; Nabi et al., 2024). It is generally accepted that the main source of MPs in freshwater is sewage treatment plants that eliminate weathered and crushed macroplastics (An et al., 2021). The incapacity of sewage treatment plants to effectively treat MP pollutants causes some of the MPs produced in the production process of the non-ferrous metal industry and small workshops to bypass the treatment process and enter the urban sewage system, which ultimately leads to rivers (Du et al., 2020; Prasad et al., 2024; Tabassum et al., 2025). Solid waste is a further significant contributor of MPs in the environment, despite the fact that it is usually ignored. Landfills are the most common method of solid waste disposal. They include between 21 and 42 percent of the world's plastics, which can decompose into secondary MPs through complex chemical processes and physical wear (Natesan et al., 2021). Additionally, a number of substantial terrestrial sources of MPs, the majority of which are directly related to human activity, are easy to overlook. Worldwide, MPs are commonly found in bottled mineral water; as groundwater drinking water is excluded, the sealing material is the primary entry point into the mineral water container (Weisser et al., 2021).

Disposable respirators may cause new plastic waste management and pollution issues as well as becoming an unavoidable source of MP pollution as a result of the widespread use and normalization of COVID-19 (Akarsu et al., 2021; Khan et al., 2023; Prasad et al., 2023; Gurjar et al., 2023). Despite being thought to be inert in biological activities, factual research has demonstrated that plastic goods have had a number of detrimental impacts through their interactions with organisms and the environment. The adsorption performance for organic pollutants is continuously changing as a result of the continuous release of monomers and additives in MPs caused by weathering, crushing, and aging (Pirsaheb et al., 2020).

Plastic trash currently stands at over 6.3 billion tons, and manufacturing is growing quickly. Plastic production reached 359 million tons in 2018 alone (Geyer et al. 2017; Plastics Europe, 2019; Galgani et al., 2015). Research shows that up to 67.5 g/kg of MPs can be found in soil in China, Australia, Switzerland, and Chile. Furthermore, the ocean receives 1.5–4.5% of the world's plastic production (Li et al., 2020; Nizzetto et al., 2016a). Every year, over 400,000 tons of MPs may be released onto land, which is four to twenty-three times more than what enters the ocean (Horton et al., 2017; Nizzetto et al., 2016b). Thus, plastic trash presents serious hazards to living things once it gets into the ecosystem (Fossi et al., 2018; Gall & Thompson, 2015). These plastic wastes are broken down into smaller bits by a variety of natural processes, such as fire, chemical oxidation, and physical damage (Hu et al., 2021; Liu et al., 2020). MPs, or particles smaller than 5 mm, make up the bulk of plastic pieces discovered in the environment (Duan et al., 2021). Thompson et al. (2004) were the first to propose the idea of MPs. MPs come in a variety of forms, such as fibrous, granular, spherical, and irregular. Polyester, polyethylene (PE), polystyrene (PS), polypropylene (PP), phenol-formaldehyde (PF), polyamide (PA), polyvinyl chloride (PVC), and polyethylene terephthalate (PET) are just a few of the various chemical compositions that they come in. Accordingly, MPs have varying mechanical strengths, hydrophobicities, melting temperatures, densities, crystallinities, and chemical stabilities (Hanun et al., 2021). Primary and secondary MPs are the two types of MPs that are categorized according to their sources. Secondary MPs are produced when bigger plastic items break down, while primary MPs are produced as tiny particles (such as tire dust and microbeads in personal care products) (Turner & Holmes, 2015). MPs can diffuse and accumulate in aquatic habitats by hydraulic migration because of their endurance, chemical reactivity, and adaptability (Werner et al., 2016). In aquatic settings, MPs weather as a result of ongoing physical and chemical processes such photooxidation, hydrodynamic flushing, and biodegradation (Andrady & Neal, 2009). When contrasted to their initial forms, these processes change the aged MPs' chemical and physical characteristics. According to studies, MPs lose crystallinity, stiffness, and thermal resilience as they get older. Furthermore, oxygen-containing functional groups form on the surface of aged MPs, raising the carbonyl index and altering their hydrophilic characteristics. Compound pollutants are created as a result of these modifications to the way MPs interact with other pollutants, including organic substances (Rochman, 2015), heavy metals (Rochman, 2015; Shruthisree et al., 2015; Amin et al., 2016, 2017) and pathogenic organisms (Malik et al., 2017; Yan et al., 2020). The surfaces of MPs may develop environmentally persistent free radicals (EPFRs) and reactive oxygen species (ROS) under specific environmental circumstances, which may be detrimental to the health of humans and other living things. Additionally, research indicates that MPs could function as photosensitive initiators, accelerating the photodegradation of organic pollutants in watery settings. MPs aging is an important phenomenon, but our understanding of it is still lacking.

2 Major Sources of MPs in Freshwater and Marine

Aquatic systems, especially the marine environment, are contaminated by a variety of sources, most often defined as originating from the land, sea, and air (Aliko et al., 2022; Auta et al., 2021). The principal pathways via which plastics go from land to sea environments are believed to be rivers (Lebreton et al., 2017). At least 80% of the plastic trash found in the sea originates on land (Andary, 2011; Mani et al., 2015). Plastic waste from beach-related tourism is discarded into the environment, while rivers carry plastic debris from sewage effluents, municipal drainage systems, and ineffective interior area management into the ocean (Barnes et al., 2009; Browne et al., 2011; Yang et al., 2021). Human mistake, unregulated waste from sewage and water treatment facilities, and the textile industry are the main reasons why plastics end up in the environment (Rastogi & Kandasubramanian, 2020). Figure 1 shows the main flows of MPs within and between the three environmental compartments: freshwater, marine and terrestrial (Horton et al., 2017). According to research done in Shaoxing, China, sewage from a textile industry region included MPs, which had an effect on the freshwater ecosystem there. MP levels in soil samples ranged from 16.7 to 1323.3 items per kilogram (dry weight), whereas those in surface water samples in the vicinity ranged from 2.1 to 71.0 items per litter (Deng et al., 2020).

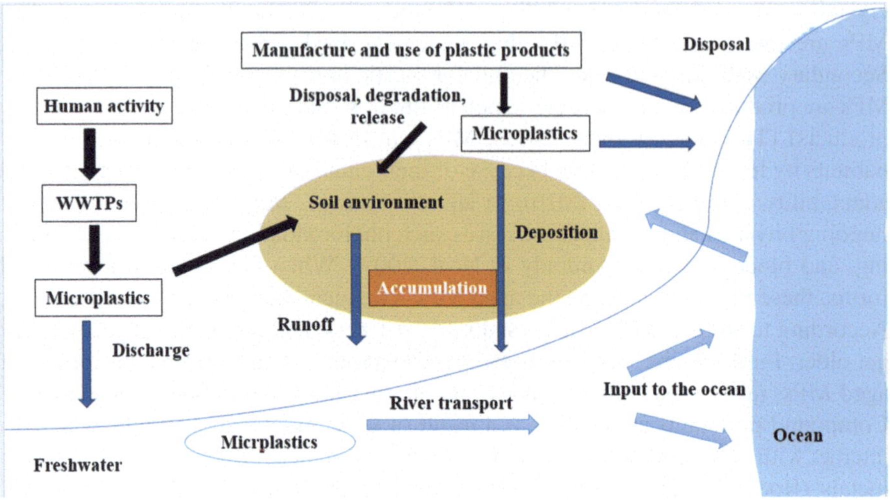

Fig. 1 Transportation of MPs particles within and between anthropogenic, terrestrial, freshwater and marine environment

Comparable research conducted on three Australian wastewater treatment facilities reveals that between 22.1^106 and 133^106 MPs are released everyday along with treated wastewater (Ziajahromi et al., 2021). MPs can also be found in effluent from the food, automobile, and pack aging industries (Gündoğdu et al., 2018). Additionally, human activity has a role in the discharge of MPs into the environment. Commonplace products like toothpaste, shampoo, and cosmetics contain these MPs (Jiang, 2018). Another important source of MPs is everyday activities like washing clothes made of synthetic fibres. According to studies, clothing composed of synthetic fibres like nylon and acrylic also releases different levels of MPs, while polyester fleece textiles shed an average of 7360 fibres per square meter per litter when washed (Carney Almroth et al., 2018). Furthermore, the terrestrial plastic accumulation ultimately seeps into the water systems due to improper landfill burial practices (Anbumani & Kakkar, 2018). According to data from 2022, 7.2 million tons of plastic garbage were dumped in EU landfills (Plastic Europe, 2023). The continuous, massive manufacturing and discharge of plastics into the maritime environment exacerbates the pollution of already polluted media (Thushari & Senevirathna, 2020). Additionally, many plastic products are used in agriculture, including things like plastic films and pipelines. As these plastics break down over time, MPs are created and subsequently carried into rivers by wind and rain. The use of 5 plastic irrigation pipes and their subsequent disposal in the environment has been identified as a major cause of MP pollution in the Ombrone River (Guerranti et al., 2017). Rivers are essential pathways for the entry of MPs into the marine environment, with an estimated 1.15 to 2.41 million metric tons of plastic waste entering rivers each year (Lebreton et al., 2017).

Moreover, a number of additional sources of plastic entering marine systems originate from tourism, commercial and recreational fishing, shipping, and the marine sector in coastal areas (Cole et al., 2011). Scientific research estimates that each year, cruise ships release over 100,000 tons of MPs into the ocean (Van Sebille et al., 2015). The World Population Review 2021 has reported that between 4.8 and 12.7 million metric tons of plastic enter the ocean annually. The Philippines, India, Malaysia, China, and Indonesia are the top five Asian countries that contribute to this, together making up around 80% of all the plastic that ends up in the seas. Even while many affluent countries produce far more plastic garbage per person than middle-class and low-income countries, the latter are responsible for a larger share of improperly managed plastic debris that ends up in the ocean. This disparity is mostly due to the safer and better plastic manufacturing methods used in high-income countries as opposed to developing ones, where significant advancements in infrastructure and technology are required. Currently, one of the biggest obstacles to reducing plastic pollution is the insufficient recycling of plastics. Only 55 million tons (Mt) of the 235 million tons of plastic garbage produced in 2019 were collected for recycling, leaving 22 Mt as residue that needed to be disposed of in the end. Just 9% of plastic garbage worldwide was recycled overall; the remaining 19% was burned, and over half was dumped in sanitary landfills. The remaining 22% was burned in open pits, spilled into the environment, or carelessly dumped in unmanaged dumpsites (OCED, 2022). Geyer et al. (2017) anticipate that by 2060, 12,000 Mt of plastic garbage will have been landfilled if the current production pattern continues.

3 Aging of MPs in the Freshwater and Marine Environment

Table 1 reviews recent research on MPs' aging process. When it comes to MP aging, biological techniques outperform chemical ones (Padervand et al., 2020). The number of MPs that have collected in the aquatic environment is significantly greater

Table 1 Recent studies of MP aging using different methods

MPs	Methods	Mass loss rate	Experimental conditions	References
PE	Mechanical fragmentation	–	The percentage of secondary MPs (1–2.4m^2) rises as the weathering duration (30 d) increases.	Resmerită et al. (2018)
14C-PS	Photo degradation	17.1 ± 0.55% in water, 6.17 ± 0.1% in air	^{14}C-PS nanoplastics exposed to UV light after 48 hours in air and water.	Tian et al. (2019)
MPs	Activated sludge	16.6%	Wastewater (influent) = 47.4 ± 7.0n/L, (effluent) = 34.1 ± 9.4n/L.	Liu et al. (2019a)
PET, -E, PS	Biodegradation	3.0%,6.2%, 5.8% for PE, PS and PET	Various MPs as the only carbon source to support Bacillus gottheilii growth in artificial medium.	Auta et al. (2017)

than what the ecosystem can clear itself of (Yang et al., 2020; Wu et al., 2020). In aquatic settings, MPs self-clean mainly through weathering or aging processes such hydrolysis, mechanical abrasion, UV photooxidation, and biodegradation.

3.1 Phyto-Oxidation

In the aquatic environment, light is one of the most significant environmental elements influencing MP aging and transformation (Andrady, 2017). Natural light exposure causes MPs to progressively change color and develop fractures, which lowers the material's elasticity, ductility, impact resistance, and other qualities and indicates MP aging (Hebner & Maurer-Jones, 2020). The functional groups on the surface of MPs have the ability to absorb light energy from natural light, particularly UV light. This helps to create free radicals, which break bonds on the backbone of the polymer and cause MPs to spontaneously depolymerize and degrade. The three stages of the photooxidation process are chain initiation, chain transfer, and chain termination (Gewart et al., 2015; Singh & Sharma, 2008). A free radical is produced during the chain initiation phase when the light energy is absorbed by unsaturated chromophoric groups on the MPs surface, breaking the C-C and C-H bonds on the polymer backbone. PE and PP's macromolecular structures are incapable of absorbing light energy because they lack unsaturated chromophoric groups in their polymer backbone. Nevertheless, non-photoreceptive MPs can absorb light and cause chain initiation if external contaminants or structural anomalies are introduced into the macromolecular structure. The chain transfer mechanism is then led by persistent free radicals that are created when the free radical reacts with O2/H2O in the aquatic environment. In addition to producing hydroperoxides, the chain transfer process also triggers intricate radical reactions. In the chain transfer reaction, olefins, aldehydes, and ketones are produced. Because they include unsaturated double bonds, they are more likely to start photooxidation. Two radicals react to produce inert products, which ends the chain. Consequently, the interaction of radicals causes the response of random chain scission, branching, crosslinking, and the creation of functional groups containing oxygen. Certain MPs, like PVC, are polymers that include chlorine; when they are exposed to light, they first go through a dechlorination reaction that yields polyene polymers and hydrochloric acid (Gewart et al., 2015). Polyene polymers then take part in the photooxidation process. UV is frequently utilized for disinfection in WWTPs. However, MPs are not significantly impacted by the UV dose that is advised for WWTPs. On the other hand, fractures, wrinkles, and protuberances are among the surface morphologies, chemical characteristics, and hydrophobicity of MPs that can be drastically changed by excessive UV irradiation (20 times) (Lin et al., 2020).

The primary adsorption processes of photo-aged MPs were partitioning, electrostatic interactions, and intermolecular hydrogen bonding; it was discovered that their headsorption capacity for hydrophilic pollutants was greater than that of virgin MPs (Yu et al., 2019). In addition to improving the adsorption capacity of certain hydrophilic contaminants, the enhanced hydrophilic and migratory properties of photo-aged MPs would also improve the adsorbed contaminants' movement capability. The findings demonstrate that PS-MP's ability to strengthen hydrogen bonds and increase polar chemicals' adsorption to their surface accounts for the notable improvement in the adsorption affinity of photo-aged PS-MP for 4-nonylphenol. Photo-aged MPs, on the other hand, improve the adsorption of non-polar contaminants like pyrene primarily because to their larger pore capacity (Liu et al., 2019c). Furthermore, compared to the initial MPs, aged MPs had a greater adsorption ability for heavy metals (such as $Cu2^{+}$ and $Zn2^{+}$) (Wang et al., 2020). This could be connected to MPs increased specific surface area and functional groups that hold oxygen with age (Yang et al., 2019). Additionally, photo-aging MPs encourage the release of chemicals, including colorants, flame retardants, and antioxidants. The oxidation of additives and adsorbed chemicals may be the cause of the MPs' dramatic increase in opacity and yellow hue. A comparison of photoaged MPs' leaching characteristics showed that there were significant differences across the leachate forms (Hermabessiere et al., 2017; Khaled et al., 2018). It was discovered that the primary releasing material following UV photoaging of PP-MPs was the antioxidant component diphosphonic acid 168 (Wu et al., 2021). A comparison of the release of hexaBDE from PS-MP surfaces in light and dark environments revealed that the release of hexaBDE from PS surfaces in light was 34.4 mg/g, higher than the release of hexaBDE in dark conditions (25.6 mg/g), indicating that light aging speeds up the release of hexaBDE (Rani et al., 2017). After photoaging MPs, it was also discovered that there was a considerable increase in the release of Ca, Cu, Zn, and total organic carbon (Bendow et al., 2017). As a result, modifications to the surface characteristics of photo-aged MPs significantly affect their capacity to adsorb environmental contaminants and release plastic additives.

3.2 *Mechanical Stress*

Photooxidation is the main cause of MP aging on sea surfaces and beaches, according to all experts. The aging process of MPs that do not come to light is yet unclear, though. Mechanical stress has been thought to have a major role in MP aging. MPs are commonly impacted by the mechanical effects of the surrounding environmental factors, such as water movement, stone and gravel wear and tear, etc. These activities will cause MPs to physically age (Enfrin et al., 2019). Physical aging can change MPs' chemical and physical properties, such as their degree of the polymerization process hydrophilia, thermal resistance, and crystallinity. MPs thus shrink and become more fragile (White & Turnbull, 1994).

When plastic products are made by processes such cold drawing, bending, cutting, rolling, etc., residual stresses may be created. These stresses can shorten the lifespan of plastic products and diminish their mechanical stability (Kim & Min, 2017; Macías et al., 2015). Following their introduction into the aquatic environment, plastic objects will fragment and produce MPs as a result of the residual stressors they contain being layered on top of mechanical stress (Hayes et al., 2015).

The molecular structure of plastic products can affect their tensile strength and density, which in turn can affect how easily they fracture when exposed to external stress. Expanded polystyrene (EPS) is prone to sand abrasion and readily breaks down into MPs, but PE and PP have been shown to fragment very little when exposed to the environmental stress of sand abrasion (Song et al., 2017).

Molecularly, MPs are composed of C-C bonds (375 kJ/mol) and C-H bonds (420 kJ/mol). The C-C and C-H bonds of plastic objects cannot be broken by mechanical abrasion because the heat energy involved is insufficient. According to a recent study, water and high-intensity ultrasonic radiation can combine to create powerful shock waves with high temperatures (5000 K) and high pressures (1000 bar), which can damage the C-C and C-H bonds and cause MPs to depolymerize and break their chains (Li et al., 2005). The aquatic environment will produce radicals due to high shear strength, which can combine with oxygen to cause radical recombination (Singh & Sharma, 2008). MPs can be effectively degraded by the recombined radical because of its increased oxidation potential. Additionally, the ball mill's mechanical stress with oxide powders including CaO, Fe2O3, SiO2, and Al2O3 promotes PVC's dechlorination, which lowers PVC's molecular weight (Inoue et al., 2004). However, high-intensity ultrasonic irradiation uses so much energy that it is not practical for use in real engineering (Li et al., 2005). Furthermore, MPs will fracture and develop additional fractures as a result of mechanical stress destroying their original surface shape, resulting in a greater specific surface area (Enfrin et al., 2019).

3.3 Advanced Oxidation Process

In WWTPs, advanced oxidation processes (AOPs) such ozone oxidation and chlorination have become commonplace recently. In WWTPs, chlorine is frequently utilized as a disinfectant. The potent oxidizing power of chlorine can alter the chemical and physical characteristics of MPs. The ability of MPs to withstand chlorination varies depending on their molecular makeup. According to reports, PP is the most chlorination-resistant material, followed by PS and high-density polyethylene (HDPE). Additionally, MPs' chemical and physical characteristics, including their toxicity and hydrophobia, can be changed by adding C-Cl bands to their polymer backbone. MPs' photosensitivity can be improved by chlorination, which can also redshift the MPs' Raman spectroscopy (Kelkar et al., 2019).

One cutting-edge tertiary treatment method for disinfection and the elimination of new pollutants is the ozone system. Since ozone has more potent oxidizing qualities, it can oxidize MPs directly and add C=O functional groups to the backbone of polymers. As ozone breaks down in the aquatic environment, hydroxyl radicals (HO), which have a greater oxidation potential than ozone and may accelerate the aging process of MPs, can also be produced (Miranda et al., 2021). Following the oxidation of ozone, ketones can be found on the surface of MPs for PE and PP. But in addition to the ketone groups that are formed, the hydroxyl group that is present on PS's surface after ozone treatment also signifies the creation of alcohols and carboxyl acids. Following ozone treatment, PE, PP, and PS's hydrophobics changed, which had an impact on pesticide adsorption (de Aragão Belé et al., 2021).

Other oxidation techniques include Fenton and persulfate treatments. The C-C and C-H bonds in plastic objects can be broken by the high standard redox potential of OH generated in the Fenton reaction (E0 (OH/H2O) = 2.80 V/SHE). According to reports, hydrophilic functional groups (C-O-H and O=C-O) can be added to PS and PC surfaces following Fenton treatment, improving surface wettability and lowering floatability. However, PVC is not significantly affected by Fenton treatment. Therefore, Fenton treatment can be used to separate harmful PVC from a mixture of polymers (Wang & Wang, 2017). Treatment with Fenton and heat-activated K2S2O8 can dramatically reduce the size of MPs. Following a 30-day Fenton and heat-activated K2S2O8 treatment, the dominant size class shrank from 40–50 m to 10–20 m, and PS and PE MPs were broken up into tiny fragments. Fenton and heat-activated $K_2S_2O_8$ treatment can reduce the contact angles of PS and PE MPs, which is explained by the addition of oxygen-containing groups (Liu et al., 2019b).

With the use of electrochemical reactions and no additional chemicals, MPs can be broken down by electrochemical advanced oxidation processes (EAOPs). On the electrode's surface, SO4 and OH can form in situ and target the MPs' surface functional group. When electrolyzed for 6 hours at a current level of 9 A, 89% of the MPs can be broken down in water. MPs undergo immediate degradation into gaseous products after anodic oxidation treatment rather than breaking down into smaller particles. EAOPs are therefore a promising technique for MPs degradation in water that doesn't result in any waste or byproducts. The surface response between MPs and electrodes should be further studied, nevertheless, because the current efficiency of EAOPs is very poor (Kiendrebeogo et al., 2021).

3.4 *Biodegradation*

In order to sustain microbial activity, organic molecules undergo a process known as biodegradation, which turns them into biogas and leftover biomass. Plastic may be utilized as a carbon source in this process (Ali & Sun, 2019; Ali et al., 2019, 2021). Numerous kinds of bacteria that break down MPs have been isolated from a variety of environmental samples, including sediments, landfills, marine settings, and activated sludge from WWTPs and plastic waste dumps. Recently, a few lepidopteran species have gained attention as extraordinary "plastivores" (Kundungal et al., 2019, 2021). It has been shown that the larger wax moth Galleria mellonella's caterpillar larvae can consume and metabolize PE at previously unheard-of speeds. According to Cassone et al.'s investigation of Galleria mellonella's gut microbiome, microbes belonging to the genus Acinetobacter seemed to have a role in the biodegradation process (Cassone et al., 2020). As of right now, there are around 56 species of bacteria and fungus from 25 genera linked to PE biodegradation (Pathak, 2017).

Bacteria degrade MPs in a complicated process involving several metabolic and regulatory processes. Bacterial extracellular enzymes must break down polymers into monomers as the initial stage of MPs degradation. This is due to MPs' extremely large particle size, which prevents them from passing through bacterial membranes. Enzyme attachment to MPs is therefore the rate-limiting step in degradation. A biofilm must grow on the MPs' surface in order to hasten the bacteria's degradation of the MPs. MPs of different shapes and chemical compositions, whether they have smooth or rough surfaces, can be used as inoculation substrates for bacteria. According to earlier studies, the surface of MPs was obscured by the biofilm produced by *Bacillus flexus* and *Pseudomonas citronellolis*, which caused PVC to appear to degrade and change (Giacomucci et al., 2019).

The molecular characteristics of MPs, including their mobility, crystallinity, molecular weight, functional groups, substituents, and modification by the addition of plasticizers or additives, all have a significant impact on the bacteria's ability to degrade MPs (Singh & Sharma, 2008). However, the external environment, which includes light, heat, humidity, and the presence of chemicals, also affects how bacteria break down MPs. The proliferation and metabolism of related bacteria are influenced by the external environment, which drastically changes the rate at which MPs biodegrade. In recent years, scientists have discovered that the intestinal habitats of insect larvae, such as Tenebrio Molitor, are ideal microenvironments for the biodegradation of plastic (Peng et al., 2019). Larvae of insects can release emulsifying agents that speed up the biodegradation of plastics and improve their bioavailability (Brandon et al., 2021).

Following their biodegradation, MPs undergo changes in their properties, such as weight loss, tensile strength, surface modification, and chemical alteration. Weight loss is caused by the breakdown of MPs into monomers as a result of the bacterially mediated hydrolysis process (Kumari et al., 2019). MPs' tensile strength decreases as a result of the degree of polymerization being diminished during biodegradation. The tensile strength of PE treated to Bacillus sp. YP1 or Enterobacter asburiae YT1 was shown to have significantly decreased (>50%) in a research of PE biodegradation by bacterial strains from the guts of plastic-eating waxworms (Yang et al., 2014). MPs' surface topography changed as a result of the emergence of cracks, pores, and holes that were attributed to the biofilm formation and the biodegrading response in the microenvironment (Awasthi et al., 2017). The surface functional group of the MPs is also altered during the biodegradation process, which causes shifts and variations in the strength of the carbonyl bands (Malachová et al., 2020). A growing amount of research has looked at various bacterial consortia's capacity to degrade MPs (Skariyachan et al., 2017, 2018; Tourova et al., 2020). According to several recent investigations, "syntrophic metabolism" occurs when the breakdown products of MPs from one species provide carbon to another (Santisi et al., 2015).

MPs can be used and adhered to by fungi as well as bacteria (Mitik-Dineva et al., 2009). Various chemical linkages, including carboxyl, carbonyl, and ester functional groups, can be formed in MPs with the help of fungi. Fungal hyphae's large surface area allows them to effectively interact and break down plastic polymers by penetrating the surface of polymeric materials (Russell et al., 2011). Moreover, fungi have the ability to facilitate the movement and change of chemicals (Mitik-Dineva et al., 2009). Research on how fungi break down MPs is currently ongoing; however, some advancements have been made.

4 Fate and Transformation of MPs

Size, crystallinity, thermal stability, surface functional groups, and other physical and chemical characteristics of MPs are altered during aging, which has a major impact on both the toxicity of MPs and the interface response between MPs and other pollutants.

4.1 Size and Morphology

Over time, MP sizes generally decrease as they age. It is known that environmental plastic particles will deteriorate further and becoming smaller and smaller until they create nanoplastics (Horton et al., 2017). MPs' surface morphology also grows rougher at this stage. Currently, MPs as small as one meter can be found in environmental samples. However, because of methodological constraints, only a small number of investigations detect plastic particles smaller than 50 m (Imhof et al., 2016). The pictures captured by an emission scanning electron microscope can be used to evaluate the size distribution of MPs. The bonds on MPs' polymer backbone can be strongly attracted by UV light, hastening MPs' disintegration. Wu et al. found that during 12 days of UV irradiation, the average size of PP-MPs dropped from 443.65 to 84.13 m. Moreover, the oxidative effect of UV irradiation on MPs was reduced with the addition of Irgafos 168 antioxidants, leading to a three-fold increase in size in comparison to MPs without antioxidants (Wu et al., 2021).

After speeding up MPs' mechanical abrasion process, Ren et al. found that the MPs' surface had broken edges, pits, and microcracks (Ren et al., 2020). Under mechanical abrasion, MPs' thickness and transmittance also reduced.

4.2 Crystallinity

Crystallinity is a measure of the orderliness of the organization of polymer monomers, and it can affect MPs' mechanical, chemical, thermal, and biodegradable properties. XRD, TGA-DSC, and ATR-FTIR can all be used to measure the crystallinity of MPs. Following the aging processes, the crystallinity changed significantly for MPs with various molecular configurations. Yu et al. examined the change in MP crystallinity following 30 years of a Shanghai landfill and found that whereas PE's crystallinity dropped from 55.6% to 20.8%, PP's grew from 24.9% to 56.8%

(Yu et al., 2022). According to a study by Bonyadinejad et al. (2022). on the impact of photooxidation on low-density polyethylene (LDPE) material qualities, LDPE's crystallinity rose from 24% to 35% following 3 weeks of UVA exposure (24 hours). Moreover, humidity hinders LDPE's ability to become more crystallinity when exposed to UVA light. The crystallinity of MPs is affected differently by various aging processes; Wu et al. discovered that after 12 days of UV oxidation, the crystallinity of PP MPs dropped from 62.04% to 56.40% (Wu et al., 2021).

4.3 Thermal Stability

The thermal characteristics of MPs are influenced by various aging processes. Using the AFM-IR approach, Luo et al. investigated the mechanical, thermal, and nanoscale infrared characteristics of TiO2-pigmented MPs both before and after aging (Luo et al., 2020a). The findings indicated that MPs' resilience to high temperatures significantly declined following aging therapy, with the average softening temperatures of aged MPs and unaged MPs being approximately 94.91 4.40 C and 209.50 11.48 C, respectively. Following the aging process, MPs' temperature at the glass transition dropped by 114.59 C. By extending the exposure duration, Ainali et al. discovered that UV radiation would cause a steady drop in the melting temperatures of LDPE, HDPE, and PP (Ainali et al., 2021). The polymer's average chain length decreased, which causes chain cleavage and lowers the average molecular mass, which was the primary cause of the melting temperature drop. Waste plastics from various landfill ages were examined by Yu et al. (2022). It was demonstrated that when landfilling years increase, PE's melting point falls and the exothermic peak shifts toward the low-temperature zone. The melting phase peak of sample PE-27, which has been in a landfill for 27 years, contains two shoulders (104.2–121 C). A polymer with a low molecular weight is created when the carbon chain of the polymer breaks. Thus, throughout the melt phase, there are two melting endothermic peaks.

4.4 Surface Functional Groups

As previously mentioned, the two primary variables that cause MPs to age in an aquatic environment are exposure to UV light and oxygen. Ainali et al. assessed how laboratory UV-B aging affected the physicochemical characteristics of LDPE, HDPE, PP, and PS (Ainali et al., 2021). The study showed that MPs' carbonyl index significantly increased as a result of the formation of novel functional groups on their surface, including vinyl, carbonyl, and hydroxyl/hydroxyperoxide. The thermal stability and crystallinity of the four types of MPs also clearly declined as a result of the chain breaking and crosslinking process brought on by UV radiation, and the MPs' surfaces were riddled with flaws. Moreover, analytical pyrolysis in conjunction with gas chromatography and mass spectrometry was used to indicate the degree of oxidation and degradation. It also demonstrated that the UV-aging process can alter the peak position of the pyrolytic marker compounds, leading to inaccurate MP quantification in aquatic environments. Luo et al. exposed PE-MPs to high temperatures and intense radiation from Xenon lamps in order to hasten the MPs' aging process. They showed that surface cracks and fragmentation made it easier for light and oxygen to reach the internal layer, which in turn sped up the aging process (Luo et al., 2020b). Moreover, MPs' toxicity was greatly elevated by the constant leakage of pigments throughout the aging process. In order to mimic natural weathering, Luo et al. subsequently exposed the MPs mixture (25 weight percent LDPE and 60 weight percent TiO2) in a Xenon lamp for 6 weeks. They also investigated the MPs mixture's mechanical, thermal, and nanoscale infrared characteristics both before and after aging (Luo et al., 2020a). The surface of the old MPs had numerous wrinkles and granular domains, according to the results. Additionally, C-O and -CH bonds were continuously forming on the MPs' surface as they aged. The hardness of aged MPs increased during the aging process, but their softening and glass transition temperatures significantly dropped after the process.

5 Future Outlook on MPs Degradation

Europe has also proposed plastics restrictions, and China has been enforcing the strictest plastic restriction order in history for almost 2 years, which has stopped or limited the production and sales of disposable plastic shopping bags, tableware, and other products (Clausen et al., 2020). As a result, the industry views mulching films and biodegradable plastic bags as substitute goods that the government is promoting, which has raised demand for them.

Biodegradable plastics are not always biodegradable, despite the fact that they are considered a significant solution to the problem of MP contamination. Because of the features of polymerization, plastics made from biological bases are far more difficult to degrade. In the natural environment, the breakdown of MPs typically occurs in two stages: oxidation and hydrolysis, followed by additional biodegradation. Furthermore, conventional petroleum-based plastics made from non-renewable resources have a very slow rate of natural deterioration (Filiciotto & Rothenberg, 2021). As a result, plastic items typically require modification to alter their performance in terms of natural or artificial degradation.

Through chemical modification techniques including grafting, doping, and modification, the majority of plastic polymers alter their characteristics, including mechanical and degrading qualities (Calvino et al., 2020). Unfortunately, most plastic polymers are employed in biomedicine rather than as engineering plastics, and chemical modification is typically expensive. Physical alteration, like structural alteration and physical mixing, is inexpensive, though. The primary way of modification for biodegradable polymers is the addition of additives.

Table 2 Natural degradation data of degradable plastics

Types of MPs	Aging process	Aging cycle	References
Butanediol succinate cofuran dicarboxylic acid	Hydrolysis	Quality loss of 1–2% after 22 weeks	Tosakul et al. (2022)
Polylactic acid	Hydrolysis	Quality loss of 5.69% in 150 days	Li et al. (2017)
Polylactic acid	Biological corrosive medium	$9.81 \wedge 10^8$ day^{-1}	Lee and Song (2011)
Polylactic acid	Enzymatic degradation	120 min	Peng et al. (2017)
Poly (3-hydroxybutyrate)	Hydrolysis	Quality loss of 0.012% after 30 days	Polyák et al. (2017)
Poly (salicylic glycolide)	Exposure to sweater	Complete degradation within 60 days	Kim et al. (2020)
VPVA-HA-Fe complexes	Placed in soil	Complete degradation in 108 days	Li et al. (2021)

Certain parameters must be met for the degradation reaction to occur in certified degradable plastics, and factors such as temperature, humidity, oxygen, particular bacteria, and so forth will influence the pace of decomposition. One of the primary types of biodegradable plastic, PLA, cannot fully biodegrade in other low-temperature settings and must meet the temperature and humidity requirements of commercial compost in order to do so. Another big substance, PBAT, degrades in water with a very poor efficiency (Lu et al., 2021). According to multiple tests, it degrades at a rate of less than 2% over the course of 365 days, meaning that, like non-biodegradable plastics, it will have a negative environmental impact on rivers, oceans, and other water habitats. It is not possible to directly promote biodegradable polymers using the experimental environment's results. Before being utilized in the appropriate environment, degradable polymers' degradation performance needs to be confirmed by their real-world application. The disintegration cycles and particular circumstances of a few degradable plastics are listed in Table 2.

Adsorbed pollutants and MPs themselves are both responsible for the toxicity of MPs. There isn't a list of biological toxicities, though. In the majority of research procedures, the exposure concentration of the MPs under investigation in organisms is typically much higher than the true value, which makes it difficult to evaluate the toxicity of MPs (Qu et al., 2023). Using computer modeling and feature information identification, a database has also been established to study and describe the origin and aging process of MPs (Lv et al., 2021). Together with data on the quantity and makeup of MPs worldwide, we may also methodically conduct life cycle studies of MPs in the soil, water, and atmosphere to uncover the influence of real environmental conditions on their destiny.

The precise toxicity of MPs to organisms needs to be elucidated based on the examination of their chemical constituents and properties. Specifically, as aging progresses, it is necessary to investigate the connection between MP toxicity and the distinctive features of aging. It is impossible to ascertain whether MPs or adsorbed pollutants are the primary cause of toxicity due to the absence of actual sample study on the subject (Repinc et al., 2022). In order to pinpoint the area of pollution control, it is necessary to ascertain the mechanism of interaction between MPs and pollutants as well as the overall toxicity and contribution to organisms (Yu et al., 2021).

Techniques for speeding up or slowing down the aging of MPs in the environment have been researched based on our understanding of the aging behavior, mechanism, and degradation of MP contaminants in the aquatic environment. The existing treatment approaches are challenging to employ in order to address the MP pollution problem since MP pollutants are stable polymers that exhibit good stability in the aquatic environment (El-Hiti et al., 2022).

6 Conclusion

MPs pose a significant and growing threat to aquatic ecosystems worldwide. This review has examined the complex lifecycle and aging processes of MPs in marine and freshwater environments, highlighting key sources, transformation mechanisms, and ecological impacts. The aging of MPs through photo-oxidation, mechanical stress, and biodegradation alters their physicochemical properties in ways that can enhance their toxicity and interactions with other pollutants. While biodegradable plastics offer promise, their environmental degradation remains limited in many aquatic conditions. Moving forward, critical research priorities include conducting more comprehensive toxicity assessments, analyzing MP lifecycles across different environmental compartments, and developing techniques to accelerate or decelerate MP degradation as needed. Addressing the pervasive issue of MPs in aquatic ecosystems will require coordinated efforts across scientific disciplines, industries, and policymakers. By advancing our understanding of MP aging and impacts, we can develop more effective strategies for mitigating this pressing environmental challenge and protecting the health of aquatic ecosystems for future generations.

References

Ainali, N. M., Bikiaris, D. N., & Lambropoulou, D. A. (2021). Aging effects on low-and high-density polyethylene, polypropylene and polystyrene under UV irradiation: An insight into decomposition mechanism by Py-GC/MS for MP analysis. *Journal of Analytical and Applied Pyrolysis, 158*, 105207.

Akarsu, C., Madenli, Ö., & Deveci, E. Ü. (2021). Characterization of littered face masks in the southeastern part of Turkey. *Environmental Science and Pollution Research, 28*(34), 47517–47527.

Ali, S. S., & Sun, J. (2019). Effective thermal pretreatment of water hyacinth (Eichhornia crassipes) for the enhancement of biomethanation: VIT® gene probe technology for microbial community analysis with special reference to methanogenic Archaea. *Journal of Environmental Chemical Engineering, 7*(1), 102853.

Ali, S. S., Al-Tohamy, R., Manni, A., Luz, F. C., Elsamahy, T., & Sun, J. (2019). Enhanced digestion of bio-pretreated sawdust using a novel bacterial consortium: Microbial community structure and methane-producing pathways. *Fuel, 254*, 115604.

Ali, S. S., Elsamahy, T., Koutra, E., Kornaros, M., El-Sheekh, M., Abdelkarim, E. A., et al. (2021). Degradation of conventional plastic wastes in the environment: A review on current status of knowledge and future perspectives of disposal. *Science of the Total Environment, 771*, 144719.

Aliko, V., Beqiraj, E. G., Qirjo, M., Cani, M., Rama, A., Bego, K., et al. (2022). Plastic invasion tolling: First evaluation of MPs in water and two crab species from the nature reserve lagoonary complex of Kune-Vain, Albania. *Science of the Total Environment, 849*, 157799.

Amin, A., Indulkar, S. T., & Pai, R. (2016). Sublethal effect of Buprofezin pesticide on carcass composition of *Cyprinus carpio* communis fingerlings. *Journal of Experimental Zoology, India, 20*(1), 269–272.

Amin, A., Ramachandra Naik, A. T., Priyadarshini, N., Nayak, H., & Sree, C. S. (2017). Toxic effect of heavy metal lead on oxygen consumption of rohu (*Labeo rohita*) fingerlings. *Biochemical & Cellular Archives, 17*(1), 225.

Amobonye, A., Bhagwat, P., Singh, S., & Pillai, S. (2021). Plastic biodegradation: Frontline microbes and their enzymes. *Science of the Total Environment, 759*, 143536.

An, W., Duan, L., Zhang, Y., Wang, B., Liu, C. S., Wang, F., et al. (2021). Occurrence, spatiotemporal distribution, seasonal and annual variation, and source apportionment of poly– and perfluoroalkyl substances (PFASs) in the northwest of Tai Lake Basin, China. *Journal of Hazardous Materials, 416*, 125784.

Anbumani, S., & Kakkar, P. (2018). Ecotoxicological effects of MPs on biota: A review. *Environmental Science and Pollution Research, 25*, 14373–14396.

Andrady, A. L. (2011). MPs in the marine environment. *Marine Pollution Bulletin, 62*(8), 1596–1605.

Andrady, A. L. (2017). The plastic in MPs: A review. *Marine Pollution Bulletin, 119*(1), 12–22.

Andrady, A. L., & Neal, M. A. (2009). Applications and societal benefits of plastics. *Philosophical Transactions of the Royal Society B: Biological Sciences, 364*(1526), 1977–1984.

Auta, H. S., Emenike, C. U., & Fauziah, S. H. (2017). Screening of Bacillus strains isolated from mangrove ecosystems in Peninsular Malaysia for MP degradation. *Environmental Pollution, 231*, 1552–1559.

Auta, H. S., Emenike, C. U., & Fauziah, S. H. (2021). Distribution and importance of MPs in the marine environment: A review of the sources, fate, effects, and potential solutions. *Environment International, 102*, 165–176.

Awasthi, S., Srivastava, P., Singh, P., Tiwary, D., & Mishra, P. K. (2017). Biodegradation of thermally treated high-density polyethylene (HDPE) by Klebsiella pneumoniae CH001. *3. Biotech, 7*(5), 332.

Bandow, N., Will, V., Wachtendorf, V., & Simon, F. G. (2017). Contaminant release from aged MP. *Environmental Chemistry, 14*(6), 394–405.

Barnes, D. K., Galgani, F., Thompson, R. C., & Barlaz, M. (2009). Accumulation and fragmentation of plastic debris in global environments. *Philosophical Transactions of the Royal Society B: Biological Sciences, 364*(1526), 1985–1998.

Bonyadinejad, G., Salehi, M., & Herath, A. (2022). Investigating the sustainability of agricultural plastic products, combined influence of polymer characteristics and environmental conditions on MPs aging. *Science of the Total Environment, 839*, 156385.

Brandon, A. M., Garcia, A. M., Khlystov, N. A., Wu, W. M., & Criddle, C. S. (2021). Enhanced bioavailability and microbial biodegradation of polystyrene in an enrichment derived from the gut microbiome of Tenebrio molitor (mealworm larvae). *Environmental Science & Technology, 55*(3), 2027–2036.

Browne, M. A., Crump, P., Niven, S. J., Teuten, E., Tonkin, A., Galloway, T., & Thompson, R. (2011). Accumulation of MP on shorelines woldwide: Sources and sinks. *Environmental Science & Technology, 45*(21), 9175–9179.

Calvino, C., Macke, N., Kato, R., & Rowan, S. J. (2020). Development, processing and applications of bio-sourced cellulose nanocrystal composites. *Progress in Polymer Science, 103*, 101221.

Carney Almroth, B. M., Åström, L., Roslund, S., Petersson, H., Johansson, M., & Persson, N. K. (2018). Quantifying shedding of synthetic fibres from textiles; a source of MPs released into the environment. *Environmental Science and Pollution Research, 25*, 1191–1199.

Cassone, B. J., Grove, H. C., Elebute, O., Villanueva, S. M., & LeMoine, C. M. (2020). Role of the intestinal microbiome in low-density polyethylene degradation by caterpillar larvae of the greater wax moth, Galleria mellonella. *Proceedings of the Royal Society B, 287*(1922), 20200112.

Clausen, L. P. W., Hansen, O. F. H., Oturai, N. B., Syberg, K., & Hansen, S. F. (2020). Stakeholder analysis with regard to a recent European restriction proposal on MPs. *PLoS One, 15*(6), e0235062.

Cole, M., Lindeque, P., Halsband, C., & Galloway, T. S. (2011). MPs as contaminants in the marine environment: A review. *Marine Pollution Bulletin, 62*(12), 2588–2597.

de Aragão Belé, T. G., Neves, T. F., Cristale, J., Prediger, P., Constapel, M., & Dantas, R. F. (2021). Oxidation of MPs by O3 and O3/H2O2: Surface modification and adsorption capacity. *Journal of Water Process Engineering, 41*, 102072.

Debnath, R., Prasad, G. S., Amin, A., Malik, M. M., Ahmad, I., Abubakr, A., Borah, S., Rather, M. A., Impellitteri, F., Tabassum, I., & Piccione, G. (2024). Understanding and addressing microplastic pollution: Impacts, mitigation, and future perspectives. *Journal of Contaminant Hydrology, 266*, 104399.

Deng, H., Wei, R., Luo, W., Hu, L., Li, B., & Shi, H. (2020). MP pollution in water and sediment in a textile industrial area. *Environmental Pollution, 258*, 113658.

Du, C., Liang, H., Li, Z., & Gong, J. (2020). Pollution characteristics of MPs in soils in southeastern suburbs of Baoding City, China. *International Journal of Environmental Research and Public Health, 17*(3), 845.

Duan, J., Bolan, N., Li, Y., Ding, S., Atugoda, T., Vithanage, M., et al. (2021). Weathering of MPs and interaction with other coexisting constituents in terrestrial and aquatic environments. *Water Research, 196*, 117011.

El-Hiti, G. A., Ahmed, D. S., Yousif, E., Al-Khazrajy, O. S., Abdallh, M., & Alanazi, S. A. (2022). Modifications of polymers through the addition of ultraviolet absorbers to reduce the aging effect of accelerated and natural irradiation. *Polymers, 14*(1), 20.

Enfrin, M., Dumée, L. F., & Lee, J. (2019). Nano/MPs in water and wastewater treatment processes–origin, impact and potential solutions. *Water Research, 161*, 621–638.

Filiciotto, L., & Rothenberg, G. (2021). Biodegradable plastics: Standards, policies, and impacts. *ChemSusChem, 14*(1), 56–72.

Fossi, M. C., Pedà, C., Compa, M., Tsangaris, C., Alomar, C., Claro, F., et al. (2018). Bioindicators for monitoring marine litter ingestion and its impacts on Mediterranean biodiversity. *Environmental Pollution, 237*, 1023–1040.

Galgani, F., Hanke, G., & Maes, T. (2015). Global distribution, composition and abundance of marine litter. In *Marine anthropogenic litter* (pp. 29–56). Springer International Publishing.

Gall, S. C., & Thompson, R. C. (2015). The impact of debris on marine life. *Marine Pollution Bulletin, 92*(1–2), 170–179.

Gewert, B., Plassmann, M. M., & MacLeod, M. (2015). Pathways for degradation of plastic polymers floating in the marine environment. *Environmental Science: Processes & Impacts, 17*(9), 1513–1521.

Geyer, R., Jambeck, J. R., & Law, K. L. (2017). Production, use, and fate of all plastics ever made. *Science Advances, 3*(7), e1700782.

Giacomucci, L., Raddadi, N., Soccio, M., Lotti, N., & Fava, F. (2019). Polyvinyl chloride biodegradation by Pseudomonas citronellolis and Bacillus flexus. *New Biotechnology, 52*, 35–41.

González-Pleiter, M., Tamayo-Belda, M., Pulido-Reyes, G., Amariei, G., Leganés, F., Rosal, R., & Fernández-Piñas, F. (2019). Secondary nanoplastics released from a biodegradable MP severely impact freshwater environments. *Environmental Science: Nano, 6*(5), 1382–1392.

Guerranti, C., Cannas, S., Scopetani, C., Fastelli, P., Cincinelli, A., & Renzi, M. (2017). Plastic litter in aquatic environments of Maremma Regional Park (Tyrrhenian Sea, Italy): Contribution by the Ombrone river and levels in marine sediments. *Marine Pollution Bulletin, 117*(1–2), 366–370.

Gündoğdu, S., Çevik, C., Güzel, E., & Kilercioğlu, S. (2018). MPs in municipal wastewater treatment plants in Turkey: A comparison of the influent and secondary effluent concentrations. *Environmental Monitoring and Assessment, 190*, 1–10.

Gurjar, U. R., Takar, S., Amin, A., & Martin Xavier, K. A. (2023). Microplastic contamination in aquatic organisms: An ecotoxicological perspective. In *Xenobiotics in aquatic animals: Reproductive and developmental impacts* (pp. 353–367). Springer Nature Singapore.

Hanun, J. N., Hassan, F., & Jiang, J. J. (2021). Occurrence, fate, and sorption behavior of contaminants of emerging concern to MPs: Influence of the weathering/aging process. *Journal of Environmental Chemical Engineering, 9*(5), 106290.

Hayes, M., Edwards, D., & Shah, A. (2015). *Fractography in failure analysis of polymers*. William Andrew.

Hebner, T. S., & Maurer-Jones, M. A. (2020). Characterizing MP size and morphology of photodegraded polymers placed in simulated moving water conditions. *Environmental Science: Processes & Impacts, 22*(2), 398–407.

Hermabessiere, L., Dehaut, A., Paul-Pont, I., Lacroix, C., Jezequel, R., Soudant, P., & Duflos, G. (2017). Occurrence and effects of plastic additives on marine environments and organisms: A review. *Chemosphere, 182*, 781–793.

Horton, A. A., Walton, A., Spurgeon, D. J., Lahive, E., & Svendsen, C. (2017). MPs in freshwater and terrestrial environments: Evaluating the current understanding to identify the knowledge gaps and future research priorities. *Science of the Total Environment, 586*, 127–141.

Hu, L., Fu, J., Wang, S., Xiang, Y., & Pan, X. (2021). MPs generated under simulated fire scenarios: Characteristics, antimony leaching, and toxicity. *Environmental Pollution, 269*, 115905.

Imhof, H. K., Laforsch, C., Wiesheu, A. C., Schmid, J., Anger, P. M., Niessner, R., & Ivleva, N. P. (2016). Pigments and plastic in limnetic ecosystems: A qualitative and quantitative study on microparticles of different size classes. *Water Research, 98*, 64–74.

Inoue, T., Miyazaki, M., Kamitani, M., Kano, J., & Saito, F. (2004). Mechanochemical dechlorination of polyvinyl chloride by co-grinding with various metal oxides. *Advanced Powder Technology, 15*(2), 215–225.

Jiang, J. Q. (2018). Occurrence of MPs and its pollution in the environment: A review. *Sustainable Production and Consumption, 13*, 16–23.

Kelkar, V. P., Rolsky, C. B., Pant, A., Green, M. D., Tongay, S., & Halden, R. U. (2019). Chemical and physical changes of MPs during sterilization by chlorination. *Water Research, 163*, 114871.

Khaled, A., Richard, C., Rivaton, A., Jaber, F., & Sleiman, M. (2018). Photodegradation of brominated flame retardants in polystyrene: Quantum yields, products and influencing factors. *Chemosphere, 211*, 943–951.

Khan, M. T., Shah, I. A., Hossain, M. F., Akther, N., Zhou, Y., Khan, M. S., et al. (2023). Personal protective equipment (PPE) disposal during COVID-19: An emerging source of MP and microfiber pollution in the environment. *Science of the Total Environment, 860*, 160322.

Kiendrebeogo, M., Estahbanati, M. K., Mostafazadeh, A. K., Drogui, P., & Tyagi, R. D. (2021). Treatment of MPs in water by anodic oxidation: A case study for polystyrene. *Environmental Pollution, 269*, 116168.

Kim, B., & Min, J. (2017). Residual stress distributions and their influence on post-manufacturing deformation of injection-molded plastic parts. *Journal of Materials Processing Technology, 245*, 215–226.

Kim, H. J., Reddi, Y., Cramer, C. J., Hillmyer, M. A., & Ellison, C. J. (2020). Readily degradable aromatic polyesters from salicylic acid. ACS Macro Letters, 9(1), 96–102.

Kumari, A., Chaudhary, D. R., & Jha, B. (2019). Destabilization of polyethylene and polyvinylchloride structure by marine bacterial strain. *Environmental Science and Pollution Research, 26*, 1507–1516.

Kundungal, H., Gangarapu, M., Sarangapani, S., Patchaiyappan, A., & Devipriya, S. P. (2019). Efficient biodegradation of polyethylene (HDPE) waste by the plastic-eating lesser waxworm (Achroia grisella). *Environmental Science and Pollution Research, 26*, 18509–18519.

Kundungal, H., Gangarapu, M., Sarangapani, S., Patchaiyappan, A., & Devipriya, S. P. (2021). Role of pretreatment and evidence for the enhanced biodegradation and mineralization of low-density polyethylene films by greater waxworm. *Environmental Technology, 42*(5), 717–730.

Lebreton, L. C., Van Der Zwet, J., Damsteeg, J. W., Slat, B., Andrady, A., & Reisser, J. (2017). River plastic emissions to the world's oceans. *Nature Communications, 8*(1), 15611.

Lee, S. H., & Song, W. S. (2011). Enzymatic hydrolysis of polylactic acid fiber. Applied Biochemistry and Biotechnology, 164, 89–102.

Li, Y., Li, J., Guo, S., & Li, H. (2005). Mechanochemical degradation kinetics of high-density polyethylene melt and its mechanism in the presence of ultrasonic irradiation. *Ultrasonics Sonochemistry, 12*(3), 183–189.

Li, J., Song, Y., & Cai, Y. (2020). Focus topics on MPs in soil: Analytical methods, occurrence, transport, and ecological risks. *Environmental Pollution, 257*, 113570.

Li, X., Chu, C., Wei, Y., Qi, C., Bai, J., Guo, C., & Chu, P. K. (2017). In vitro degradation kinetics of pure PLA and Mg/PLA composite: Effects of immersion temperature and compression stress. *Acta Biomaterialia, 48*, 468–478

Li, Y., Li, S., & Sun, J. (2021). Degradable Poly (vinyl alcohol)-based supramolecular plastics with high mechanical strength in a watery environment. *Advanced Materials, 33*(13), 2007371.

Lin, J., Yan, D., Fu, J., Chen, Y., & Ou, H. (2020). Ultraviolet-C and vacuum ultraviolet inducing surface degradation of MPs. *Water Research, 186*, 116360.

Liu, X., Yuan, W., Di, M., Li, Z., & Wang, J. (2019a). Transfer and fate of MPs during the conventional activated sludge process in one wastewater treatment plant of China. *Chemical Engineering Journal, 362*, 176–182.

Liu, Y., Deng, Y., Chen, P., Duan, M., Lin, X., & Zhang, Y. (2019b). Biodegradation analysis of polyvinyl alcohol during the compost burial course. *Journal of Basic Microbiology, 59*(4), 368–374.

Liu, J., Zhang, T., Tian, L., Liu, X., Qi, Z., Ma, Y., et al. (2019c). Aging significantly affects mobility and contaminant-mobilizing ability of nanoplastics in saturated loamy sand. *Environmental Science & Technology, 53*(10), 5805–5815.

Liu, P., Zhan, X., Wu, X., Li, J., Wang, H., & Gao, S. (2020). Effect of weathering on environmental behavior of MPs: Properties, sorption and potential risks. *Chemosphere, 242*, 125193.

Lu, G., Zhu, J., Yu, H., Jin, M., Abdalkarim, S. Y. H., & Wei, Y. (2021). Degradation mechanism of green biopolyester nanocomposites with various cellulose nanocrystal based nanohybrids. *Cellulose, 28*, 7735–7748.

Luo, H., Xiang, Y., Zhao, Y., Li, Y., & Pan, X. (2020a). Nanoscale infrared, thermal and mechanical properties of aged MPs revealed by an atomic force microscopy coupled with infrared spectroscopy (AFM-IR) technique. *Science of the Total Environment, 744*, 140944.

Luo, H., Zhao, Y., Li, Y., Xiang, Y., He, D., & Pan, X. (2020b). Aging of MPs affects their surface properties, thermal decomposition, additives leaching and interactions in simulated fluids. *Science of the Total Environment, 714*, 136862.

Lv, L., Yan, X., Feng, L., Jiang, S., Lu, Z., Xie, H., et al. (2021). Challenge for the detection of MPs in the environment. *Water Environment Research, 93*(1), 5–15.

Ma, J., Nawarathna, H. M., & Hesp, S. A. (2022). On the sustainable use of recycled plastics in flexible asphalt pavements. *Journal of Cleaner Production, 359*, 132081.

Macías, C., Meza, O., & Pérez, E. (2015). Relaxation of residual stresses in plastic cover lenses with applications in the injection molding process. *Engineering Failure Analysis, 57*, 490–498.

Malachová, K., Novotný, Č., Adamus, G., Lotti, N., Rybková, Z., Soccio, M., et al. (2020). Ability of Trichoderma hamatum isolated from plastics-polluted environments to attack petroleum-based, synthetic polymer films. *PRO, 8*(4), 467.

Malik, M., Balkhi, M. H., & Abubakr, A. (2017). Comparative study on the contamination level of Dal Lake, Kashmir based on Total Viable Bacterial Counts over a period of time. *Ecology, Environment, and Conservation, 23*(4), 2145–2147.

Mani, T., Hauk, A., Walter, U., & Burkhardt-Holm, P. (2015). MPs profile along the Rhine River. *Scientific Reports, 5*(1), 17988.

Miranda, M. N., Sampaio, M. J., Tavares, P. B., Silva, A. M., & Pereira, M. F. R. (2021). Aging assessment of MPs (LDPE, PET and uPVC) under urban environment stressors. *Science of the Total Environment, 796*, 148914.

Mitik-Dineva, N., Wang, J., Truong, V. K., Stoddart, P., Malherbe, F., Crawford, R. J., & Ivanova, E. P. (2009). Escherichia coli, Pseudomonas aeruginosa, and Staphylococcus aureus attachment patterns on glass surfaces with nanoscale roughness. *Current Microbiology, 58*, 268–273.

Nabi, N., Ahmad, I., Amin, A., Rather, M. A., Ahmed, I., Hajam, Y. A., Khursheed, S., Malik, M. M., & Abubakr, A. (2024). Understanding the sources, fate and effects of microplastics in aquatic environments with a focus on risk profiling in aquaculture systems. *Reviews in Aquaculture, 16*(4), 1947–1980.

Natesan, U., Vaikunth, R., Kumar, P., Ruthra, R., & Srinivasalu, S. (2021). Spatial distribution of MP concentration around landfill sites and its potential risk on groundwater. *Chemosphere, 277*, 130263.

Nizzetto, L., Bussi, G., Futter, M. N., Butterfield, D., & Whitehead, P. G. (2016a). A theoretical assessment of MP transport in river catchments and their retention by soils and river sediments. *Environmental Science: Processes & Impacts, 18*(8), 1050–1059.

Nizzetto, L., Futter, M., & Langaas, S. (2016b). Are agricultural soils dumps for MPs of urban origin?.

OECD. (2022). Global primary plastic production by polymer, 1990 to 2019. https://www.oecd.org/en/publications/global-plastics-outlook_aa1edf33-en.html

Padervand, M., Lichtfouse, E., Robert, D., & Wang, C. (2020). Removal of MPs from the environment. A review. *Environmental Chemistry Letters, 18*(3), 807–828.

Pathak, V. M. (2017). Review on the current status of polymer degradation: A microbial approach. *Bioresources and Bioprocessing, 4*(1), 1–31.

Peng, S., Wu, B., Wu, L., Li, B. G., & Dubois, P. (2017). Hydrolytic degradation of biobased poly (butylene succinate-co furandicarboxylate) and poly (butylene adipate-co furandicarboxylate) copolyesters under mild conditions. *Journal of Applied Polymer Science, 134*(15).

Peng, B. Y., Su, Y., Chen, Z., Chen, J., Zhou, X., Benbow, M. E., et al. (2019). Biodegradation of polystyrene by dark (Tenebrio obscurus) and yellow (Tenebrio molitor) mealworms (Coleoptera: Tenebrionidae). *Environmental Science & Technology, 53*(9), 5256–5265.

Pirsaheb, M., Hossini, H., & Makhdoumi, P. (2020). Review of MP occurrence and toxicological effects in marine environment: Experimental evidence of inflammation. *Process Safety and Environmental Protection, 142*, 1–14.

Plastics Europe. (2023). Plastics—the fast Facts 2023. Plastics – the fast Facts 2023 • Plastics Europe

PlasticsEurope, E. P. R. O. (2019). Plastics—the facts 2019. An analysis of European plastics production, demand and waste data. *PlasticEurope*.

Polyák, P., Szemerszki, D., Vörös, G., & Pukánszky, B. (2017). Mechanism and kinetics of the hydrolytic degradation of amorphous poly (3-hydroxybutyrate). *Polymer Degradation and Stability, 140*, 1–8.

Prasad, G. S., Rout, S. K., Malik, M. M., Karmakar, S., Amin, A., & Ahmad, I. (2023). Occurrence of xenoestrogen alkylphenols (Octylphenols and Nonylphenol) and its impact on the aquatic ecosystem. In *Xenobiotics in aquatic animals: Reproductive and developmental impacts* (pp. 275–284). Springer Nature Singapore.

Prasad, G. S., Rout, S. K., Amin, A., Malik, M. M., Ahmad, I., Rather, M. A., Karmakar, S., Khalil, H. S., & Abubakar, A. (2024). Impact of COVID-19 on the aquatic environment and implications for fisheries with special emphasis to Indian perspective. *Proceedings of the Indian National Science Academy, 90*(4), 821–839.

Qu, H., Diao, H., Han, J., Wang, B., & Yu, G. (2023). Understanding and addressing the environmental risk of MPs. *Frontiers of Environmental Science & Engineering, 17*(1), 12.

Rani, M., Shim, W. J., Jang, M., Han, G. M., & Hong, S. H. (2017). Releasing of hexabromocyclododecanes from expanded polystyrenes in seawater-field and laboratory experiments. *Chemosphere, 185*, 798–805.

Rastogi, S., & Kandasubramanian, B. (2020). Processing trends of silk fibres: Silk degumming, regeneration and physical functionalization. *The Journal of The Textile Institute, 111*(12), 1794–1810.

Ren, S. Y., Sun, Q., Ni, H. G., & Wang, J. (2020). A minimalist approach to quantify emission factor of MP by mechanical abrasion. *Chemosphere, 245*, 125630.

Repinc, S. K., Bizjan, B., Budhiraja, V., Dular, M., Gostiša, J., Humar, B. B., et al. (2022). Integral analysis of hydrodynamic cavitation effects on waste activated sludge characteristics, potentially toxic metals, microorganisms and identification of MPs. *Science of the Total Environment, 806*, 151414.

Resmeriță, A. M., Coroaba, A., Darie, R., Doroftei, F., Spiridon, I., Simionescu, B. C., & Navard, P. (2018). Erosion as a possible mechanism for the decrease of size of plastic pieces floating in oceans. *Marine Pollution Bulletin, 127*, 387–395.

Rochman, C. M. (2015). The complex mixture, fate and toxicity of chemicals associated with plastic debris in the marine environment. In *Marine anthropogenic litter* (pp. 117–140). Springer International Publishing.

Russell, J. R., Huang, J., Anand, P., Kucera, K., Sandoval, A. G., Dantzler, K. W., et al. (2011). Biodegradation of polyester polyurethane by endophytic fungi. *Applied and Environmental Microbiology, 77*(17), 6076–6084.

Santisi, S., Cappello, S., Catalfamo, M., Mancini, G., Hassanshahian, M., Genovese, L., et al. (2015). Biodegradation of crude oil by individual bacterial strains and a mixed bacterial consortium. *Brazilian Journal of Microbiology, 46*(2), 377–387.

Singh, B., & Sharma, N. (2008). Mechanistic implications of plastic degradation. *Polymer Degradation and Stability, 93*(3), 561–584.

Skariyachan, S., Setlur, A. S., Naik, S. Y., Naik, A. A., Usharani, M., & Vasist, K. S. (2017). Enhanced biodegradation of low and high-density polyethylene by novel bacterial consortia formulated from plastic-contaminated cow dung under thermophilic conditions. *Environmental Science and Pollution Research, 24*, 8443–8457.

Skariyachan, S., Patil, A. A., Shankar, A., Manjunath, M., Bachappanavar, N., & Kiran, S. (2018). Enhanced polymer degradation of polyethylene and polypropylene by novel thermophilic consortia of Brevibacillus sps. and Aneurinibacillus sp. screened from waste management landfills and sewage treatment plants. *Polymer Degradation and Stability, 149*, 52–68.

Song, Y. K., Hong, S. H., Jang, M., Han, G. M., Jung, S. W., & Shim, W. J. (2017). Combined effects of UV exposure duration and mechanical abrasion on MP fragmentation by polymer type. *Environmental Science & Technology, 51*(8), 4368–4376.

Sruthisree, C., Gowda, G., Nayak, H., Harish, B. D., & Amin, A. (2015). Effect of heavy metal (lead) on oxygen consumption and ammonia excretion of Iridescent Shark (*Pangusius hypohthalmus*). *Journal of Experimental Zoology, India, 18*(2), 637.

Tabassum, I., Hassan, N., Amin, A., Malik, M. M., Abubakr, A., Prasad, G. S., Latief, F., & Rather, S. A. (2025). COVID-19 and global environmental change: A threat to food security and nutrition. In *Food security, nutrition and sustainability through aquaculture technologies* (pp. 437–445). Springer Nature Switzerland.

Thompson, R. C., Olsen, Y., Mitchell, R. P., Davis, A., Rowland, S. J., John, A. W., et al. (2004). Lost at sea: Where is all the plastic? *Science, 304*(5672), 838–838.

Thushari, G. G. N., & Senevirathna, J. D. M. (2020). Plastic pollution in the marine environment. *Heliyon, 6*(8), e04709.

Tian, L., Chen, Q., Jiang, W., Wang, L., Xie, H., Kalogerakis, N., et al. (2019). A carbon-14 radiotracer-based study on the phototransformation of polystyrene nanoplastics in water versus in air. *Environmental Science: Nano, 6*(9), 2907–2917.

Tosakul, T., Suetong, P., Chanthot, P., & Pattamaprom, C. (2022). Degradation of polylactic acid and polylactic acid/natural rubber blown films in aquatic environment. *Journal of Polymer Research, 29*(6), 242.

Tourova, T., Sokolova, D., Nazina, T., Grouzdev, D., Kurshev, E., & Laptev, A. (2020). Biodiversity of microorganisms colonizing the surface of polystyrene samples exposed to different aqueous environments. *Sustainability, 12*(9), 3624.

Turner, A., & Holmes, L. A. (2015). Adsorption of trace metals by MP pellets in fresh water. *Environmental Chemistry, 12*(5), 600–610.

Van Sebille, E., Wilcox, C., Lebreton, L., Maximenko, N., Hardesty, B. D., Van Franeker, J. A., et al. (2015). A global inventory of small floating plastic debris. *Environmental Research Letters, 10*(12), 124006.

Wang, J. C., & Wang, H. (2017). Fenton treatment for flotation separation of polyvinyl chloride from plastic mixtures. *Separation and Purification Technology, 187*, 415–425.

Wang, Q., Wangjin, X., Zhang, Y., Wang, N., Wang, Y., Meng, G., & Chen, Y. (2020). The toxicity of virgin and UV-aged PVC MPs on the growth of freshwater algae Chlamydomonas reinhardtii. *Science of the Total Environment, 749*, 141603.

Weisser, J., Beer, I., Hufnagl, B., Hofmann, T., Lohninger, H., Ivleva, N. P., & Glas, K. (2021). From the well to the bottle: Identifying sources of MPs in mineral water. *Water, 13*(6), 841.

Werner, S., Budziak, A., Van Franeker, J. A., Galgani, F., Hanke, G., Maes, T., et al. (2016). Harm caused by marine litter.

White, J. R., & Turnbull, A. (1994). Weathering of polymers: Mechanisms of degradation and stabilization, testing strategies and modelling. *Journal of Materials Science, 29*, 584–613.

World Population Review. (2021). *Plastic pollution in the ocean*. Plastic Pollution by Country 2024 .

Wu, N., Zhang, Y., Zhao, Z., He, J., Li, W., Li, J., et al. (2020). Colonization characteristics of bacterial communities on MPs compared with ambient environments (water and sediment) in Haihe Estuary. *Science of the Total Environment, 708*, 134876.

Wu, X., Liu, P., Shi, H., Wang, H., Huang, H., Shi, Y., & Gao, S. (2021). Photo aging and fragmentation of polypropylene food packaging materials in artificial seawater. *Water Research, 188*, 116456.

Xu, Z., Sui, Q., Li, A., Sun, M., Zhang, L., Lyu, S., & Zhao, W. (2020). How to detect small MPs (20–100 μm) in freshwater, municipal wastewaters and landfill leachates? A trial from sampling to identification. *Science of the Total Environment, 733*, 139218.

Yan, W., Hamid, N., Deng, S., Jia, P. P., & Pei, D. S. (2020). Individual and combined toxicogenetic effects of MPs and heavy metals (Cd, Pb, and Zn) perturb gut microbiota homeostasis and gonadal development in marine medaka (Oryzias melastigma). *Journal of Hazardous Materials, 397*, 122795.

Yang, J., Yang, Y., Wu, W. M., Zhao, J., & Jiang, L. (2014). Evidence of polyethylene biodegradation by bacterial strains from the guts of plastic-eating waxworms. *Environmental Science & Technology, 48*(23), 13776–13784.

Yang, J., Cang, L., Sun, Q., Dong, G., Ata-Ul-Karim, S. T., & Zhou, D. (2019). Effects of soil environmental factors and UV aging on Cu 2+ adsorption on MPs. *Environmental Science and Pollution Research, 26*, 23027–23036.

Yang, Y., Liu, W., Zhang, Z., Grossart, H. P., & Gadd, G. M. (2020). MPs provide new microbial niches in aquatic environments. *Applied Microbiology and Biotechnology, 104*, 6501–6511.

Yang, H., Chen, G., & Wang, J. (2021). MPs in the marine environment: Sources, fates, impacts and microbial degradation. *Toxics, 9*(2), 41.

Yu, F., Yang, C., Zhu, Z., Bai, X., & Ma, J. (2019). Adsorption behavior of organic pollutants and metals on micro/nanoplastics in the aquatic environment. *Science of the Total Environment, 694*, 133643.

Yu, Z. F., Song, S., Xu, X. L., Ma, Q., & Lu, Y. (2021). Sources, migration, accumulation and influence of MPs in terrestrial plant communities. *Environmental and Experimental Botany, 192*, 104635.

Yu, F., Wu, Z., Wang, J., Li, Y., Chu, R., Pei, Y., & Ma, J. (2022). Effect of landfill age on the physical and chemical characteristics of waste plastics/MPs in a waste landfill sites. *Environmental Pollution, 306*, 119366.

Ziajahromi, S., Neale, P. A., Silveira, I. T., Chua, A., & Leusch, F. D. (2021). An audit of MP abundance throughout three Australian wastewater treatment plants. *Chemosphere, 263*, 128294.

COVID-19 and Microplastics: Analysing the Role of Medical Waste in Environmental Pollution

Shigufta Ali

1 Introduction

Plastics have transformed the world in many ways and it has been a boon for mankind since they facilitated the advancement of computers, mobile phones, and other life-saving innovations in contemporary medicine. It is used widely for its malleability, adaptability, and durability, and it has diverse applications; nonetheless, its misuse has resulted in detrimental effects on the environment. The COVID-19 pandemic has caused a substantial increase in the demand for plastic products. Coronavirus disease 2019 (COVID-19) was declared a pandemic by the World Health Organization (WHO) on March 11, 2020 (World Health Organization, 2020).

As of October 2023, there were over 700 million reported cases worldwide, with more than 6 million documented deaths (dela Cruz et al., 2024). While the pandemic's devastating impacts on human health and the global economy have been widely acknowledged, its profound environmental implications also warrant significant attention. The pandemic resulted in a significant rise in municipal waste, largely due to the extensive utilisation of single-use plastic products, including gloves, masks, tissues, and various forms of personal protective equipment (PPE) (Patrício Silva, 2021). The WHO estimated a monthly requirement of 89 million face masks, 76 million gloves, 30 million gowns, 1.6 million goggles, and 2.9 million hand sanitisers to assist frontline healthcare workers in the fight against COVID-19 therefore it required a 40% augmentation in the worldwide manufacturing of medical safety products. Consequently, the usage of various plastic products, including face masks (e.g., N95 or FFP2), boots, water bottles, food containers, and plastic bags, increased significantly (Benson et al., 2021). For a global population of 7.8 billion, the monthly consumption of face masks and gloves was estimated to be 129 billion

S. Ali (✉)
Fish Nutrition and Research Laboratory, Department of Zoology, Aligarh Muslim University, Aligarh, India

A. Abubakr et al. (eds.), *Microplastics: Ecological and Food Security Implications*,
https://doi.org/10.1007/978-3-031-94480-2_13

and 65 billion, respectively (Kalina et al., 2022). Specific instances underscore the scale of this demand. In February 2020, China's daily production of face masks surged to 116 million, over 12 times its usual level ("Coronavirus: China's manufacturing supply chain pummelled from all sides in efforts to restart | South China Morning Post," n.d.). Similarly, in the United Kingdom, 39,500 facial masks, 11,500 medical gloves, 1500 gowns, and 4200 FFP3 masks were used in the same month (Duer, 2021). This unprecedented reliance on PPE and other healthcare products resulted in a significant increase in biomedical and municipal solid waste globally (Masud et al., 2023; Prasad et al., 2023). The extensive utilisation of plastic materials, produced from polystyrene (PS), polypropylene (PP), polyethylene (PE), polyacrylonitrile (PAN), and polyurethane (PU), was crucial for manufacturing vital products such as face masks, gloves, aprons, and facial coverings (Ardusso et al., 2021). However, the significant reliance on such materials has raised environmental concerns, particularly regarding their improper disposal and accumulation. The growing consumption and mismanagement of plastic-based PPE and other single-use products during the pandemic have posed considerable risks to terrestrial, aquatic, and atmospheric systems (Mohana et al., 2023). Additionally, the emergence of microplastics as a byproduct of this plastic waste is of particular concern. This chapter analyses the environmental repercussions of the COVID-19 epidemic, specifically addressing the increase in plastic waste and its effects. The discourse seeks to emphasise the urgent necessity for sustainable waste management solutions and to offer perspectives for tackling analogous difficulties in the future (Fig. 1).

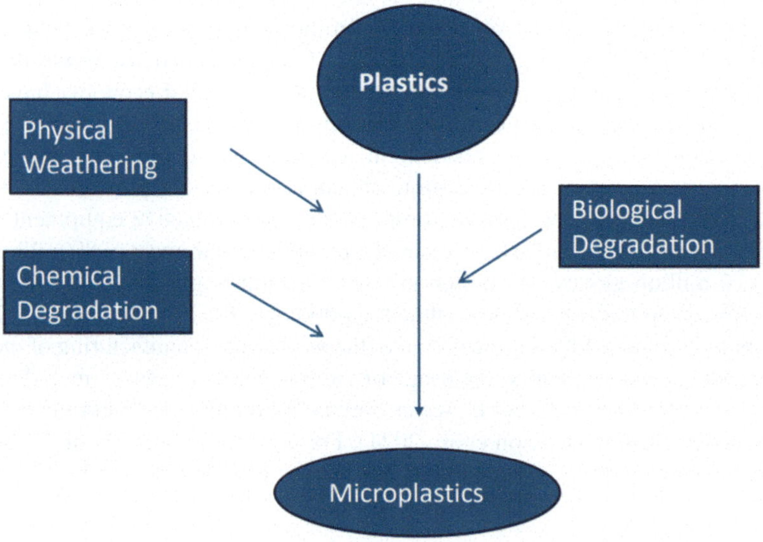

Fig. 1 Disintegration of plastics into microplastics

2 Covid 19 and Emergence of Biomedical Waste

The COVID-19 pandemic led to a significant surge in global biomedical waste production (Behera, 2024; Prasad et al., 2024; Tabassum et al., 2025), primarily driven by the unprecedented rise in single-use plastics that reshaped global consumption patterns. The exponential increase in personal protective equipment (PPE) demand was a crucial factor, with the World Health Organization (WHO) reporting monthly global consumption of approximately 129 billion face masks and 65 billion gloves at the height of the crisis (Prata, 2018). While essential for the safety of healthcare workers and the public, the extensive use of these materials created significant environmental challenges. As of May 20, 2021, the pandemic's global impact was apparent, with verified COVID-19 cases totalling 163,738,674 and a death toll of 3,384,750 ("The European Respiratory Virus Surveillance Summary (ERVISS)," n.d.). Recurring waves of infection placed immense strain on healthcare systems worldwide (Derraik et al., 2020). Although responses varied by socioeconomic context, most health authorities recommended PPE use for healthcare workers and other at-risk personnel (Nguyen et al., 2020) and encouraged widespread public mask-wearing to mitigate transmission (Tso & Cowling, 2020). Despite debates regarding mask efficacy and societal tensions resulting in discrimination against non-wearers, face masks became normalized and often legally mandated. In 2020 alone, single-use mask consumption reached an estimated 129 billion per month (Ren et al., 2021). In 2020 alone, single-use mask consumption reached an estimated 129 billion per month (Ren et al., 2021). To meet the surging PPE requirements, many countries rapidly scaled up production. China, as the leading global manufacturer, increased daily mask output from 20 million in January 2020 to 200 million by March ("Coronavirus: China's manufacturing supply chain pummelled from all sides in efforts to restart South China Morning Post," n.d.). Similar expansions took place in various countries as governments emphasised public health initiatives rather than environmental factors. Simultaneously, there was a notable shift in consumer behaviour, characterised by substantial increases in food delivery services and online shopping. These trends drove increased use of disposable plastic items, including food containers, bubble wrap, and other packaging, which, while maintaining hygiene standards, intensified the already burgeoning plastic waste crisis. The pandemic underscored a significant contrast between immediate public health needs and the goals of long-term environmental sustainability. Urban areas with elevated infection rates have become the epicentre of plastic pollution, with masks, gloves, and other PPE items noticeably accumulating on streets and waterways.

Hence, the persistence of plastics with single-use face masks and PPE kits will probably lead to the prevalence of these materials in the environment in the upcoming years (Sajjad et al., 2022a).

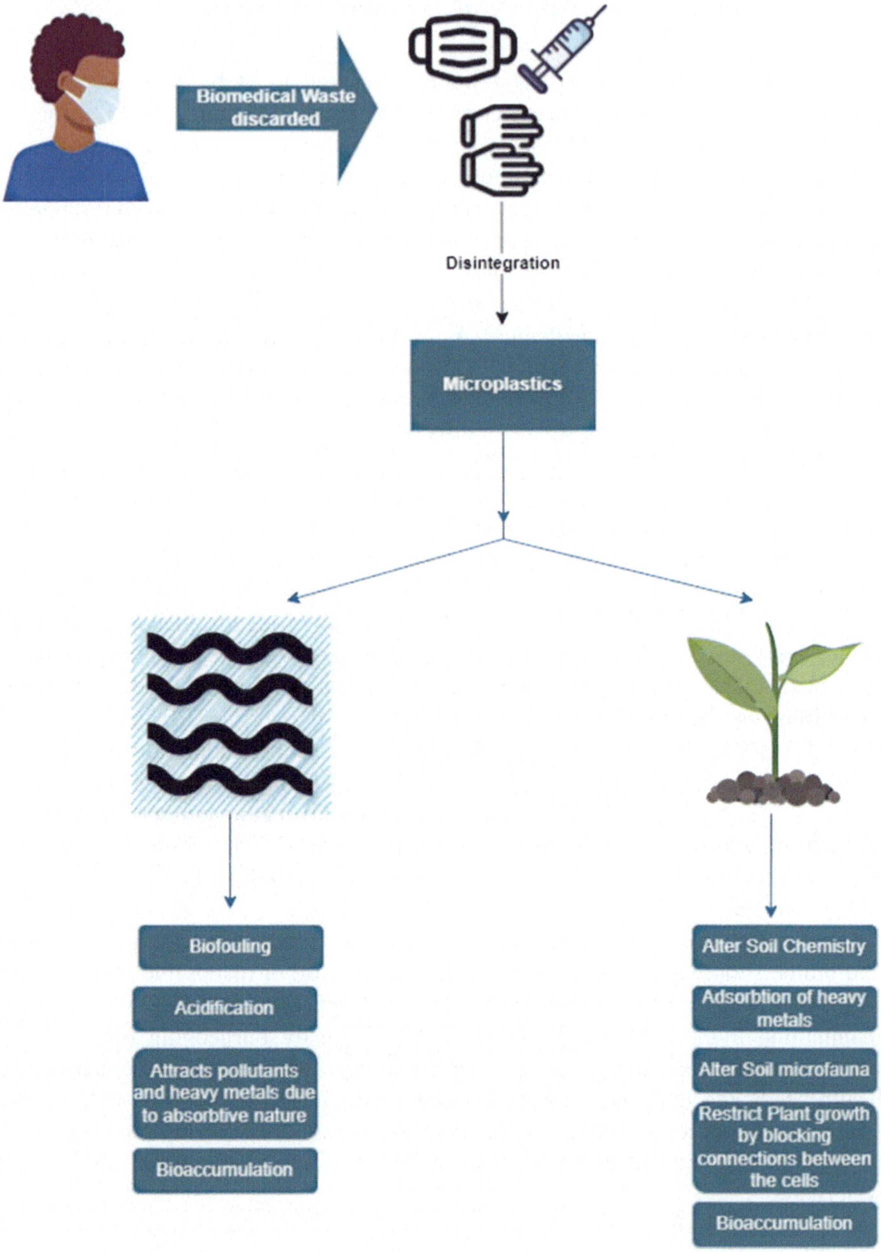

Fig. 2 Pathways from which microplastics enter the aquatic and terrestrial environment

3 Biomedical Waste as a Source of Microplastics

PPE like face masks and gloves were most used during the pandemic and their improper disposal led to its accumulation in landfills and water bodies (Rathinamoorthy & Raja Balasaraswathi, 2023). Since these are produced from polypropylene (PP), polyurethane (PU), polyacrylonitrile (PAN), polystyrene (PS), polycarbonate (PC), polyethylene (PE), and polyethylene terephthalate (PET) are (Chellamani et al., 2013). These compounds, when exposed to ultraviolet light and mechanical wear and tear, they break up into micro/nanoparticles (Fadare & Okoffo, 2020; Kutralam-Muniasamy et al., 2023; Mhiret Gela & Aragaw, 2022) which is a source of secondary plastic in the environment. It is predicted that a mask in ambient circumstances of a coastal ecosystem can discharge approximately 173,000 microfibers daily (Saliu et al., 2021). According to the research conducted by (Sullivan et al., 2021) it was found that masks subjected to water emitted heavy metals, including lead, cadmium, and antimony (Sorasan et al., 2022), which is further hazardous for aquatic and terrestrial environments.

4 Environmental Impact of Microplastics

Microplastics act as carriers for harmful chemicals, including heavy metals and pesticides, amplifying their toxic effects (Hasan & Tarannum, 2025; Nabi et al., 2024). Their ability to facilitate the spread of invasive species and pathogens further destabilizes ecosystems (Zhao et al., 2023). With the emergence of nanoplastics compounds these challenges due to their smaller size allows them to penetrate biological membranes and interact with cells at a molecular level, potentially causing significant harm (Shruti et al., 2020; Gurjar et al., 2023; Debnath et al., 2024).

4.1 Aquatic Systems

Water currents transport microplastics that ultimately aggregate in oceans and rivers. Less dense plastics such as polythene accumulate on the water's surface, but denser materials sink. Such materials affect all marine life, both species at the surface and those at the bottom of a sea body. The polypropylene composition of masks, together with other PPE debris, is a significant contributor to marine pollution, have been observed in marine flora and fauna (Aragaw, 2020; De-la-Torre et al., 2022). Over 20% of examined crustaceans have been demonstrated to contain microplastics linked to masks and personal protective equipment (Ray et al., 2022). Single-use masks and other forms of plastic personal protective equipment (PPE) contribute to the toxicity of the world's oceanic waters through plastic pollution.This mechanism can result in the poisoning of aquatic organisms that ingest plastic particles (Ray et al., 2022). The impact of microplastics on the ecological environment is undeniably harmful. Marine organisms, including fish and other species, ingest microplastics, leading to physical blockages, as well as toxic repercussions and bioaccumulation at higher trophic levels. Consuming contaminated seafood containing microorganisms presents a significant threat to the health of apex predators, including humans (Andrady, 2011). Over time, the ocean will further acidify due to accumulation of microplastics.

4.2. Terrestrial Systems Terrestial systems comes in contact with biomedical wastes through improper disposal and microplastics released from it disrupt soil health by altering its structure and microbial activity. Studies have revealed that these particles reduce soil fertility and water retention, potentially impacting agricultural productivity (Kwak & An, 2021). The presence of high concentrations of microplastics (MPs) in soil cannot be ignored due to their potential environmental threats, including altering soil properties, impacting microbial activity, harming soil fauna, and hindering plant growth (Li et al., 2021). Due to their potentially harmful effects, together with their longevity and persistence, the presence of microplastics poses a hazard to soil ecology and ecosystems (Hidalgo-Ruz et al., 2012). Microplastics present in the soil contribute to the deterioration of soil structure, impair the soil's infiltration capacity for rain and irrigation, and negatively affect its water retention ability. Agricultural and urban soils are identified as the most significant sinks for microplastics (MPs) (Hurley & Nizzetto, 2018). Plastic residues in the soil gradually break down into microplastics (MPs) over time through various environmental processes. These MPs disperse throughout the soil and interact with other contaminants, such as heavy metals, pesticides, and persistent organic pollutants (POPs), leading to combined toxic effects on soil organisms and ecosystems (Shruthisree et al., 2015; Amin et al., 2016, 2017). These MPs may be ultimately transferred to underground water via seepage and by agricultural water runoff to rivers, oceans, and other water causing water pollution (Sajjad et al., 2022b). In terrestrial ecosystems, soil-dwelling organisms such as earthworms have demonstrated reduced growth and reproductive capacity when exposed to microplastics. These particles also disrupt microbial communities essential for nutrient cycling, undermining soil health and productivity (Lakshmi Kavya et al., 2020).

4.3. Atmospheric Systems Microplastics also become airborne, carried by wind across continents. Urban centres and remote regions alike have reported microplastic deposition, illustrating the pervasiveness of this pollution (Selvam et al., 2021). Inhalation of microplastics cause many respiratory problems such as asthma, bronchitis and pneumonia (Bhat, 2024).

5 Human Health Concerns

Humans are exposed to microplastics through ingestion, inhalation, and dermal contact. Alarmingly, studies have detected microplastics in human tissues, including the lungs, intestines, and even placentas. The potential health implications are vast including cellular damage like inducing oxidative stress, inflammation, and DNA damage, heightening the risk of chronic illnesses and cancers (Phu et al., 2023). Microplastics also carry endocrine-disrupting chemicals (EDCs) that interfere with hormonal balance, potentially causing reproductive and developmental issues (Tiwari et al., 2023). Likewise inhaled microplastics contribute to respiratory conditions, including inflammation and fibrosis of lungs posing long-term health risks) also microplastics induce contact dermatitis (Wang et al., 2021).

6 Challenges in Waste Management

The global waste management systems were flooded by the huge volume of plastic waste, which underscored critical gaps in infrastructure and policy. Developing countries, in particular, encountered severe obstacles. For example, Manila, Philippines, reported a 280-ton daily increase in hazardous medical waste, while Thailand experienced a 62% increase in plastic waste between 2019 and 2021 (Godbole, 2021; Masud et al., 2023). Even nations with sophisticated systems encountered difficulties in managing the influx, frequently resorting to landfill disposal or incineration, both of which have their own environmental disadvantages. The degradation of plastics into microplastics and nano plastics was expedited by improper disposal practices, including open dumping and burning. Subsequently, these minute particulates penetrated ecosystems via soil, water, and air pathways, resulting in long-term remediation challenges. The new and urgent dimension of the global plastic pollution crisis can be seen by the widespread visibility of PPE debris in oceans, rivers, and even remote forests crisis (Sangkham, 2020).

7 Mitigation Strategies

7.1 Transition to Circular Economy and Improved Waste Management Techniques

The circular economy encompasses the concept of consumption and production, which involves reusing, recycling, refurbishing, and repairing existing materials as long as possible. This could be achieved by improving the waste management system, advancing waste recycling technologies such as chemical recycling which can recover high value materials for example pyrolysis convert plastic waste into value added products like char, oil and gas which further can be used as the source of fuel and building materials. Likewise enhanced segregation of waste at the source can improve recycling efficiency and reduce contamination (Zhao & Zhang, 2023).

7.2 *Development of Alternative*

Due to the non-biodegradable nature of plastics and their persistence in the environment, it is very important to find alternatives to it. Polyhydroxyalkanoates (PHA), plant-based plastics also known as bioplastics, are sustainable alternatives to petroleum-based plastics. They are often biodegradable and are not as deleterious to environment as their counterparts, therefore scaling up the production and adoption of these material by government incentives can significantly reduce reliance on traditional plastics (Fadare & Okoffo, 2020). However, their production is currently costly and limited in scale compared to conventional plastics therefore efforts must be made to lower down their cost.

7.3 *Policy and Public Engagement*

Educational Campaigns: Engaging the public about the necessity of reducing plastic consumption and adopting environmentally friendly practices will facilitate a transformation in habits. Campaigns must encourage people to restrict single-use plastics, for recycling, and support local initiatives.

Community Engagement: Connecting local communities to waste management initiatives and enabling them to execute their own programs can foster a sense of ownership and enhance involvement in recycling activities. Collaboration models can enhance the efficacy of these programs.

Policies that focus on the 3Rs and ban any form of waste are also needed. Campaigns of this sort can help people to be socially responsible, as well as seek interactive alternatives to mitigate the plastic pollution problem (United Nations Environment Programme, 2018).

7.4 *Discussion*

The interplay between the realms of public health and environmental sustainability has been brought to light during the COVID-19 crisis. The reliance on single-use plastics during the pandemic has shown how plastic generated safety equipments can lead to adverse repercussions for nature, even if they are useful in the short term. This two-sidedness requires a thorough comprehension of the nature of the problem and the means to ameliorate it. The need for resilient waste management systems cannot be over emphasized, as this is a core lesson learnt during the pandemic. Countries which had developed efficient mechanisms for recycling and waste segregation were able to limit some negative impacts, whilst those which did not have the requisite mechanisms faced challenges. This difference of practices highlights the need for adequate resources allocation to global waste management practices especially in developing countries where the situation with respect to plastic pollution is at its worst. The scientific community alongside the policymaking bodies have raised great concerns regarding microplastics and nanoplastics being an evolving environmental threat. The COVID-19 pandemic has in fact, widened the scope for their use and distribution which showcases how changes in material usage can worsen pollution problems The omnipresence of microplastics in terrestrial, aquatic, and atmospheric systems indicates the existence of a global problem that requires all nations to work collaboratively and to set suitable management policies in order to solve. Furthermore, the pandemic has stimulated the search for alternative materials like biodegradable compounds like PHA and bio-based plastics are forecasting future trends. Nevertheless, there are problems of economic viability and performance, and scaling-up remains a challenge. Research and development work must be undertaken by governments and industries jointly to make sustainable alternatives competitive for the market. Public awareness also emerged as a critical factor in addressing the problem of plastic waste. Nonetheless, to address the long-term necessary changes in social behaviour such as minimization of consumption of single-use plastic, extensive public education campaigns supported by policy measures are needed. In this regard, the coronavirus pandemic has turned out to be a perfect time to reconsider our dependence on plastics. We can help in making the world more sustainable through appropriate measures with the understanding of the people and using relevant technologies to promote healthy societies. Such a challenge requires collective action and resolve from the governments, businesses, researchers, and society as a whole in order to foster a balanced and sustainable future.

8 Conclusion

The COVID-19 global health crisis revealed a multi-faceted tension between the public health requirements and the need to sustain the planet's environment. The issue of the pollution of the oceans by plastic has to be engagements at the level of the global community coupled with appropriate scientific innovations and lasting policies. If we are concerned with the longer-term negative impacts of the pandemic on the environment, then the way out is to change our way and to promote global collaboration between countries and regions of the world.

References

Amin, A., Indulkar, S. T., & Pai, R. (2016). Sublethal effect of Buprofezin pesticide on carcass composition of *Cyprinus carpio* communis fingerlings. *Journal of Experimental Zoology, India, 20*(1), 269–272.

Amin, A., Ramachandra Naik, A. T., Priyadarshini, N., Nayak, H., & Sree, C. S. (2017). Toxic effect of heavy metal lead on oxygen consumption of rohu (*Labeo rohita*) fingerlings. *Biochemical & Cellular Archives, 17*(1), 225.

Andrady, A. L. (2011). Microplastics in the marine environment. *Marine Pollution Bulletin, 62*(8), 1596–1605.

Aragaw, T. A. (2020). Surgical face masks as a potential source for microplastic pollution in the COVID-19 scenario. *Marine Pollution Bulletin, 159*, 111517.

Ardusso, M., Forero-López, A. D., Buzzi, N. S., Spetter, C. V., & Fernández-Severini, M. D. (2021). COVID-19 pandemic repercussions on plastic and antiviral polymeric textile causing pollution on beaches and coasts of South America. *Science of the Total Environment, 763*, 144365. https://doi.org/10.1016/J.SCITOTENV.2020.144365

Behera, S. K. (2024). Microbial remediation of microplastics and integration into the circular economy. In *Sustainable microbial technology for synthetic and cellulosic microfiber bioremediation* (pp. 207–220). Springer Nature Switzerland. https://doi.org/10.1007/978-3-031-62600-5_11

Benson, N. U., Bassey, D. E., & Palanisami, T. (2021). COVID pollution: Impact of COVID-19 pandemic on global plastic waste footprint. *Heliyon, 7*, e06343. https://doi.org/10.1016/j.heliyon.2021.e06343

Bhat, M. A. (2024). A comprehensive characterization of indoor ambient microplastics in households during the COVID-19 pandemic. *Air Quality, Atmosphere and Health, 17*, 2017–2033. https://doi.org/10.1007/s11869-024-01559-6

Chellamani, K. P., Veerasubramanian, D., & Balaji, R. V. (2013). Surgical face masks: manufacturing methods and classification. *Journal of Academia and Industrial Research, 2*(6), 320–324.

Coronavirus: China's manufacturing supply chain pummelled from all sides in efforts to restart | South China Morning Post [WWW Document]. (n.d.). https://www.scmp.com/economy/china-economy/article/3051534/coronavirus-chinas-manufacturing-supply-chain-pummelled-all?firstTimeRegister=true. Accessed 1.12.25.

De-la-Torre, G. E., Dioses-Salinas, D. C., Dobaradaran, S., Spitz, J., Nabipour, I., Keshtkar, M., ... & Javanfekr, F. (2022). Release of phthalate esters (PAEs) and microplastics (MPs) from face masks and gloves during the COVID-19 pandemic. *Environmental Research, 215*, 114337.

Debnath, R., Prasad, G. S., Amin, A., Malik, M. M., Ahmad, I., Abubakr, A., Borah, S., Rather, M. A., Impellitteri, F., Tabassum, I., & Piccione, G. (2024). Understanding and addressing microplastic pollution: Impacts, mitigation, and future perspectives. *Journal of Contaminant Hydrology, 266*, 104399.

dela Cruz, J., Lammel, D., Kim, S. W., Bi, M., & Rillig, M. (2024). COVID-19 pandemic-related drugs and microplastics from mask fibers jointly affect soil functions and processes. *Environmental Science and Pollution Research, 31*, 50630–50641. https://doi.org/10.1007/s11356-024-34587-x

Derraik, J. G., Anderson, W. A., Connelly, E. A., & Anderson, Y. C. (2020). Rapid review of SARS-CoV-1 and SARS-CoV-2 viability, susceptibility to treatment, and the disinfection and reuse of PPE, particularly filtering facepiece respirators. *International Journal of Environmental Research and Public Health, 17*(17), 6117.

Duer, J. (2021, May). Single-use plastics are on the rise due to COVID-19. In World Econ. Forum. https://www.weforum.org/agenda/2020/07/plastic-waste-management-covid19-ppe/. Accessed (Vol. 9).

The European Respiratory Virus Surveillance Summary (ERVISS) [WWW Document]. (n.d.). https://www.ecdc.europa.eu/en/publications-data/european-respiratory-virus-surveillance-summary-erviss. Accessed 1.12.25.

Fadare, O. O., & Okoffo, E. D. (2020). Covid-19 face masks: A potential source of microplastic fibers in the environment. *Science of the Total Environment, 737*, 140279. https://doi.org/10.1016/J.SCITOTENV.2020.140279

Godbole, T., (2021). Coronavirus Aggravates Thailand's Plastic Waste Crisis, pp. 2020–2022.

Gurjar, U. R., Takar, S., Amin, A., & Martin Xavier, K. A. (2023). Microplastic contamination in aquatic organisms: An ecotoxicological perspective. In *Xenobiotics in aquatic animals: Reproductive and developmental impacts* (pp. 353–367). Springer Nature Singapore.

Hasan, M. M., & Tarannum, M. N. (2025). Adverse impacts of microplastics on soil physicochemical properties and crop health in agricultural systems. *Journal of Hazardous Materials Advances, 17*, 100528. https://doi.org/10.1016/J.HAZADV.2024.100528

Hidalgo-Ruz, V., Gutow, L., Thompson, R. C., & Thiel, M. (2012). Microplastics in the marine environment: A review of the methods used for identification and quantification. *Environmental Science & Technology, 46*, 3060–3075. https://doi.org/10.1021/ES2031505/ASSET/IMAGES/MEDIUM/ES-2011-031505_0006.GIF

Hurley, R. R., & Nizzetto, L. (2018). Fate and occurrence of micro(nano)plastics in soils: Knowledge gaps and possible risks. *Current Opinion in Environmental Science & Health, 1*, 6–11. https://doi.org/10.1016/J.COESH.2017.10.006

Kalina, M., Kwangulero, J., Ali, F., & Tilley, E. (2022). "You need to dispose of them somewhere safe": Covid-19, masks, and the pit latrine in Malawi and South Africa. *PLoS One, 17*, e0262741. https://doi.org/10.1371/journal.pone.0262741

Kutralam-Muniasamy, G., Shruti, V. C., Pérez-Guevara, F., & Roy, P. D. (2023). Microplastic diagnostics in humans: "The 3Ps" Progress, problems, and prospects. *Science of the Total Environment, 856*, 159164. https://doi.org/10.1016/J.SCITOTENV.2022.159164

Kwak, J. I., & An, Y. J. (2021). Post COVID-19 pandemic: Biofragmentation and soil ecotoxicological effects of microplastics derived from face masks. *Journal of Hazardous Materials, 416*, 126169. https://doi.org/10.1016/J.JHAZMAT.2021.126169

Lakshmi Kavya, N. V., Sundarrajan, A., & Ramakrishna, S. (2020). Identification and characterization of micro-plastics in the marine environment: A mini review. *Marine Pollution Bulletin, 160*, 111704. https://doi.org/10.1016/J.MARPOLBUL.2020.111704

Li, J., Lai, S., Gao, G. F., & Shi, W. (2021). The emergence, genomic diversity and global spread of SARS-CoV-2. *Nature, 600*(7889), 408–418.

Masud, R. I., Suman, K. H., Tasnim, S., Begum, M. S., Sikder, M. H., Uddin, M. J., & Haque, M. N. (2023). A review on enhanced microplastics derived from biomedical waste during the COVID-19 pandemic with its toxicity, health risks, and biomarkers. *Environmental Research, 216*, 114434. https://doi.org/10.1016/J.ENVRES.2022.114434

Mhiret Gela, S., & Aragaw, T. A. (2022). Abundance and characterization of microplastics in main urban ditches across the Bahir Dar City, Ethiopia. *Frontiers in Environmental Science, 10*. https://doi.org/10.3389/FENVS.2022.831417/

Mohana, A. A., Islam, M. M., Rahman, M., Pramanik, S. K., Haque, N., Gao, L., & Pramanik, B. K. (2023). Generation and consequence of nano/microplastics from medical waste and household plastic during the COVID-19 pandemic. *Chemosphere, 311*, 137014. https://doi.org/10.1016/j.chemosphere.2022.137014

Nabi, N., Ahmad, I., Amin, A., Rather, M. A., Ahmed, I., Hajam, Y. A., Khursheed, S., Malik, M. M., & Abubakr, A. (2024). Understanding the sources, fate and effects of microplastics in aquatic environments with a focus on risk profiling in aquaculture systems. *Reviews in Aquaculture, 16*(4), 1947–1980.

Nguyen, L. H., Drew, D. A., Graham, M. S., Joshi, A. D., Guo, C. G., Ma, W., ... & Zhang, F. (2020). Risk of COVID-19 among front-line health-care workers and the general community: a prospective cohort study. *The Lancet Public Health, 5*(9), e475–e483.

Patrício Silva, A. L. (2021). New frontiers in remediation of (micro)plastics. *Current Opinion in Green and Sustainable Chemistry, 28*, 100443. https://doi.org/10.1016/J.COGSC.2020.100443

Phu, H., Han, H. T. N., & Thao, N. L. N. (2023). Sources of microplastic pollution in the Saigon-Dong Nai rivers, potential risks affecting human health and recommendations for mitigation solutions. *IOP Conference Series: Earth and Environmental Science, 1226*, 012017. https://doi.org/10.1088/1755-1315/1226/1/012017

Prasad, G. S., Rout, S. K., Malik, M. M., Karmakar, S., Amin, A., & Ahmad, I. (2023). Occurrence of xenoestrogen alkylphenols (Octylphenols and Nonylphenol) and its impact on the aquatic ecosystem. In *Xenobiotics in aquatic animals: Reproductive and developmental impacts* (pp. 275–284). Springer Nature Singapore.

Prasad, G. S., Rout, S. K., Amin, A., Malik, M. M., Ahmad, I., Rather, M. A., Karmakar, S., Khalil, H. S., & Abubakar, A. (2024). Impact of COVID-19 on the aquatic environment and implications for fisheries with special emphasis to Indian perspective. *Proceedings of the Indian National Science Academy, 90*(4), 821–839.

Prata, J. C. (2018). Airborne microplastics: Consequences to human health? *Environmental Pollution, 234*, 115–126. https://doi.org/10.1016/J.ENVPOL.2017.11.043

Rathinamoorthy, R., & Raja Balasaraswathi, S. (2023). Characterization of microfibers originated from the textile screen printing industry. *Science of the Total Environment, 874*, 162550. https://doi.org/10.1016/J.SCITOTENV.2023.162550

Ray, R. L., Singh, V. P., Singh, S. K., Acharya, B. S., & He, Y. (2022). What is the impact of COVID-19 pandemic on global carbon emissions?. *Science of The Total Environment, 816*, 151503.

Ren, Z., Gui, X., Wei, Y., Chen, X., Xu, X., Zhao, L., Qiu, H., & Cao, X. (2021). Chemical and photo-initiated aging enhances transport risk of microplastics in saturated soils: Key factors, mechanisms, and modeling. *Water Research, 202*, 117407. https://doi.org/10.1016/J.WATRES.2021.117407

Sajjad, M., Huang, Q., Khan, S., Khan, M. A., Liu, Y., Wang, J., Lian, F., Wang, Q., & Guo, G. (2022a). Microplastics in the soil environment: A critical review. *Environmental Technology and Innovation, 27*, 102408. https://doi.org/10.1016/j.eti.2022.102408

Sajjad, M., Huang, Q., Khan, S., Khan, M. A., Liu, Y., Wang, J., Lian, F., Wang, Q., & Guo, G. (2022b). Microplastics in the soil environment: A critical review. *Environmental Technology and Innovation, 27*, 102408. https://doi.org/10.1016/J.ETI.2022.102408

Saliu, F., Veronelli, M., Raguso, C., Barana, D., Galli, P., & Lasagni, M. (2021). The release process of microfibers: From surgical face masks into the marine environment. *Environmental Advances, 4*, 100042. https://doi.org/10.1016/J.ENVADV.2021.100042

Sangkham, S. (2020). Face mask and medical waste disposal during the novel COVID-19 pandemic in Asia. *Case Studies in Chemical and Environmental Engineering, 2*, 100052. https://doi.org/10.1016/J.CSCEE.2020.100052

Selvam, S., Manisha, A., Roy, P. D., Venkatramanan, S., Chung, S. Y., Muthukumar, P., Jesuraja, K., Elgorban, A. M., Ahmed, B., & Elzain, H. E. (2021). Microplastics and trace metals in fish species of the Gulf of Mannar (Indian Ocean) and evaluation of human health. *Environmental Pollution, 291*, 118089. https://doi.org/10.1016/J.ENVPOL.2021.118089

Shruti, V. C., Pérez-Guevara, F., Elizalde-Martínez, I., & Kutralam-Muniasamy, G. (2020). First study of its kind on the microplastic contamination of soft drinks, cold tea and energy drinks-Future research and environmental considerations. *Science of the Total Environment, 726*, 138580.

Single-use plastics: A roadmap for sustainability | UNEP - UN Environment Programme. (2018). https://www.unep.org/resources/report/single-use-plastics-roadmap-sustainability.

Sorasan, C., Ortega-Ojeda, F. E., Rodríguez, A., & Rosal, R. (2022). Modelling the Photodegradation of marine microplastics by means of infrared spectrometry and chemometric techniques. *Microplastics, 1*, 198–210. https://doi.org/10.3390/MICROPLASTICS1010013/S1

Sruthisree, C., Gowda, G., Nayak, H., Harish, B. D., & Amin, A. (2015). Effect of heavy metal (lead) on oxygen consumption and ammonia excretion of Iridescent Shark (*Pangusius hypohthalmus*). *Journal of Experimental Zoology, India, 18*(2), 637.

Sullivan, G. L., Delgado-Gallardo, J., Watson, T. M., & Sarp, S. (2021). An investigation into the leaching of micro and nano particles and chemical pollutants from disposable face masks - linked to the COVID-19 pandemic. *Water Research, 196*, 117033. https://doi.org/10.1016/J.WATRES.2021.117033

Tabassum, I., Hassan, N., Amin, A., Malik, M. M., Abubakr, A., Prasad, G. S., Latief, F., & Rather, S. A. (2025). COVID-19 and global environmental change: A threat to food security and nutrition. In *Food security, nutrition and sustainability through aquaculture technologies* (pp. 437–445). Springer Nature Switzerland.

Tiwari, B. R., Lecka, J., Pulicharla, R., & Brar, S. K. (2023). Microplastic pollution and associated health hazards: Impact of COVID-19 pandemic. *Current Opinion in Environmental Science & Health, 34*, 100480. https://doi.org/10.1016/j.coesh.2023.100480

Tso, R. V., & Cowling, B. J. (2020). Importance of face masks for COVID-19: A call for effective public education. *Clinical Infectious Diseases, 71*, 2195–2198. https://doi.org/10.1093/cid/ciaa593

Wang, C., Zhao, J., & Xing, B. (2021). Environmental source, fate, and toxicity of microplastics. *Journal of Hazardous Materials, 407*, 124357. https://doi.org/10.1016/J.JHAZMAT.2020.124357

World Health Organization. (2020). Laboratory testing for 2019 novel coronavirus (2019-nCoV) in suspected human cases: Interim guidance, 17 January 2020. World Health Organization.

Zhang, Y., Kang, S., Allen, S., Allen, D., Gao, T., & Sillanpää, M. (2020). Atmospheric microplastics: A review on current status and perspectives. *Earth-Science Reviews, 203*, 103118.

Zhao, S., & Zhang, J. (2023). Microplastics in soils during the COVID-19 pandemic: Sources, migration and transformations, and remediation technologies. *Science of the Total Environment, 883*, 163700. https://doi.org/10.1016/j.scitotenv.2023.163700

Zhao, X., Liu, Z., Cai, L., & Han, J. (2023). Occurrence and distribution of microplastics in surface sediments of a typical river with a highly eroded catchment, a case of the Yan River, a tributary of the Yellow River. *Science of the Total Environment, 863*, 160932. https://doi.org/10.1016/J.SCITOTENV.2022.160932

Microplastic Contamination in Aquaculture: Sources, Ecological Effects, and Future Solutions

Samiya Majeed and Ifra Tabassum

1 Introduction

The world's population is increasing and wild fish populations are declining; under such circumstances, aquaculture has become a sustainable option to provide a stable food supply. Aquaculture, often referred to as aquatic agriculture, involves the cultivation, breeding, and harvesting of various aquatic organisms, including fish, shellfish, algae, and more, within diverse water environments. Beyond its role in food production, aquaculture serves multiple purposes, such as generating commercial products, restoring habitats, replenishing wild populations, and aiding in the recovery of endangered and threatened species (NOAA, 2023). This practice encompasses the controlled cultivation of aquatic animals throughout significant portions or the entirety of their life cycles. Its advancement holds the potential to address future food demands and reduce the strain on natural resources. (USDA, 2023) Aquaculture has contributed significantly to food production and the availability of essential nutrients in recent years (Wang et al., 2015; Boyd et al., 2020). In 2022, the worldwide production of aquatic animals achieved a record level of 185 million tonnes, with approximately 164.6 million tonnes (89%) designated for human consumption, translating to an average of 20.7 kg per person. The production of farmed aquatic animals through global aquaculture has shown a steady upward trend in recent years, culminating in a total of 94 million tonnes in 2022 (FAO, 2024).

Regretfully, the aquaculture environment today faces many difficulties out of which Microplastics is a major concern (Kaleem & Sabi, 2021). The term

S. Majeed (✉)
Division of Aquaculture, Faculty of Fisheries Rangil, Ganderbal, SKUAST-Kashmir, Srinagar, India

I. Tabassum
Division of Aquatic Environmental Management, Faculty of Fisheries, Sher-e-Kashmir University of Agricultural Science and Technology of Kashmir, Shalimar, Srinagar, India

A. Abubakr et al. (eds.), *Microplastics: Ecological and Food Security Implications*,
https://doi.org/10.1007/978-3-031-94480-2_14

"microplastics" originally appeared in the mid-2000s. Nowadays, it is commonly used to refer to plastic pieces that have a maximum diameter of 5 mm (UNEP, 2019). According to Lusher et al. (2017), microplastics can be divided into fragments, filaments, beads, sheets, and pellets. There are two types of microplastics that are currently causing problems in the environment: primary microplastics, which are made smaller than 5 mm or are created during deliberate use, and secondary microplastics, which are created when bigger plastics break apart during environmental weathering (Fred-Ahmadu et al., 2020; Debnath et al., 2024). According to numerous researches, secondary microplastics often make up the majority of the environment (Zhang et al., 2018; Meng et al., 2020).

Microplastics have been detected across diverse aquatic environments, such as rivers, lakes, coastal regions, and the deep ocean. The use of plastic mulch and sludge in agricultural practices, along with their presence in textile manufacturing, common consumer products, cleaning supplies, and health and personal care items, plays a substantial role in the rising levels of microplastic pollution within aquatic ecosystems (Nguyen et al., 2023; Prasad et al., 2024; Tabassum et al., 2025). The ability of microplastics to adsorb poisons makes them more likely to contain a variety of environmental pollutants (Padervand et al., 2020).

Although the aquaculture business finds plastic items to be very convenient, with time, these products break down and decompose, releasing microplastics into sediment layers and water columns (Vázquez-Rowe et al., 2021). The issue of microplastic contamination has become a major worldwide environmental concern, leading to heightened study focus on aquaculture (Wu et al., 2023; Gurjar et al., 2023). These microplastics are an increasingly prevalent aquaculture pollutant that has been found in fish diets and cultured fish (Iheanacho et al., 2023). Numerous studies have repeatedly emphasized the issue of microplastic accumulation in a wide range of aquaculture environments (Bordós et al., 2019). According to Korez et al. (2019), microplastic pollution was common in areas around aquaculture farms and facilities. Additionally, the evolving use of chemicals in aquaculture and agriculture is affecting the environment and endangering food safety (Burridge et al., 2010; Prasad et al., 2023; Nabi et al., 2024), and there is evidence that farmed fish generally include more microplastic than wild-caught fish (Wootton et al., 2021). Aiguo et al. (2022) highlight that the existence of microplastics represents a significant threat to the safety and quality of aquatic products. These microplastics adversely affect fish productivity, energy consumption, and the utilization of water resources. This chapter intends to examine the prevalence, origins, and effects of microplastics on aquaculture species.

2 Sources of Microplastics in Aquaculture

2.1 Industrial Wastewater

The primary industrial wastewater source of fibrous microplastics in rivers is textile wastewater (Wu et al., 2023). Sewage originating from a textile industrial zone in

Shaoxing, China, has been identified as a source of microplastics that have adversely impacted the local freshwater ecosystem. The concentration of microplastics in surface water samples from the adjacent aquatic environment varied between 2.1 and 71.0 items per liter, whereas sediment samples exhibited a range of 16.7–1323.3 items per kilogram (d.w.) (Deng et al., 2020). Microplastics can also be found in wastewater from the automotive, packaging, and food industries (Gundogdu et al., 2018). Human activities contribute to the release of microplastics into the environment. These microplastics are present in common consumer products, including cosmetics, toothpaste, and shampoo (Jiang, 2018). Additionally, the laundering of synthetic fiber garments on a regular basis generates microplastics. Research indicates that clothing made from synthetic fibers such as acrylic and nylon releases different quantities of microplastics during washing, while polyester fleece fabrics can release an average of 7360 fibers per square meter per liter in a single wash. (Jiang, 2018).

2.2 *Wastewater Treatment Plants*

A primary contributor to microplastics in aquatic ecosystems is the effluent discharged from wastewater treatment facilities, which is identified as a distinct source of aquatic contaminants (Habib et al., 2020; Zeri et al., 2021). Wastewater treatment plants typically process a significant portion of the sewage generated by urban centers and extensive industrial regions. As a result, these facilities have accumulated complicated sources of microplastics and are now a point source of their emissions. WWTP treatment reduces microplastics in wastewater, however discharged effluents still include them.

Despite a significant decrease in microplastics in wastewater following treatment by wastewater treatment plants (WWTPs), the aquatic ecosystem is still seriously threatened by effluents released from WWTPs (Liu et al., 2019). Aquaculture facilities, especially those involved in coastal aquaculture and mariculture, may be exposed to MPs from this source. According to Wang et al. (2022a, b, c), MPs from wastewater treatment facilities attach themselves to chemicals employed in the production process to create poisonous and persistent compounds (PBTs). Even though MPs contain PBTs and polycyclic aromatic hydrocarbon (PAHs), they have a high affinity for the compounds and are capable of efficiently absorbing them (Koelmans et al., 2016; Gomiero et al., 2018). When these harmful elements are consumed by aquatic species, they bioaccumulate in them. They are crucial to the anticipated hazard and risk of MPs in the aquatic environment. According to Ziajahromi et al. (2021), three wastewater treatment plants in Australia discharge between 22.1 × 106 and 133 × 106 microplastics per day.

2.3 Environmental Transport

Geographical and meteorological elements significantly influence the distribution and accumulation of microplastics within aquaculture settings (Chen et al., 2018a; Wang et al., 2019). Most plastic waste initially enters the terrestrial ecosystem as a result of improper disposal, littering, or leakage. (Chia et al., 2021). There, it undergoes various weathering processes, including chemical, biological, and photodegradation, or it is carried by wind or water flow due to its low weight (Zhang et al., 2021).

The main route through which macro- and microplastics transition from land to water bodies is through surface runoff caused by rainfall (Liu et al., 2019; Ma et al., 2020; Pinon-Colin et al., 2020). Another common way that microplastics are transported is by wind (Chen et al., 2018a). Typhoons have been shown to alter the forms, colours, and kinds of microplastics as well as raise the average concentrations of these particles by about 40% in seawater and sediments (Wang et al., 2019).

2.4 Atmospheric Deposition

Recent findings indicate that atmospheric deposition significantly contributes to the influx of microplastics, especially in remote regions (Zhang et al., 2020). Research by Cai et al. (2017) revealed that the concentration of microplastics in airborne fallout from Dongguan, China, ranged from 175 to 313 items per square meter per day. Furthermore, Wright et al. (2020) reported that the deposition of microplastics in the air over urban London varied between 575 and 1008 items per square meter per day. In protected areas across the United States, the average daily deposition rate of microplastics was found to be 132 items per square meter per day (Brahney et al., 2020). The significance of air deposition to microplastic contamination in aquaculture ponds must be considered in light of research findings from both urban and distant places, even though atmospheric deposition of microplastics in aquaculture settings has not been studied before.

2.5 Microplastics Produced During Aquaculture Practices

Fish feed, fertilizer, and pesticide plastic packets and containers are frequently left on the pond's shore after usage, which can cause secondary microplastics to form in the pond. Furthermore, the aquaculture environment will unavoidably be contaminated with microplastics due to the use of fishing gear. The nets, fishing ropes, floating balls, and other plastic goods used in aquaculture can produce microplastics due to prolonged immersion, erosion, wear, and collision (FAO, 2018). UV catalytic degradation will speed up this process (Song et al., 2017).

According to Wright et al. (2021), twisted rope, braided rope, and filament—all of which are commonly used in aquaculture—may contribute 0.78 to 6.39 $\pm$ 2.33 cm^3 of estimated plastic volume per meter of beach and may release 300 to 1277 $\pm$ 431 microplastic fragments per meter of beach. Meanwhile, fishing gear would be lost in aquaculture for a number of reasons, including falling off the ropes (Chen et al., 2018b) or being swept away by waves (Chaves and Silveira, 2016). These fishing gears have the potential to unintentionally show up in aquaculture areas and pollute the industry with microplastics (FAO, 2018; Huntington, 2019). In the waters of the Norwegian commercial fishing route, the quantity of lost or discarded fishing gear-including trawls, purse seines, Danish seines, gillnets, longlines, and traps—was recorded at under 500 tonnes in 2007, while it surged to over 4000 tonnes by 2016 (Deshpande et al., 2020). Additionally, it is likely that the main contributors to microplastics in aquaculture stem from the equipment used in breeding facilities, which encompasses plastic tools, bait machines, aerators, water pumps, electric wiring, and other related items.

The packaging process represents the concluding phase for aquaculture products prior to their departure from the production environment. These products are commonly contained in plastic trays, corrugated plastic boxes, and expanded polystyrene containers (Margeirsson et al., 2011; Skirtun et al., 2022). Studies have shown that microplastic fibers can be released from plastic packaging composed of various materials, with polystyrene plastics demonstrating the highest levels of release (Du et al., 2020). It has also been discovered that rainbow trout are susceptible to microplastic pollution from polystyrene packing (Alak et al., 2021). Microplastic pollution in aquaculture areas will also be made worse by the deterioration and improper use of plastic items like rubber gloves, shoes, and aprons that are frequently used by aquaculture professionals.

2.6 Aquatic Feeds

Fish and shrimp meal are often used as aquatic feeds (Luthada-Raswiswi et al., 2021). However, microplastics (MPs) have been commonly discovered in fish and shrimp from natural environments (Jabeen et al., 2017; Curren et al., 2020), raising concerns about their introduction into aquatic feeds. Previous studies have confirmed the presence of microplastics (MPs) in commercial fish feeds and have demonstrated a positive correlation between the concentration of MPs in these feeds and the accumulation of MPs in farmed common carp that ingest them (Hanachi et al., 2019). In a separate investigation, Yao et al. (2021) analyzed seafood-based feeds and found that the abundance of MPs varied from 0.1 to 0.54 items per kilogram in fish meals, while shrimp meals had an average of 1.07 items per kilogram.

Aquatic feeds containing MPs have the potential to contaminate the environment by being consumed by cultivated aquatic species or discharged into aquaculture ponds. MP contamination has been found to be considerable in studies on fish meals used in aquafeeds. In a global research, MPs were found in fish meal samples from 10 countries with China, Peru, and Myanmar having the highest quantities (5.5 ± 1.6 items/g) (Wang et al., 2022a, b, c). The high MP content in fish meal is due to high contamination levels in both the fish and their surroundings.

Several fish species used to produce fish meal have been shown to contain MPs such as include *Scomber japonicus*, *Sardinops pilchardus*, and *Sardinella aurita* (253.3 ± 43.4 n/kg) (Gündoğdu et al., 2021); *Trichiurus lepturus* (46.00 n/g) (Hasan et al., 2022); *Heteropneustes fossilis* (0.33–1.57 g/m, 7–15 μm) (Rahman et al., 2022); and *Sardinops sagax* (5.50 ± 1.6 items/g, 500–1000 μm) (Wang et al., 2022a, b, c). Furthermore, it has been observed that fish meals from Italy contain notable quantities of MP, with 50–100 mg/kg of polystyrene, 50–100 mg/kg of highly oxidized polyolefins, and 12.9 mg/kg of polyester (Castelvetro et al., 2021a, b).

In aquaculture, artificial compound feeds are another important source of MPs (Wright et al., 2013b; Van Cauwenberghe et al., 2015). Due to the complexity of the constituents in these feeds, MPs can infiltrate at various stages during the production, processing, storage, transit, and sales processes. Furthermore, the concentration of MPs in processed fish meal is notably higher than in raw fish, indicating that MPs enter fish meal products during the production process (Thiele et al., 2021). An examination of three commercial fish meal products from Malaysia, for instance, found 336 MP particles, of which over 50% were plastic polymers and 25% were pigment particles. Plastic fragments were the most common MP form (Karbalaei et al., 2020).

Beyond feed sources, fish medication and animal protection products bring MPs into aquaculture systems. The longer environmental persistence of MPs is caused by the adsorption of several pesticides and antibiotics that are frequently employed in the early phases of aquaculture (Koelmans et al., 2016; Li et al., 2018). Plastics and MPs infiltrate the aquaculture ecosystem through their use in disease control, water quality management, anti-stress therapies, and growth promotion.

3 Impact on Aquatic Organisms

Fish populations, both domestic and wild, have been greatly harmed by plastic pollution, which also introduces possible health hazards from harvested products (Drillet, 2020). Aquaculture species face significant physical and chemical risks from microplastics (MPs) in aquatic habitats, which can impact their survival, reproduction, and metabolism.

Generally speaking, MP consumption and buildup may result in oxidative (inflammation) and mechanical (abrasion) harm to biological systems such the circulatory, neurological, reproductive, respiratory, and digestive systems (Talvitie et al., 2015; Wright et al., 2013a; Jovanovic et al., 2018; Espinosa et al., 2018; Kim et al., 2020; Iheanacho et al., 2020a, b; Yu et al., 2020; Hamed et al., 2021; Parker et al., 2021; Bobori et al., 2022). Unfortunately, the information provided may be limited due to a knowledge gap about organismal responses to MPs in aquaculture systems (Halsband & Galloway, 2015; Lyu et al., 2020).

3.1 Biological and Physiological Effects of Microplastics

3.1.1 Digestive System

The consumption of microplastics leads to obstruction and physical damage to the gastrointestinal tract, potentially resulting in decreased food consumption, a sensation of false fullness, and in severe cases, perforations of the stomach and intestines, which can be life-threatening (Malik et al., 2017; Farrell & Nelson, 2013; Kühn et al., 2015; Sharma & Chatterjee, 2017). MPs can move to other biological systems; however, they are primarily found in fish digestive tracts (da Costa et al., 2016; Seltenrich, 2015).

3.1.2 Metabolic and Oxidative Stress

The exposure to microplastics (MP) influences the lipid metabolism, oxidative stress levels, carbohydrate metabolism, and toxin elimination in fish (Jacob et al. 2020). The increased production of reactive oxygen species (ROS) may lead to oxidative stress, adversely affecting the functionality of enzymes associated with lipid metabolism (Cedervall et al., 2012; Shruthisree et al., 2015; Amin et al., 2016, 2017; Karami et al., 2016; Wan et al., 2019).

3.1.3 Swimming and Feeding

Ingestion of MP impairs fish eating and swimming ability. Polyethylene MP exposure was found to reduce swimming and predation skills in the common goby (*Pomatoschistus microps*) (Oliveira et al., 2013) and swimming and feeding capacities in *S. schlegelii* (Yin et al., 2018)

3.1.4 Neurological and Brain Function

Microplastic particles as small as 24–27 nm can harm brain development, producing structural damage and impaired acetylcholinesterase function (Mattsson et al., 2015; LeMoine et al., 2018; Barboza et al., 2018). The accumulation of larger MPs (>500 nm) in the tissues and blood vessels around the brain can cause metabolic and immunological abnormalities (Borre et al., 2014; Chen et al., 2017).

3.1.5 Cardiovascular and Respiratory Effects

MPs have an adverse effect on heart and gill function. According to Pitt et al. (2018), zebrafish exposed to 51 nm polystyrene MPs experienced oxidative stress, which resulted in a reduction in heart rate. *Clarias gariepinus* gills suffered tissue injury from MPs that were 60 μm in size (Karami et al., 2016).

3.2 Growth and Reproductive Impacts

MP consumption influences energy storage and growth in aquaculture species. For example, in edible mussels, PET MPs decrease growth rates while increasing energy consumption (Détrée & Gallardo-Escárate, 2018). The growth rate of loach juveniles was considerably slowed by MPs (Wang et al., 2022a, b, c).

Reproductive health is negatively impacted by MPs and their additives. Oyster reproductive health and energy absorption were adversely affected by MP exposure (Sussarellu et al., 2016). Fish and shrimp gonadal development is impacted by BPA released by MPs (Mandich et al., 2007; Han et al., 2022). There have also been reports of decreased fertilization success in bivalves as a result of MPs (Martinez-Gomez et al., 2017; Shi et al., 2022).

3.3 Mortality and Economic Consequences

Aquaculture species had higher mortality rates when contaminated by MP. MPs and other inorganic materials were detected in one-third of deceased fish, and 50% of young *Thunnus orientalis* perished within 30 days of cage culture (Okada et al., 2014). Adult dagger blade shrimp exposed to MPs had mortality rates ranging from 5 to 40% (Gray & Weinstein, 2017). Aquaculture experiences financial losses as a result of increased death rates and decreased reproductive efficiency. The elevated mortality rate of oysters in France from 2001 to 2008 led to a 28% decline in shellfish aquaculture enterprises and a 16% reduction in related employment (Girard & Pérez Agúndez, 2014).

Currently, a number of research have investigated ways to address microplastic pollution, including membrane bioreactors (Lares et al., 2018), electro-coagulation (Perren et al., 2018), and a foam material filtering technique for saltwater based on zirconium metal-organic frameworks (Chen et al., 2020). With a purifying effectiveness of over 95%, these innovative technologies can efficiently minimize the amount of microplastics that reach the environment.

Nevertheless, the majority of these techniques have not been extensively implemented and are unable to directly lower the introduction of microplastics into aquaculture; therefore, alternative strategies are required to directly manage the microplastic pollution in aquaculture (Table 1).

Table 1 Impact of microplastics on commercially important aquaculture species

Species	Exposure condition	Impacts	References
Ctenopharyngodon idella	PS (32–40 μm)	Altered metabolism Reduced survival Oxidative stress	Jia et al. (2022)
Oreochromis mossambicus	PP, PE (<1 mm)	Accumulation in the GIT	Li et al. (2021)
Murgil cephalus	PP, PE (<2 mm)	Less effects on aquaculture Mullets than those in the wild.	Cheung et al. (2018)
Carassius auratus	EVA fiber (0.7–5.0 mm), PS fragment (2.5–3.0 mm), PEA pellets (4.9–5.0 mm)	Food with fiber causes Damage in liver; Food with fragment cause damage in jaws	Jabeen et al. (2018)
Lates calcarifer	PS (97 μm)	Fish behaviour altered, Erratic movement and Lower swimming speed.	Guven et al. (2018)
Oreochromis niloticus	Microplastic particles with irregularshaped (>100 nm, 1, 10, 100 mg/L)	Anemia Perturbations in hemato-biochemical parameters	Hamed et al. (2019)
Chanos chanos	Fibres, granules	Reduced growth.	Priscilla and Patria (2019)
Cyprinus carpio	PVC, PE, PS (<5 mm, 452 ± 161 μm)	Altered protein level, Gastrointestinal damage, Induced CAT, SOD, GPX Reduced MDA, Decreased growth.	Xia et al. (2020), Hanachi et al. (2019)
Clarias gariepinus	PVC (95.41 μm)	Reduce catalase activity in the brain. Oxidative stress. Altered glutathione peroxidase activity in the brain and gill. Elevated lipid peroxidation levels in the brain Reduced neutrophil counts Decreased haemoglobin values Reduced Hematocrit values	Iheanacho and Odo (2020a, b)
Danio rerio	PE (100–1000 μg/Ml, 106–125 μm)	Inflammation and Lipid accumulation in Liver Oxidative stress Changes in metabolic profiles within the liver. Disruption in lipid and energy metabolism	Yuan et al. (2023), Bobori et al. (2022)

(continued)

Table 1 (continued)

Species	Exposure condition	Impacts	References
Mytilus galloprovincialis	PS 3–45 μm PE 20–25 μm PE 4.5–45 μm	Accumulation in the haemolymph Activated multixenobiotic resistance system activity Oxidative damage. Neurotoxicity Altered immune system Accumulation in the gill, haemolymph, digestive tissues. Accumulation in the gills and digestive gland Histological distortions in the digestive gland	Franzellitti et al. (2019), Pittura et al. (2018), Hellfeld et al. (2022)
Carassius carassius	Secondary PE film, 90% <848.37 μm, 6.38, 12.18, and 22.33 mg microplastics/fish/day	Decrease in growth rate at high concentration; Liver damage; Changes in gut microbiome composition	Hu et al. (2022)
Macrobrachium rosenbergii	PS 0.5–1.0 μm PE 30.-150 μm	Elevated MDA and antioxidant enzymes (CAT, SOD, GST, GSH) activities Oxidative stress Reduced GPT, GOT Heat shock protein (HSP70) gene Inhibited growth	Jaikumar et al. (2019)
Penaeus monodon,	Fiber 1–5 mm Filament 250–500 μm	Accumulation in the GIT	Hossain et al. (2020)
Crassostrea gigas	PS PE, 6 μm PP <400 μm	Higher mortality Accumulation in the GIT, digestive tubules. Decreased lysosomal membrane activity. Altered immune system DNA damage Accumulation/bioaccumulation in the gills, digestive glands.	Thomas et al. (2020), Revel et al. (2020)

4 Strategies to Mitigate Microplastic Pollution in Aquaculture

Microplastic pollution represents a considerable danger to aquatic ecosystems, particularly in aquaculture settings (Gavigan et al., 2020; Ilechukwu et al., 2021; Loctier, 2021; Prata et al., 2019; Taniguchi et al., 2016; Wang et al., 2017). A variety of studies have recently explored potential solutions to combat microplastic pollution, including the use of membrane bioreactors (Lares et al., 2018), electrocoagulation methods (Perren et al., 2018), and a foam material filtering technique for saltwater that utilizes zirconium metal-organic frameworks. (Chen et al., 2020). While these technologies have demonstrated purification efficiencies exceeding 95%, their widespread application remains limited. Nevertheless, the majority of these techniques have not been extensively implemented and are unable to directly lower the introduction of microplastics into aquaculture; therefore, alternative strategies are required to directly manage the microplastic pollution in aquaculture.

4.1 Blocking External Sources of Microplastics in Aquaculture-Ecological Interception

Ecological interception is the process of using aquatic vegetation to reduce water velocity and capture pollutants in order to reduce microplastic pollution. Due to their capacity to absorb and collect pollutants in their plants and soils, mangroves are frequently utilized for this purpose (Liu et al., 2022). Moreover, mangroves lower pollution levels in shrimp production and preserve aquaculture stability (Do & Thuy, 2022). It is commonly known that they serve as nursery environments for offshore aquaculture (Barbier et al., 2011). Studies indicate that the concentration of microplastics at the edges of mangroves significantly exceeds that found in adjacent regions (Duan et al., 2021), with surface water concentrations varying between 620 and 13,100 n/m^3, and sediment levels ranging from 142 to 488 n/kg (Liu et al., 2022).

According to Tang et al. (2018), these values are higher than those found close to the coast (103–2017 n/m^3 in surface seawater and 76–333 n/kg in sediments). The biodegradation rates of polyethylene plastics in mangrove soils range from 3.77% to 4.21% over a nine-month period, demonstrating that mangroves also aid in the breakdown of microplastics (Kathiresan, 2003). When isolated from mangrove sediments, microorganisms including *Bacillus* and *Rhodococcus* have shown 6.4% and 4.0% breakdown rates for polypropylene microplastics over a 40-day period (Auta et al., 2018).

Reeds are similarly significant in fishpond ecosystems, as they offer habitat and nourishment while also affecting biogeochemical cycles. Their capacity to adsorb microplastics (Plestenjak et al., 2021) results in the accumulation of these particles at the edges of reed beds (Yao et al., 2019). The mean concentration of microplastics found in the sediments of reed farms was recorded at 511.2 ± 295.0 items/kg, surpassing the levels detected in the adjacent water (Yin et al., 2021). Despite its advantages, ecological interception is limited by the presence of suitable aquatic plants in specific aquaculture situations.

4.2 Reducing Microplastic Production in Aquaculture: Sustainable Fishing Gear and Waste Management

Fishing gear is the principal cause of microplastic pollution in aquaculture (Food and Agriculture Organization, 2018). Biodegradable materials in fishing gear, such as biodegradable stockings for mussel farming (Skirtun et al., 2022) and robust biodegradable monofilaments (Deroine et al., 2019), offer promise in minimizing microplastic pollution. Though their effects on aquaculture products are less severe, even biodegradable plastics can provide microplastic hazards.

Enhancing the design of fishing gear can help cut down on the release of microplastics. For example, Shetland mussel farming has reduced gear loss and microplastic emissions by substituting continuous lines for mussel pegs (Raedemaecker et al., 2020).

Further, Consumer plastic items like gloves, protective gear, and food packaging contribute to microplastic contamination, making efficient plastic waste management in aquaculture essential (Sandra et al., 2019). Recycling and reusing fishing gear reduces pollution while also providing economic rewards, encouraging aquaculture practitioners to embrace sustainable techniques.

4.3 Reducing Microplastic Contamination in Aquaculture Products: Purification and Packaging Improvements

Despite protective precautions, aquaculture products are nevertheless susceptible to microplastic contamination. A possible approach is to relocate items to clean aquatic habitats with strong microplastic purification procedures, which has already been used to combat dangerous algal blooms (Mardones et al., 2021). Water exchange and recirculating pond systems improve aquaculture conditions by replacing contaminated water with clean water (Wang et al., 2022a, b, c). A 28.95% drop in farmed mussels shows that purification is an effective way to reduce microplastic contamination (Birnstiel et al., 2019). Furthermore, it was demonstrated that purification improved the oxidative stress and inflammatory reactions caused by microplastics in *Sparus aurata* (Solomando et al., 2020, 2021).

Packaging also has a substantial impact on microplastic contamination. Food packaging made of plastic can allow microplastics to seep into it (Kedzierski et al., 2020). According to Alak et al. (2021), rainbow trout fillets kept in polystyrene and wrapped films for aquaculture products had a high level of microplastic contamination, primarily fibers, shards, and pellets. Sustainable packaging solutions, such as biodegradable mesh bags for clams, hardwood trays for oysters, and recycled vacuum-sealed mussel bags, have been developed. Through a 20% reduction in tray weights for product transportation, aquaculture companies in Belgium have reduced their yearly plastic usage by 96 tons (Raedemaecker et al., 2020).

4.4 Policy Interventions and Wastewater Treatment for Microplastic Reduction

It is imperative that microplastic pollution be addressed at the policy level. Regulations aimed at minimizing plastic waste have been established in various countries, including restrictions on single-use plastics and bans on plastic microbeads in personal care products (Huntington, 2019; Schmaltz et al., 2020; Wagner, 2017). Nations such as Canada, France, the United Kingdom, and New Zealand have implemented similar bans to mitigate the influx of microplastics into aquatic ecosystems (Guerranti et al., 2019; Xanthos & Walker, 2017). In the United States, the Microbead-Free Waters Act, enacted in 2015 and effective from 2017, prohibits the inclusion of plastic microbeads in personal care items. Additionally, Kenya has enacted one of the most rigorous laws against the manufacture and possession of plastic bags (Biswas & Hartley, 2017; Lam et al., 2018).

4.5 Alternative Protein Sources and Biodegradable Materials in Aquaculture

One possible tactic to reduce microplastic contamination in aquaculture feed is to substitute alternate protein sources for fishmeal. Microorganisms like bacteria, fungi, yeast, and algae are the source of single-cell proteins (SCPs), which have a high nutritional value and lessen reliance on conventional fishmeal that might be tainted with microplastics (Buttle et al., 2024). The increasing need for premium protein-rich feeds has led to the identification of SCPs as sustainable and economical protein sources for aquaculture (Glencross et al., 2020).

Along with feed innovations, microplastic pollution can be greatly decreased by substituting natural or biodegradable materials for plastic-based aquaculture equipment. Materials such as coconut shells and bamboo have been investigated as sustainable substitutes for synthetic elements in aquaculture systems. The creation of a sponge from cotton and squid bone, for example, has shown great effectiveness in removing microplastics from water, indicating possible uses in aquaculture facilities (Perkins, 2024). To enable the broad use of these materials, however, issues including price, durability, and particular degrading circumstances need to be resolved. The environmental effect and biodegradability of the goods produced by the bioplastics sector are also being scrutinized, underscoring the necessity of thorough assessment prior to widespread use (Rawnsley, 2024).

5 Conclusion

Global food security and the *sustainability* of fish farming are at risk due to the growing problem of microplastic pollution in aquaculture. Aquaculture animals suffer serious ecological and physiological effects as a result of the permanence and toxicity of plastic garbage, which keeps finding its way into aquatic environments. The various sources and vast distribution of microplastics in aquatic settings make this a difficult issue to handle. The negative impacts on the physiology, behaviour, and survival of cultured species underscore the critical need for mitigation techniques. Advancing towards a sustainable future necessitates a comprehensive strategy that encompasses enhanced waste management practices, the creation of biodegradable substitutes for traditional plastics, advanced filtration technologies, and the implementation of more stringent regulations regarding plastic usage and disposal within the aquaculture sector. Additionally, further research is essential to fully understand the long-term impacts of microplastics on aquaculture species and to develop innovative solutions for addressing the existing microplastic pollution in aquatic ecosystems.

References

Aiguo, Z., Di, S., Chong, W., Yuliang, C., Shaolin, X., Peiqin, L., et al. (2022). Characteristics and differences of microplastics ingestion for farmed fish with different water depths, feeding habits and diets. *Journal of Environmental Chemical Engineering, 10*(2), 107189.

Alak, G., Köktürk, M., & Atamanalp, M. (2021). Evaluation of different packaging methods and storage temperature on MPs abundance and fillet quality of rainbow trout. *Journal of Hazardous Materials, 420*, 126573.

Amin, A., Indulkar, S. T., & Pai, R. (2016). Sublethal effect of Buprofezin pesticide on carcass composition of *Cyprinus carpio* communis fingerlings. *Journal of Experimental Zoology, India, 20*(1), 269–272.

Amin, A., Ramachandra Naik, A. T., Priyadarshini, N., Nayak, H., & Sree, C. S. (2017). Toxic effect of heavy metal lead on oxygen consumption of rohu (*Labeo rohita*) fingerlings. *Biochemical & Cellular Archives, 17*(1), 225.

Auta, H. S., Emenike, C. U., Jayanthi, B., & Fauziah, S. H. (2018). Growth kinetics and biodeterioration of polypropylene microplastics by Bacillus sp. and Rhodococcus sp. isolated from mangrove sediment. *Marine Pollution Bulletin, 127*, 15–21.

Barbier, E. B., Hacker, S. D., Kennedy, C., Koch, E. W., Stier, A. C., & Silliman, B. R. (2011). The value of estuarine and coastal ecosystem services. *Ecological Monographs, 81*(2), 169–193.

Barboza, L. G. A., Vieira, L. R., & Guilhermino, L. (2018). Single and combined effects of microplastics and mercury on juveniles of the European seabass (*Dicentrarchus labrax*): Changes in behavioural responses and reduction of swimming velocity and resistance time. *Environmental Pollution, 236*, 1014–1019.

Birnstiel, S., Soares-Gomes, A., & da Gama, B. A. (2019). Depuration reduces microplastic content in wild and farmed mussels. *Marine Pollution Bulletin, 140*, 241–247.

Biswas, A. K., & Hartley, K. (2017). *Real solutions for plastic problems: Tackling microplastics require big policy proposals*. Asia and the pacific policy society, policy forum. 28 september. https://www.policyforum.net/real-solutions-plastic-problems/. Accessed 29 March 2022.

Bobori, D. C., Dimitriadi, A., Feidantsis, K., Samiotaki, A., Fafouti, D., Sampsonidis, I., Kalogiannis, S., Kastrinaki, G., Lambropoulou, D. A., Kyzas, G. Z., Koumoundouros, G., Bikiaris, D. N., Kaloyianni, M., et al. (2022). Differentiation in the expression of toxic effects of polyethylene-microplastics on two freshwater fish species: Size matters. *Science of the Total Environment*. https://doi.org/10.1016/j.scitotenv.2022.154603

Bordós, G., Urbányi, B., Micsinai, A., Kriszt, B., Palotai, Z., Szabó, I., et al. (2019). Identification of microplastics in fish ponds and natural freshwater environments of the Carpathian basin, Europe. *Chemosphere, 216*, 110–116.

Borre, Y. E., Moloney, R. D., Clarke, G., Dinan, T. G., & Cryan, J. F. (2014). The impact of microbiota on brain and behavior: Mechanisms & therapeutic potential. In *Microbial endocrinology: The microbiota-gut-brain axis in health and disease* (pp. 373–403).

Boyd, C. E., D'Abramo, L. R., Glencross, B. D., Huyben, D. C., Juarez, L. M., Lockwood, G. S., et al. (2020). Achieving sustainable aquaculture: Historical and current perspectives and future needs and challenges. *Journal of the World Aquaculture Society, 51*(3), 578–633.

Brahney, J., Hallerud, M., Heim, E., Hahnenberger, M., & Sukumaran, S. (2020). Plastic rain in protected areas of the United States. *Science, 368*(6496), 1257–1260.

Burridge, L., Weis, J. S., Cabello, F., Pizarro, J., & Bostick, K. (2010). Chemical use in salmon aquaculture: A review of current practices and possible environmental effects. *Aquaculture, 306*(1–4), 7–23. https://doi.org/10.1016/j.aquaculture.2010.05.020

Buttle, L., Noorman, H., Roa Engel, C., & Santigosa, E. (2024). Bridging the protein gap with single-cell protein use in aquafeeds. *Frontiers in Marine Science, 11*, 1384083.

Cai, L., Wang, J., Peng, J., Tan, Z., Zhan, Z., Tan, X., & Chen, Q. (2017). Characteristic of microplastics in the atmospheric fallout from Dongguan city, China: Preliminary research and first evidence. *Environmental Science and Pollution Research, 24*, 24928–24935.

Castelvetro, V., Corti, A., Bianchi, S., Giacomelli, G., Manariti, A., & Vinciguerra, V. (2021a). Microplastics in fish meal: Contamination level analyzed by polymer type, including polyester (PET), polyolefins, and polystyrene. *Environmental Pollution, 273*, 115792.

Castelvetro, V., Corti, A., Bianchi, S., Giacomelli, G., Manariti, A., & Vinciguerra, V. (2021b). Development and optimization of a standard method for extraction of microplastics in mussels by enzyme digestion of soft tissues. *Environmental Toxicology and Chemistry, 36*(4), 947–951.

Cedervall, T., Hansson, L. A., Lard, M., Frohm, B., & Linse, S. (2012). Food chain transport of nanoparticles affects behaviour and fat metabolism in fish. *PLoS One, 7*(2), e32254.

Chaves, P. D. T., & Silveira, B. E. D. (2016). Artisanal fishing net float loss and a proposal for a float design solution. *Brazilian Journal of Oceanography, 64*, 89–94.

Chen, Q., Gundlach, M., Yang, S., Jiang, J., Velki, M., Yin, D., & Hollert, H. (2017). Quantitative investigation of the mechanisms of microplastics and nanoplastics toward zebrafish larvae locomotor activity. *Science of the Total Environment, 584*, 1022–1031.

Chen, C. L., Kuo, P. H., Lee, T. C., & Liu, C. H. (2018a). Snow lines on shorelines: Solving Styrofoam buoy marine debris from oyster culture in Taiwan. *Ocean & Coastal Management, 165*, 346–355.

Chen, M., Jin, M., Tao, P., Wang, Z., Xie, W., Yu, X., & Wang, K. (2018b). Assessment of microplastics derived from mariculture in Xiangshan Bay, China. *Environmental Pollution, 242*, 1146–1156.

Chen, B., Fan, Y., Huang, W., Rayhan, A. S., Chen, K., & Cai, M. (2020). Observation of microplastics in mariculture water of Longjiao Bay, Southeast China: Influence by human activities. *Marine Pollution Bulletin, 160*, 111655.

Cheung, L. T. O., Lui, C. Y., & Fok, L. (2018). Microplastic contamination of wild and captive flathead grey mullet (*Mugil cephalus*). *International Journal of Environmental Research and Public Health, 15*(4). https://doi.org/10.3390/ijerph15040597

Chia, R. W., Lee, J. Y., Kim, H., & Jang, J. (2021). Microplastic pollution in soil and groundwater: A review. *Environmental Chemistry Letters, 19*(6), 4211–4224.

Curren, E., Leaw, C. P., Lim, P. T., & Leong, S. C. Y. (2020). Evidence of marine microplastics in commercially harvested seafood. *Frontiers in Bioengineering and Biotechnology, 8*, 562760.

Da Costa, J. P., Santos, P. S., Duarte, A. C., & Rocha-Santos, T. (2016). Nano plastics in the environment–sources, fates and effects. *Science of the Total Environment, 566*, 15–26.

Debnath, R., Prasad, G. S., Amin, A., Malik, M. M., Ahmad, I., Abubakr, A., Borah, S., Rather, M. A., Impellitteri, F., Tabassum, I., & Piccione, G. (2024). Understanding and addressing microplastic pollution: Impacts, mitigation, and future perspectives. *Journal of Contaminant Hydrology, 266*, 104399.

Deng, H., Wei, R., Luo, W., Hu, L., Li, B., & Shi, H. (2020). Microplastic pollution in water and sediment in a textile industrial area. *Environmental Pollution, 258*, 113658.

Deroiné, M., Pillin, I., Le Maguer, G., Chauvel, M., & Grohens, Y. (2019). Development of new generation fishing gear: A resistant and biodegradable monofilament. *Polymer Testing, 74*, 163–169.

Deshpande, P. C., Philis, G., Brattebø, H., & Fet, A. M. (2020). Using Material Flow Analysis (MFA) to generate the evidence on plastic waste management from commercial fishing gears in Norway. *Resources, Conservation & Recycling: X, 5*, 100024.

Détrée, C., & Gallardo-Escárate, C. (2018). Single and repetitive microplastics exposures induce immune system modulation and homeostasis alteration in the edible mussel *Mytilus galloprovincialis*. *Fish & Shellfish Immunology, 83*, 52–60.

Do, H. L., & Thuy, T. D. (2022). Productivity response and production risk: A study of mangrove forest effects in aquaculture in the Mekong River Delta. *Ecological Economics, 194*, 107326.

Drillet, G. (2020). *Plastics in aquaculture – the WAS-IMarEST roundtables*. World Aquaculture Society Magazine Accessed 1/4/22 https://www.was.org/articles/Plastics-in-Aquaculture-The-WAS-IMarEST-Roundtables.aspx#.Yk aASKgo9LM

Du, F., Cai, H., Zhang, Q., Chen, Q., & Shi, H. (2020). Microplastics in take-out food containers. *Journal of Hazardous Materials, 399*, 122969.

Duan, J., Han, J., Cheung, S. G., Chong, R. K. Y., Lo, C. M., Lee, F. W. F., et al. (2021). How mangrove plants affect microplastic distribution in sediments of coastal wetlands: Case study in Shenzhen Bay, South China. *Science of the Total Environment, 767*, 144695.

Espinosa, C., Garcia, B. J. M., Esteba, M. A., & Cuesta, A. (2018). Invitro effects of virgin microplastic on fish Head- kidney leucocytes activities. *Environmental Pollution, 235*, 30–38.

FAO. (2018). *The state of world fisheries and aquaculture 2018*. United Nations.

FAO. (2024). *The State of World Fisheries and Aquaculture 2024 – Blue Transformation in action*. Rome. https://doi.org/10.4060/cd0683en.

Farrell, P., & Nelson, K. (2013). Trophic level transfer of microplastic: *Mytilus edulis* (L.) to *Carcinus maenas* (L.). *Environmental Pollution, 177*, 1–3.

Food and Agriculture Organization of the United Nations and the sustainable development goals (2018). Sustainable development, 22, 1–450.

Franzellitti, S., Capolupo, M., Wathsala, R. H. G. R., Valbonesi, P., & Fabbri, E. (2019). The Multixenobiotic resistance system as a possible protective response triggered by microplastic ingestion in Mediterranean mussels (*Mytilus galloprovincialis*): Larvae and adult stages. *Comparative Biochemistry and Physiology Toxicology & Pharmacology: CB, 219*, 50–58. https://doi.org/10.1016/j.cbpc.2019.02.005

Fred-Ahmadu, O. H., Ayejuyo, O. O., & Benson, N. U. (2020). Microplastics distribution and characterization in epipsammic sediments of tropical Atlantic Ocean, Nigeria. *Regional Studies in Marine Science, 38*, 101365. https://doi.org/10.1016/j.rsma.2020.101365

Gavigan, J., Kefela, T., Macadam-Somer, I., Suh, S., & Geyer, R. (2020). Synthetic microfiber emissions to land rival those to waterbodies and are growing. *PLoS One, 15*(9), Article e0237839. https://doi.org/10.1371/journal.pone.0237839

Girard, S., & Agundez, J. A. P. (2014). The effects of the oyster mortality crisis on the economics of the shellfish farming sector: Preliminary review and prospects from a case study in Marennes-Oleron Bay (France). *Marine Policy, 48*, 142–151.

Glencross, B. D., Huyben, D., & Schrama, J. W. (2020). The application of single-cell ingredients in aquaculture feeds—A review. *Fishes, 5*(3), 22.

Gomiero, A., Strafella, P., Pellini, G., Salvalaggio, V., & Fabi, G. (2018). Comparative effects of ingested PVC micro particles with and without adsorbed benzo(a)pyrene vs. Spiked sediments on the cellular and sub cellular processes of the benthic organism *Hediste diversicolor. Frontiers in Marine Science, 5*, 99. https://doi.org/10.3389/fmars.2018.00099

Gray, A. D., & Weinstein, J. E. (2017). Size-and shape-dependent effects of microplastic particles on adult daggerblade grass shrimp (*Palaemonetes pugio*). *Environmental Toxicology and Chemistry, 36*(11), 3074–3080.

Guerranti, C., Martellini, T., Perra, G., Scopetani, C., & Cincinelli, A. (2019). Microplastics in cosmetics: Environmental issues and needs for global bans. *Environmental Toxicology and Pharmacology, 68*, 75–79. https://doi.org/10.1016/j.etap.2019.03.007. Accessed 10 December 2020.

Gündoğdu, S., Çevik, C., Güzel, E., & Kilercioğlu, S. (2018). Microplastics in municipal wastewater treatment plants in Turkey: A comparison of the influent and secondary effluent concentrations. *Environmental Monitoring and Assessment, 190*, 1–10.

Gündogdu, S., Eroldogan, O. T., Evliyaoglu, E., Turchini, G. M., & Wu, X. G. (2021). Fish out, plastic in: Global pattern of plastics in commercial fishmeal. *Aquaculture, 534*, Article 736316. https://doi.org/10.1016/j.aquaculture.2020.736316

Gurjar, U. R., Takar, S., Amin, A., & Martin Xavier, K. A. (2023). Microplastic contamination in aquatic organisms: An ecotoxicological perspective. In *Xenobiotics in aquatic animals: Reproductive and developmental impacts* (pp. 353–367). Springer Nature Singapore.

Guven, O., Ach, L., Munk, P., Dinh, K. V., Mariani, P., & Nielsen, T. G. (2018). Microplastic does not magnify the acute effect of PAH pyrene on predatory performance of a tropical fish (*Lates calcarifer*). *Aquatic Toxicology, 198*, 287–293.

Hamed, M., Soliman, H. A., Osman, A. G., & Sayed, A. E. D. H. (2019). Assessment the effect of exposure to microplastics in Nile Tilapia (Oreochromis niloticus) early juvenile: I. blood biomarkers. Chemosphere, 228, 345–350.

Habib, R. Z., Thiemann, T., & Al Kendi, R. (2020). Microplastics and wastewater treatment plants—A review. *Journal of Water Resource and Protection, 12*, 1–35. https://doi.org/10.4236/jwarp.2020.121001

Halsband, C., & Galloway, T. S. (2015). The impact of polystyrene microplastics on feeding, function and fecundity in the marine copepod *Calanus helgolandicus*. *Environmental Science and Technology, 49*, 1130–1137.

Hamed, M., Soliman, H. A. M., Badrey, A. E. A., & Osman, A. G. M. (2021). Microplastics induced histopathological lesions in some tissues of tilapia (*Oreochromis niloticus*) early juveniles. *Tissue and Cell, 71*, Article 101512.

Han, Y., Shi, W., Tang, Y., Zhou, W., Sun, H., Zhang, J., et al. (2022). Microplastics and bisphenol A hamper gonadal development of whiteleg shrimp (*Litopenaeus vannamei*) by interfering with metabolism and disrupting hormone regulation. *Science of the Total Environment, 810*, 152354.

Hanachi, P., Karbalaei, S., Walker, T. R., Cole, M., & Hosseini, S. V. (2019). Abundance and properties of microplastics found in commercial fish meal and cultured common carp (*Cyprinus carpio*). *Environmental Science and Pollution Research, 26*, 23777–23787.

Hasan, J., Islam, S. M., Alam, M. S., Johnson, D., Belton, B., Hossain, M. A. R., & Shahjahan, M. (2022). Presence of microplastics in two common dried marine fish species from Bangladesh. *Marine Pollution Bulletin, 176*, 113430.

Hellfeld, V., R., Zarzuelo, M., Zaldibar, B., Cajaraville, M. P., & Orbea, A. (2022). Accumulation, depuration, and biological effects of polystyrene microplastic spheres and adsorbed cadmium and benzo (a) pyrene on the mussel Mytilus galloprovincialis. *Toxics, 10*(1), 18.

Hossain, M. S., Rahman, M. S., Uddin, M. N., Sharifuzzaman, S. M., Chowdhury, S. R., et al. (2020). Microplastic contamination in penaeid shrimp from the northern bay of bengal. *Chemosphere, 238*, Article 124688.

Hu, J., Zuo, J., Li, J., Zhang, Y., Ai, X., Zhang, J., et al. (2022). Effects of secondary polyethylene microplastic exposure on crucian (Carassius carassius) growth, liver damage, and gut microbiome composition. *Science of the Total Environment, 802*, 149736.

Huntington, T. (2019). *Marine Litter and Aquaculture Gear: White Paper*. Report produced by poseidon aquatic resources management Ltd for the aquaculture stewardship Council. 20.

Iheanacho, S. C., & Odo, G. E. (2020a). Neurotoxicity, oxidative stress biomarkers and hematological responses in African catfish (*Clarias gariepinus*) exposed to polyvinyl chloride microparticles. *Comparative Biochemistry and Physiology - Part C: Toxicology & Pharmacology, 232*, Article 108741.

Iheanacho, S. C., & Odo, G. E. (2020b). Dietary exposure to polyvinyl chloride microparticles induced oxidative stress and hepatic damage in *Clarias gariepinus* (Burchell, 1822). *Environmental Science and Pollution Research, 27*(17), 21159–21173.

Iheanacho, S., Ogbu, M., Bhuyan, M. S., & Ogunji, J. (2023). Microplastic pollution: An emerging contaminant in aquaculture. *Aquaculture and Fisheries, 8*(6), 603–616.

Ilechukwu, I., Ndukwe, G. I., Ehigiator, B. E., Ezeh, C. S., & Asogwa, S. L. (2021). Microplastics in silver catfish (*Chrysichthys nigrodigitatus*) from new Calabar river in Niger delta, Nigeria. *Ghana Journal of Science, 62*(2), 16–24.

Jabeen, K., Su, L., Li, J., Yang, D., Tong, C., Mu, J., & Shi, H. (2017). Microplastics and mesoplastics in fish from coastal and fresh waters of China. *Environmental Pollution, 221*, 141–149.

Jabeen, K., Li, B., Chen, Q., Su, L., Wu, C., Hollert, H., & Shi, H. (2018). Effects of virgin microplastics on goldfish (Carassius auratus). *Chemosphere, 213*, 323–332.

Jacob, H., Besson, M., Swarzenski, P. W., Lecchini, D., & Metian, M. (2020). Effects of virgin micro-and nanoplastics on fish: Trends, meta-analysis, and perspectives. *Environmental Science & Technology, 54*(8), 4733–4745.

Jaikumar, G., Brun, N. R., Vijver, M. G., & Bosker, T. (2019). Reproductive toxicity of primary and secondary microplastics to three cladocerans during chronic exposure. *Environmental Pollution, 249*, 638–646.

Jia, X., Liu, Y., He, Y., Yu, H., Liu, Y., Shen, Y., Xu, X., & Li, J. (2022). Exposure to microplastics induces lower survival, oxidative stress, disordered microbiota and altered metabolism in the intestines of grass carp (Ctenopharyngodon idella). *Aquaculture and Fisheries*. https://doi.org/10.1016/j.aaf.2022.09.004

Jiang, J. Q. (2018). Occurrence of microplastics and its pollution in the environment: A review. *Sustainable Production and Consumption, 13*, 16–23.

Jovanovic, B., Gokdag, K., Guven, O., Emre, Y., Whitely, E. M., & Kideys, A. E. (2018). Virgin microplastics are not causing imminent harm to fish after dietary exposure. *Marine Pollution Bulletin, 130*, 123–131.

Kaleem, O., & Sabi, A. F. B. S. (2021). Overview of aquaculture systems in Egypt and Nigeria, prospects, potentials, and constraints. *Aquaculture and Fisheries, 6*(6), 535–547.

Karami, A., Romano, N., Galloway, T., & Hamzah, H. (2016). Virgin microplastics cause toxicity and modulate the impacts of phenanthrene on biomarker responses in African catfish (*Clarias gariepinus*). *Environmental Research, 151*, 58–70.

Karbalaei, S., Golieskardi, A., Watt, D. U., Boiret, M., Hanachi, P., Walker, T. R., & Karami, A. (2020). Analysis and inorganic composition of microplastics in commercial Malaysian fish meals. *Marine Pollution Bulletin, 150*, 110687.

Kathiresan, K. (2003). Polythene and plastics-degrading microbes from the mangrove soil. *Revista de Biologia Tropical, 51*(3–4), 629–633.

Kedzierski, M., Lechat, B., Sire, O., Le Maguer, G., Le Tilly, V., & Bruzaud, S. (2020). Microplastic contamination of packaged meat: Occurrence and associated risks. *Food Packaging and Shelf Life, 24*, 100489.

Kim, J., Poirier, D. G., Helm, P. A., Bayoumi, M., & Rochman, C. M. (2020). No evidence of spherical microplastics (10-300 μm) translocation in adult rainbow trout (*Oncorhynchus mykiss*) after a two-week dietary exposure. *PLoS One, 15*(9), Article e0239128. https://doi.org/10.1371/journal.pone.0239128

Koelmans, A. A., Bakir, A., Burton, G. A., & Janssen, C. R. (2016). Microplastic as a vector for chemicals in the aquatic environment: Critical review and model-supported reinterpretation of empirical studies. *Environmental Science and Technology, 50*, 3315–3326.

Korez, Š., Gutow, L., & Saborowski, R. (2019). Microplastics at the strandlines of Slovenian beaches. *Marine Pollution Bulletin, 145*, 334–342.

Kühn, S., Bravo Rebolledo, E. L., & Van Franeker, J. A. (2015). Deleterious effects of litter on marine life. In *Marine anthropogenic litter* (pp. 75–116).

Lam, C. S., Ramanathan, S., Carbery, M., Gray, K., Vanka, K. S., & Maurin, C. (2018). A comprehensive analysis of plastics and microplastic legislation worldwide. *Water, Air, and Soil Pollution, 229*(11), 1–19. https://doi.org/10.1007/s11270-018-4002-z

Lares, M., Ncibi, M. C., Sillanpää, M., & Sillanpää, M. (2018). Occurrence, identification and removal of microplastic particles and fibers in conventional activated sludge process and advanced MBR technology. *Water Research, 133*, 236–246.

LeMoine, C. M., Kelleher, B. M., Lagarde, R., Northam, C., Elebute, O. O., & Cassone, B. J. (2018). Transcriptional effects of polyethylene microplastics ingestion in developing zebrafish (*Danio rerio*). *Environmental Pollution, 243*, 591–600.

Li, H. X., Ma, L. S., Lin, L., Ni, Z. X., Xu, X. R., Shi, H. H., & Rittschof, D. (2018). Microplastics in oysters Saccostrea cucullata along the Pearl River estuary, China. Environmental Pollution, 236, 619–625.

Li, Y., Zhang, Y., Chen, G., Xu, K., Gong, H., Huang, K., et al. (2021). Microplastics in surface waters and sediments from Guangdong Coastal Areas, South China. *Sustainability, 13*, 2691. https://doi.org/10.3390/su13052691

Liu, X., Yuan, W., Di, M., Li, Z., & Wang, J. (2019). Transfer and fate of microplastics during the conventional activated sludge process in one wastewater treatment plant of China. *Chemical Engineering Journal, 362*, 176–182.

Liu, X., Liu, H., Chen, L., & Wang, X. (2022). Ecological interception effect of mangroves on microplastics. *Journal of Hazardous Materials, 423*, 127231.

Loctier, D. (2021). Reducing plastic in fishing and aquaculture: What alternatives to protect our oceans?. https://www.euronews.

Lusher, A., Hollman, P., & Mendoza-Hill, J. (2017). *Microplastics in fisheries and aquaculture: Status of knowledge on their occurrence and implications for aquatic organisms and food safety*. FAO.

Luthada-Raswiswi, R., Mukaratirwa, S., & O'brien, G. (2021). Animal protein sources as a substitute for fishmeal in aquaculture diets: A systematic review and meta-analysis. *Applied Sciences, 11*(9), 3854.

Lyu, W., Chen, Q., Cheng, L., & Zhou, W. (2020). Microplastics in aquaculture systems and their transfer in the food chain. In *Microplastics in Terrestrial Environments: Emerg.Contam. Major Chall* (pp. 357–e382). Springer.

Ma, J., Niu, X., Zhang, D., Lu, L., Ye, X., Deng, W., et al. (2020). High levels of microplastic pollution in aquaculture water of fish ponds in the Pearl River Estuary of Guangzhou, China. *Science of the Total Environment, 744*, 140679.

Malik, M., Balkhi, M. H., & Abubakr, A. (2017). Comparative study on the contamination level of Dal Lake, Kashmir based on total viable bacterial counts over a period of time. *Ecology, Environment, and Conservation, 23*(4), 2145–2147.

Mandich, A., Bottero, S., Benfenati, E., Cevasco, A., Erratico, C., Maggioni, S., et al. (2007). In vivo exposure of carp to graded concentrations of bisphenol A. *General and Comparative Endocrinology, 153*(1–3), 15–24.

Mardones, J. I., Paredes, J., Godoy, M., Suarez, R., Norambuena, L., Vargas, V., et al. (2021). Disentangling the environmental processes responsible for the world's largest farmed fish-killing harmful algal bloom: Chile, 2016. *Science of the Total Environment, 766*, 144383.

Margeirsson, B., Gospavic, R., Pálsson, H., Arason, S., & Popov, V. (2011). Experimental and numerical modelling comparison of thermal performance of expanded polystyrene and corrugated plastic packaging for fresh fish. *International Journal of Refrigeration, 34*(2), 573–585.

Martínez-Gómez, C., León, V. M., Calles, S., Gomáriz-Olcina, M., & Vethaak, A. D. (2017). The adverse effects of virgin microplastics on the fertilization and larval development of sea urchins. *Marine Environmental Research, 130*, 69–76.

Mattsson, K., Ekvall, M. T., Hansson, L. A., Linse, S., Malmendal, A., & Cedervall, T. (2015). Altered behavior, physiology, and metabolism in fish exposed to polystyrene nanoparticles. *Environmental Science & Technology, 49*(1), 553–561.

Meng, Y., Kelly, F. J., & Wright, S. L. (2020). Advances and challenges of microplastic pollution in freshwater ecosystems: A UK perspective. *Environmental Pollution, 256*, 113445.

Nabi, N., Ahmad, I., Amin, A., Rather, M. A., Ahmed, I., Hajam, Y. A., Khursheed, S., Malik, M. M., & Abubakr, A. (2024). Understanding the sources, fate and effects of microplastics in aquatic environments with a focus on risk profiling in aquaculture systems. *Reviews in Aquaculture, 16*(4), 1947–1980.

Nguyen, M. K., Lin, C., Nguyen, H. L., Le, V. R., Priya, K. L., Singh, J., et al. (2023). Emergence of microplastics in the aquatic ecosystem and their potential effects on health risks: The insights into Vietnam. *Journal of Environmental Management, 344*, 118499. https://doi.org/10.1016/j.jenvman.2023.118499

NOAA. (2023). What is aquaculture? National Ocean Service Website. https://oceanservice.noaa.gov/facts/aquaculture.html

Okada, T., Uchida, S., & Nakata, K. (2014, July). Direct joining of aluminum alloy and plastic sheets by friction lap processing. *In Materials Science Forum* (Vol. 794, pp. 395–400). Trans Tech Publications Ltd.

Oliveira, M., Ribeiro, A., Hylland, K., & Guilhermino, L. (2013). Single and combined effects of microplastics and pyrene on juveniles (0+ group) of the common goby *Pomatoschistus microps* (Teleostei, Gobiidae). *Ecological Indicators, 34*, 641–647.

Padervand, M., Lichtfouse, E., Robert, D., & Wang, C. (2020). Removal of microplastics from the environment. A review. *Environmental Chemistry Letters, 18*(3), 807–828.

Parker, B., Andreou, D., Green, I. D., & Britton, J. R. (2021). Microplastics in freshwater fishes: Occurrence, impacts and future perspectives. *Fish and Fisheries, 22*(3), 467–488.

Perkins, T. (2024, December 10). *Cotton-and-squid-bone sponge can soak up 99.9% of microplastics, scientists say*. The Guardian. https://www.theguardian.com/environment/2024/dec/10/microplastics-pollution-sponge-cotton-squid-bone.

Perren, W., Wojtasik, A., & Cai, Q. (2018). Removal of microbeads from wastewater using electrocoagulation. *ACS Omega, 3*(3), 3357–3364.

Piñon-Colin, T., Rodriguez-Jimenez, R., Rogel-Hernandez, E., Alvarez-Andrade, A., & Wakida, F. T. (2020). Microplastics in stormwater runoff in a semiarid region, Tijuana, Mexico. *Science of the Total Environment, 704*, 135411.

Pitt, J. A., Kozal, J. S., Jayasundara, N., Massarsky, A., Trevisan, R., Geitner, N., et al. (2018). Uptake, tissue distribution, and toxicity of polystyrene nanoparticles in developing zebrafish (*Danio rerio*). *Aquatic Toxicology, 194*, 185–194.

Pittura, L., Avio, C. G., Giuliani, M. E., d'Errico, G., Keiter, S. H., Cormier, B., & Regoli, F. (2018). Microplastics as vehicles of environmental PAHs to marine organisms: combined chemical and physical hazards to the Mediterranean mussels, Mytilus galloprovincialis. *Frontiers in marine science, 5*, 103.

Plestenjak, G., Eler, K., Mihelič, R., Ferlan, M., Ogrinc, N., Krajnc, B., & Vodnik, D. (2021). Can additional air supply enhance decomposition processes in sludge treatment reed beds? *Journal of Environmental Management, 277*, 111511.

Prasad, G. S., Rout, S. K., Malik, M. M., Karmakar, S., Amin, A., & Ahmad, I. (2023). Occurrence of xenoestrogen alkylphenols (Octylphenols and Nonylphenol) and its impact on the aquatic ecosystem. In *Xenobiotics in aquatic animals: Reproductive and developmental impacts* (pp. 275–284).

Prasad, G. S., Rout, S. K., Amin, A., Malik, M. M., Ahmad, I., Rather, M. A., Karmakar, S., Khalil, H. S., & Abubakar, A. (2024). Impact of COVID-19 on the aquatic environment and implications for fisheries with special emphasis to Indian perspective. *Proceedings of the Indian National Science Academy, 90*(4), 821–839.

Prata, J. C., Silva, A. L. P., da Costa, J. P., Mouneyrac, C., Walker, T. R., Duarte, A. C., & Rocha-Santos, T. (2019). Solutions and integrated strategies for the control and mitigation of plastic and microplastic pollution. *International Journal of Environmental Research and Public Health, 16*, 2411. https://doi.org/10.3390/ijerph16132411

Priscilla, V., & Patria, M. P. (2019). Comparison of microplastic abundance in aquaculture ponds of milkfish Chanos chanos (Forsskål, 1775) at Muara Kamal and Marunda, Jakarta Bay IOP Conference Series: Earth and Environmental Science. In *The 4th EMBRIO international symposium and the 7th international symposium of East Asia Fisheries and Technologists Association 5–6 August 2019, Bogor* (Vol. 404).

Raedemaecker, F. D., Sandra, M., Devriese, L., Gin, I., Altvater, S., Rigo, M. V., Alomar, C., Deudero, S., Lara, M. M. (2020). *Learning Lab report North Sea.* D3.4 Aqua-Lit project 72 pp. Flanders Marine Institute (VLIZ), Oostende.

Rahman, T., Mustakima, S., Ferdous, Z., Tabassum, T., Mahamud, A. S. U., Siddika, M., & Haque, M. N. (2022). Properties and abundance of microplastics found in fish feed, tissues, and culture water of catfish (*Heteropneustes fossilis*). *International Journal of Aquatic Biology, 10*(1), 1–10. https://doi.org/10.22034/ijab.v10i1

Rawnsley, J. (2024, November 15). *Bioplastics start-ups gain ground but green credentials face scrutiny.* Financial Times. https://www.ft.com/content/39ec3f72-5840-4127-9c8f-f1f8d5bf036d

Revel, M., Chatel, A., Perrein-Ettajani, H., Bruneau, M., Akcha, F., Sussarellu, R., et al. (2020). Realistic environmental exposure to microplastics does not induce biological effects in the Pacific oyster (*Crassostrea gigas*). *Marine Pollution Bulletin, 150*, Article 110627.

Sandra, M., Devriese, L., De Raedemaecker, F., Lonneville, B., Lukic, I., Altvater, S., et al. (2019). Knowledge wave on marine litter from aquaculture sources: D2. 2 Aqua-Lit project.

Schmaltz, E., Melvin, E. C., Diana, Z., Gunady, E. F., Rittschof, D., Somarelli, J. A., & Dunphy-Daly, M. M. (2020). Plastic pollution solutions: emerging technologies to prevent and collect marine plastic pollution. *Environment international, 144*, 106067.

Seltenrich, N. (2015). New link in the food chain? Marine plastic pollution and seafood safety. *Environmental Health Perspectives*. https://doi.org/10.1289/ehp.123-A34

Sharma, S., & Chatterjee, S. (2017). Microplastic pollution, a threat to marine ecosystem and human health: A short review. *Environmental Science and Pollution Research, 24*, 21530–21547.

Shi, W., Sun, S., Han, Y., Tang, Y., Zhou, W., Zhang, W., et al. (2022). Microplastics hamper the fertilization success of a broadcast spawning bivalve through reducing gamete collision and gamete fusion efficiency. *Aquatic Toxicology, 242*, 106049.

Skirtun, M., Sandra, M., Strietman, W. J., van den Burg, S. W., De Raedemaecker, F., & Devriese, L. I. (2022). Plastic pollution pathways from marine aquaculture practices and potential solutions for the North-East Atlantic region. *Marine Pollution Bulletin, 174*, 113178.

Solomando, A., Capó, X., Alomar, C., Álvarez, E., Compa, M., Valencia, J. M., et al. (2020). Long-term exposure to microplastics induces oxidative stress and a pro-inflammatory response in the gut of Sparus aurata Linnaeus, 1758. *Environmental Pollution, 266*, 115295.

Solomando, A., Capó, X., Alomar, C., Compa, M., Valencia, J. M., Sureda, A., & Deudero, S. (2021). Assessment of the effect of long-term exposure to microplastics and depuration period in *Sparus aurata* Linnaeus, 1758: Liver and blood biomarkers. *Science of the Total Environment, 786*, 147479.

Song, Y. K., Hong, S. H., Jang, M., Han, G. M., Jung, S. W., & Shim, W. J. (2017). Combined effects of UV exposure duration and mechanical abrasion on microplastic fragmentation by polymer type. *Environmental Science & Technology, 51*(8), 4368–4376.

Sruthisree, C., Gowda, G., Nayak, H., Harish, B. D., & Amin, A. (2015). Effect of heavy metal (lead) on oxygen consumption and ammonia excretion of Iridescent Shark (*Pangusius hypohthalmus*). *Journal of Experimental Zoology, India, 18*(2), 637.

Sussarellu, R., Suquet, M., Thomas, Y., Lambert, C., Fabioux, C., Pernet, M. E. J., et al. (2016). Oyster reproduction is affected by exposure to polystyrene microplastics. *Proceedings of the National Academy of Sciences, 113*(9), 2430–2435.

Tabassum, I., Hassan, N., Amin, A., Malik, M. M., Abubakr, A., Prasad, G. S., Latief, F., & Rather, S. A. (2025). COVID-19 and global environmental change: A threat to food security and nutrition. In *Food security, nutrition and sustainability through aquaculture technologies* (pp. 437–445). Springer Nature Switzerland.

Talvitie, J., Heinonen, M., Paakkonen, J. P., Vahtera, E., Mikola, A., Setala, O., & Vahala, R. (2015). Do wastewater treatment plants act as a potential point source of microplastics? Preliminary study in the coastal gulf of Finland, baltic sea. *Water Science and Technology, 72*, 1495–1505. https://doi.org/10.2166/wst.2011.079

Tang, G., Liu, M., Zhou, Q., He, H., Chen, K., Zhang, H., et al. (2018). Microplastics and polycyclic aromatic hydrocarbons (PAHs) in Xiamen coastal areas: Implications for anthropogenic impacts. *Science of the Total Environment, 634*, 811–820.

Taniguchi, S., Colabuono, F. I., Dias, P. S., Oliveira, R., Fisner, M., Turra, A., Izar, G. M., Abessa, D. M. S., Saha, M., Hosoda, J., Yamashita, R., Takada, H., Lourenco, R. A., Magalhaes, C. A., Bicego, M. C., & Montone, R. C. (2016). Spatial variability in persistent organic pollutants and polycyclic aromatic hydrocarbons found in beach– Stranded pellets along the coast of the state of Sao Paulo, Southeastern Brazil. *Marine Pollution Bulletin, 106*(1–2), 87–94.

Thiele, C. J., Hudson, M. D., Russell, A. E., Saluveer, M., & Sidaoui-Haddad, G. (2021). Microplastics in fish and fishmeal: An emerging environmental challenge? *Scientific Reports, 11*(1), 2045.

Thomas, M., Jon, B., Craig, S., Edward, R., Ruth, H., John, B., Dick, V. A., Heather, L. A., & Matthew, S. (2020). The world is your oyster: Low-dose, long-term microplastic exposure of juvenile oysters. *Heliyon, 6*(1), Article e03103. https://doi.org/10.1016/j.heliyon.2019.e03103

UNEP. (2019). https://www.unep.org/resources/report/microplastics

USDA. (2023). Aquaculture. https://www.usda.gov/topics/farming/aquaculture

Van Cauwenberghe, L., Devriese, L., Galgani, F., Robbens, J., & Janssen, C. R. (2015). Microplastics in sediments: A review of techniques, occurrence and effects. *Marine Environmental Research, 111*, 5–17.

Vázquez-Rowe, I., Ita-Nagy, D., & Kahhat, R. (2021). Microplastics in fisheries and aquaculture: Implications to food sustainability and safety. *Current Opinion in Green and Sustainable Chemistry, 29*, 100464.

Wagner, R. (2017). La obra de arte del futuro. Universitat de Valéncia.

Wan, Z., Wang, C., Zhou, J., Shen, M., Wang, X., Fu, Z., & Jin, Y. (2019). Effects of polystyrene microplastics on the composition of the microbiome and metabolism in larval zebrafish. *Chemosphere, 217*, 646–658.

Wang, Q., Cheng, L., Liu, J., Li, Z., Xie, S., & De Silva, S. S. (2015). Freshwater aquaculture in Pr China: Trends and prospects. *Reviews in Aquaculture, 7*(4), 283–302.

Wang, J., Peng, J., Tan, Z., Gao, Y., Zhan, Z., Chen, Q., & Cai, L. (2017). Microplastics in the surface sediments from the Beijiang River littoral Zone: Composition, abundance, surface textures and interaction with heavy metals. *Chemosphere, 171*, 248–258.

Wang, J., Lu, L., Wang, M., Jiang, T., Liu, X., & Ru, S. (2019). Typhoons increase the abundance of microplastics in the marine environment and cultured organisms: A case study in Sanggou Bay, China. *Science of the Total Environment, 667*, 1–8.

Wang, C., Jiang, C., Gao, T., Peng, X., Ma, S., Sun, Q., et al. (2022a). Improvement of fish production and water quality in a recirculating aquaculture pond enhanced with bacteria-microalgae association. *Aquaculture, 547*, 737420.

Wang, Q., Li, J., Zhu, X., Sun, C., Teng, J., Chen, L., Shan, E., & Zhao, J. (2022b). Microplastics in fish meals: An exposure route for aquaculture animals. *Science of the Total Environment, 807*(3), Article 151049. https://doi.org/10.1016/j.scitotenv.2021.151049

Wang, X., Jian, S., Zhang, S., Wu, D., Wang, J., Gao, M., et al. (2022c). Enrichment of polystyrene microplastics induces histological damage, oxidative stress, Keap1-Nrf2 signaling pathway-related gene expression in loach juveniles (*Paramisgurnus dabryanus*). *Ecotoxicology and Environmental Safety, 237*, 113540.

Wootton, N., Reis-Santos, P., & Gillanders, B. M. (2021). Microplastic in fish–a global synthesis. *Reviews in Fish Biology and Fisheries, 31*(4), 753–771. https://doi.org/10.1007/s11160-021-09684-6

Wright, S. L., Rowe, D., Thompson, R. C., & Galloway, T. S. (2013a). Microplastic ingestion decreases energy reserves in marine worms. *Current Biology, 23*, 1031–1033.

Wright, S. L., Thompson, R. C., & Galloway, T. S. (2013b). The physical impacts of microplastics on marine organisms: A review. *Environmental Pollution, 178*, 483–492.

Wright, S. L., Ulke, J., Font, A., Chan, K. L. A., & Kelly, F. J. (2020). Atmospheric microplastic deposition in an urban environment and an evaluation of transport. *Environment International, 136*, 105411.

Wright, L. S., Napper, I. E., & Thompson, R. C. (2021). Potential microplastic release from beached fishing gear in Great Britain's region of highest fishing litter density. *Marine Pollution Bulletin, 173*, 113115.

Wu, H., Hou, J., & Wang, X. (2023). A review of microplastic pollution in aquaculture: Sources, effects, removal strategies and prospects. *Ecotoxicology and Environmental Safety, 252*, 114567.

Xanthos, D., & Walker, T. R. (2017). International policies to reduce plastic marine pollution from single-use plastics (plastic bags and microbeads): A review. *Marine Pollution Bulletin, 118*(1–2), 17–26. https://doi.org/10.1016/j.marpolbul.2017.02.048

Xia, X., Sun, M., Zhou, M., Chang, Z., & Li, L. (2020). Polyvinyl chloride microplastics induce growth inhibition and oxidative stress in *Cyprinus carpio* var. larvae. *Science of the Total Environment, 716*, Article 136479.

Yao, W., Di, D., Wang, Z., Liao, Z., Huang, H., Mei, K., et al. (2019). Micro-and macroplastic accumulation in a newly formed Spartina alterniflora colonized estuarine saltmarsh in Southeast China. *Marine Pollution Bulletin, 149*, 110636.

Yao, C., Liu, X., Wang, H., Sun, X., Qian, Q., & Zhou, J. (2021). Occurrence of microplastics in fish and shrimp feeds. *Bulletin of Environmental Contamination and Toxicology, 107*, 684–692.

Yin, L., Chen, B., Xia, B., Shi, X., & Qu, K. (2018). Polystyrene microplastics alter the behavior, energy reserve and nutritional composition of marine jacopever (*Sebastes schlegelii*). *Journal of Hazardous Materials, 360*, 97–105.

Yin, L., Wen, X., Huang, D., Zeng, G., Deng, R., Liu, R., et al. (2021). Microplastics retention by reeds in freshwater environment. *Science of the Total Environment, 790*, 148200.

Yu, J., Tian, J. Y., Xu, R., Zhang, Z. Y., Yang, G. P., Wang, X. D., & Chen, R. (2020). Effects of microplastics exposure on ingestion, fecundity, development, and dimethylsulfide production in *Tigriopus japonicus* (Harpacticoida, copepod). *Environmental Pollution, 267*, Article 115429.

Yuan, Y., Sepulveda, M. S., Bi, B., Huang, Y., Kong, L., Yan, H., & Gao, Y. (2023). Acute polyethylene microplastic (PE-MPs) exposure activates the intestinal mucosal immune network pathway in adult zebrafish (*Danio rerio*). *Chemosphere, 311*(1), Article 137048.

Zeri, C., Adamopoulou, A., Koi, A., Koutsikos, N., Lytras, E., Dimitriou, E., & Rivers, A. (2021). Wastewater-treatment plants as microplastic pathways to eastern Mediterranean Waters: First records for the aegean sea, Greece. *Sustainability, 13*(5328), 2021. https://doi.org/10.3390/su13105328

Zhang, K., Shi, H., Peng, J., Wang, Y., Xiong, X., Wu, C., & Lam, P. K. (2018). Microplastic pollution in China's inland water systems: A review of findings, methods, characteristics, effects, and management. *Science of the Total Environment, 630*, 1641–1653.

Zhang, Y., Kang, S., Allen, S., Allen, D., Gao, T., & Sillanpää, M. (2020). Atmospheric microplastics: A review on current status and perspectives. *Earth-Science Reviews, 203*, 103118.

Zhang, C., Wang, J., Pan, Z., Wang, S., Zhang, L., Wang, Q., et al. (2021). A dosage-effect assessment of acute toxicology tests of microplastic exposure in filter-feeding fish. *Fish & Shellfish Immunology, 113*, 154–161.

Ziajahromi, S., Neale, P. A., Silveira, I. T., Chua, A., & Leusch, F. D. (2021). An audit of microplastic abundance throughout three Australian wastewater treatment plants. *Chemosphere, 263*, 128294.

Impact of Microplastics (MPs) on Aquatic Ecosystems: A Sustainable Approach to Bioremediation and Food Security

Rohit Kumar Gautam, Ashish Sahu, Mahender Singh, and Abha Mishra

1 Introduction

Microplastics (MPs) are plastic fragments smaller than 5 mm in diameter (Arthur et al., 2009) and have emerged as a pervasive pollutant in ecosystems globally (Lamichhane et al., 2023). These tiny particles originate from two primary sources fragmentation of larger plastic debris through environmental factors (sunlight, wind, and mechanical forces), and direct use of microplastic materials in consumer products (cosmetics, cleaning agents, and synthetic fibres) (Nawalage & Bellanthudawa, 2022). Over the past few decades, the exponential growth in plastic production has resulted in the substantial accumulation of microplastics across terrestrial and aquatic habitats. The history of plastic production traces back to the mid-twentieth century, an era when plastic materials revolutionized industries and transformed daily life (Quye, 2017). However, several properties that make plastics desirable durability and resistance to biodegradation have resulted in environmental hazards. Plastics, including microplastics, persist in ecosystems for decades, fragmenting into

R. K. Gautam
Department of Zoology, Babasaheb Bhimrao Ambedkar University, Lucknow, UP, India

Faculty of Fisheries Science, Department of Fisheries Resources Management, Kerala University of Fisheries and Ocean Studies, Cochin, Kerala, India

A. Sahu
Faculty of Fisheries Science, Department of Fisheries Resources Management, Kerala University of Fisheries and Ocean Studies, Cochin, Kerala, India

ICAR-National Bureau of Fish Genetic Resources, Lucknow, Uttar Pradesh, India

M. Singh
ICAR-National Bureau of Fish Genetic Resources, Lucknow, Uttar Pradesh, India

A. Mishra (✉)
Department of Zoology, Babasaheb Bhimrao Ambedkar University, Lucknow, UP, India

A. Abubakr et al. (eds.), *Microplastics: Ecological and Food Security Implications*,
https://doi.org/10.1007/978-3-031-94480-2_15

progressively smaller particles that infiltrate soil, water, and organisms, leading to widespread ecological disruption. In aquatic habitats, microplastics are particularly prevalent, with millions of tons entering rivers, lakes, and oceans annually (Hale et al., 2020; Gurjar et al., 2023; Debnath et al., 2024). These particles are ingested by a diverse array of organisms, ranging from plankton to fish, and ultimately enter the human food chain through the consumption of finfish and shellfish. The accumulation of microplastics within the food chain raises significant concerns regarding both human health and the stability of ecosystems. Microplastics can cause physical harm to aquatic species, disrupt feeding and reproduction biology, and alter ecological functions, all of which jeopardize food security (De-la-Torre, 2020). This chapter addresses the escalating hurdles of microplastic pollution, focusing on sustainable bioremediation approaches to mitigate its impact. By exploring innovative methods such as microbial degradation and phytoremediation, we aim to highlight solutions that not only reduce the ecological footprint of plastics but also contribute to safeguarding food security and promoting ecological health in the face of this widespread environmental threat.

2 How Do MPs Accumulate in Aquatic Ecosystems and Transfer to Organisms Through the Food Web?

Aquatic ecosystems are significantly affected by microplastic debris (synthetic organic polymers), which changes habitats, introduces harmful compounds, and causes damage to organisms across all trophic levels (Huang et al., 2021; Nabi et al., 2024). The base of aquatic food webs, plankton, is particularly vulnerable to microplastic pollutants. Due to their resemblance to natural food in both size and appearance, these organisms inadvertently ingest the microplastics, exposing them to potential harm. Plankton populations are threatened by physical obstructions, reduced nutrient absorption, and impaired health resulting from microplastic ingestion. The decline in plankton poses a significant risk to hydrophilic fauna and birds, as it serves as a primary food source for several organisms. Direct ingestion or predation on contaminated prey also exposes fish and mid-trophic organisms to microplastics. Beyond inducing oxidative stress and damaging internal organs, microplastics can disrupt behaviours such as feeding and reproduction efficiency. Moreover, studies indicate that exposure to microplastics negatively impacts fish reproductive systems, leading to declining population sizes (Wu et al., 2024). Existing pressures, such as overfishing and habitat degradation, are further exacerbated by the cascading effects of microplastic pollution on aquatic life. Large fish, marine animals, and seabirds, as apex predators, face heightened risks due to bioaccumulation and biomagnification of microplastics (Fig. 1). These particles adhere to their surfaces, acting as carriers for toxic contaminants such as hydrocarbons, heavy metals, and persistent organic pollutants (POPs). These pollutants accumulate in organisms at lower trophic levels and become more concentrated as

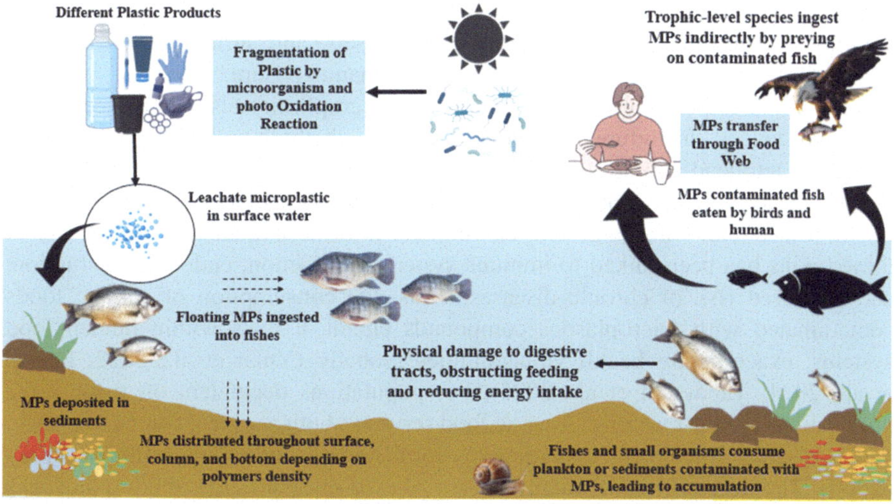

Fig. 1 Accumulation and transfer of microplastics in aquatic biomes: ecological impacts on Flora and Fauna, as well as food web disruptions

they move up the food chain. When apex predators consume contaminated prey, they are exposed to these toxins over time, potentially leading to immune system dysfunction, hormonal imbalances, and elevated mortality rates (Sruthisree et al., 2015; Amin et al., 2016, 2017; Prasad et al., 2023; Hasan et al., 2024).

Microplastics not only exert biological effects but also physically and chemically disrupt habitats. By suffocating species and blocking sunlight, they damage vital habitats such as coral reefs, sponges, seaweeds, and seagrass beds, alter sediment composition, and deteriorate water quality. Chemical additives, including phthalates and bisphenol A (BPA), leach into the water, acting as endocrine disruptors and intensifying environmental degradation (Wang et al., 2024). These multifaceted impacts jeopardize food security, environmental stability, and biodiversity, highlighting the urgent need for effective mitigation strategies.

3 Implications of Microplastic Contamination for Food and Nutrient Security

The contamination of microplastics in aquatic food products could pose a threat to food safety security and human health. Now it is regarded as a global issue due to its toxic effects on fish and humans. Fish, shellfish, and other seafood species frequently ingest microplastics directly from water or indirectly through prey, leading to their accumulation in the tissues of edible animals. Consequently, microplastics are introduced into human diets via fish food consumption (Smith et al., 2018). It reaches up to gastrointestinal tract which causes oxidative stress, cytotoxicity, and translocation to other tissues after absorbing. Studies have detected microplastics in commonly consumed species such as fish, mussels, and shrimp, raising serious concerns about long-term health impacts and human exposure.

In addition to being physical contaminants, microplastics also act as vectors for toxic substances, including bisphenol A (BPA), diethylhexyl phthalate, dibutyl phthalate, tetralead trioxide sulphate, nonylphenol, heavy metals, and POPs, etc. (Wang et al., 2024). These hazardous compounds adhere to the surface of microplastics and, when ingested, can transfer into the human body. Exposure to these toxins has been linked to immune system dysfunction, endocrine disruption, and increased risk of chronic diseases. Thus, the consumption of aquatic foods contaminated with microplastics compounds chemical risks within human food systems, exacerbating health vulnerabilities globally (Yuan et al., 2022; Prasad et al., 2024; Tabassum et al., 2025). For populations dependent on fishing, the implications are far-reaching. Aquatic food serves as both a critical source of protein and a key economic resource, particularly for fishermen communities in developing nations. The degradation of water resources due to microplastic pollution threatens not only nutritional security but also economic stability and livelihoods. Moreover, diminished confidence in the safety and quality of seafood could disrupt international trade, with cascading effects on both local economies and global markets. Addressing microplastic contamination is therefore essential to safeguard food security, public health, and economic resilience.

4 Human Health Implications of Microplastic Ingestion via Aquatic Food Consumption

Aquatic food (especially seafood) consumption has increasingly been linked to the ingestion of microplastics, raising significant concerns about potential human health impacts (Fig. 2). Studies have estimated that individuals consume thousands of microplastic particles annually through commonly consumed seafood, such as mussels and oysters. Research conducted in Europe has confirmed detectable levels of microplastics in these species, highlighting the extent of contamination in marine food products. In Asian countries, where seafood forms a dietary staple, higher levels of microplastic exposure have been correlated with an elevated risk of toxin-related health issues. Microplastics not only infiltrate human diets but also act as carriers for harmful substances. This raises serious concerns about the chronic health implications of consuming contaminated seafood, including endocrine disruption, immune system dysfunction, and increased susceptibility to diseases. These findings underscore the urgent need for research into the long-term effects of microplastic-laden seafood on human health. As the prevalence of microplastic contamination continues to rise, understanding its implications is critical to protecting public health and ensuring the safety of global food systems.

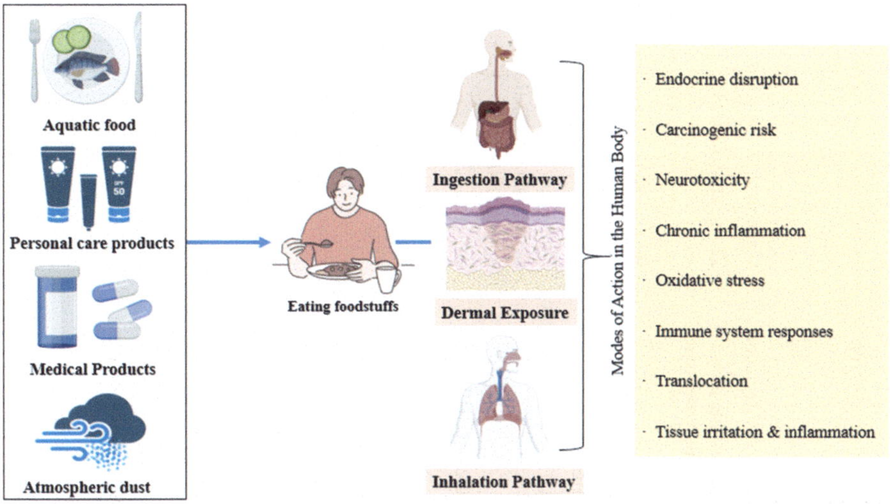

Fig. 2 Potential threat to human health due to environmental exposure to MPs

Aquatic food consumption is a significant route for human exposure to microplastics. By 2015, aquatic food accounted for 6.7% of total protein intake globally and approximately 17% of the protein derived from animal sources. Aquatic food consumption in the gastrointestinal tract (GIT) or inhalation into the lungs (Revel et al., 2018), with endocytosis being a common entry mechanism for both routes. Ingestion is the primary exposure pathway, where MPs pass through gaps in the villus tip epithelium to access the GIT circulatory system. Their translocation to secondary organs depends on particle size, hydrophobicity, surface charge, functionalization, and associated protein corona. Studies (Revel et al., 2018) reveal the presence of MPs (20 μm) in human blood and organs, although it remains uncertain if larger MPs are similarly transported. Research on polyethylene (PE) MPs suggests that MPs (<50 μm) may cross the GIT barrier and reach lymph nodes, liver, and spleen (Lithner et al., 2011), triggering inflammation, macrophage activation, and cytokine production. Additionally, MPs (<2.5 μm) can penetrate respiratory barriers, accumulating in lungs and linking air pollution to respiratory and cardiovascular toxicity (Liao et al., 2020).

5 Traditional Methods for Microplastic (MP) Mitigation Strategies

Microplastic mitigation strategies have traditionally relied on three primary methods mechanical (Advanced filtration systems), chemical (Certain chemical treatments), and biological (Bacteria and Fungi) each offering unique approaches to address microplastic pollution in the environment (Table 1);

Table 1 Summarized different types of microplastic degradation approaches

Strategy	Method	Degradation agents	Types of polymers degraded	Citation
Mechanical methods	Multi-separation approach (filtration, sieving, and grinding)	Physical filters, ultrafiltration systems	General microplastic removal; not specific to polymer type	Zhang et al. (2021)
Chemical methods	Oxidation, hydrolysis, Solvolysis	Acid/base catalysts, advanced oxidants (e.g., Fenton's reagent)	Polyethylene (PE), Polypropylene (PP), Polyethylene Terephthalate (PET)	Weber (2024) and De Clercq et al. (2017)
Thermal methods	Pyrolysis, thermal Depolymerization	High temperature and pressure	Polystyrene (PS), PET, and other thermoplastics	Kijo-Kleczkowska and Gnatowski (2022) and Kasar et al. (2020)
Bioremediation	Microbial degradation	Bacteria (e.g., Ideonella sakaiensis), Fungi (Aspergillus sp., Pestalotiopsis sp.)	PET, Polyurethane (PU), Polylactic Acid (PLA)	Ali et al. (2021)
In vivo degradation	Degradation by insect	Lesser waxworm (*Achroia grisella*), dark (*Tenebrio obscurus*) and yellow (*Tenebrio molitor*) mealworms.	High-density polyethylene (HDPE), PS	Kundungal et al. (2019) and Peng et al. (2019)
Enzymatic degradation	Enzyme-based treatment	PETase, Cutinase, Lipase	PET, PU, PLA	Tournier et al. (2023) and Sahu et al. (2023)
Photo-degradation	UV light exposure	UV radiation, solar radiation	PE, PP, PS	Andrady et al. (2022)
Advanced treatments	Nanotechnology-assisted degradation	Metal oxide nanoparticles (e.g., TiO2, ZnO), Catalysts	Various thermoplastics and elastomers	Remya et al. (2022)
Biodegradable polymers	Synthetic bioplastics	Microorganisms, natural enzymes	PLA, Polyhydroxyalkanoates (PHA), Starch-Based Plastics	Meereboer et al. (2020)

Mechanical and Physical Methods (Filtration and Screening Techniques) Filtration and screening are widely recognized mechanical and physical methodologies employed to mitigate microplastic pollution, particularly in industrial and wastewater treatment facilities. These systems utilize advanced technologies such as membrane technology, sand filters, and mesh screens of different sizes to capture MP particles effectively. By efficiently intercepting larger plastic fragments, these techniques significantly reduce their release into aquatic ecosystems. For surface water cleanup and sediment remediation, sieving techniques are often deployed to remove visible plastic debris (Ahmed et al., 2024).

Filtration techniques for microplastic removal involve different filter media, equipment, and supportive tools. Filter media include microfilters (pore sizes 0.1–10 μm), ultrafiltration membranes (0.01–0.1 μm), nanofilters for extremely fine particles, sand filters for larger plastic particles, and activated carbon filters for adsorbing smaller particles. Filtration equipment includes rotary drum filters for continuous industrial filtration, plate, and frame filter presses for high-pressure filtration, bag filters for trapping microplastics in flow streams, cartridge filters with precise pore sizes, and disc filters for compact multi-layered water treatment. Supportive tools like vacuum filtration systems are used in lab-scale research, while high-pressure filters are applied in municipal and industrial water treatment systems.

Screening techniques for plastic removal involve the use of specialized screens, nets, and equipment. Fine mesh screens (stainless steel or nylon) with sizes down to 0.05 mm, grizzly screens for coarse plastics, polyurethane screens for durability, trommel screens for larger pieces, and micro-nets for field research are commonly used. Screening equipment includes vibrating screens for size separation, drum screens (trommels) for bulk removal, disc screens for fine particles, hydrocyclones for smaller microplastics, and air classifiers that use airflow to separate lightweight plastics. Field sampling tools like manta trawls for surface microplastics, neuston nets for surface-level sampling, plankton nets for small particles, and bongo nets for depth-specific sampling are essential for research and monitoring.

Combined systems for plastic removal incorporate multiple methods to enhance efficiency. Multistage filtration systems combine coarse screening, sand filtration, and membrane filtration to target plastics of varying sizes. Electrostatic filtration uses electrical charges to capture microplastics effectively. Hybrid systems integrate biological treatment with physical filtration, offering a comprehensive approach to plastic removal from water.

Despite their efficacy, these methods exhibit limitations, particularly in addressing nanoscale polymers. Conventional filtration devices often fail to retain microplastics smaller than a few micrometres due to their minute size and buoyant properties. Moreover, in complex aquatic environments where microplastics are embedded in sediments or dispersed in deep-water zones, sieving methods prove less effective. Although reverse osmosis and nanofiltration are examples of advanced filtration technologies that have increased the effectiveness of microplastic removal, they are frequently associated with significant operating costs and energy demands. These drawbacks underscore the critical need for complementary and innovative solutions capable of targeting nanoscale polymers while minimizing the environmental and economic impacts of existing approaches (Poerio et al., 2019). Addressing these challenges is essential to developing sustainable, scalable strategies for combating microplastic pollution, particularly in resource-limited regions.

Chemical Treatment Methods Chemical treatments offer innovative pathways for addressing microplastic pollution by targeting, degrading, or eliminating plastic particles. Coagulation, a widely employed method, uses chemical agents to aggregate microplastic particles into larger clusters that are more easily removed through filtration or sedimentation. Advanced oxidation processes (AOPs), leveraging highly reactive species such as hydroxyl radicals, break down microplastics into simpler molecules or smaller fragments (Jain et al., 2023). These methods have demonstrated particular effectiveness in industrial wastewater and effluent treatment.

However, chemical approaches come with substantial environmental trade-offs. Coagulation depends on chemical additives, which, if mismanaged, can introduce secondary pollutants into aquatic habitats. Similarly, advanced oxidation techniques, while effective, are energy-intensive and may produce hazardous byproducts during degradation (De Clercq et al., 2017). The high cost of chemical treatments further limits their accessibility and scalability, particularly in low-income regions. Although these approaches hold promise, their environmental and economic implications necessitate a balanced approach, integrating them with complementary strategies to reduce unforeseen repercussions.

6 Bioremediation: A Sustainable Approach for Environmental Cleanup

Bioremediation is the biological process that uses different biological agents (bacteria, fungi, and plants) to eliminate or reduce the concentrations of hazardous pollutants. It is an environmentally friendly and cost-effective alternative to conventional chemical and mechanical cleanup methods. This approach capitalizes on natural metabolic pathways to break down contaminants such as hydrocarbons, pesticides, and heavy metals. Key mechanisms include enzymatic reactions, symbiotic interactions within ecosystems, and microbial activity under aerobic or anaerobic conditions. Historically, bioremediation has been effectively employed in various scenarios, including wastewater treatment, remediation of contaminated soils, and oil spill cleanup. Notable examples include the use of hydrocarbon-degrading bacteria to mitigate the Exxon Valdez oil spill and phytoremediation techniques to extract heavy metals from polluted soils. While these successes underscore the potential of bioremediation, its application in addressing microplastic pollution remains in its early stages. Recent advancements in microbial ecology and molecular biology are enhancing our understanding of how bioremediation might be adapted to target microplastics. Emerging research on microbial consortia capable of degrading synthetic polymers presents promising opportunities for tackling this pervasive environmental challenge. As this field evolves, bioremediation could play a pivotal role in developing innovative and sustainable solutions to mitigate microplastic contamination.

In Vivo, Degradation of Microplastics Microplastic pollutants are highly persistent in the environment, with degradation spanning decades to centuries due to their resilient structural properties. However, certain organisms in nature can degrade MPs, offering a natural pathway for mitigation. While often overlooked, these organisms are now gaining attention for their potential role in addressing MP pollution. Recent studies have highlighted in vivo degradation by specific species as a promising solution. For example, *Achroia grisella* (lesser waxworm) has been shown to degrade high-density polyethylene (HDPE) effectively. Remarkably, this process does not disrupt the waxworm's normal life cycle, suggesting its feasibility for practical applications (Kundungal et al., 2019). Exploring the mechanisms of in vivo degradation could provide sustainable and ecologically integrated solutions to combat plastic pollution. Further research into these natural processes is essential to unlock their full potential for mitigating MP contamination.

Microbial Degradation of Microplastics Microorganisms, including fungi and bacteria, play a vital role in the degradation of microplastics. Enzymes secreted by these organisms break down polymer chains into smaller fragments, which can then be metabolized as energy sources. Certain fungi, such as *Aspergillus* and *Penicillium* species, and bacteria, like *Ideonella sakaiensis*, have demonstrated remarkable potential in degrading plastics such as polyethylene terephthalate (PET) and polyethylene (PE) (Ali et al., 2021). Furthermore, microbial consortia diverse communities of microorganisms frequently exhibit enhanced degradation capabilities due to synergistic interactions among species. *Ideonella sakaiensis* has been found to significantly degrade PET films in laboratory experiments. Similarly, marine fungi *Aspergillus sp*. and *Zalerion maritimum* (Zhou et al., 2024) can degrade plastics in aquatic environments.

Enzymatic Degradation Microplastics Enzymatic degradation aims to break down plastic polymers into smaller, biodegradable components using specialized enzymes. Enzymes such as PETase and Leaf-branch Compost Cutinase (LCC) are targeting polyethylene terephthalate (PET), a widely used plastic. These enzymes catalyze the hydrolysis of polymer chains, converting plastics into monomers that can be either recycled into new materials or further degraded by microbial action. Due to its efficiency and selectivity, enzymatic degradation presents a promising approach for managing plastic waste. Advancements in enzymology are driving progress in this field. Techniques such as protein engineering and directed evolution have enhanced the stability and activity of enzymes (PETase), enabling them to target a broader range of plastic types under diverse environmental conditions (Tournier et al., 2023). Despite these breakthroughs, challenges remain, including the high production costs of enzymes and the scalability of enzymatic processes for large-scale applications. To fully realize the potential of enzymatic degradation in mitigating plastic pollution, further research and technological innovations are essential. Addressing these obstacles will pave the way for integrating enzymatic solutions into sustainable waste management systems.

7 Biodegradable Polymers: A Sustainable Alternative

Compared to traditional plastics, biodegradable polymers are specifically designed to decompose more rapidly and naturally, reducing their environmental persistence. Examples such as polylactic acid (PLA) and polyhydroxyalkanoates (PHAs) are engineered to break down under specific conditions, including composting and microbial activity. These materials offer improved degradability while retaining properties comparable to conventional plastics, making them ideal for applications in packaging, agriculture, and medicine.

Biodegradable polymers also enhance natural degradation processes by providing substrates that microbes and enzymes can more easily metabolize. However, their degradation efficiency is influenced by environmental factors such as temperature, humidity, and microbial activity. Advances in polymer formulation, such as incorporating natural fibres or additives, have shown promise in accelerating degradation and improving compatibility with existing waste management systems. As the demand for sustainable materials grows, biodegradable polymers are poised to play a crucial role in reducing plastic pollution and supporting the transition to a circular economy. Continued innovation and integration into global waste management strategies will further their potential to address environmental challenges.

8 Factors Influencing the Bioremediation Process

The efficiency and effectiveness of bioremediation are governed by several interconnected factors that influence the metabolic activities of microorganisms and their ability to degrade pollutants. These factors can be broadly categorized into environmental conditions, Microplastic Characteristics, and adaptation of biological agents.

Environmental Factors Environmental factors have a big impact on how well bioremediation procedures that target microplastics work. Microbial activity and enzymatic efficiency are significantly influenced by variables such as temperature, pH, and nutrition availability. Ideal temperatures accelerate the breakdown of microplastics by increasing the metabolic rates of microorganisms. Since many microbial species thrive in particular pH ranges can also affect the stability and activity of enzymes involved in polymer breakdown. Carbon, nitrogen, and phosphorus are among the nutrients that are available to support microbial development and their capacity to generate the enzymes needed for breakdown. Microbial colonization and activity can be impacted by variations in salinity, oxygen concentrations, and water movement. For example, poor nutrient availability and excessive salt concentrations in marine environments may restrict microbial efficiency.

Microplastic Characteristics The properties of microplastics, including their size, shape, surface area, structure, and polymer composition, play a critical role in determining their biodegradation rate. Polymers such as polyethylene (PE) and polypropylene (PP) are highly resistant to microbial degradation due to their large molecular weight and hydrophobic properties. In contrast, materials like polyethylene terephthalate (PET) and polylactic acid (PLA), which contain ester linkages, are more susceptible to enzymatic breakdown, making them more amenable to bioremediation efforts.

In addition to polymer type, chemical additives incorporated into plastics—such as stabilizers, colorants, and plasticizers can influence degradation rates. These additives may either facilitate or inhibit microbial activity, depending on their nature and concentration. Consequently, plastics containing such additives can exhibit varying levels of biodegradability. The physical attributes of microplastics significantly affect how biological agents interact with them. Smaller microplastics, with their larger surface-area-to-volume ratio, are more readily colonized by microorganisms and are more accessible to enzymatic degradation. Furthermore, irregularly shaped or rough particles provide greater surface area for microbial attachment compared to smooth, spherical particles. The formation of biofilms on microplastic surfaces also plays a dual role in the degradation process biofilms can either hinder microbial degradation by acting as a protective layer or promote it by fostering microbial activity. To optimize bioremediation strategies, it is essential to tailor techniques that consider these diverse physical and chemical properties of microplastics.

Adaptation of Biological Agents The success of bioremediation efforts to address microplastic pollution largely depends on the adaptation of biological agents, particularly microorganisms. Microbial consortia, communities of diverse microorganisms, often evolve to enhance their capacity for plastic degradation. Through natural selection and environmental exposure, microorganisms can learn to metabolize specific polymers or adapt to challenging conditions such as elevated salinity or nutrient scarcity. The diversity within microbial consortia fosters synergistic interactions, wherein multiple species collaborate to break down complex plastic polymers into simpler, more biodegradable compounds.

To further enhance microbial degradation efficiency, two promising strategies are bioaugmentation and genetic engineering. Genetic modification can augment the production of plastic-degrading enzymes or expand the range of polymers that microorganisms can degrade. Bioaugmentation involves introducing naturally occurring or engineered microbial strains into contaminated environments to accelerate plastic breakdown. Additionally, immobilizing microorganisms on substrates like alginate beads or activated charcoal can boost their activity and stability across diverse environmental conditions. These approaches offer significant potential for improving bioremediation processes and addressing the global challenge of microplastic pollution, particularly when integrated with further research into microbial adaptation mechanisms.

9 Different Challenges in the Bioremediation Process

Although microplastics have numerous detrimental effects across various environmental sectors, their management remains a formidable challenge. Every stage of MP monitoring ranging from sampling and extraction to isolation and detection presents significant obstacles to large-scale surveillance efforts. Several significant challenges are associated with the identification, quantification, and management of microplastics, as illustrated in Fig. 3.

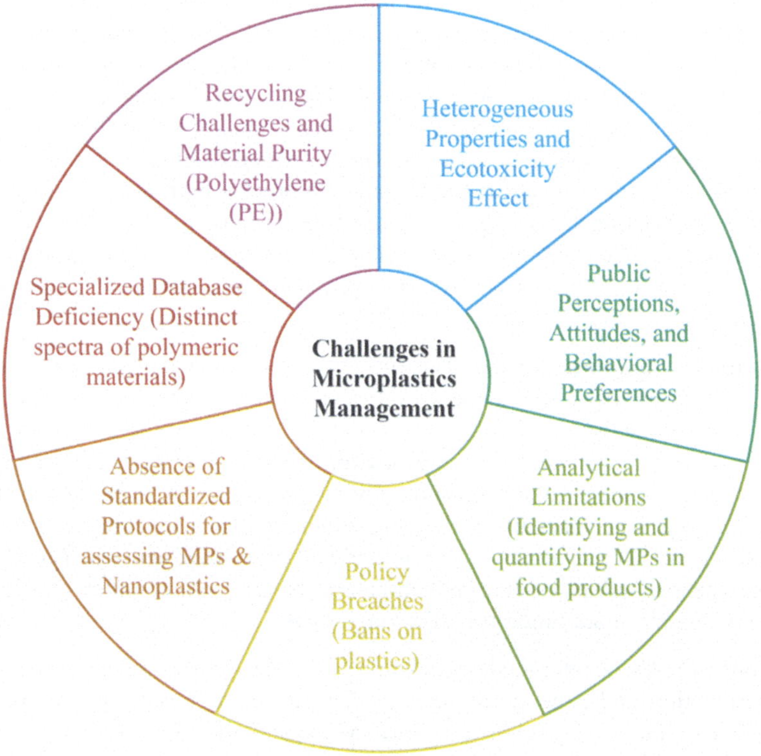

Fig. 3 Different key challenges in microplastics (MPs) management

10 Conclusion

Microplastic pollution has emerged as a critical environmental issue, significantly affecting human health, food security, and ecosystems. These minute plastic particles infiltrate both terrestrial and aquatic environments, disrupting food chains and causing toxic accumulation that impacts all forms of life, from humans to apex predators and plankton. Traditional mitigation methods, such as mechanical filtration, chemical treatments, and regulatory measures, have demonstrated limited effectiveness, highlighting the urgent need for innovative and sustainable solutions to address this growing crisis. Bioremediation, utilizing microorganisms and enzymes to break down microplastics into harmless byproducts, offers a promising ecological restoration approach. Leveraging natural processes, this method is both environmentally friendly and adaptable to various ecosystems. The potential of bioremediation is further bolstered by advances in biotechnology, including the development of biodegradable polymers and enzyme engineering. While challenges related to efficiency and scalability persist, this approach can be optimized by integrating emerging technologies, such as nanotechnology and hybrid mitigation strategies. To ensure a sustainable future, it is essential to reduce plastic production, promote recycling, and transition towards a circular economy. Major initiatives involve such as increasing research funding, fostering community involvement, and strengthening policy frameworks. By prioritizing bioremediation and systemic reforms, humanity can mitigate microplastic pollution, restore ecosystems, and secure food and environmental safety for generations to come.

References

Ahmed, S. F., Islam, N., Tasannum, N., Mehjabin, A., Momtahin, A., Chowdhury, A. A., & Mofijur, M. (2024). Microplastic removal and management strategies for wastewater treatment plants. *Chemosphere, 347*, 140648. https://doi.org/10.1016/j.chemosphere.2023.140648

Ali, S. S., Elsamahy, T., Al-Tohamy, R., Zhu, D., Mahmoud, Y. A. G., Koutra, E., & Sun, J. (2021). Plastic wastes biodegradation: Mechanisms, challenges and future prospects. *Science of The Total Environment, 780*, 146590. https://doi.org/10.1016/j.scitotenv.2021.146590

Amin, A., Indulkar, S. T., & Pai, R. (2016). Sublethal effect of Buprofezin pesticide on carcass composition of *Cyprinus carpio* communis fingerlings. *Journal of Experimental Zoology, India, 20*(1), 269–272.

Amin, A., Ramachandra Naik, A. T., Priyadarshini, N., Nayak, H., & Sree, C. S. (2017). Toxic effect of heavy metal lead on oxygen consumption of rohu (*Labeo rohita*) fingerlings. *Biochemical & Cellular Archives, 17*(1), 225–228.

Andrady, A. L., Barnes, P. W., Bornman, J. F., Gouin, T., Madronich, S., White, C. C., & Jansen, M. A. K. (2022). Oxidation and fragmentation of plastics in a changing environment; from UV-radiation to biological degradation. *Science of the Total Environment, 851*, 158022. https://doi.org/10.1016/j.scitotenv.2022.158022

Arthur, C., Baker, J. E., & Bamford, H. A. (2009). *Proceedings of the international research workshop on the occurrence, effects, and fate of microplastic marine debris, September 9–11, 2008*. University of Washington Tacoma. NOAA technical memorandum NOS-OR&R 30.

De Clercq, R., Dusselier, M., & Sels, B. F. (2017). Heterogeneous catalysis for bio-based polyester monomers from cellulosic biomass: advances, challenges and prospects. *Green Chemistry, 19*(21), 5012–5040. https://doi.org/10.1039/C7GC02040F

Debnath, R., Prasad, G. S., Amin, A., Malik, M. M., Ahmad, I., Abubakr, A., Borah, S., Rather, M. A., Impellitteri, F., Tabassum, I., & Piccione, G. (2024). Understanding and addressing microplastic pollution: Impacts, mitigation, and future perspectives. *Journal of Contaminant Hydrology, 266*, 104399.

De-la-Torre, G. E. (2020). Microplastics: an emerging threat to food security and human health. *Journal of Food Science and Technology, 57*(5), 1601–1608. https://doi.org/10.1007/s13197-019-04138-1

Gurjar, UR., Martin,K.A., Xavier, S. P., Shukla, S., Takar, A.K., Jaiswar, G., Deshmukhe, G., Nayak, B. B., (2023) Seasonal distribution and abundance of microplastics in the coastal sediments of north eastern. *Arabian Sea Marine Pollution Bulletin, 187*, 114545. https://doi.org/10.1016/j.marpolbul.2022.114545

Hale, R. C., Seeley, M. E., La Guardia, M. J., Mai, L., & Zeng, E. Y. (2020). A global perspective on microplastics. *Journal of Geophysical Research: Oceans, 125*(1), e2018JC014719. https://doi.org/10.1029/2018JC014719

Hasan, A. M., Hamed, M., Hasan, J., Martyniuk, C. J., Niyogi, S., & Chivers, D. P. (2024). A review of the neurobehavioural, physiological, and reproductive toxicity of microplastics in fishes. *Ecotoxicology and Environmental Safety, 282*, 116712. https://doi.org/10.1016/j.ecoenv.2024.116712

Huang, W., Song, B., Liang, J., Niu, Q., Zeng, G., Shen, M., & Zhang, Y. (2021). Microplastics and associated contaminants in the aquatic environment: A review on their ecotoxicological effects, trophic transfer, and potential impacts to human health. *Journal of Hazardous Materials, 405*, 124187. https://doi.org/10.1016/j.jhazmat.2020.124187

Jain, R., Gaur, A., Suravajhala, R., Chauhan, U., Pant, M., Tripathi, V., & Pant, G. (2023). Microplastic pollution: Understanding microbial degradation and strategies for pollutant reduction. *Science of The Total Environment, 905*, 167098. https://doi.org/10.1016/j.scitotenv.2023.167098

Kasar, P., Sharma, D. K., & Ahmaruzzaman, M. (2020). Thermal and catalytic decomposition of waste plastics and its co-processing with petroleum residue through pyrolysis process. *Journal of Cleaner Production, 265*, 121639. https://doi.org/10.1016/j.jclepro.2020.121639

Kijo-Kleczkowska, A., & Gnatowski, A. (2022). Recycling of plastic waste, with particular emphasis on thermal methods. *Energies, 15*(6), 2114. https://doi.org/10.3390/en15062114

Kundungal, H., Gangarapu, M., Sarangapani, S., Patchaiyappan, A., & Devipriya, S. P. (2019). Efficient biodegradation of polyethylene (HDPE) waste by the plastic-eating lesser waxworm (Achroia grisella). *Environmental Science and Pollution Research, 26*, 18509–18519. https://doi.org/10.1007/s11356-019-05038-9

Lamichhane, G., Acharya, A., Marahatha, R., Modi, B., Paudel, R., Adhikari, A., & Parajuli, N. (2023). Microplastics in environment: global concern, challenges, and controlling measures. *International Journal of Environmental Science and Technology, 20*(4), 4673–4694. https://doi.org/10.1007/s13762-022-04261-1

Liao, Y. L., & Yang, J. Y. (2020). Microplastic serves as a potential vector for Cr in an in-vitro human digestive model. *Science of the Total Environment, 703*, 134805. https://doi.org/10.1016/j.scitotenv.2019.134805

Lithner, D., Larsson, Å., & Dave, G. (2011). Environmental and health hazard ranking and assessment of plastic polymers based on chemical composition. *Science of the total environment, 409*(18), 3309–3324. https://doi.org/10.1016/j.scitotenv.2011.04.038

Meereboer, K. W., Misra, M., & Mohanty, A. K. (2020). Review of recent advances in the biodegradability of polyhydroxyalkanoate (PHA) bioplastics and their composites. *Green Chemistry, 22*(17), 5519–5558. https://doi.org/10.1039/D0GC01647K

Nabi, N., Ahmad, I., Amin, A., Rather, M. A., Ahmed, I., Hajam, Y. A., Khursheed, S., Malik, M. M., & Abubakr, A. (2024). Understanding the sources, fate and effects of microplastics in aquatic environments with a focus on risk profiling in aquaculture systems. *Reviews in Aquaculture, 16*(4), 1947–1980.

Nawalage, N. S. K., & Bellanthudawa, B. K. A. (2022). Synthetic polymers in personal care and cosmetics products (PCCPs) as a source of microplastic (MP) pollution. *Marine Pollution Bulletin, 182*, 113927. https://doi.org/10.1016/j.marpolbul.2022.113927

Peng, B. Y., Su, Y., Chen, Z., Chen, J., Zhou, X., Benbow, M. E., & Zhang, Y. (2019). Biodegradation of polystyrene by dark (Tenebrio obscurus) and yellow (Tenebrio molitor) mealworms (Coleoptera: Tenebrionidae). *Environmental Science & Technology, 53*(9), 5256–5265. https://doi.org/10.1021/acs.est.8b06963

Poerio, T., Piacentini, E., & Mazzei, R. (2019). Membrane processes for microplastic removal. *Molecules, 24*(22), 4148. https://doi.org/10.3390/molecules24224148

Prasad, G. S., Rout, S. K., Malik, M. M., Karmakar, S., Amin, A., & Ahmad, I. (2023). Occurrence of xenoestrogen alkylphenols (Octylphenols and Nonylphenol) and its impact on the aquatic ecosystem. In M. A. Rather, A. Amin, Y. A. Hajam, A. Jamwal, & I. Ahmad (Eds.), *Xenobiotics in aquatic animals* (pp. 275–284). Springer.

Prasad, G. S., Rout, S. K., Amin, A., Malik, M. M., Ahmad, I., Rather, M. A., Karmakar, S., Khalil, H. S., & Abubakar, A. (2024). Impact of COVID-19 on the aquatic environment and implications for fisheries with special emphasis to Indian perspective. *Proceedings of the Indian National Science Academy, 90*(4), 821–839.

Quye, A. (2017). Quality matters for historical plastics: The past-making of cellulose nitrates for future preservation. *Cahiers François Viète, III-2*, 45–65. https://doi.org/10.4000/cahierscfv.799

Remya, R. R., Julius, A., Suman, T. Y., Mohanavel, V., Karthick, A., Pazhanimuthu, C., & Muhibbullah, M. (2022). Role of nanoparticles in biodegradation and their importance in environmental and biomedical applications. *Journal of Nanomaterials, 2022*(1), 6090846. https://doi.org/10.1155/2022/6090846

Revel, M., Châtel, A., & Mouneyrac, C. (2018). Micro (nano) plastics: A threat to human health? *Current Opinion in Environmental Science & Health, 1*, 17–23. https://doi.org/10.1016/j.coesh.2017.10.003

Sahu, S., Kaur, A., Khatri, M., Singh, G., & Arya, S. K. (2023). A review on cutinases enzyme in degradation of microplastics. *Journal of Environmental Management, 347*, 119193. https://doi.org/10.1016/j.jenvman.2023.119193

Smith, M., Love, D. C., Rochman, C. M., & Neff, R. A. (2018). Microplastics in seafood and the implications for human health. *Current Environmental Health Reports, 5*, 375–386. https://doi.org/10.1007/s40572-018-0206-z

Sruthisree, C., Gowda, G., Nayak, H., Harish, B. D., & Amin, A. (2015). Effect of heavy metal (lead) on oxygen consumption and ammonia excretion of Iridescent Shark (*Pangusius hypohthalmus*). *Journal of Experimental Zoology, India, 18*(2).

Tabassum, I., Hassan, N., Amin, A., Malik, M. M., Abubakr, A., Prasad, G. S., Latief, F., & Rather, S. A. (2025). COVID-19 and global environmental change: A threat to food security and nutrition. In *Food security, nutrition and sustainability through aquaculture technologies* (pp. 437–445). Springer.

Tournier, V., Duquesne, S., Guillamot, F., Cramail, H., Taton, D., Marty, A., & André, I. (2023). Enzymes' power for plastics degradation. *Chemical Reviews, 123*(9), 5612–5701. https://doi.org/10.1021/acs.chemrev.2c00644

Wang, Z., Liang, G., Jiang, S., Wang, F., Li, H., Li, B., & Gong, W. (2024). Understanding the environmental impact and risks of organic additives in plastics: a call for sustained research and sustainable solutions. *Emerging Contaminants, 10*, 100388. https://doi.org/10.1016/j.emcon.2024.100388

Weber, R. S. (2024). Electrochemical upvaluing of waste plastic. *Current Opinion in Electrochemistry, 46*, 101493. https://doi.org/10.1016/j.coelec.2024.101493

Wu, B., Yu, H., Yi, J., Lei, P., He, J., Ruan, J., & Zhang, X. (2024). Behavioral studies of zebrafish reveal a new perspective on the reproductive toxicity of micro-and nanoplastics. *Toxics, 12*(3), 178. https://doi.org/10.3390/toxics12030178

Yuan, Z., Nag, R., & Cummins, E. (2022). Human health concerns regarding microplastics in the aquatic environment-From marine to food systems. *Science of the Total Environment, 823*, 153730. https://doi.org/10.1016/j.scitotenv.2022.153730

Zhang, Y., Jiang, H., Bian, K., Wang, H., & Wang, C. (2021). A critical review of control and removal strategies for microplastics from aquatic environments. *Journal of Environmental Chemical Engineering, 9*(4), 105463. https://doi.org/10.1016/j.jece.2021.105463

Zhou, J., Chen, M., Li, Y., Wang, J., Chen, G., & Wang, J. (2024). Microbial bioremediation techniques of microplastics and nanoplastics in the marine environment. *TrAC Trends in Analytical Chemistry, 180*, 117971. https://doi.org/10.1016/j.trac.2024.117971

Navigating the Plastic Crisis: Global Legislation and Policy Responses to Plastic Waste

Naila M. Bhat, Ifrah Rashid, Mohd Ashraf Malik, Ubair Nisar, Saba Nabi, Mohammad Talib, Gowhar Iqbal, and Shakir Ahmad Mir

1 Introduction

Plastic pollution ranks as the most significant environmental issues the twenty-first century has grappled with. The widespread use of plastic in daily life, combined with its durability and resistance to degradation, has led to widespread accumulation of plastic waste in both terrestrial and aquatic ecosystems. In terrestrial ecosystems, plastic pollution contaminates soil through improper disposal, municipal waste, and agricultural activities. This contamination affects soil health and biodiversity, as microplastics disrupt soil organisms like earthworms and potentially harm plants, microorganisms, and insects, leading to ecological imbalances (Chae & An, 2018; Debnath et al., 2024). Plastic pollution affects both ecosystems and human health through the release of harmful chemicals into soil and water, contaminating food and water supplies. Single-use plastics and microplastics harm wildlife by causing ingestion, entanglement, and poisoning, thereby disrupting ecosystems and biodiversity. Moreover, chemical additives in plastics can leach into the environment, which may affect human health through absorption or consumption (Reddy et al.,

N. M. Bhat · I. Rashid
Faculty of Fisheries, Sher-e- Kashmir University of Science and Technology, Baramulla, Kashmir, India

M. A. Malik (✉)
College of Fisheries Science, Birsa Agricultural University, Gumla-835207, India

U. Nisar
Business and Policy Research (IBPR), Sher-e- Kashmir University of Science and Technology, Baramulla, Kashmir, India

S. Nabi
College of Fisheries, Kishanganj, Bihar, India

M. Talib · G. Iqbal · S. A. Mir
ICAR-Central Institute of Fisheries Education, Mumbai, India

A. Abubakr et al. (eds.), *Microplastics: Ecological and Food Security Implications*,
https://doi.org/10.1007/978-3-031-94480-2_16

2014; Gurjar et al., 2023; Nabi et al., 2024). In marine ecosystems, plastic pollution occurs through entanglement, ingestion, and the toxic effects of plastic on marine life. These disruptions cause starvation, suffocation, and significant biodiversity loss. Microplastics are found to accumulate in water, sediments, and organisms, thereby threatening food webs and essential ecosystem services. Besides environmental impacts, plastic pollution also affects socio-economic factors like fisheries, tourism, and human health, which requires immediate mitigation efforts (Thushari & Senevirathna, 2020).

A crisis is a sudden event requiring urgent attention, threatening high-priority goals, and surprising decision-makers with limited time for response (Ladi & Tsarouhas, 2020). Crises often prompt collective action and can drive positive change but may also be exploited politically, disrupting government routines and policy agendas (McConnell, 2020a). While some crises emerge suddenly, others, known as "creeping crises," develop gradually over time. Marine plastic pollution is one example of a creeping crisis-a "wicked problem" characterized by its lack of clear start or end (Mæland & Staupe-Delgado, 2020; Vince & Stoett, 2018; Nielsen et al., 2020). While urgent, plastic pollution has been delayed in its management due to slow adoption and implementation of policies (Belontz et al., 2019; Landon-Lane, 2018). Plastic pollution grows because oceanic plastic accumulation tracks global production volumes well (Wilcox et al., 2020). Efforts to reduce plastic waste in waterways have largely failed, underscoring the urgency of the issue (Borrelle et al., 2020). However, governments often frame the breaking point of this crisis as a future concern, reflecting a preference for low-risk strategies in managing complex problems (McConnell, 2018).

The global plastic recycling system is inadequate, with less than 10% of plastics being recycled. Industries often justify increasing plastic production under the guise of recycling, producing cheaper virgin material over recycled material (Sullivan, 2020; Parashar & Hait, 2021). Governments often make policy traps whereby they adopt the most superficial measures appearing proactive but which do not actually have the resource or commitment needed to solve them (McConnell, 2020b). The COVID-19 pandemic has heightened plastic pollution since single-use plastics and personal protective equipment for health and safety have derailed the sustainable efforts of humankind (Johansson, 2021; Silva et al., 2020, 2021; Hu et al., 2022).

Plastic politics covers every stage of its life cycle encompassing it's production, consumption, waste management and environment pollution. Investments in conventional feedstocks expand the fossil fuel-dependent plastic production process. Meanwhile, other materials such as bio-based plastics encounter economic and sociocultural hurdles. Reducing reliance on single-use plastics is thus being approached in three ways: *changing behavior, changing regulation, and changing products.* However, waste management continues to be an important issue, for recycling rates have not kept pace with growth in plastic production. Down-cycling and the controversies over global waste trade remain significant hurdles. Marine plastic pollution, in particular, attracts the most attention due to its wide environmental, economic, and health implications, which underscores the need for integrated policies on plastic flows holistically.

Currently, there is no centralized global authority that deals with the more comprehensive plastic problem. Most of the efforts focus on marine pollution through initiatives like the Global Partnership on Marine Litter (GPML) and the Joint Group of Experts on the Scientific Aspects of Marine Environmental Protection (GESAMP).

The lack of consensus on defining "the plastic problem" leads to varied focus areas, solutions, and sustainability visions. Prominent approaches include promoting a circular economy, developing bioplastics, and reducing plastic use alongside other emerging strategies (Nielsen et al., 2020).

2 Global Legislative Frameworks

Various agreements and frameworks have been established to address the escalating plastic pollution crisis, reflecting global efforts to mitigate its impacts. These include international treaties, regional strategies, and national policies aimed at reducing plastic waste and promoting sustainable management practices. Some of the international agreements have been given in Fig. 1.

3 Multilateral Environmental Agreements

Multilateral Environmental Agreements (MEAs) are international treaties where several countries come together to tackle shared environmental issues, whether on a global or regional scale. They provide a framework for cooperation, setting legally binding commitments to protect and manage environmental resources, biodiversity, and ecosystems. Multilateral Environmental Agreements (MEAs) addressing plastic pollution focus primarily on regulating waste management, marine pollution, and the lifecycle impacts of plastics (English, 2023).

3.1 Basel Convention

The Basel Convention, established in 1989 and enforced since 1992, is a global agreement with almost universal participation. Its primary goal is to safeguard human health and the environment from the dangers posed by hazardous waste. It regulates transboundary movements of toxic, flammable, infectious, and other dangerous wastes and ensures their environmentally sound management and disposal. Parties are obligated to minimize waste generation, transport, and ensure disposal near the source to reduce environmental impacts. In 2019, the convention covered plastic wastes within its governing scope and regulated international movements for environmental protection purposes (UNEP, 2011).

Fig. 1 Agreements on plastic pollution at global level

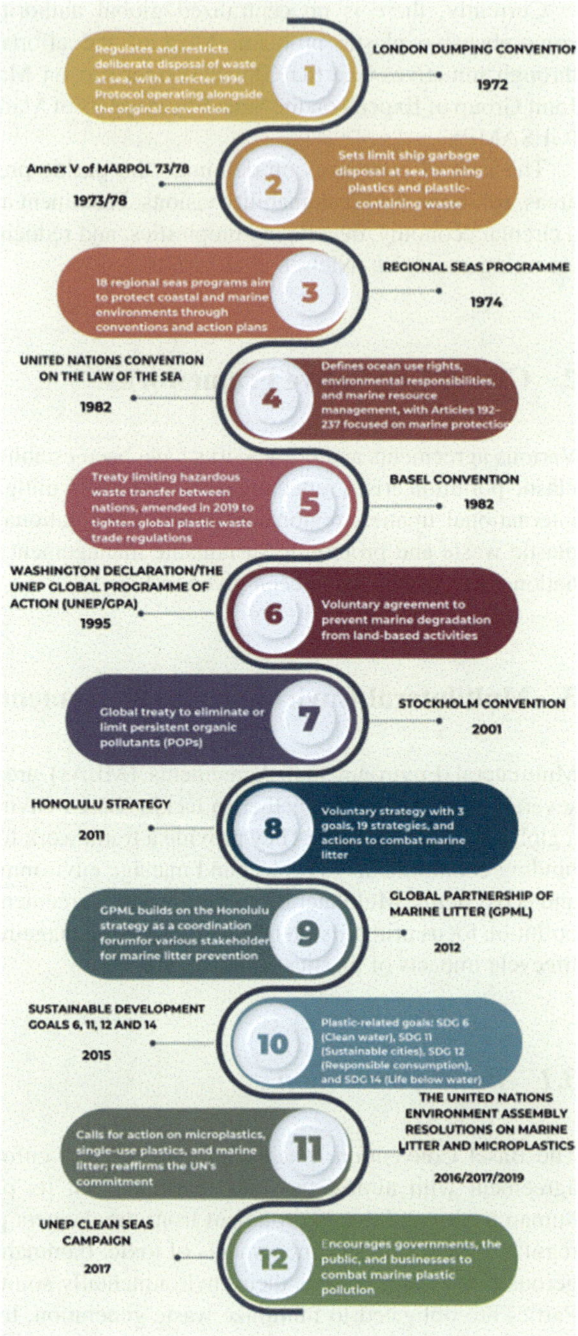

3.2 United Nations Environment Assembly Resolution

The resolution of the United Nations Environment Assembly on plastic pollution represents global will and action towards addressing the issue. The UNEP report *Turning off the Tap: How the World Can End Plastic Pollution and Create a Circular Economy* highlights that plastic pollution can be reduced by 80% by 2040 through three key market shifts: *reuse, recycling, and reorienting production toward alternatives*. Such measures as promoting refillable containers, improving recycling systems by removing fossil fuel subsidies, and replacing problematic plastics with sustainable materials can significantly cut waste. However, managing 100 million metric tons of residual plastics and legacy pollution will involve setting design and safety standards and enforcing producer responsibility for microplastics; indeed, mobilizing investments into circular infrastructure, supported by policies such as levies on virgin plastics and *Extended Producer Responsibility* schemes, might be some of these steps (IISD, 2023; Islam et al., 2024).

3.3 Global Plastic Treaty

As of March 2022, the UN Environmental Assembly met in Nairobi, Kenya, in a discussion over the global plastic crisis. In a historic turn of events, 175 countries voted to have a global agreement on plastic pollution-agreeing on an expedited timeline such that the agreement could be into effect as soon as 2025 (WWF, 2023). About 175 UN Member States adopted Resolution *5/14 "End Plastic Pollution: Towards an International Legally Binding Instrument*, mandating creation of a legally binding treaty. This has been supported by initiatives such as the High Ambition Coalition to End Plastic Pollution by 2040, which negotiations aim to focus on the lifecycle of plastics, lessons drawn from previous international environmental agreements, such as the Montreal Protocol on ozone-depleting substances and the Stockholm Convention on persistent organic pollutants. The treaty attempts to integrate national and regional measures such as the directives of the EU on plastic bags and single-use plastics, as well as policies in African countries and behavioral interventions in the U.S., towards reducing, replacing, recycling, and managing plastic waste. By learning from the successes and limitations of previous IEAs and leveraging economic, political, and scientific insights, the treaty aspires to create a coordinated, effective framework to combat plastic pollution globally (Aanesen et al., 2024).

3.4 Legislation on Marine Plastic Pollution

Marine plastic pollution is a worldwide issue, with an estimated 4.8 to 12.7 million tons of plastic waste flowing into the oceans annually. In total, this contributes to approximately 150 million tons of plastic currently polluting marine environments (Haward, 2018; Worm et al., 2017).

Plastic pollution kills marine life, enters the food chain, accumulates toxins, and hosts unique microbial communities in marine ecosystems (Chen, 2015; Haward, 2018; Hugo, 2018). Most Microplastic debris (MPD) comes from land-based sources, carried by rivers, wastewater, wind, and tides, while a smaller percentage comes from ocean-based activities such as fishing and shipping (Nyka, 2019; Worm et al., 2017). Land-based contributions account for 80%–90% of MPD, projected to double by 2030 and again by 2050 (Worm et al., 2017). International efforts include the *United Nations Convention on the Law of the Sea (UNCLOS), MARPOL*, and the *London Convention*, which establish broad obligations to protect marine environments from pollution. However, none of these specifically target MPD (Wu, 2022). The *Basel Convention* also plays a role in preventing the illegal trafficking of plastic waste, indirectly helping to reduce ocean plastic pollution (Wu, 2022). Despite these frameworks, targeted legal measures are required to effectively address the growing challenge of MPD.

4 Regional and National Policies

4.1 European Union

Europe, the third-largest global plastic producer, manufactured 60 Mt. of plastic resins in 2017, with demand driven by packaging (39.7%), automotive (19.8%), building and construction (10.1%) industries (Plastics Europe, 2017, 2018). Of the 27.1 Mt. of post-consumer plastic waste collected, only 31.1% was recycled, with 27.3% landfilled and 41.6% incinerated. Packaging waste recycling was higher at 40.8%, but challenges remain with long-term use plastics in construction and automotive sectors due to data fragmentation (Plastics Europe, 2017). Through the *Strategy for Plastics in a Circular Economy*, the European Commission has led the way with the goal to have 100% reusable or recyclable plastic packaging by 2030 as well as ambitions such as to recycle more than half of the plastic waste that Europe generates as well as increasing sorting and recycling capacity four-fold since 2015, further creating 200,000 new jobs (European Commission, 2018a). The Action Plan for Circular Economy (EC, 2015) focuses on moving Europe toward becoming a circular economy, with improvements through recycling and the reuse and resource management agenda towards environmental advantages. It calls for durable, repairable, and upgradeable designs especially in the electrical and electronic sectors; efficiency in production; and information accessibility to consumers such as environmental footprints, energy efficiency labeling, and repair options coupled with economic incentives like adjusted taxation. In addition, the strategy promotes high-quality recycling and energy recovery with minimum landfill use while facilitating market penetration of secondary raw materials and water reuse, "closing the loop" of product life-cycles (EC, 2015). The *2018 Circular Economy Package* outlined four areas for improvement on the path to a circular economy: lack of information for substances of concern, existence of legacy substances within recycled products, uncertainty in transition from waste to material, and classification problems for hazardous waste and chemicals. Its operational impacts are yet to be known (IEEP, 2018a, b).

Action such as *Extended Producer Responsibility* (where producers need to assume the costs of plastic waste recycling) have been pivotal in promoting accountability and reducing environmental burdens. By shifting the responsibility for waste management from governments to manufacturers, EPR policies encourage sustainable product design, improved recycling systems, and the reduction of plastic pollution at its source. In the European Union, rules around Extended Producer Responsibility (EPR) for packaging have been in place since 1994 (Cho, 2020). This system is widely regarded as one of the most effective ways to manage municipal waste, helping to implement the European waste hierarchy (Milios, 2018). Thanks to EPR, many countries like those in the EU, China, and Africa have seen a noticeable boost in recycling rates for packaging. Under this approach, it's the responsibility of producers or importers to cover the costs of sorting, recycling, or disposing of packaging when recycling isn't an option (NERC, 2020). While the specifics of EPR programs can differ, Canada and Europe often use a model that works through producer responsibility organizations (PROs). Market-based instruments are encouraged to align stakeholders to reduce plastic pollution, while specific actions address single-use plastics (SUPs), which constitute half of marine litter, with design improvements, market interventions, and consumer behavior changes targeting the ten main SUPs impacting marine environments. Restrictions have been placed on items like cotton bud sticks, balloon sticks, cutlery, plates, stirrers, straws, and oxo-degradable plastic food containers (Foschi & Bonoli, 2019). To address the environmental problems caused by single-use plastics, the European Commission has introduced several initiatives. One prominent example is lightweight plastic carrier bags, whose usage increased from 0.75 million tons in 2010 to an estimated 0.84 million tons by 2020 (BIO Intelligence Service, 2011). Directive 2015/720 introduced consumption reduction targets, aiming for no more than 90 lightweight bags per person annually by 2019 and 40 by 2025, alongside restrictions on very lightweight bags (European Parliament, 2019). The lack of clear guidelines for biodegradable and oxo-degradable plastics has complicated waste management (da Cruz et al., 2014). The Commission responded by proposing further restrictions on oxo-degradable plastics and concepts to harmonized rules regarding labeling of biodegradable plastics towards proper disposal (European Commission, 2018b).

4.2 United States

Landfilling remains the most common municipal solid waste (MSW) disposal in the USA, accounting for 49.9% of total MSW and 77% of plastic waste landfilled in 2018 (EPA, 2018). The recycling rate has significantly declined from 8.7% in 2018 to 5–6% in 2021 due to high costs and consumer inconvenience (Weforum, 2022). When the COVID-19 pandemic emerged, more waste was generated. Residential waste increased by 15–25%, and municipalities faced high costs for managing the waste (Trentmann, 2020). Economic problems such as unstable fuel prices make it difficult to maintain profitability in recycling services (Corkery, 2019; Wastedive, 2019). *Extended Producer Responsibility (EPR)* programs, which hold producers accountable for their products from design to end-of-life recycling, are a proposed solution to improve waste management efficiency and reduce costs (Valderrama, 2021). The *United States Microbead-Free Waters Act of 2015*, enacted with bipartisan support, bans the use of intentionally manufactured plastic microbeads (<5 mm) in rinse-off personal care products like facewash and toothpaste to reduce microplastic pollution. However, it has been criticized as addressing only a minor source of microplastics, not including secondary microplastics (e.g., through fabric washing or plastic weathering) and microbeads in some cosmetics. Moreover, the Act broadly prohibits all plastic microbeads, including innovative bioplastics that have lower environmental impact, without comprehensive standards on environmental safety, toxicity, and bioaccumulation. This legislative compromise calls for a more subtle approach in policy implementation to effectively address the plastic pollution problem at large. In the United States, laws concerning plastics have been enacted in cities, states, and federally, with stronger laws generally applicable unless superseded by federal law.

While several states, such as California and Hawaii, have led initiatives to ban single-use plastic bags and regulate microplastic (MP) levels in drinking water, comprehensive federal policies addressing macroplastics or MPs are lacking. Current federal regulations often grandfather plastic polymers as safe or focus on limited aspects, like microbeads in personal care products. Local initiatives, from plastic straw prohibitions to larger bans on use, have moved forward but faced criticism for affecting minor contributors towards plastic pollution when packaging, now 36% of global plastics produced, would be left almost untouched. Combined with the failure to standardize methodologies for detection and regulation and the challenges raised by the Covid-19, legislative progress still lags at a slow place. Moving forward, it involves focusing on packaged plastics and incorporates emerging concerns including nanoplastics (NPs) in a regulatory framework with the goal to effectively deal with plastic pollution as recommended by Sorensen et al. (2023). Extended producer responsibility (EPR) laws shift post-consumer product management from municipalities to manufacturers and encourage circular economy practices through careful product design and material selection (OECD, 2001). The first U.S. packaging EPR law was passed in Maine as LD 1541; 2021 with the goal of strengthening municipal recycling programs (Mozak, 2022). EPR can help to better manage waste and support environmental sustainability.

4.3 China

In 2021, China produced an astounding 80.1 million tons of plastic (Furst & Feng, 2022). While estimates of its share of plastic pollution in the world's oceans differ, China is widely recognized as one of the largest contributors to this global issue (Chen et al., 2019). This highlights the vital role China must play in tackling the challenge of plastic pollution on a global scale.

China, the world's largest producer and consumer of plastics, generates the most plastic waste globally, with rapid e-commerce growth and increased online food delivery during the COVID-19 pandemic exacerbating plastic pollution (Prasad et al., 2024; Tabassum et al., 2025). The 2017 import ban on plastic waste has added new challenges to managing this escalating issue. Since the introduction of plastic fast-food containers in 1986, China has faced serious challenges with plastic pollution, known as "white pollution." In the last two decades, the country has taken a number of measures to address plastic waste, including a nationwide ban on thin plastic shopping bags in 2008, which reduced their use by 49%. Despite this, design flaws in policies, limited public awareness, and challenges in the agricultural sector have hindered the effective implementation of policies (Wang & Fu, 2022; Xing et al., 2009). The specific policies, such as plastic bans, taxation, and restrictions on single-use plastics, is a step forward; however, issues like insufficient recycling programs and gaps in public engagement remain unaddressed. Marine plastic policies are underdeveloped and lack national-level strategies or regulations for microplastics. While the regulatory framework for plastics in China has become more diverse and comprehensive, covering various life cycle stages and products, questions about the overall efficacy of these measures in mitigating plastic pollution remain unresolved.

China's plastic pollution control strategy revolves around four key aspects. Firstly, the Five-Year Plan, officially *The Outline of the Five-Year Plan for National Economic and Social Development of the People's Republic of China*, is a general guideline for mid- and long-term economic, cultural, and environmental development, which could determine the direction of decisions and evaluate the achievement of policy. Secondly, the 2017 import ban of foreign waste marked a turning point in the global and national plastic waste management as attention is slowly shifted to the closed domestic recycling and treatment systems. Thirdly, tremendous efforts have been put into developing plastic recycling and treatment systems, such as renewable resource recycling initiatives, circular economy frameworks, and upgrading urban solid waste management infrastructure. Finally, China has implemented specific policies to deal with emerging challenges, such as single-use plastics, microplastics, and waste from fast-growing industries such as e-commerce and food delivery. These steps show a comprehensive approach to addressing plastic pollution and promoting sustainable waste management (Liu & Liu, 2023). China has strived to prevent and control plastic waste pollution through an integrative and detailed legal system and innovative measures. Starting from the *1995 Law on Prevention and Control of Environmental Pollution by Solid Wastes*, the nation has incrementally enhanced its rule-of-law system while following the ideas of "*reduce, reuse, recycle*," full-process management, and differentiated waste management.

Some of the significant policy milestones include the *2019 Opinions on Further Strengthening the Treatment of Plastic Pollution*, which was supported by General Secretary Xi Jinping, and the *2020 directive* issued by the National Development and Reform Commission and the Ministry of Ecology and Environment, which focused on regulated recycling, industrialization, and technological innovation in waste management. Other initiatives include "*waste-free cities*" and "*beautiful villages*," which encourage sustainable waste practices and improve the recycling industry, ensuring that China deals with plastic pollution effectively (Wang et al., 2019). China's fast-growing economy has attracted the spotlight towards Municipal Solid Waste Management (MSWM). Hence, MSWM regulations were amended with new rules in the form of 2020 MSW classification law and a policy such as Shanghai's four-category mandatory sorting. All these would lead to proper waste segregation with organic waste recycling being 35% and city-area recycling at 90%. However, problems such as low participation rates, mixed waste collection, and contamination are not so effective, as China's plastic recycling rate in 2020 was 17.1% (Tumu et al., 2023).

4.4 India

India's total volume of trade for plastic-related products and goods stood at USD 64.8 billion in 2022, or 5.5% of the total volume of the country's products trade and 2.3% of the world's total trade volume, as reported by United Nations Conference on Trade and Development.

India is a net importer of plastic-related goods and products, having a negative trade balance of nearly USD 16.8 billion in 2022 (SWITCH, 2024). In 2019, India annually produced around 9.4 million tonnes of plastic waste, which accounted for 3.1% of the global plastic waste generation of over 380 million tonnes (Rafey & Siddiqui, 2023). While the global packaging sector consumes 42% of plastics and the construction sector 17%, in India, these sectors account for 35% and 23% of consumption, respectively.

The Indian plastic industry expanded from 8.3 MMTPA (Million Metric Tons Per Annum) in 2010 to 13.4 MMTPA in 2015, with a projected growth to 22 MMTPA by 2020 due to diverse applications (FICCI, 2014). While there has been growth in per capita plastic usage, in 2016 India stood at a consumption of only 11 kg per year compared with the world average of 28 kg/year; however, estimates suggest this could increase to 20 kg by 2022 (Bhattacharya et al., 2018; Singh & Ruj, 2015). Plastic waste is mainly generated by Maharashtra, Tamil Nadu, and Gujarat among other regions, clearly evidencing the variability in regions between plastic use and waste disposal habits. India's Plastic Waste Management (PWM) framework functions across three levels: central, state, and urban local bodies (ULBs). These tiers are interconnected to ensure waste management practices are effectively implemented at every level. Municipalities and local authorities are key players, especially in managing the flow of municipal solid waste (MSW) and coordinating door-to-door collection of household packaging waste (Rafey & Siddiqui, 2023).

Consumers are the primary agents for the segregation of plastic waste in the process of efficient collection and treatment by the municipalities, assisted by the unorganized sector, such as *kabadi walas* (scrap dealers), which works quite effectively at the level of households. The government acts as the prime mover behind plastic waste management through regulations like the *Plastic Waste Management Rules, 2016*, which complements other movements as well-the *Swachh Bharat Mission* and the Sustainable Development Goals. Government has introduced various laws for tackling plastic pollution which are given in Fig. 2. Efforts to reduce plastic waste include a partnership between the United Nations Development Programme (UNDP) and Hindustan Coca-Cola Beverages Private Limited, which aims to strengthen existing systems and promote sustainable practices like segregation, recycling, and reuse to advance a circular economy (UNDP, 2020). As one of the largest plastic consumers globally, India has implemented and amended regulations to minimize waste generation and ensure industry compliance. The Plastic Waste Management (PWM) Rules, 2016, were brought into India to reduce plastic waste generation and promote sustainable waste management. According to these rules, companies as generators of waste have developed an obligation to minimize waste generation at the source, segregation, storage, and forwarding the waste for processing to authorized agencies, and payment of waste management charges to local bodies of plastic waste management (Rafey & Siddiqui, 2023). These are complemented by *the Swachh Bharat Abhiyan* (SBA), which creates awareness in the public domain regarding the treatment and disposal of plastic waste and eradicates plastic littering, as cited by (Bura, 2019). India has a well-rounded policy framework addressing waste management and plastic pollution, built on a long history of binding laws, regulations, and procedures. The key national policies include Environment (Protection) Act, 1986 and the Environment Protection Rules, 1986, along with their amendments in 2021. There are waste-specific rules, for example, Municipal Solid Wastes (Management and Handling) Rules, 2000 that were later substituted by the Solid Waste Management Rules, 2016, and the Plastic Waste Management Rules, 2016, followed by several amendments in 2018, 2021, 2022, and 2024 along with.

These guidelines are based on the previous legislations such as the Recycled Plastics Usage Rules, 1988, the Plastics Manufacture, Sale and Usage Rules, 1999, and the Plastic Waste Management and Handling Rules, 2011. Other than this, India has progressively put forward roadmaps, handbooks, and plans at regional and local levels to implement these policies by progressing toward a circular economy and better waste management practices (SWITCH, 2024). Extended producer responsibility is the first of a key market-based mechanism introduced by the Plastic Waste

Fig. 2 Evolution of plastic pollution management laws in India

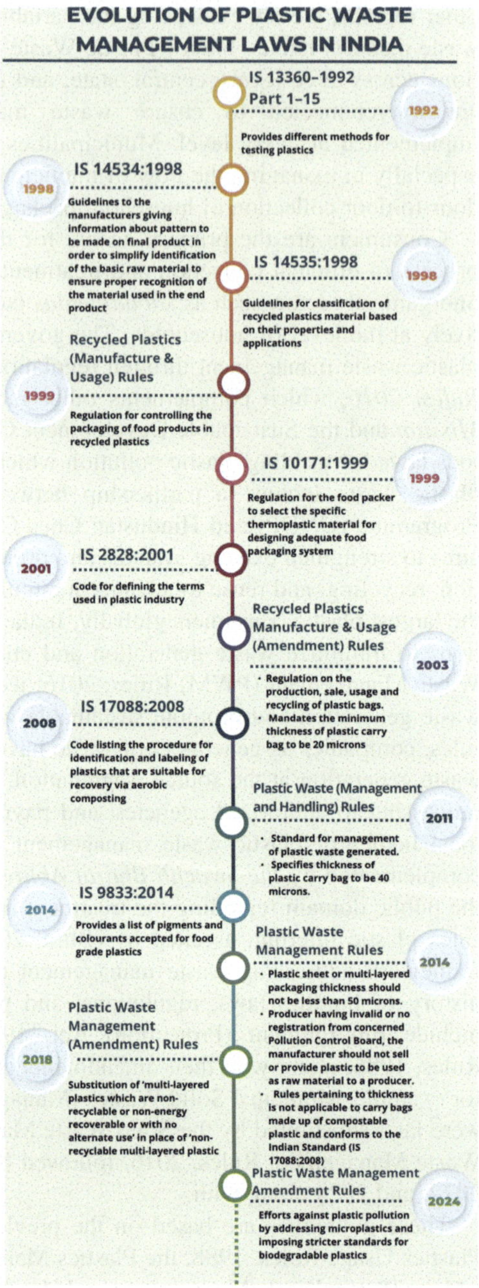

Management (PWM) Amendment Rules 2022. Mentioned in PWM Rules 2016 for the first time, EPR was called the 'polluter pays principle'. This has more elaborate guidance on plastic packaging for the amendment rules 2022.

5 Economic Instruments

Economic instruments and incentives are also vital in reducing plastic waste and promoting sustainability. The main approaches to reducing environmental waste under the legislation are command-and-control measures and market-based economic instruments (Oosterhuis & Peeters, 2014). Command-and-control measures regulate direct activities, including banning plastic microbeads or single-use plastic bags. Economic instruments include Container Deposit Legislation (CDL), plastic bag levies, or disposal taxes and work based on financial incentives or disincentives for behaviour. Container deposit legislation (CDL) is one of the several legislative actions that lawmakers have introduced and offered to reduce the debris that finds its way to the oceans (Schuyler et al., 2018). Though incentives are expensive to enforce, but they also avoid associated problems with disincentives, such as illegal dumping (Walls, 2011). Refund systems and positive incentives have been shown to generate higher material recovery rates for materials, including beverage containers and old cars (Lerner, 2011).

Plastic taxes, such as the UK's Plastic Packaging Tax, discourage the use and production of virgin plastics by creating a financial disincentive for producing virgin plastics while making recycled alternatives more appealing. Subsidies for alternatives can be given in the form of financial incentives for developing biodegradable and sustainable materials to encourage innovation in eco-friendly solutions. Such policies align economic incentives with environmental goals, thus nudging towards a circular economy (Oosterhuis et al., 2014).

Building on this market-based approach, the EU's 2021–2027 budget introduces a plastics-related tax that will require Member States to contribute €0.80 per kg of non-recycled plastic packaging waste. The measure is framed as a modern funding source for the EU budget and is estimated to rise up to €7 billion in revenue over the period, further solidifying the economic incentives for recycling and waste reduction (IEEP, 2018a, b).

6 Grassroots Initiatives

Grassroots, community-based efforts and corporate social responsibility are also key efforts in fighting plastic pollution. At the community level, one of the most impactful activities is the beach cleanup, such as those done by the International Coastal Cleanup, which organizes millions of volunteers across the globe to remove plastic and other waste from shorelines, waterways, and beaches. This program, initiated by *Ocean Conservancy*, has cleaned up more than 350 million pounds of litter since its creation, demonstrating the strength of collective power (Trice et al., 2021). In addition, campaigns such as the "*Beat Plastic Pollution*" project by the United Nations Environment Programme (UNEP) are trying to educate the public and the community about how plastics affect the environment and discourage the use of single-use plastics. Such initiatives have been effective in creating awareness through events, social media outreach, and educational resources, which lead to long-term behavioural change (UNEA, 2022).

Corporations have also made significant strides in reducing plastic pollution. Companies such as Unilever and Coca-Cola have pledged to make all their packaging recyclable, reusable, or compostable by 2025, which is a sign of a shift towards sustainable business practices (Murphy & Murphy, 2018; Coca-Cola, 2020). There is also the idea of plastic offsetting, where companies sponsor projects that gather and recycle plastic waste to offset their plastic use. For instance, Plastic Collective's programs are not only removing existing plastic but also generating local economic benefits by aligning waste collection with circular economy models (Plastic Collective, 2021). This synergy of grassroots action, public awareness, and corporate accountability underscores the importance of a collaborative approach to address the global issue of plastic pollution.

7 Global Governance Challenges and Technological Innovations in Plastic Pollution

A huge difference between the developed and developing countries exists in their systems of management of plastic waste. The difference ranges from infrastructure, technology, and governance. Challenges in developing countries include unorganized source separation, low effectiveness of waste collection efficiency, incompetence in the transportation system, facilities for recycling, and unsecured landfilling. These factors lead to increased plastic litter, with the majority of this waste being either dumped in the open or dumped into water. Developed countries, on the other hand, have well-organized waste collection mechanisms, efficient recycling processes, and secure landfilling procedures. However, even in developed nations, a vast amount of plastic waste is sent to countries where the waste management systems are much less stringent. In a collaborative manner, infrastructure, regulations enforcement, and practices of responsible waste management all need to be improved in regions worldwide (Rafey & Siddiqui, 2023).

The persistence of plastic pollution remains a pressing global issue despite various policy interventions (Dauvergne, 2018; Jameck et al., 2015). The governance of the matter is often fragmented between several national and local jurisdictions where the business and political-economy often overshadow the capacity of the marine system to cope. Most international institutions have not been very effective in closing these gaps. It should be clear that the lifetime and wide dispersal of plastics, ambiguities about the actual volumes flowing into the oceans, and difficulties in accountability further compound these challenges. All of these have been complicated by a booming plastics industry that uses obscure trade and channels waste to areas with inefficient waste disposal systems (Dauvergne, 2018).

In fact, existing global regulatory structures, such as MARPOL Annex V and the 2011 Honolulu Strategy, have shown to fail in addressing marine plastic litter (Dauvergne, 2018; Gold et al., 2013). The current regulations, along with state-level policies and non-state interventions, will also fall short of consumer expectations and liability categories regarding global environmental protection (DeSomre, 2018). The problems resulting from fragmented governance are loopholes that create opportunities for noncompliance and cost evasion for plastic pollution (Dauvergne, 2018). Solution strategies would entail governments collaborating with private enterprise and employing technological innovations to comprehensively address this global problem. As Gold et al. (2013) put it; investments in new technologies should enhance marine litter cleanup operations efficiency.

Innovation in technology has taken a swift advancement in dealing with plastic pollution by private individuals and NGOs. For instance, Ubintoo is a web-based database that collects and disseminates solutions emanating from private individuals, local organisations, governments, and academia. The involvement of companies including SAP, Modis, Cermaq, and Wilhelmsen, in projects like UN's Reboot the Ocean Challenge, a global challenge to inspire innovators across the world to come up with ideas on how to address marine plastic pollution through a technological approach, is proof of that argument (Schmaltz et al., 2020). Solutions targeting different stages of the plastic lifecycle, such as production, consumption and waste, including landfilling, recycling, and repurposing should be comprehensively addressed (Nielsen et al., 2020; Prata et al., 2019).

Land-based sources are responsible for at least 80% of marine plastic pollution (Li et al., 2016; Ritchie & Roser, 2018) and flow into oceans mainly through littering, inappropriate waste disposal, or leakage from open, uncontrolled landfills. Microplastics are finding their way into aquatic environments through wastewater, stormwater runoff, and extreme rainfall events, which open up pathways for a range of pollutants to enter marine systems (Law, 2017). Technologies designed to tackle these challenges either prevent plastic from leaking into waterways or collect existing pollution. Emerging recycling technologies such as those employed to convert plastic to fuel and bioremediation offer potential solutions to this problem (Mohanraj et al., 2017; Sheth et al., 2019; Tournier et al., 2020; Yoshida et al., 2016). These innovations act in synergy with the policies designed to address reduction in marine plastic pollution (Cordier & Uehara, 2019; Gold et al., 2013; Worm et al., 2017).

The UNEA Resolution 2/11 emphasizes regional and international cooperation to address hotspots of marine litter and to develop eco-friendly removal and disposal systems (Resolution 2/11, 2016; ten Brink et al., 2018). This research contributes towards this goal by developing the Technology Inventory for Prevention and Collection of Plastic Pollution, a resource for stakeholders addressing marine plastic pollution (Schmaltz et al., 2020). The inventory concerns technology for either preventing plastic leakage to waters or collection of already released pollution associated with oceans. It aims to facilitate future research and decision making, as well as comparative assessments of current solutions allowing innovators and policymakers to direct future efforts effectively. The collection broadens its horizons by complementing earlier studies to include marine environments and provide a searchable database. It allows for deeper understanding of the strengths and weaknesses of technologies and the gaps between the technology and its implementation. It is expected that the inventory will inspire innovations and policy integration toward a more holistic response to the global crisis in plastic pollution.

8 Conclusion

The global plastic crisis demands urgent and coordinated action, with legislation and policy serving as critical tools to mitigate its widespread impacts. This chapter has highlighted the diverse strategies employed worldwide, from international treaties and national policies to local community-driven initiatives, all aimed at addressing plastic production, consumption, and waste. While progress has been made, challenges such as enforcement gaps, unequal resource availability, and varying policy effectiveness persist. To navigate this crisis effectively, the global community must prioritize harmonized regulations, innovative technologies, and inclusive governance frameworks that balance environmental sustainability with economic considerations. Collaborative efforts among governments, industries, and civil society are essential to transition toward a circular economy and ensure the responsible management of plastic waste. By fostering systemic change and long-term commitment, the world can move closer to a sustainable future free from the devastating impacts of plastic pollution.

References

Aanesen, M., Ahi, J. C., Abate, T. G., Khan, F. R., de Vries, F. P., Kite-Powell, H., & Beaumont, N. J. (2024). Insights from international environmental legislation and protocols for the global plastic treaty. *Scientific Reports, 14*(1), 2750.

Belontz, S. L., Corcoran, P. L., Davis, H., Hill, K. A., Jazvac, K., Robertson, K., & Wood, K. (2019). Embracing an interdisciplinary approach to plastics pollution awareness and action. *Ambio, 48*, 855–866. https://doi.org/10.1007/s13280-018-1126-8

Bhattacharya, R. R. N., Chandrasekhar, K., Roy, P., & Khan, A., (2018). Challenges and opportunities: Plastic waste management in India.

BIO Intelligence Service. (2011). *Assessment of impacts of options to reduce the use of single-use plastics bags*. Final report prepared for the european commission—DG environment 2011. Available online: http://ec.europa.eu/environment/waste/packaging/pdf/report_options.pdf. Accessed on 9 Jan 2025.

Borrelle, S. B., Ringma, J., Law, K. L., Monnahan, C. C., Lebreton, L., McGivern, A., Murphy, E., Jambeck, J., Leonard, G. H., & Hilleary, M. A. (2020). Predicted growth in plastic waste exceeds efforts to mitigate plastic pollution. *Science, 369*, 1515–1518. https://www.science.org/doi/full/10.1126/science.aba3656.

Bura, N., (2019). An overview of plastic waste management in India. Waste management and resource efficiency. In: Proceedings of 6th IconSWM 2016, pp. 935–943.

Chae, Y., & An, Y. J. (2018). Current research trends on plastic pollution and ecological impacts on the soil ecosystem: A review. *Environmental Pollution, 240*, 387–395.

Chen, C. L. (2015). Regulation and management of marine litter. In M. Bergmann, L. Gutow, & M. Klages (Eds.), *Marine Anthropogenic Litter* (pp. 395–428). Springer.

Chen, Y., Cui, Z., Cui, X., Liu, W., Wang, X., & Li, X. (2019). Life cycle assessment of end-of-life treatments of waste plastics in China. *Resources, Conservation and Recycling, 146*, 348–357. https://doi.org/10.1016/j.resconrec.2019.03.011

Cho, R., (2020). Recycling in the U.S. is broken. How do we fix it? https://news.climate.columbia.edu/2020/03/13/fix-recycling-america/. Accessed 9 Jan 2025.

Coca Cola. (2020). *Packaging solutions*. https://www.coca-colacompany.com/sustainability/packaging

Cordier, M., & Uehara, T. (2019). How much innovation is needed to protect the ocean from plastic contamination? *Science of the Total Environment, 670*, 789–799. https://doi.org/10.1016/j.scitotenv.2019.03.258

Corkery, M. (2019). As costs skyrocket, more U.S. cities stop recycling. *The New York Times Company.*. https://www.nytimes.com/2019/03/16/business/local-recycling-costs.html

da Cruz, N. F., Ferreira, S., Cabral, M., Simões, P., & Marques, R. C. (2014). Packaging waste recycling in Europe: Is the industry paying for it? *Waste Management, 34*(2), 298–308.

Dauvergne, P. (2018). The power of environmental norms: Marine plastic pollution and the politics of microeads. *Environmental Politics., 27*, 579–597. https://doi.org/10.1080/09644016.2018.1449090

Debnath, R., Prasad, G. S., Amin, A., Malik, M. M., Ahmad, I., Abubakr, A., Borah, S., Rather, M. A., Impellitteri, F., Tabassum, I., & Piccione, G. (2024). Understanding and addressing microplastic pollution: Impacts, mitigation, and future perspectives. *Journal of Contaminant Hydrology, 266*, 104399.

DeSomre, E. R. (2018). Ocean governance. In P. Dauvergne & J. Alger (Eds.), *A research agenda for Gloal environmental politics* (pp. 114–125). Edward Elgar Pulishing, Ltd.. https://doi.org/10.4337/9781788110952

EC. (2015). *Closing the loop – an EU action plan for the circular economy*. https://eur-lex.europa.eu/legal-content/EN/TXT/?uri=CELEX:52015DC0614

English, C. (2023). Pushing out plastic: Lessons for the unea to consider in drafting its newest multilateral environmental agreement. *Vanderbit Journal of Transnational Law, 56*, 1131.

EPA. (2018). *Facts and figures about materials, waste and recycling*. Retrieved from https://www.epa.gov/facts-and-figures-about-materials-waste-and-recycling/glass-material-specific-data.

European Commission. (2018a). A *European strategy for plastics in a circular economy*. Available online: https://publications.europa.eu/en/publication-detail/-/publication/e7904416-72a2-11e8-9483-01aa75ed71a1. Accessed on 15 Dec 2018.

European Commission. (2018b). *Report on the impact of the use of oxo-degradable plastic, including oxo-degradable plastic carrier bags, on the environment*.

European-parliament. (2019). *Directive (Eu) 2019/904 of the european parliament and of the council*. Retrieved from. https://eur-lex.europa.eu/legal-content/EN/TXT/HTML/?uri=CELEX%3A32019L0904.

FICCI. (2014). Potential of plastics industry in northern India with special focus on Plasticulture and food processing – 2014: A report on plastic industry.

Foschi, E., & Bonoli, A. (2019). The commitment of packaging industry in the framework of the European strategy for plastics in a circular economy. *Administrative Sciences, 9*(1), 18.

Furst, K., & Feng, Y. (2022). China's regulatory respond to plastic pollution: Trends and trajectories. *Frontiers in Marine Science, 9*, 982546.

Gold, M., Mika, K., Horowitz, C., Herzog, M., & Leitner, L. (2013). *Stemming the tide of plastic marine litter*. A Gloal Action Agenda.

Gurjar, U. R., Takar, S., Amin, A., & Martin Xavier, K. A. (2023). Microplastic contamination in aquatic organisms: An ecotoxicological perspective. In *Xenobiotics in aquatic animals: Reproductive and developmental impacts* (pp. 353–367). Springer.

Haward, M. (2018). Plastic pollution of the world's seas and oceans as a contemporary challenge in ocean governance. *Nature Communications, 9*(1), 667.

Hu, X., Davies, R., Morrissey, K., Smith, R., Fleming, L. E., Sharmina, M., Clair, R., & Hopkinson, P. (2022). Single-use plastic and Covid-19 in the NHS: Barriers and opportunities. *Journal of Public Health Research, 11*, 2483. https://doi.org/10.4081/jphr.2021.2483

Hugo, T. G. (2018). *The case for a treaty on marine plastic pollution*. Norwegian Academy of International Law.

IEEP. (2018a). *Institute for European Environmental Policy*. https://ieep.eu/wp-content/uploads/2022/12/Think-2030-A-circular-economy-for-plastics-by-2030-1.pdf

IEEP. (2018b). *Moving towards a circular economy for plastics in the EU by 2030*. https://ieep.eu/wp-content/uploads/2022/12/Think-2030-A-circular-economy-for-plastics-by-2030-1.pdf

IISD. (2023). https://sdg.iisd.org/news/unep-report-identifies-market-shifts-needed-toreduce-plastic-pollution/

Islam, M. S., Lee, Z., Shaleh, A., & Soo, H. S. (2024). The United Nations environment assembly resolution to end plastic pollution: Challenges to effective policy interventions. *Environment, Development and Sustainability, 26*(5), 10927–10944.

Jameck, J., Geyer, R., Wilcox, C., Siegler, T. R., Perryman, M., Andrady, A., Narayan, R., & Law, K. L. (2015). Plastic waste inputs from land into the ocean. *Science, 347*(6223), 768–771. https://doi.org/10.1126/science.1260352

Johansson, N. (2021). *Intervention–disaster capitalism, COVID-19, and single-use plastic*. Antipode Online. https://antipodeonline.org/2021/01/26/covid-19-and-single-use-plastic/.

Ladi, S., & Tsarouhas, D. (2020). EU economic governance and Covid-19: Policy learning and windows of opportunity. *Journal of European Integration, 42*(8), 1041–1056.

Landon-Lane, M. (2018). Corporate social responsibility in marine plastic debris governance. *Marine Pollution Bulletin, 127*, 310–319. https://doi.org/10.1016/j.marpolbul.2017.11.054

Law, K. L. (2017). Plastics in the marine environment. *Annual Review of Marine Science, 9*(1), 205–229. https://doi.org/10.1146/annurev-marine-010816-060409

Lerner, M. L. (2011). Cash for clunkers, dimes for Duracells: An effective model to motivate the proper disposal of household toxic waste. *Jurimetrics.*, 141–179.

Li, W. C., Tse, H. F., & Fok, L. (2016). Plastic waste in the marine environment: A review of sources, occurrence and effects. *Science of the Total Environment, 566–567*, 333–349. https://doi.org/10.1016/j.scitotenv.2016.05.084

Liu, C., & Liu, C. (2023). Exploring plastic-management policy in China: Status, challenges and policy insights. *Sustainability, 15*(11), 9087.

Mæland, C. E., & Staupe-Delgado, R. (2020). Can the global problem of marine litter be considered a crisis? *Risk, Hazards & Crisis in Public Policy, 11*, 87–104. https://doi.org/10.1002/rhc3.12180

McConnell, A. (2018). Rethinking wicked problems as political problems and policy problems. *Policy & Politics, 46*, 165–180. https://doi.org/10.1332/030557317X15072085902640

McConnell, A. (2020a). The politics of crisis terminology. In *Oxford Research Encyclopedia of Politics*. Oxford University Press. https://doi.org/10.1093/acrefore/9780190228637.013.1590

McConnell, A. (2020b). The use of placebo policies to escape from policy traps. *Journal of European Public Policy, 27*, 957–976. https://doi.org/10.1080/13501763.2019.1662827

Milios, L. (2018). Advancing to a circular economy: Three essential ingredients for a comprehensive policy mix. *Sustainability Science, 13*(3), 861–878. https://doi.org/10.1007/s11625-017-0502-9

Mohanraj, C., Senthilkumar, T., & Chandrasekar, M. (2017). A review on conversion techniques of liquid fuel from waste plastic materials: Waste plastics, conversion techniques, CI engine operations. *International Journal of Energy Research, 41*(11), 1534–1552. https://doi.org/10.1002/er.3720

Mozak, M. (2022). It's time to trash consumer responsibility for plastics: An analysis of extended producer responsibility laws' success in Maine. *The Journal of Sustainable Development Law and Policy, 23*, 12.

Murphy, P. E., & Murphy, C. E. (2018). *Sustainable living: Unilever. Progressive business models: Creating sustainable and pro-social enterprise* (pp. 263–286). Palgrave.

Nabi, N., Ahmad, I., Amin, A., Rather, M. A., Ahmed, I., Hajam, Y. A., Khursheed, S., Malik, M. M., & Abubakr, A. (2024). Understanding the sources, fate and effects of microplastics in aquatic environments with a focus on risk profiling in aquaculture systems. *Reviews in Aquaculture, 16*(4), 1947–1980.

NERC. (2020). Extended producer responsibility (EPR) for packaging and paper products. https://nerc.org/documents/EPR%20for%20Packaging%20&%20Paper%20Fact%20Sheet.pdf.

Nielsen, T. D., Hasselalch, J., Holmerg, K., & Stripple, J. (2020). Politics and the plastic crisis: A review throughout the plastic life cycle. *WIREs Energy and Environment, 9*(1), e360. https://doi.org/10.1002/wene.360

Nyka, M., 2019. Legal approaches to the problem of pollution of marine environment with plastic. Zeszyty Naukowe Akademii Morskiej w Szczecinie.

OECD. (2001). *Extended producer responsibility*. OECD. https://doi.org/10.1787/9789264189867-en

Oosterhuis, F., & Peeters, M. (2014). Limits to integration in pollution prevention and control. In *EU environmental legislation* (pp. 91–115). Edward Elgar Publishing.

Oosterhuis, F., Papyrakis, E., & Boteler, B. (2014). Economic instruments and marine litter control. *Ocean & Coastal Management, 102*, 47–54.

Parashar, N., & Hait, S. (2021). Plastics in the time of COVID-19 pandemic: Protector or polluter? *Science of the Total Environment, 759*, 144274.

Plastic Collective. (2021). *Joins UN Sea of Solutions 2021*. https://www.plasticcollective.co/plastic-collective-joins-un-sea-of-solutions-2021/

Plastics Europe. (2017). Plastic—The facts 2017. Plastics Europe. Available online: https://www.plasticseurope.org/application/files/5715/1717/4180/Plastics_the_facts_2017_FINAL_for_website_one_page.pdf. Accessed on 15 Dec 2024.

Plastics Europe. (2018). Plastic—The facts 2018. : Plastics Europe, Available online: https://www.plasticseurope.org/application/files/6315/4510/9658/Plastics_the_facts_2018_AF_web.pdf. Accessed on 30 Dec 2024.

Prasad, G. S., Rout, S. K., Amin, A., Malik, M. M., Ahmad, I., Rather, M. A., Karmakar, S., Khalil, H. S., & Abubakar, A. (2024). Impact of COVID-19 on the aquatic environment and implications for fisheries with special emphasis to Indian perspective. *Proceedings of the Indian National Science Academy, 90*(4), 821–839.

Prata, J. C., Silva, A. L. P., da Costa, J. P., Mouneyrac, C., Walker, T. R., Duarte, A. C., & RochaSantos, T. (2019). Solutions and integrated strategies for the control and mitigation of plastic and microplastic pollution. *International Journal of Environmental Research and Pulic Health, 16*(13), 2411. https://doi.org/10.3390/ijerph16132411

Rafey, A., & Siddiqui, F. Z. (2023). A review of plastic waste management in India–challenges and opportunities. *International Journal of Environmental Analytical Chemistry, 103*(16), 3971–3987.

Reddy, M. S., Reddy, P. S., Subbaiah, G. V., & Subbaiah, H. V. (2014). Effect of plastic pollution on environment. *Journal of Chemical and Pharmaceutical Sciences, 14*, 28–29.

Resolution 2/11. (2016). Marine plastic litter and microplastics, United Nations Environment Assemly. https://nicholasinstitute.duke.edu/plastics-policies/unea-resolution-211-marine-plastic-litter-and-microplastics.

Ritchie, H., & Roser, M., (2018). Plastic pollution. Our World in Data. https://ourworldindata.org/plastic-pollution.

Schmaltz, E., Melvin, E. C., Diana, Z., Gunady, E. F., Rittschof, D., Somarelli, J. A., Virdin, J., & Dunphy-Daly, M. M. (2020). Plastic pollution solutions: Emerging technologies to prevent and collect marine plastic pollution. *Environment International, 144*, 106067.

Schuyler, Q., Hardesty, B. D., Lawson, T. J., Opie, K., & Wilcox, C. (2018). Economic incentives reduce plastic inputs to the ocean. *Marine Policy, 96*, 250–255.

Sheth, M. U., Kwartler, S. K., Schmaltz, E. R., Hoskinson, S. M., Martz, E. J., Dunphy-Daly, M. M., Schultz, T. F., Read, A. J., Eward, W. C., & Somarelli, J. A. (2019). Bioengineering a future free of marine plastic waste. *Frontiers in Marine Science., 6*, 624. https://www.frontiersin.org/articles/10.3389/fmars.2019.00624/full

Silva, A. L., Patrício, J. C., Prata, T. R., Walker, D. C., Duarte, A. C., Soares, A. M. V. M., Barcelò, D., & Rocha-Santos, T. (2020). Rethinking and Optimising plastic waste management under Covid-19 pandemic: Policy solutions based on redesign and reduction of single-use plastics and personal protective equipment. *Science of the Total Environment, 742*, 140565. https://doi.org/10.1016/j.scitotenv.2020.140565

Silva, A. L., Patrício, J. C., Prata, T. R., Walker, A. C., Duarte, W. O., Barcelò, D., & Rocha-Santos, T. (2021). Increased plastic pollution due to Covid-19 pandemic: Challenges and recommendations. *Chemical Engineering Journal, 405*, 126683. https://doi.org/10.1016/j.cej.2020.126683

Singh, R. K., & Ruj, B. (2015). Plasticwaste management and disposal techniques-Indian scenario. *International Journal of Plastics Technology, 19*(2), 211–226.

Sorensen, R. M., Kanwar, R. S., & Jovanovi, B. (2023). Past, present, and possible future policies on plastic use in the United States, particularly microplastics and nanoplastics: A review. *Integrated Environmental Assessment and Management, 19*(2), 474–488.

Sullivan, L. (2020). *How big oil misled the public into believing plastic would be recycled*. NPR. https://www.npr.org/2020/09/11/897692090/how-big-oil-misled-the-public-into-believing-plastic-would-be-recycled

SWITCH. (2024). Plastic policies in India. Country Profile.

Tabassum, I., Hassan, N., Amin, A., Malik, M. M., Abubakr, A., Prasad, G. S., Latief, F., & Rather, S. A. (2025). COVID-19 and global environmental change: A threat to food Security and nutrition. In *Food security, nutrition and sustainability through aquaculture technologies* (pp. 437–445). Springer.

Ten Brink, P., Schweitzer, J. -P., Watkins, E., Janssens, C., Smet, M. D., Leslie, H., & Galgani, F., (2018). Circular economy measures to keep plastics and their value in the economy, avoid waste and reduce marine litter. Economics: The Open-Access, https://www.econstor.eu/itstream/10419/173128/1/1011145367.pdf.

Thushari, G. G. N., & Senevirathna, J. D. M. (2020). Plastic pollution in the marine environment. *Heliyon, 6*(8), e04709.

Tournier, V., Topham, C. M., Gilles, A., David, B., Folgoas, C., Moya-Leclair, E., Kamionka, E., Desrousseaux, M.-L., Texier, H., Gavalda, S., Cot, M., Guémard, E., Daliey, M., Nomme, J., Cioci, G., Bare, S., Chateau, M., André, I., Duquesne, S., & Marty, A. (2020). An engineered PET depolymerase to reak down and recycle plastic ottles. *Nature, 580*(7802), 216–219. https://doi.org/10.1038/s41586-020-2149-4

Trentmann, N., (2020). Waste management wants consumers to pay more as it moves more trash. *The Wall Street Journal*. https://www.wsj.com/articles/waste-management-wants-consumers-to-pay-more-as-it-moves-more-trash-11589115601. Accessed 4 Dec 2024.

Trice, A., Robbins, C., Philip, N., & Rumsey, M. (2021). *Challenges and opportunities for ocean data to advance conservation and management*. Ocean Conservancy.

Tumu, K., Vorst, K., & Curtzwiler, G. (2023). Global plastic waste recycling and extended producer responsibility laws. *Journal of Environmental Management, 348*, 119242.

UNDP. (2020). Plastic waste management programme: A partnership 2018–2024, Central Pollution Control Board. https://eprplastic.cpcb.gov.in/#/plastic/home

UNEA. (2022). End plastic pollution: Towards an international legally binding instrument. Draft resolution. United Nations environment assembly of the United Nations environment Programme. 28 February- 2 march 2022. https://wedocs.unep.org/bitstream/handle/20.500.11822/38522/k2200647_-_unep-ea-5-l-23-rev-1_-_advance.pdf?sequence=1&isAllowed=y. Accessed on 4 Dec 2024.

UNEP. (2011). Report. Basel convention on the control of transboundary movements of hazardous wastes.

Valderrama, G., (2021). Pioneering EPR policies in the United States.

Vince, J., & Stoett, P. (2018). From problem to crisis to interdisciplinary solutions: Plastic marine debris. *Marine Policy, 96*, 200–203. https://doi.org/10.1016/j.marpol.2018.05.006

Walls, M., (2011). Deposit-refund systems in practice and theory. Resources for the future discussion paper, (11–47).

Wang, Q. H., & Fu, K. M. (2022). Status and thinking of plastic wastes recycling industry in China under the new situation. *China Plas India, 50*, 38–42.

Wang, M. H., He, Y., & Sen, B. (2019). Research and management of plastic pollution in coastal environments of China. *Environmental Pollution, 248*, 898–905.

Wastedive. (2019). *How recycling has changed in all 50 states.* Wastedive. Retrieved from. https://www.wastedive.com/news/what-chinese-import-policies-mean-forall-50-states/510751/.

Weforum. (2022). *How is the growing plastic waste problem impacting America?* Retrieved from. https://www.weforum.org/agenda/2022/05/plastic-waste-generation-new-highs-us-report/#:~:text=The%20plastic%20recycling%20rate%20is,and%20Turkey%20banning%20such%20imports

Wilcox, C., Hardesty, B. D., & Law, K. L. (2020). Abundance of floating plastic particles is increasing in the Western North Atlantic Ocean. *Environmental Science and Technology, 54*, 790–796. https://doi.org/10.1021/acs.est.9b04812

World Wildlife Fund (WWF). (2023). https://www.worldwildlife.org/pages/global-plastics-treaty

Worm, B., Lotze, H. K., Juinville, I., Wilcox, C., & Jameck, J. (2017). Plastic as a persistent marine pollutant. *Annual Review of Environment and Resources, 42*(1), 1–26. https://doi.org/10.1146/annurev-environ-102016-060700

Wu, H. H. (2022). A study on transnational regulatory governance for marine plastic debris: Trends, challenges, and prospect. *Marine Policy, 136*, 103988.

Xing, X. F., Wang, Q., & Tao, G. D. (2009). Analysis and extended thinking on the effectiveness of the "plastic restriction order" policy. *Ecological Economics, 3*(184–187), 190.

Yoshida, S., Hiraga, K., Takehana, T., Taniguchi, I., Yamaji, H., Maeda, Y., Toyohara, K., Miyamoto, K., Kimura, Y., & Oda, K. (2016). A acterium that degrades and assimilates poly (ethylene terephthalate). *Science, 351*, 1196–1199.

Addressing Plastic Pollution: A Strategic Framework for Solving the Global Crisis

Neha Wajahat Qureshi and Saba N Reshi

1 Introduction

Plastics have become an indispensable part of modern society, reshaping industries and transforming daily life with their unparalleled versatility, durability, and affordability (United Nations, 2023). From extending the shelf life of food to revolutionizing industrial production, plastics have significantly contributed to societal advancements. This indispensable material, however, presents a profound paradox: while enabling convenience and innovation, it has simultaneously catalyzed one of the most pressing environmental burden of our time (Anthony et al., 2015). Plastics, emblematic of technological progress and human ingenuity, now represent one of the defining environmental challenges of the twenty-first century.

Plastics, once celebrated for their adaptability, are now among the most enduring pollutants. They infiltrate ecosystems ranging from the summit of Mount Everest to the ocean's deepest trenches, and even the human body (Eriksen et al., 2014; Ragusa et al., 2021; Leslie et al., 2022; Yang et al., 2023; Prasad et al., 2024; Tabassum et al., 2025). The origins of plastics trace back to 1907, when Bakelite emerged as the first synthetic polymer, revolutionizing human life by providing a material of unparalleled utility. However, by the late twentieth century, plastics had become pervasive pollutants.

Macro-plastics (pieces larger than 0.5 mm) account for 88% of global plastic leakage, with single-use items such as bottles, caps, shopping bags and straws dominating waste streams (Reisser et al., 2013). Carried ashore by ocean currents or settling into seabeds, these materials create persistent environmental hazards. Once discarded, plastics neither biodegrade nor naturally decompose. Instead, they

N. W. Qureshi
ICAR-Central Institute of Fisheries Education, Mumbai, India

S. N. Reshi (✉)
College of Fisheries, Kishanganj, Bihar, India

A. Abubakr et al. (eds.), *Microplastics: Ecological and Food Security Implications*,
https://doi.org/10.1007/978-3-031-94480-2_17

fragment into microplastics and nanoplastics that persist for centuries, polluting all ecosystems they encounter (Phanisankar et al., 2020; Debnath et al., 2024; Nabi et al., 2024).

Microplastics (fragments smaller than 5 mm) and nanoplastics (smaller than 100 nm) infiltrate food systems and marine organisms, posing poorly understood risks to human and animal health (Reisser et al., 2013; Gurjar et al., 2023). These particles are either shed from synthetic products like textiles and tires or formed through the breakdown of larger plastics under the influence of solar radiation, wind, and ocean currents. Nanoplastics, in particular, represent a silent but insidious threat, capable of crossing cell membranes and accumulating within living organisms (Reisser et al., 2013).

The disproportionate impact of plastic pollution on the Global South, juxtaposed with policies largely driven by the consumption patterns of the Global North, further highlights systemic inequities (Brandon et al., 2023). Disparities in plastic waste management are also starkly evident between high-income and low- and middle-income countries. High-income nations, despite possessing superior waste collection and recycling systems, remain significant contributors to global plastic pollution due to their high per capita plastic consumption (Brandon et al., 2023). Conversely, low- and middle-income nations face acute challenges stemming from unregulated waste streams, insufficient infrastructure, and informal recycling sectors. India epitomizes this dichotomy. Existing regulatory frameworks in India, such as the Plastics Manufacture, Sale and Usage Rules, 1999, and the Coastal Regulation Zone (CRZ) Notification of 1991, have introduced measures to curb plastic pollution. However, these policies lack specificity and enforcement mechanisms, limiting their effectiveness in addressing the catastrophe comprehensively (Woods et al., 2021).

The informal and fragmented nature of India's plastic waste management system further exacerbates the problem, with significant amounts of waste remaining unprocessed and unmanaged. Moreover, behavioral shifts have been minimal, as the unparalleled convenience of plastics undermines targeted initiatives such as single-use plastic bans.

This chapter provides a comprehensive analysis of plastic pollution as a global environmental crisis, examining its implications for sustainability and public health, while assessing the effectiveness of existing measures. By aligning global insights with regional contexts, it seeks to chalk out a path toward a sustainable future where human convenience and environmental conservation coexist in harmony.

2 Current Scenario of Plastic Waste

The scale of plastic production illustrates the enormity of this impasse. The global production and consumption of plastics have expanded at an extraordinary pace since the mid-twentieth century, rendering plastics both indispensable and deeply problematic to modern society. Between 1950 and 2015, global plastic production skyrocketed from a mere two million metric tons to an astounding 380 million metric tons, an unprecedented 190-fold increase (Geyer et al., 2017). Since 2004, the world has produced as much plastic as it did in the entire preceding half-century, reflecting the escalating shift toward single-use plastics and a pervasive disposable consumer culture. Annual production surged from 234 million tonnes in 2000 to 460 million tonnes in 2019, with projections exceeding 700 million tonnes by 2040 (OECD, 2022).

Plastics, primarily high-molecular-mass synthetic organic polymers derived from hydrocarbons extracted from crude oil and natural gas, serve myriad purposes. Over one-third of global plastic production is devoted to packaging, most of it designed for single-use. This over-reliance on disposable materials has created a waste stream with minimal recycling and significant contributions to environmental pollution. While plastic products persist anywhere between 20 to 500 years (WWF, 2021), less than 10% of global plastic waste is recycled, leaving vast amounts to accumulate in landfills, rivers, and oceans (Landrigan et al., 2023). These trends have significantly intensified the accumulation of plastic waste in natural environments, leading to widespread environmental degradation across oceans, rivers, and terrestrial ecosystems (Reisser et al., 2013).

Marine ecosystems bear a disproportionate brunt of this crisis with approximately 20 million tonnes entering oceans annually. This influx disrupts ecosystems, endangers biodiversity, and threatens the livelihoods of millions reliant on coastal resources. Mismanaged plastic waste, often originating from inadequate terrestrial systems, is transported through waterways to oceans, where it disrupts biodiversity and threatens sustainability. Rivers like the Ganges-Brahmaputra serve as major conduits of plastic waste into marine environments, exacerbating ocean pollution (Barnes, 2019).

Despite a per capita plastic consumption of only 11 kg, well below the global average of 28 kg (Bhattacharya et al., 2018), India produced over 3.3 million metric tonnes of plastic in 2019, an increase of over one million tonnes compared to 2017–18 (CPCB, 2019). Key contributors include Maharashtra (12%), Tamil Nadu (12%), Gujarat (11%), West Bengal (9%), Uttar Pradesh (8%), Karnataka (8%), and Delhi (7%) (CPCB, 2019).

India's rapidly expanding plastic sector, comprising 15 major polymer manufacturers, 30,000 processing units, and over 7000 recycling facilities, operates within a poorly integrated and largely informal framework (Sivadas et al., 2022). This fragmented nature of plastic waste management leaves substantial quantities of waste unprocessed or mismanaged, exacerbating environmental challenges and undermining sustainability efforts.

Sectoral analysis highlights that the packaging industry remains the largest contributor to plastic waste, accounting for substantial portions of unprocessed and unrecycled materials. Beyond packaging, industrial, agricultural, and consumer sectors also contribute significantly to plastic pollution, with the extent varying according to economic activities and the stringency of regulatory frameworks. The urgency of this crisis demands transformative solutions at local, national, and international levels to reconcile human advancement with environmental sustainability.

3 Impact of Plastic Pollution

The proliferation of plastics in the environment has precipitated a cascade of adverse consequences across ecological, human, and economic dimensions.

Environmental Consequences Plastic pollution has inflicted substantial damage on marine ecosystems, terrestrial environments, and soil health. An estimated eight million tonnes of plastic enters the oceans annually, the volume roughly equivalent to one garbage truck per minute. Without intervention, this figure is projected to quadruple by 2050 (Fletcher et al., 2023). The accumulation of plastics in marine environments disrupts habitats, introduces persistent toxic substances, and jeopardizes biodiversity. Ghost gear which is abandoned, lost, or discarded fishing equipment accounts for approximately 10% of all marine litter (Debris, 2020). It devastates aquatic habitats, ensnares marine life, and perpetuates ecological degradation. Microplastics, the result of plastic fragmentation, further infiltrate trophic levels, contaminating food webs and undermining the resilience of marine ecosystems (Ragusa et al., 2021). Plastics also impair soil health through contamination by additives and leachates, hindering microbial activity and fertility. These environmental ramifications underscore the interconnectedness of ecosystems and the cascading effects of plastic pollution on ecological balance.

Human Health Risks The human health implications of plastic pollution are deeply concerning, albeit still incompletely understood. Microplastics, ubiquitous in aquatic systems and increasingly found in seafood, are ingested by humans, raising alarms about their potential biological effects. These particles, alongside chemical additives such as phthalates, bisphenols, and flame retardants, may act as endocrine disruptors and carcinogens, compounding public health risks (Yang et al., 2023). Although long-term exposure and cumulative effects remain difficult to quantify, mounting evidence links plastic-related contaminants to adverse health outcomes, warranting urgent research and policy responses (Leslie et al., 2022). In fisheries-dependent communities, particularly in the Global South, the risks are magnified. Contaminated seafood not only threatens public health but also exacerbates food insecurity and economic precarity. These vulnerabilities disproportionately affect marginalized populations per se, underscoring the social inequities inherent in plastic pollution.

Economic Costs Plastic pollution exacts a mammoth economic toll, draining resources through cleanup expenses, diminished tourism revenues, and losses in critical industries such as fisheries. Plastic packaging alone generates an estimated social cost of $40 billion annually, a figure poised to escalate under a business-as-usual trajectory. Marine debris, particularly ghost gear, imposes significant costs on fisheries, reducing productivity, damaging equipment, and increasing expenditures for replacement and cleanup. For millions of fishers reliant on marine resources, these losses threaten livelihoods and deepen economic vulnerabilities. The tourism industry, too, suffers as plastic-laden beaches and polluted waters deter visitors, undermining the economic potential of coastal regions. Governments and municipalities face mounting cleanup costs, diverting resources from other essential priorities. Collectively, these economic burdens illustrate the pervasive nature of plastic pollution and its capacity to erode societal and financial stability.

Climate Synergies The intersection of plastic pollution and climate change compounds these challenges. Plastics, derived predominantly from fossil fuels, contribute to greenhouse gas emissions throughout their lifecycle, from production to degradation (Cottom et al., 2024). These emissions intensify ocean warming, acidification, and habitat loss, further undermining marine biodiversity and fisheries' adaptive capacity. As plastics degrade, they release methane and ethylene, potent greenhouse gases that makes this crisis indelible.

4 Global and National Efforts to Address Plastic Pollution

The complexity, transboundary nature, and global scale of plastic pollution necessitate international cooperation, as no single state can address the issue in isolation (Raubenheimer & McIlgorm, 2018; Simon et al., 2021). The global response is characterized by a fragmented array of international frameworks, national initiatives, and enduring challenges in enforcement and implementation (Kumar et al., 2021; UNEP, 2021). While there is growing recognition of plastic pollution as a critical environmental and developmental concern, existing policies highlight both progress and persistent shortcomings.

Existing International Frameworks At the international level, several legal frameworks and strategies have emerged to combat plastic pollution. The United Nations Environment Assembly (UNEA) has played a pivotal role, adopting key resolutions on marine litter and microplastics since its inaugural session in 2014 (Lau et al., 2020). Significant milestones include the establishment of the Ad-Hoc Open-Ended Expert Group on Marine Litter and Microplastics at UNEA-3 and heightened focus on single-use plastics (SUPs) at UNEA-4 (De Kock et al., 2020). These initiatives underscore the growing systemic understanding of plastic pollution and the need for coordinated global action.

Despite these advances, existing frameworks exhibit critical limitations. Participation is inconsistent, and approaches remain fragmented, lacking cohesive compliance mechanisms, dedicated funding systems, and effective implementation support. The transboundary nature of plastic waste exacerbated by global trade in plastic scrap and natural dispersal processes further complicates governance efforts. Accountability is notably absent, emphasizing the urgent need for a legally binding global treaty. Advocates such as WWF have called for such an agreement, proposing clear state responsibilities, shared accountability measures, and equitable solutions to curtail production, consumption, and environmental leakage of plastics (Lindebjerg, 2019).

Recent international initiatives reflect an encouraging shift toward inclusivity and sustainability. At the fifth session of the Intergovernmental Negotiating Committee

(INC-5) in Busan, South Korea, India proposed the establishment of a multilateral fund modeled on the Montreal Protocol (Indian Express, 2024). This proposal highlights India's leadership in promoting equitable global solutions through grant-based financial assistance for developing nations, engagement with the private sector to diversify funding sources, and technology transfer to promote innovation. Such initiatives underscore the increasing emphasis on collective governance and capacity-building to address the scale and complexity of plastic pollution.

National Initiatives National governments are increasingly implementing policies to address plastic pollution, with a predominant focus on curbing single-use plastics. Regulatory measures, including bans, taxes, and extended producer responsibility (EPR) frameworks, have gained traction worldwide (Rafey & Siddiqui, 2023). Regions such as Europe and North America have made notable progress, particularly in reducing the consumption of plastic bags through bans and levies, providing evidence of their efficacy.

India presents a unique case in the global context. India has pledged to eliminate all single-use plastics by 2022, underscoring its commitment through initiatives aimed at promoting cleanliness and aesthetics. However, these efforts often prioritize littering concerns over systemic issues, such as inadequate waste collection, insufficient source segregation, and the toxic emissions associated with plastic burning. India's evolving policy framework reflects a nuanced attempt to balance development imperatives with environmental stewardship.

At the national level, enforcement challenges, resource constraints, and competing interests hinder progress. In India, for example, efforts to reduce litter often overshadow more structural issues, such as enhancing waste management infrastructure, decreasing reliance on fossil fuels, and addressing health concerns associated with alternatives to single-use plastics. Industry lobbying, health and safety considerations, and social equity challenges further complicate the implementation of ambitious regulatory measures (Knoblauch & Mederake, 2021).

To bridge these gaps, systemic approaches that address the entire plastic value chain are imperative. Policies must transcend superficial measures, emphasizing circular economy principles, rigorous enforcement, and inclusive strategies that balance environmental, economic, and social objectives. The establishment of a cohesive, legally binding global treaty, complemented by targeted national interventions and strengthened enforcement mechanisms, remains essential to mitigating the escalating crisis of plastic pollution.

5 Prospects and Strategies for Tackling Plastic Pollution

In overcoming these impacts, an opportunity beckons: using the plastics innovation engine to move the industry into a positive spiral of value capture, stronger economics and better environmental outcomes.

Increased Environmental Awareness Over the years, a significant rise in environmental consciousness has paved the way for stronger advocacy for plastic pollution regulations and enforcement. This heightened awareness presents an opportunity to establish a comprehensive legal framework to combat plastic pollution. By leveraging this momentum, strategies can include:

Targeted Public Awareness Campaigns: Educating citizens on the impacts of plastic pollution and promoting eco-conscious behaviors.
Community Engagement: Mobilizing local communities to actively participate in waste management initiatives.
Incentives for Eco-Friendly Practices: Encouraging businesses and individuals to adopt sustainable solutions through rewards.
Strengthened Environmental Education: Integrating environmental education into school curriculums to cultivate long-term behavioral change.
Partnerships with NGOs: Collaborating with non-profits to implement localized solutions and raise awareness.
Mandatory Environmental Impact Assessments: Enforcing transparency and accountability in industrial operations to ensure pollution reduction.

These combined efforts, driven by heightened public consciousness, hold immense potential to foster a cleaner and healthier environment.

Government Commitment Government commitment forms the cornerstone of building a robust legal and institutional framework for addressing plastic pollution. The Indian government has taken notable strides toward tackling this issue:

Policy Initiatives: The Government of India's pledge to eliminate single-use plastics (SUPs) by 2022 reflects its intention to curb pollution at the source.
Programs and Guidelines: Regulatory measures such as the Plastic Waste Management Rules have introduced key mandates, including extended producer responsibility (EPR) and waste collection targets.
Financial Incentives: Proposals to integrate economic incentives such as subsidies for eco-friendly materials can encourage industries to transition to sustainable alternatives.
Global Leadership: At the recent Intergovernmental Negotiating Committee (INC-5) session in Busan, India proposed a Montreal Protocol-like multilateral fund to support equitable transitions for developing nations, underscoring its commitment to balancing development goals with environmental stewardship.

Public-Private Partnerships (PPPs) Collaboration between the government and private sector entities can amplify the framework's effectiveness. Key aspects of PPPs in combating plastic pollution include:

Shared Expertise: Combining public sector oversight with private sector innovation ensures practical, scalable solutions.
Resource Mobilization: Leveraging private sector investments to supplement government funding for pollution control measures.

Technology Adoption: Facilitating the use of advanced technologies for waste segregation, recycling, and alternative material production.
Awareness Campaigns: Driving large-scale educational initiatives and community engagement programs to foster shared responsibility.
Capacity-Building Initiatives: Enhancing skills and knowledge in local communities to support sustainable practices.

Strategic Innovations for Plastic Pollution Mitigation in Fisheries Having identified the dangers of plastic pollution, it is essential to take strategic steps to correct the damaging effects that have been caused and to prevent more damage from been done. To do this, India may need to take a cue from other countries that have successfully transitioned from the use of plastic to other sustainable means of packaging.

Technological Innovations

Biodegradable Fishing Gear: Innovations such as Japan's cellulose-based fishing nets provide viable, sustainable alternatives to conventional plastic gear, significantly reducing "ghost fishing" caused by abandoned and discarded nets.
Marine Debris Retrieval Systems: Norway's Clean Sweep program integrates waste collection into fishing operations, showcasing how environmental restoration can align with economic objectives. Such innovations present scalable models for fisheries worldwide, especially in coastal economies like India's.

Regulatory Interventions

Marine Litter Policies: Germany's Deposit Return Scheme (DRS) exemplifies how well-structured regulatory policies can achieve recycling rates exceeding 90%, proving the effectiveness of incentivized waste management.
Community-Based Waste Management: Programs like the Philippines' Plastic Bank empower communities by providing tradeable rewards for plastic collection. This grassroots engagement serves as a model for India's coastal communities, fostering economic participation while mitigating marine pollution.

Behavioral and Educational Campaigns

Grassroots Awareness: Programs such as India's National Green Corps engage youth in environmental stewardship, promoting long-term behavioral change through education. Reshi et al. (2022) found that coastal communities demand increased awareness for responsible management and critically need a circular economy.
Behavioral Nudges: Sweden's interactive recycling stations demonstrate the potential of behavioral science to influence responsible waste disposal, an approach that can be replicated in urban fishing hubs across India.

Collaborative R&D Innovations such as South Korea's seawater-degradable polymers highlight the transformative role of global partnerships in fostering technological solutions to plastic waste.

India, as a major player in global fisheries, stands to benefit immensely from these innovations. Collaborative R&D, combined with localized implementation, can strengthen India's capacity to mitigate plastic pollution in its marine ecosystems.

Economic Incentives for Sustainable Practices Economic incentives are a critical driver for promoting sustainable practices and establishing a comprehensive legal framework to reduce plastic pollution in India. Key approaches include:

Tax Breaks and Subsidies: Offering financial incentives to industries adopting biodegradable materials or cleaner technologies.
Grants for Innovation: Supporting startups and research institutions working on sustainable alternatives to plastics.
Preferential Treatment: Providing benefits, such as procurement priority or certifications, to businesses complying with plastic reduction regulations.

By aligning economic incentives with environmental goals, India can stimulate investments in the green economy, fostering job creation in sectors like renewable energy, waste management, and pollution control. This approach creates a win-win scenario, promoting economic growth while safeguarding environmental health.

6 Conclusion

Plastics, while indispensable to modern lifestyles due to their durability, versatility, and affordability, have emerged as one of the most significant environmental challenges of our time. The inadequacy of current measures to address the escalating plastic pollution crisis highlights the urgent need for transformative action. With plastic production steadily rising and its adverse implications for human health and ecosystems becoming increasingly apparent, the next 3–5 years represent a critical juncture for systemic intervention.

The path forward lies in embracing bold, multidimensional solutions. Governments must prioritize waste management as a cornerstone of sustainable development, aligning efforts with the waste management hierarchy to tackle plastic pollution at its source. Public-private partnerships and economic incentives can catalyze innovation and amplify the impact of these measures, while international collaborations can facilitate the exchange of best practices and advanced technologies.

Furthermore, addressing this crisis necessitates a paradigm shift in societal behavior and consumption patterns. Grassroots awareness campaigns, regulatory interventions, and technological innovations must converge to foster a culture of responsibility and sustainability. While transitioning to eco-friendly alternatives may pose economic challenges, the long-term benefits to public health, biodiversity, and the global economy far outweigh the costs.

Ultimately, the drive to eliminate plastic pollution is not merely an environmental imperative rather it is a moral obligation to ensure a livable planet for future generations. By acting decisively within this narrow window of opportunity, humanity can avert the dire consequences of inaction and chart a path toward a resilient and sustainable future.

References

Anthony, A., Melody, B., Dale, S., Fossi, M. C., & Arne, H., (2015). Plastics in a sustainable society-background paper. In: *Plastics in a sustainable society*. MISTRA-The Swedish Foundation for Strategic Environmental Research.

Barnes, S. J. (2019). Understanding plastics pollution: The role of economic development and technological research. *Environmental Pollution, 249*, 812–821.

Bhattacharya, R. R. N., Chandrasekhar, K., Roy, P., & Khan, A. (2018). Challenges and opportunities: Plastic waste management in India.

Brandon, A., Vanapalli, K. R., Martin, O. V., Dijkstra, H., De la Torre, G. E., Hartmann, N. B., Meier, M. A., Pathak, G., Busch, P.-O., & Ma, D. (2023). Charting success for the plastics treaty. *One Earth, 6*(6), 575–581.

Cottom, J. W., Cook, E., & Velis, C. A. (2024). A local-to-global emissions inventory of macroplastic pollution. *Nature, 633*(8028), 101–108.

CPCB. (2019). Annual report for the year 2018–19 on implementation of plastic waste management rules., 1–18. Retrieved from chrome-extension://efaidnbmnnnibpcajpcglclefindmkaj/https://cpcb.nic.in/uploads/plasticwaste/Annual_Report_2019-20_PWM.pdf.

De Kock, L., Sadan, Z., Arp, R., & Upadhyaya, P. (2020). A circular economy response to plastic pollution: Current policy landscape and consumer perception. *South African Journal of Science, 116*(5–6), 1–2.

Debnath, R., Prasad, G. S., Amin, A., Malik, M. M., Ahmad, I., Abubakr, A., Borah, S., Rather, M. A., Impellitteri, F., Tabassum, I., & Piccione, G. (2024). Understanding and addressing microplastic pollution: Impacts, mitigation, and future perspectives. *Journal of Contaminant Hydrology, 266*, 104399.

Debris, M. P. (2020). Stop ghost gear.

Eriksen, M., Lebreton, L. C. M., Carson, H. S., Thiel, M., Moore, C. J., Borerro, J. C., Galgani, F., Ryan, P. G., & Reisser, J. (2014). Plastic pollution in the world's oceans: More than 5 trillion plastic pieces weighing over 250,000 tons afloat at sea. *PLoS One, 9*(12), e111913.

Fletcher, S., March, A. L. A., Roberts, K., Shirian, Y., i Canals, L. M., Cairns, A., et al. (2023). *Turning off the tap: How the world can end plastic pollution and create a circular economy*. United Nations Environment Programme.

Geyer, R., Jambeck, J. R., & Law, K. L. (2017). Production, use, and fate of all plastics ever made. *Science Advances, 3*(7), e1700782.

Gurjar, U. R., Takar, S., Amin, A., & Martin Xavier, K. A. (2023). Microplastic contamination in aquatic organisms: An ecotoxicological perspective. In *Xenobiotics in aquatic animals: Reproductive and developmental impacts* (pp. 353–367). Singapore.

Indian Express. (2024). *Why the world needs a global plastic treaty*. Retrieved from., https://indianexpress.com/article/explained/explained-climate/global-plastic-treaty-9686703/. Accessed on 15 Dec 2024

Knoblauch, D., & Mederake, L. (2021). Government policies combatting plastic pollution. *Current Opinion in Toxicology, 28*, 87–96.

Kumar, R., Verma, A., Shome, A., Sinha, R., Sinha, S., Jha, P. K., et al. (2021). Impacts of plastic pollution on ecosystem services, sustainable development goals, and need to focus on circular economy and policy interventions. *Sustainability, 13*(17), 9963.

Landrigan, P., Symeonides, C., Raps, H., & Dunlop, S. (2023). The global plastics treaty: Why is it needed? *The Lancet, 402*(10419), 2274–2276.

Lau, W. W., Shiran, Y., Bailey, R. M., Cook, E., Stuchtey, M. R., Koskella, J., et al. (2020). Evaluating scenarios toward zero plastic pollution. *Science, 369*(6510), 1455–1461.

Leslie, H., et al. (2022). Discovery and quantification of plastic particle pollution in human blood. *Environment International, 163*, 107199.

Lindebjerg, E. (2019). *WWF policy paper: Tackling marine plastic pollution*. WWF.

Nabi, N., Ahmad, I., Amin, A., Rather, M. A., Ahmed, I., Hajam, Y. A., Khursheed, S., Malik, M. M., & Abubakr, A. (2024). Understanding the sources, fate and effects of microplastics in aquatic environments with a focus on risk profiling in aquaculture systems. *Reviews in Aquaculture, 16*(4), 1947–1980.

OECD. (2022). *Plastic pollution is growing relentlessly as waste management and recycling fall short*. Retrieved from: https://www.oecd.org/en/about/news/press-releases/2022/02/plastic-pollution-is-growing-relentlessly-as-waste-management-and-recycling-fall-short.html

Phanisankar, B. S. S., Vasudeva Rao, N., & Manikanta, J. E. (2020). Conversion of waste plastic to fuel products. *Materials Today: Proceedings, 33*(8), 5190–5195. https://doi.org/10.1016/j.matpr.2020.02.880

Prasad, G. S., Rout, S. K., Amin, A., Malik, M. M., Ahmad, I., Rather, M. A., Karmakar, S., Khalil, H. S., & Abubakar, A. (2024). Impact of COVID-19 on the aquatic environment and implications for fisheries with special emphasis to Indian perspective. *Proceedings of the Indian National Science Academy, 90*(4), 821–839.

Rafey, A., & Siddiqui, F. Z. (2023). A review of plastic waste management in India–challenges and opportunities. *International Journal of Environmental Analytical Chemistry, 103*(16), 3971–3987.

Ragusa, A., Svelato, A., Santacroce, C., Catalano, P., Notarstefano, V., Carnevali, O., Papa, F., Rongioletti, M. C. A., Baiocco, F., Draghi, S., D'Amore, E., Rinaldo, D., Matta, M., & Giorgini, E. (2021). Plasticenta: First evidence of microplastics in human placenta. *Environment International, 146*, 106274.

Raubenheimer, K., & McIlgorm, A. (2018). Can the Basel and Stockholm conventions provide a global framework to reduce the impact of marine plastic litter? *Marine Policy, 96*, 285–290. https://doi.org/10.1016/j.marpol.2018.01.013

Reisser, J., Shaw, J., Wilcox, C., Hardesty, B. D., Proietti, M., Thums, M., & Pattiaratchi, C. (2013). Marine plastic pollution in waters around Australia: Characteristics, concentrations, and pathways. *PLoS One, 8*(11), e80466.

Reshi, S. N., Qureshi, N. W., & Krishnan, M. (2022). Awareness, perception and adaptation strategies of fisher community towards marine plastic pollution along Mumbai coast, Maharashtra, India. *India Journal of Fisheries, 69*(3), 135–143.

Simon, N., Raubenheimer, K., Urho, N., Unger, S., Azoulay, D., Farrelly, T., Sousa, J., van Asselt, H., Carlini, G., Sekomo, C., Schulte, M. L., Busch, P. O., Wienrich, N., & Weiand, L. (2021). A binding global agreement to address the life cycle of plastics. *Science, 373*(6550), 43–47. https://doi.org/10.1126/science.abi9010

Sivadas, S. K., Mishra, P., Kaviarasan, T., Sambandam, M., Dhineka, K., Murthy, M. R., et al. (2022). Litter and plastic monitoring in the Indian marine environment: A review of current research, policies, waste management, and a roadmap for multidisciplinary action. *Marine Pollution Bulletin, 176*, 113424.

Tabassum, I., Hassan, N., Amin, A., Malik, M. M., Abubakr, A., Prasad, G. S., Latief, F., & Rather, S. A. (2025). COVID-19 and global environmental change: A threat to food Security and nutrition. In *Food security, nutrition and sustainability through aquaculture technologies* (pp. 437–445). Springer.

UNEP. (2021, October 21). From pollution to solution: A global assessment of marine litter and plastic pollution. United Nations Environment Programme, Nairobi. Available online: https://www.unep.org/interactives/beat-plastic-pollution/?gad_source=1&gclid=CjwKCAiApY-7BhBjEiwAQMrrEe9SgbGFulbFBN_DnRKH6Ux38hSE_v5gvq7c4Z-XPlQdjoG8ORsO1hoCre4QAvD_BwE. Accessed on 15 Dec 2024.

United Nations. (2023). *Understanding plastic pollution and its impact on lives.* Retrieved from https://www.un.org/africarenewal/magazine/may-2023/understanding-plastic-pollution-and-its-impact-lives

Woods, J. S., Verones, F., Jolliet, O., Vázquez-Rowe, I., & Boulay, A. M. (2021). A framework for the assessment of marine litter impacts in life cycle impact assessment. *Ecological Indicators, 129*, 107918.

WWF-Australia. (2021). The lifecycle of plastics. Available online: https://wwf.org.au/blogs/the-lifecycle-of-plastics/ Accessed on 15 Dec 2024

Yang, Y., Xie, E., Du, Z., Peng, Z., Han, Z., Li, L., Zhao, R., Qin, Y., Xue, M., Li, F., Hua, K., & Yang, X. (2023). Detection of various microplastics in patients undergoing cardiac surgery. *Environmental Science & Technology, 57*(30), 10911–10918.

Index

A. Abubakr et al. (eds.), *Microplastics: Ecological and Food Security Implications*,
https://doi.org/10.1007/978-3-031-94480-2

The manufacturer's authorised representative in the EU is Springer Nature Customer Service Centre GmbH, Europaplatz 3, 69115 Heidelberg, Germany. If you have any concerns regarding our products, please contact ProductSafety@springernature.com

Printed and bound by CPI Group (UK) Ltd, Croydon, CR0 4YY

30/09/2026

02225590-0018